BIOSYNTHESES

Georges N. Cohen

BIOSYNTHESES

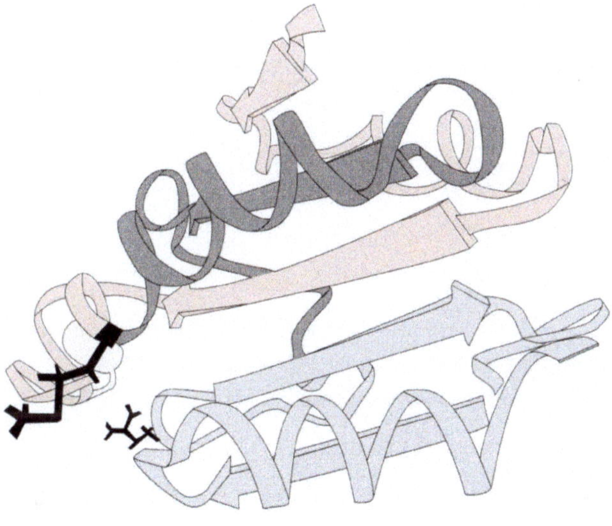

HERMANN PUBLISHERS IN ART AND SCIENCES

CHAPMAN & HALL • NEW YORK • LONDON

Published by Hermann, publishers in art and sciences, 293 rue Lecourbe, 75015 Paris, France.
Chapman & Hall, One Penn Plaza, New York, NY 10119.

ISBN 2 7056 6260 8 (Hermann)
ISBN 0 – 412 – 99551 – 4 (Chapman & Hall)

Printed in France.

Library of Congress Cataloging-in-Publication Data

Cohen, Georges N.
Biosyntheses / by Georges Cohen.
428 p. 19,5 x 25,5 cm.
Includes bibliographical references and index.
ISBN 0 – 412 – 99551 – 4
1. Biosynthesis. I. Title.
QP517. B57C64 1994
574. 19'29 – – dc 20

Table

JACQUES MONOD (1910–1976)

Foreword

For nearly fifty years I have been living and working with micro-organisms, mainly *Escherichia coli.* I have spent most of my scientific life at the Institut Pasteur in Paris, where many concepts of modern molecular biology were born or developed.

The fields of microbial physiology, biochemistry, and genetics have made considerable advances in the last twenty years through the introduction of gene cloning and sequencing, and by the exponential development of physical methods such as X-ray crystallography of proteins. Although the younger generation of biochemists is, legitimately, interested in the problems raised by differentiation and development in higher organisms, and in the neurosciences, it is still my feeling that the study of prokaryotes will long remain the best introduction to general biology.

Particular attention has been paid here to certain systems, extensively studied from physiological, enzymological, structural, genetic, and evolutionary points of view. I present my apologies to those who may find this choice too personal, reflecting my interest in subjects to which I myself have made a contribution, or where important results have been obtained by some of my good friends.

My thanks are due to my wife, Louisette Cohen, for her patience and help, not only while this book was being written, but during our more than fifty years of life together.

Much of the book was written while I was visiting professor at the University of Osaka Medical School, Institute of Molecular and Cellular Biology: I wish to thank Professor Kenichi Matsubara for his hospitality. This work is dedicated to the memory of my beloved colleagues; my mentor Jacques Monod, the late Dean B. Cowie, Michael Doudoroff, Ben Nisman, Roger Y. Stanier, Huguette and Kissel Szulmajster.

Paris, March 1993 Georges N. Cohen

Introduction

The ability of cells to multiply, leading to a net increase in mass, is due to a network of chemical reactions which can be classified as anabolic. Their study forms a chapter in biochemistry to which the name biosynthesis can be given. Biosynthetic reactions require energy; this is provided by another set of chemical reactions which are called catabolic.

A study of cellular metabolism must therefore concern itself with the reactions which produce energy and with the reactions of biosynthesis. This distinction, useful in a didactic way, must not obscure the fact that many intermediates involved in the classical degradation processes, glycolysis and the tricarboxylic acid cycle, are branch points from which purely biosynthetic pathways arise. The degradation sequences are therefore not only important insofar as they provide energy in the form of ATP, but also because they provide the carbon atoms which are necessary for the synthesis of cellular constituents.

Furthermore, if we consider the growth of a bacterium such as *Escherichia coli* on succinate as sole source of carbon, it is evident that this organism must be able to carry out the reactions of glycolysis in the reverse direction in order to obtain, for example, glucose 6-phosphate, which, when transformed to erythrose 4-phosphate, is required in the biosynthesis of the aromatic amino acids. In this instance, the glycolytic reactions have a purely biosynthetic role. The term "amphibolic" has been introduced to describe such reactions which function in both catabolism and anabolism.

Let us consider the bacterium *E. coli*: it can grow exponentially on a mineral medium containing a usable carbon source. Table I gives the composition of one such medium, and table II is a very incomplete list of carbon sources for this bacterium. Such a cell will first of all have to be brought into contact with the carbon source of the medium. We shall see that this is generally achieved not by simple diffusion but by means of proteins localized in the cytoplasmic membrane, which are responsible for the ingress of metabolites into the intracellular space.

Table I. Phosphate is in large excess, in order to buffer the medium against pH variations. Metals such as zinc, molybdenum cobalt, etc., found in proteins and certain vitamins exist as trace impurities in the commercial salts used.

KH_2PO_4	13.6 g
$(NH_4)_2SO_4$	2.0 g
$MgSO_4$, 7 H_2O	0.2 g
$FeSO_4$, 7 H_2O	0.0005 g
Bring to pH 7.0 with KOH	
H_2O to 1 liter	

Table II. Some carbon sources used by *Escherichia coli*.

Monosaccharides	Disaccharides	Acids	Polyols
Glucose	Maltose	Acetate	Glycerol
Fructose	Lactose	Succinate	Mannitol
Mannose			Dulcitol
Galactose			Sorbitol
Arabinose			
Xylose			
Rhamnose			

CHAPTER I

Bacterial growth

Lag phase

When cells are inoculated from an agar slant into a synthetic medium, a certain time elapses before cells grow at a constant rate. This time is called the *lag* phase during which at first no growth is detected, and then cells experience an accelerating growth rate before the constant rate is established.

Most authors agree that the lag phase is the expression of phenomena which are not linked to the growth proper, and which are due to contingent reasons:

a) After growth has stopped, the number of ribosomes per cell diminishes and protein synthesis cannot resume at a convenient rate until new ribosomes have been produced.

b) In a culture where a part of the population is no longer viable, the dead cells continue to diffuse light. Since the bacterial mass is often measured by nephelometry rather than by viable count, a certain time will be necessary before a change in optical density becomes significant. Actually, the lag phase determined by viable count is shorter than the lag phase determined by nephelometry.

c) A transfer from a certain medium to a different one may create a lag time if the utilization, say, of the carbon source of the second medium requires the synthesis of one or more inducible enzymes, or if growth on the first medium has caused the repression of some enzymes which must be synthesized in the second one.

d) Finally, if exponentially growing bacteria are transferred in a fresh medium identical to the initial one, all other conditions being equal, no lag time is observed.

Exponential phase

When no essential nutrient is limiting, the culture grows at a constant rate, and the growth rate is proportional to the culture density. The growth curve is therefore an exponential.

After n generations, the population of an exponential culture equal to x_0 at zero time becomes:

$$x = x_0 2^n.$$

If μ is the number of divisions per unit time, x the density at time t, and x_0 the density at time t_0, one obtains:

$$x = x_0\, 2^{\mu t}$$

from which the value of μ is extracted:

$$\mu = \frac{\log x - \log x_0}{(t - t_0) \log 2}.$$

Consequently, if the size of the population is known at two times in the exponential phase, the value of μ, the growth rate, can be derived. In practice, the time unit chosen is the hour and μ is the number of divisions per hour.

The best graphic representation of exponential growth is obtained by putting as ordinates the optical density of the culture and as abscissas the time at which the measures were made. The simple inspection of the straight line obtained on semilogarithmic paper gives the value of μ. A simple interpolation gives the time necessary to obtain a doubling of the population.

If an essential nutrient disappears during growth, if oxygen becomes limiting, or if the medium becomes too acid or too basic, the growth rate decreases and finally reaches zero. This phase is ill-defined: its characteristics and duration depend on the cause which has limited growth. The number of viable cells and the bacterial mass cease to increase almost simultaneously. On the other hand, the accumulation of toxic substances can inhibit cell division (the number of viable cells remaining constant), but in this case the optical density of the cultures may continue to increase.

Linear growth

When certain amino acid analogues are added to a culture of *E. coli* growing exponentially, growth does not stop altogether, but the exponential growth is transformed into arithmetic (linear) growth, *i.e.* where the increase in mass becomes directly proportional to time, according to

$$x - x_0 = kt.$$

Everything happens as if an essential component ceases to be synthesized, at least under its active form; growth becoming, as a result, directly proportional to the amount of the essential component present before the addition of the analogue.

Microscopic examination of bacteria grown under these conditions show long filaments due to an anomaly in cell division. After a certain increase in cell mass, growth stops altogether, but the bacteria remain in general viable and division resumes upon removal of the analogue.

Yield of growth

It is defined by the following ratio:

$$\frac{\text{Total growth (mg dry weight per unit volume)}}{\text{Carbon source concentration (mg per unit volume)}}.$$

Examination of table I, taken from Monod's classical work, shows that total growth is directly proportional to the initial concentration of the carbon source in the medium; in other terms, the growth yield is independent of the concentration of the limiting nutrient, within the limits of the concentrations studied. Obviously, it has been verified that all the glucose was used up in this experiment.

Table I. Total growth and yield of *E. coli* cultures in function of the concentration of glucose in the synthetic medium.

Glucose (mg/l)	Total growth (mg dry wt/l)	Yield	Deviation from the mean value (%)
200	47.0	0.235	+ 0.9
180	42.4	0.236	+ 1.3
160	37.4	0 234	+ 0.43
140	33.0	0.236	+ 1.3
120	28.0	0.234	+ 0.43
90	20.8	0.230	− 1.3
70	16.0	0.228	− 2.1
50	11.5	0.230	− 1.3
25	5.4	0.258	+ 10.7
		Mean yield: 0.233	

Careful experiments have shown that growth yield is totally independent of growth rate, when the variations of growth rate are caused by oxygen limitation, for example. Figure 1, taken from the same work by Monod, shows the growth of two cultures of *E. coli* on the same synthetic medium. One culture is shaken, the other not. In the second, oxygen becomes limiting and growth is slower. However, the two cultures reach the same maximum.

The growth yield measured as a function of different carbon sources is highly variable and seems roughly to reflect the different ATP yields characteristic of each catabolic pathway. However, if the growth yield is calculated as a function of the amount of ATP synthesized, it is roughly constant and equal to 10 g of dry matter per "mole" of energy-rich bond.

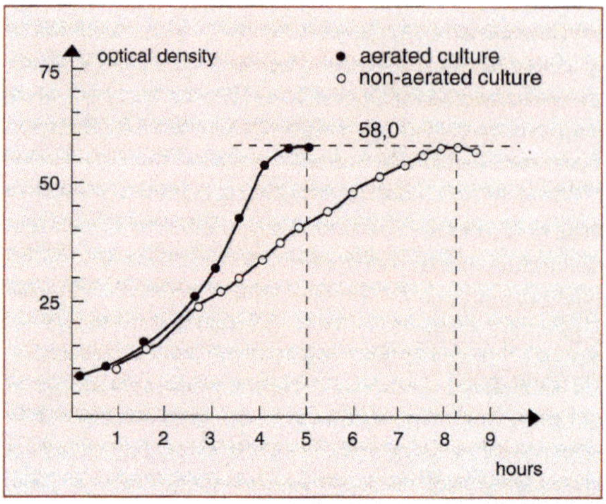

Figure 1. Growth of two *E. coli* cultures growing under two different aeration conditions.

This view has been seriously challenged by Marr in a 1991 review. Marr provides evidence that the growth rate of *E. coli* is set by the flux of a precursor metabolite and of the monomers derived from it rather than by the flux of ATP or, with the possible exception of the maximal growth rate, the rate of protein synthesis. The physiology of *E. coli* at low growth rates is clearly different from that at moderate to high growth rates. Although one can account for the rate of protein synthesis only in terms of rate of peptide elongation, it seems likely that initiation of translation is modulated to prevent the depletion of pools of amino acids and aminoacyl-tRNAs. Modulation of initiation might be one of a set of controls which confer resistance to starvation. Marr speculates that the signal for regulation of protein synthesis is the ratio of uncharged tRNA to charged aminoacyl-tRNA, that this signal controls the concentration of guanosine tetraphosphate which itself controls the transcription of the ribosomal RNA genes. The protein synthesizing system is an important but not singular example of synthesis on demand, which is also charac-ter-istic of peptidoglycan, phospholipids, and even DNA. Unfortunately, in no case is the mechanism of such modulation well understood.

Variation of the growth rate at limiting carbon source concentrations

The growth rate is constant only above a certain concentration of the carbon source. Below that concentration, it decreases considerably. The curves describing the variation of the growth rate of *E. coli* in a synthetic medium in function of the glucose concentration of glucose, mannitol or lactose are hyperbolas expressing the relation:

$$\mu = \mu_0 \frac{C}{C_l + C},$$

μ being the growth rate, μ_0 the maximal growth rate, C the concentration of the limiting nutrient. The ratio $C/C_1 + C$ being equal to 1/2 when $C = C_1$. Constant C_1 is therefore the concentration of the carbon source for which the growth rate is half of the maximum rate. The above equation has a form identical to a Langmuir isotherm. What limits the growth rate is the existence of a saturable system limiting the utilization of the carbon source. This system has been shown to be the active transport of the carbon source (see next chapter).

Continuous growth. The chemostat

Under the usual batch conditions of growth, the exponential phase can last only during a limited number of generations, a nutrient or the aeration of the medium becoming limiting. It can be prolonged by successive transfers of the bacteria into a fresh medium. Monod, and Novick and Szilard have independently introduced the more practical method of continuous cultures which, in addition, has proven extremely interesting both conceptually and experimentally.

The principle is the following: bacterial growth takes place in a vessel B connected to a reservoir R containing fresh medium which can be continuously delivered at the desired rate; on the other hand, vessel B is equipped with an exit tubing placed in such a way that a volume of culture equal to the volume of medium delivered from the reservoir be withdrawn. A non-limiting aeration and an efficient stirring are necessary in vessel B. If the bacteria grown under these conditions are required, they can be instantaneously taken from vessel P, and cooled so as to prevent further growth (fig. 2).

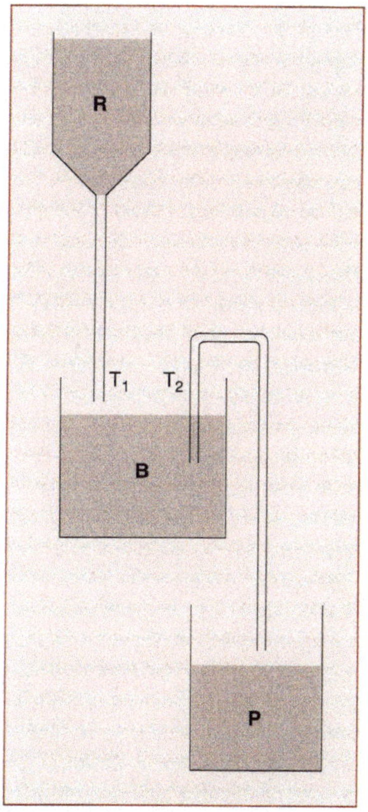

Figure 2. The chemostat, an apparatus for continuous culture. The actual chemostats are much more elaborate. The figure only explains the principle on which they are built.

How does N, the bacterial population in the growing vessel B, vary as a function of time? By derivation of the exponential growth equation

$$N_t = N_0\, \mathrm{e}^{\mu t}$$

one obtains the instantaneous growth rate:

$$\frac{\mathrm{d}N}{\mathrm{d}t} = \mu N.$$

Let V be the volume of the medium in the growth flask and v the volume delivered from the reservoir per unit time (equal to the volume withdrawn from flask B per unit time). The rate of removal of cells is given by

$$\frac{\mathrm{d}N}{\mathrm{d}t} = -N\,\frac{v}{V_t}.$$

Let us call $\frac{v}{V_t} = D$, the dilution rate.

The net population change in vessel B is given by the algebraic sum of these two functions:

$$\frac{dN}{dt} = \mu N - ND = N(\mu - D).$$

We have previously noted that, for a given microorganism, μ cannot exceed a maximal value μ_0 under experimental conditions where the value of μ depends exclusively on the concentration of a limiting factor. If the dilution rate D becomes higher than μ, the expression dN/dt becomes negative and the population in the growth flasks is gradually depleted. If D is smaller than μ, dn/dt becomes positive and the size of the population increases in vessel B. The increase, however, cannot be indefinite and will continue up to a moment when a nutrient of the reservoir becomes limiting. Then μ decreases, and when it becomes equal to *D, the exponentially growing culture of vessel B attains a constant size.*

We are thus in presence of an autoregulated system, the *chemostat,* where the growth rate will adjust automatically to its equilibrium value and remain constant indefinitely. The only way, all other conditions being equal, to interfere with the value of μ is to increase or decrease the rate of admission of the medium from the reservoir and in consequence to act on the dilution rate.

When equilibrium is reached, after one generation, we can write

$$N = 2N_0 = N_0 \, e \, \mu t,$$

from which the following ensues:

$$e \, \mu t = 2 \text{ and } \mu t = ln\, 2 = 0.69.$$

Since, at equilibrium,

$$\mu N = -N \frac{v}{V_t}$$

then

$$\mu t = \frac{v}{V}$$

and the generation time is defined by the value $v/V = 0.69$, *i.e.* when a volume equal to 0.69 V has circulated through the growth flask.

Advantages of the continuous exponential culture

The theory and the practice of the simplified chemostat we have just described show that it is possible to obtain a cultured maintained in the exponential phase by continuous dilution calculated in order that the growth of the microorganisms is exactly compensated. In principle, such a culture grows indefinitely, at a constant rate, under constant conditions. Since the growth rate can be varied at will, one can study its specific effects on cell size and chemical composition. These are very marked effects: the DNA, RNA and protein content vary quite significantly (table II).

Table II. Relative contents of macromlecules in *E. coli* grown at different growth rates.

Growth rate μ	Cells/g dry weight	DNA	RNA	Protein
		mg/g dry weight		
2.4	1.3 10^{12}	30	310	670
1.2	3.1 10^{12}	35	220	740
0.6	4.8 10^{12}	37	180	780
0.2	6.3 10^{12}	40	120	830

Using auxotrophic mutants, the essential metabolite required as the result of the mutation can be made easily limiting with the chemostat, and the effect of this limitation on the synthesis of more specific components can then be studied, as we shall see in a later chapter.

Diauxic growth

In 1941, while studying the growth of cultures of *B. subtilis* in a synthetic medium containing mixtures of carbohydrates, Monod observed a phenomenon to which he gave the name "*diauxie*". Whereas in certain mixtures the growth curve is normal (with a growth rate value intermediate between those obtained with the two isolated carbohydrates), certain other mixtures gave rise to a curve which could be decomposed in two distinct exponential phases separated by a *lag* phase.

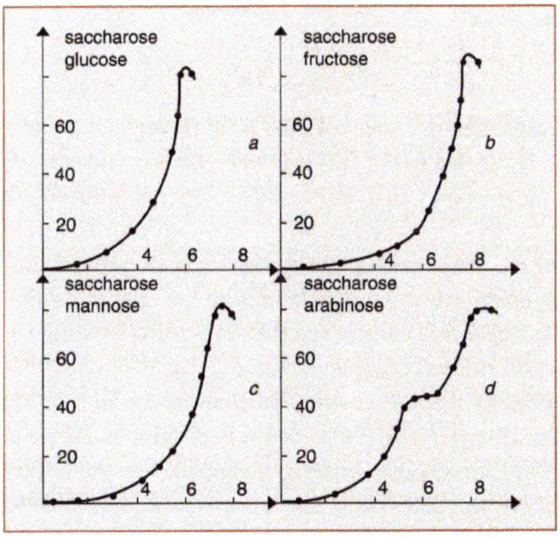

Figure 3. Abscissa: time (h). Ordinate: optical density.

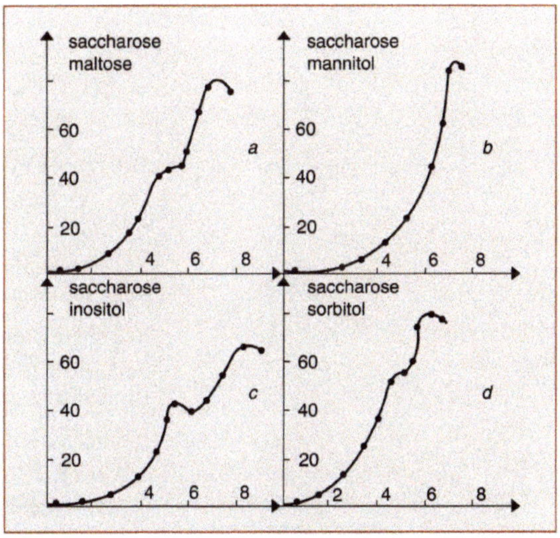

Figure 4. Abscissa: time (h). Ordinate: optical density.

If mixtures are studied where one carbohydrate, *e.g.* saccharose, is always present and the other is varied, the results show that certain mixtures grow in the usual way (fig. 3 *a, b, c* & 4 *b)*, whereas in certain other mixtures, the diauxie is clearly visible (fig. 3 *d,* & 4 *a, c, d*).

If the sugars which, associated to saccharose, give rise or not to a diauxie, are grouped into two lists, one obtains:

"Non diauxic" sugars with saccharose *A*	"Diauxic" sugars with saccharose *B*
Glucose	Sorbitol
Fructose	Inositol
Mannose	Arabinose
Mannitol	Maltose
Saccharose	Dextrin

No chemical similarity is specific to one or the other group. Monod made all the possible binary combinations with any sugar of the first list and any sugar of the second: he *always* observed a diauxic growth; such a curve was *never* observed with two sugars belonging to the same list.

A similar phenomenon is observed with *E. coli* or *Salmonella typhimurium* where the two lists are as follows:

A	*B*
Glucose	Arabinose
Mannose	Xylose
Fructose	Rhamnose
Mannitol	Sorbitol
	Dulcitol
	Maltose
	Lactose

The lists are similar to the ones obtained with *B. subtilis*, the few differences observed being due to the fact that enterobacteria do not utilize saccharose or inositol and that other sugars are not utilized by *B. subtilis.*

An analysis of the phenomenon reveals that, in both types of bacteria, the first growth phase is always associated with the total utilization of a sugar from list *A*. Sugar *B* is never, even partially, utilized during the first growth cycle. The cultures appear unable to metabolize simultaneously the constituents of the *A-B* couple. We shall return later to the mechanism of diauxie (Chapter VIII). It is useful to discuss its nature at this point. Monod has dismissed, by convincing experiments, the improbable hypothesis that the second cycle is due to the selection of mutants able to metabolize *B* sugars.

Another hypothesis for explaining diauxie is based on observations made by Emile Duclaux and reported in his *Traité de Microbiologie* (1899).

He observed that in *Aspergillus niger,* certain enzyme activities were always present whereas the appearance of other enzymes required the presence of their substrates in the growth medium. It was only in 1930 that this phenomenon was carefully studied in Karström's doctoral thesis. We shall give an example of the numerous experiments performed by Karström, which all show that the type of "adaptation" suspected by Duclaux is a direct temporary answer of the cell to the constituents of the growth medium, which disappears when the microorganism is grown in the absence of the stimulating agent: a xylose-fermenting strain of *Aerobacter aerogenes* is grown on whey and the washed suspension of the organism added to xylose in the presence of chalk; no fermentation occurs in 15 hours. A source of nitrogen is then added: fermentation sets in only two hours later. If the organism had been grown in xylose broth and treated as before, fermentation set in immediately, while the fermentation of glucose, as opposed to xylose, occurred whether glucose was present in the growth medium or not. The glucose-fermenting enzyme, being apparently a constant constituent of the cell was called "constitutive" by Karström, whilst he called the xylose-fermenting enzyme "adaptive".

Monod was aware of Karström's thesis and hypothesized that the conclusions of the Finnish scientist could plausibly explain the diauxic growth curves; bacteria would always be equipped with the enzymes necessary for degrading sugars *A*. The enzymes responsible for the degradations of sugars *B* would appear only after a beforehand adaptation, slow enough to explain the two successive cycles characteristic of diauxie. In addition, sugars *A* should inhibit the adaptation to sugars *B*.

SELECTED REFERENCES

Bacterial growth. Diauxie

J. Monod, *Recherches sur la croissance des cultures bactériennes*, Thesis, 1942, second edition, Hermann, Paris (1958).

Linear growth

G. N. Cohen & R. Munier, *Biochim. Biophys. Acta* **31**, 347-356 (1959).

Continuous growth. The chemostat

J. Monod, *Ann. Institut Pasteur* **79**, 390-410 (1950).

Influence of growth rate on cellular constituents

O. Maaloe & N. O. Kjeldgaard, *Control of Macromolecular Synthesis*. W. A. Benjamin, Inc., New York and Amsterdam (1966).

A. J. Marr, *Microbiological Reviews* **55**, 316-333 (1991).

Adaptive (inducible) enzymes. Prehistory

H. Karström, Thesis, Helsingfors (1930).

CHAPTER II

Cellular permeability

Gram-negative bacteria have a complex cell surface consisting of three layers: an outer membrane composed of protein, phospholipid, and lipopolysaccharide; the cell wall proper or peptidoglycan; and an inner, or cytoplasmic, membrane composed of protein and phospholipid. Nutrients, therefore, have to pass through this rather formidable architectural and protective structure. The outer membrane constitutes an important barrier against a variety of assaults, such as enzymatic and detergent attack in the digestive tract, and antibiotic entry. Limited permeability to small solutes is available through the outer membrane by way of proteinaceous channels, constituted of a special class of proteins which can be substrate-non specific (the proteins in this case are named porins) or substrate-specific. The cell wall proper is commonly regarded as an entirely permeable layer, conferring rigidity to the cell while forming a widely open network through which nutrient diffusion occurs readily. The cytoplasmic membrane, on the other hand, is impermeable to almost every solute unless a special transport system is provided.

The selective permeation of certain molecular species across certain tissues, or into certain cells, has been recognized since the beginning of the century as a phenomenon of fundamental importance in animal physiology. The situation was, up to 1955-1957, quite different in the field of microbiology. Although the importance of recognizing and studying selective permeability effects had been frequently emphasized, the available evidence appeared ambiguous, and the very concept of selective permeation was looked upon with suspicion by many microbiologists who believed that, in the absence of direct proof, it served mostly as a verbal "explanation" of otherwise obscure results.

In the second part of the 1950s, however, definite proof was obtained of the existence in bacteria of stereospecific permeation systems, functionally specialized and distinct from metabolic enzymes. We know now that the entry into a given type of bacterial cell of most of the organic and inorganic nutrilites it is able to metabolize is in fact mediated by such specific permeation systems. Many of these systems have been characterized as composed of one or more proteins, and defined by a combination of highly characteristic properties. The generic name "permeases" has been suggested for these systems. Although this designation has been criticized, it has the overwhelming advantage that its general meaning and scope are immediately understood.

Accumulation, crypticity and selective permeability

That the entry of organic substrates into bacterial cells may be mediated by more or less selective permeation systems has been suggested primarily by two kinds of observations concerning, respectively, (a) the capacity of certain cells to accumulate certain nutrilites; (b) the state of "crypticity" of certain cells toward certain substrates, their incapacity to metabolize a given substrate, even though they possess the relevant enzyme system.

Let us see why both accumulation and crypticity phenomena were strongly suggestive, yet inconclusive, as evidence of selective permeation systems.

The classical work of Gale on the uptake of amino acids by staphylococcal cells posed the problem of the accumulation mechanism as early as 1947. Gale and his associates found that these cells grown on casein hydrolyzate contain large amounts of glutamic acid, lysine, and other amino acids which could be extracted by water from crushed, but not from intact, cells. These observations appeared to indicate that the cells were highly impermeable to amino acids. If this were true, then the entry of amino acids could not occur by simple diffusion, which is by definition a reversible process: it had to be mediated by some special, uni-directional transfer mechanism. This conclusion was also suggested by the fact that glutamic acid enters the cells only in the presence of glucose. However, lysine, which is accumulated to a similar extent as glutamic acid, and is equally retained by intact cells, does not require glucose for its entry. An alternative mechanism therefore has to be considered, namely that the amino acids are retained within staphylococcal cells by intermolecular forces, for instance by some kind of intracellular receptors. If so, no permeable barrier, or any permeation mechanism, need to be assumed to account for the accumulation. These two alternatives must both be considered, and weighed against each other, whenever attempting to interpret the mechanism of accumulation of a compound by a cell. A choice between them is always difficult; all the more so since they are not mutually exclusive: proving a contribution to the accumulation process by one of these mechanisms does not in itself disprove a contribution by the other.

The paradoxical finding that enzymes active against a given substrate may, in some cases, be extracted from cells which, when intact, are inert toward the same substrate, has been noted many times and diversely interpreted by puzzled microbiologists. There is of course no paradox when the "cryptic" state of the cells concerns a whole class of chemical compounds, since there is no difficulty in assuming that the solubility and/or electrical properties of a class of compounds may forbid their passage through the cell membrane. The phosphorylated metabolites (nucleotides, hexose phosphates) provide classical examples.

The paradox arises when interpretations in terms of non-specific forces or properties become inadequate; that is to say, when crypticity is highly stereospecific. The study of the metabolism of disaccharides by yeast has provided some of the earliest cases of specific crypticity. For instance, intact baker's yeast does not ferment maltose, although autolyzates of the same yeast contain maltase (α-glucosidase). Analogous observations have been made with other yeasts for cellobiose and cellobiase (β-glucosidase), sucrose and sucrase, etc. Similarly, it was observed in 1939 that dried preparations of a strain of *E. coli* which did not ferment lactose possessed lactase (β-galactosidase).

An essential point is that the cells which are cryptic towards a given carbohydrate nevertheless behave as a rule quite normally towards other carbohydrates. For instance, an *E. coli* cell which is cryptic towards lactose metabolizes glucose, maltose and other carbohydrates at a high rate. Now, if crypticity to a particular carbohydrate is attributed to the impermeability of the cell membrane, then the membrane must be impermeable to all compounds presenting similar solubility properties and molecular weight, that is to say to virtually all the carbohydrates. Therefore, those carbohydrates that do enter the cell and are metabolized at a high rate must use some highly specific stratagem for getting through the barrier.

The actual demonstration and identification of a specific permeation system, as distinct from other similar systems and from intracellular metabolic enzymes, has rested upon a sort of operational isolation *in vivo*, which has required a combination of several approaches, and we shall now examine it in some detail.

β-galactoside permease

Before introducing the system, it should be recalled that *E. coli* metabolizes lactose and other galactosides *via* the inducible system β-galactosidase. Analogues of β-galactosides where the oxygen atom of the glycosidic linkage is replaced by sulfur (fig. 1) are not split by β-galactosidase, nor are they used by *E. coli* as a source of energy, carbon or sulfur.

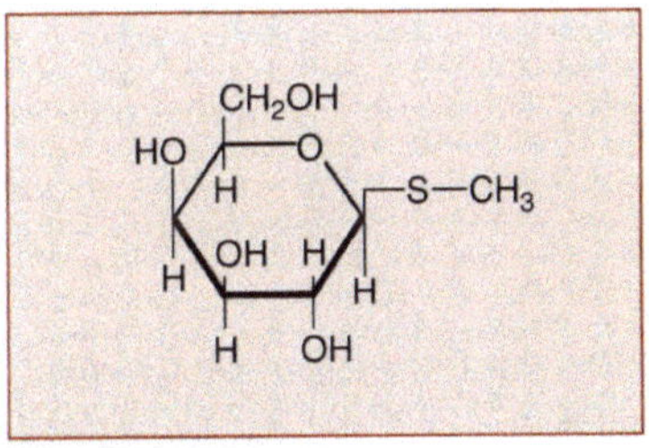

Figure 1. A thiogalactoside, methyl-β-D-thiogalactoside (IPTG).

ACCUMULATION IN INDUCED CELLS: KINETICS AND SPECIFICITY

When a suspension of *E. coli,* previously induced by growth in the presence of a galactoside, is shaken for a few minutes with an ^{35}S-labeled thiogalactoside, and the cells are rapidly separated from the suspending fluid by membrane filtration, they are found to retain an amount of radioactivity corresponding to an intracellular concentration of galactoside which may exceed by 100-fold or more its concentration in the external medium. Non-induced cells, *i.e.*, cells grown in the absence of a galactoside do not accumulate any significant amount of radioactivity.

The accumulated radioactivity is quantitatively extracted by boiling water. Chromatographic analysis of extracts show a major spot, which by all criteria corresponds to the free unchanged thiogalactoside. A minor compound accumulates if the incubation is prolonged. It corresponds to 6-acetylthiogalactoside, and its formation may be disregarded in discussing the kinetics of accumulation.

The accumulation is reversible: when the uptake of galactoside is followed as a function of time, a stable maximum is seen to be reached gradually (within 5 to 20 min. at 34 °C, depending on the galactoside used). If, at this point, an unlabeled galactoside is added to the medium at a suitable concentration, the radioactivity flows out of the cell (fig. 2).

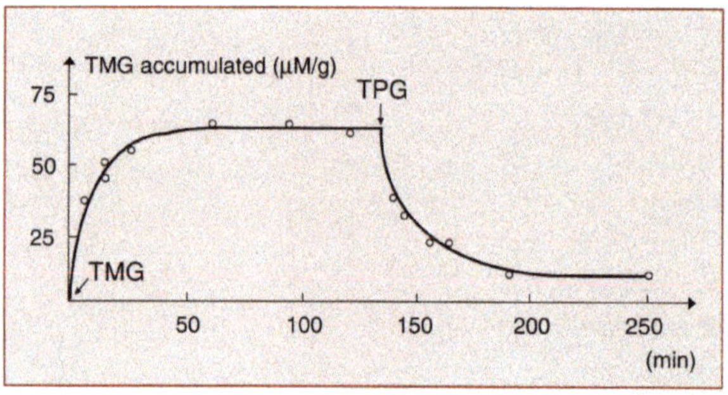

Figure 2. Accumulation of methyl-β-D-galactoside (TMG) by induced bacteria. Non-radioactive β-D-phenylthiogalactoside (TPG) was added at the point shown by an arrow. The experiment was carried out at 0 °C to enable the kinetics to be followed. (From G. N. Cohen & J. Monod, with permission of *Bacteriological Reviews*.)

The amount of intracellular galactoside at equilibrium, in presence of increasing external concentrations of galactoside, follows quite accurately an adsorption isotherm (fig. 3).

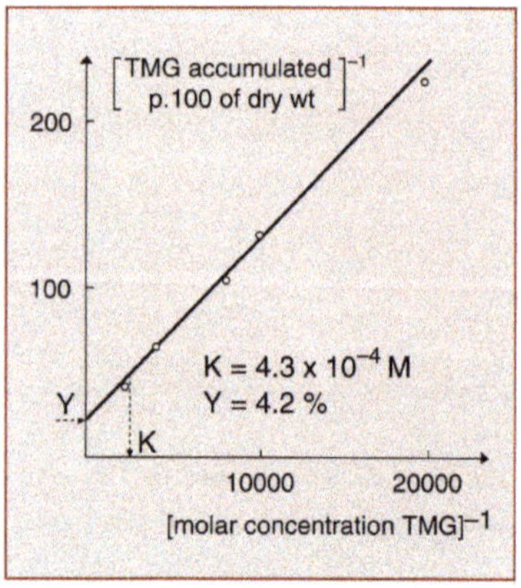

Figure 3. Accumulation of radioactive thiogalactoside as a function of external concentration.
Y and K are the constants of Equation 1.

Calling the external concentration of galactoside G_{ex}, and the amount taken up by the cells at equilibrium ${}^{eq}G_{in}$, one may write:

$$ {}^{eq}G_{in} = Y \frac{G_{ex}}{G_{ex} + K} \qquad (1) $$

K is the dissociation constant of the "bacterium-galactoside complex" and Y is another constant, called capacity, which expresses the maximal amount of galactoside which the cells take up at saturating concentration[1].

The displacement of a labeled galactoside by another, unlabeled, galactoside also follows quite accurately the classical laws of competition for a common site (fig. 4).

This allows the determination of specific affinity constants for any competitive compound. The specificity of the system proves very strict: only those compounds which possess an unsubstituted galactosidic residue (in either α or β linkage) present detectable affinity for the competition site. Glucosides or other carbohydrates, even though they may differ from galactosides only by the position of a single hydroxyl, do not com-

1. The "total capacity" is defined as the capacity per unit volume of cell suspension. The "specific capacity" is the capacity per unit weight of organisms. It may be expressed in per cent dry weight or, preferably, in moles per unit dry weight.

pete with the galactosides. Moreover, all the effective competitors which have been tested have also proved to be accumulated within the cells. The affinity constant for each can then be determined either directly, by measurement of accumulation, or indirectly by displacement of another galactoside. The two values agree reasonably well.

The results show that the accumulation of galactosides within induced cells is due to, and limited by, stereospecific sites able to form a reversible complex with α- and β-galactosides.

A stoechiometric model, according to which galactosides are accumulated within the cells in combination with specific receptor sites is dismissed mainly by a common-sense argument: in induced cells, the level of galactoside accumulation may be very high, and actually exceed 5% of the dry weight of the cells. If the intracellular galactoside was adsorbed onto stereospecific sites (presumably associated with cellular proteins) there would have to exist, in highly induced cells, one such site for each fraction of cellular protein of molecular weight 2,000, an assumption which seems quite unreasonable.

The second interpretation (catalytic or permease model) assigns to the specific "permease" sites the role of catalyzing the accumulation of the galactosides into the cell, rather than of serving as final acceptors. In order to account for the properties of the system, one is then led to the following scheme:

$$G_{ex} \xrightarrow[\text{(entry)}]{(y)} G_{in} \xrightarrow[\text{(exit)}]{} G_{ex}$$

according to which the intracellular galactoside (G_{in}) is a steady-state intermediate between an entry reaction, catalyzed by the stereospecific sites, and an independent exit reaction. The entry reaction involves the transitory formation of a specific complex between the sites and the galactoside and should follow the kinetics of enzyme reactions. The exit reaction is assumed not to involve the sites, and its rate to be proportional to the amount of intracellular galactoside G_{in}. According to these assumptions, the rate of increase of the intracellular galactoside is given by:

$$\frac{dG_{in}}{dt} = y \frac{G_{ex}}{G_{ex} + K} - cG_{in} \qquad (2)$$

If $\frac{y}{c} = Y$, equation (2) reduces to equation (1) for equilibrium conditions, when $\frac{dG_{in}}{dt} = 0$. The constant K again corresponds to the dissociation constant of the galactoside-site complex, while the capacity Y is now the ratio of the permease activity y to the exit rate constant c. Insofar as the latter remains constant, the level of intracellular galactoside is proportional to the activity of the permease.

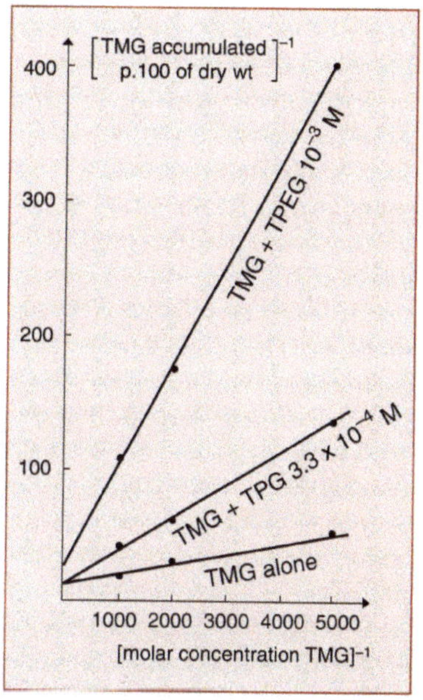

Figure 4. Competitive displacement of radioactive thiomethylgalactoside by thiophenylethylgalactoside (TPEG), and thiophenylgalactoside (TPG), in *Escherichia coli*.

The kinetics of intracellular galactoside accumulation also prove incompatible with the stoechiometric model while in good agreement with the permease model. But the most decisive reason for adopting the permease interpretation is the evidence that cells genetically or otherwise devoid of permease are specifically cryptic towards galactosides. This evidence will be reviewed later in this chapter.

The permease model assigns the binding essentially to a high degree of impermeability of the cell membrane (or other osmotic membrane) toward carbohydrates. The permease sites catalyzing the entry must then be assumed to be associated with the osmotic barrier itself.

The following experiment depends on the fact that protoplasts of *E. coli* retain the permeability properties of intact cells while their resistance to osmotic pressure is very greatly reduced. Differences in osmotic pressure between the intra- and extracellular phases may result from either a decrease of the external osmolality or an increase of the internal pressure resulting, for example, from intracellular accumulation of a compound in an osmotically active state, *i.e.*, in a *free* state. The bursting of protoplasts is easily measured as a decrease of the optical density of the suspension, so that protoplasts can be used as sensitive indicators of variation of their own osmotic pressure. By comparing the extent of lysis which occurs when the internal pressure increases as a result of permease activity with that caused by a known decrease in external pressure, one can determine approximately the intracellular concentration of free permease substrate at the time of lysis.

Using this experimental principle, the lysis caused by accumulation of various galactosides in *E. coli* protoplasts has been studied. The strain used was a mutant possessing an inducible galactoside permease and devoid of galactosidase, *i. e.*, able to accumulate, but unable to hydrolyse, lactose or other galactosides. It was observed that protoplasts prepared from induced cells (*i.e.*, cells grown in presence of an inducer of galactoside permease) and suspended in 0.1 M phosphate buffer underwent lysis in presence of 10^{-2} M lactose, while protoplasts from uninduced cells were insensitive to the addition of lactose. Comparisons with lysis provoked by decreasing the molarity of the suspending buffer showed that the induced protoplasts had accumulated *free* lactose to the extent of about 22% of their dry weight, a figure closely approximating direct estimations made on intact cells. The addition of a rapidly metabolizable carbohydrate, namely glucose, did not result in any significant lysis.

These experiments demonstrate that the bulk, if not the totality, of the substrates of galactoside permease are accumulated within the cells in a free form. Therefore the accumulating mechanism must be catalytic, and the retention is due to a permeability barrier, not to binding or other intermolecular forces.

INDUCED SYNTHESIS OF GALACTOSIDE PERMEASE

The fact that galactoside permease is an inducible system has been of particular value for its study and characterization. As mentioned above, the system is active only in cells previously grown in the presence of a compound possessing a free unsubstituted galactosidic residue. No other carbohydrates show any inductive activity. Not even all galactosides are inducers. The specificity of induction can best be studied using thiogalactosides which are not hydrolyzed or "transgalactosylated" in the cells. The specificity pattern of induction is strikingly parallel to that of β-galactosidase, although there are some minor differences which might be significant (table I) in the relative inducing activity of different compounds.

Table I. Induction of galactoside permease and β-galactosidase by various thiogalactosides.
The specific activities are, respectively, μmoles TMG concentrated/g dry weight and mμmoles ONPG hydrolyzed per mn and per mg dry weight. The cultures were on synthetic medium, in presence of 10^{-3} M inducer. Permease was measured with 10^{-3} M TMG.

Inducer added during growth	Permease (specific activity)	β-galactosidase (specific activity)
None	< 2	2
Methyl-β-D-thiogalactoside	176	8,500
Propyl-β-D-thiogalactoside	181	8,500
Isopropyl-β-D-thiogalactoside	124	9,500
Hexyl-β-D-thiogalactoside	14	8.5
Phenyl-β-D-thiogalactoside	5	2
Benzyl-β-D-thiogalactoside	10	16
Phenylethyl-β-D-thiogalactoside	50	110
Galactoside-β-D-thiogalactoside	10	4

Since all galactosides are concentrated by the permease, all inducers are also "substrates" of the system. However, certain substances known to be actively concentrated (phenyl-β-D-thiogalactoside; thio-di-β-galactoside) show little or no inducing activity.

The induction is effective only under conditions allowing the synthesis of protein; it is blocked by chloramphenicol or in the absence of a required amino acid. The kinetics of the induced synthesis show that the increase in capacity corresponds to a *de novo* synthesis of protein.

FUNCTIONAL SIGNIFICANCE OF GALACTOSIDE PERMEASE; SPECIFIC CRYPTICITY

If the permease model is correct, *i.e.*, if (a) the cells are virtually impermeable to galactosides, except for the specific activity of the permease, (b) the permease is distinct from galactosidase, (c) galactosidase is strictly intracellular, one should expect at least two phenotypes among *E. coli* mutants unable to metabolize galactosides, one type corresponding to the loss of the permease, the other to the loss of the galactosidase. In fact, the two predicted types have been found.

Cells of the first mutant type ("absolute-negative"), grown in presence of a suitable inducer (thiomethylgalactoside), accumulate normal amounts of galactoside, but they form no detectable trace of β-galactosidase. These organisms, called thereafter z^-, accumulate in particular large amounts of lactose (up to 20% dry weight), while the induced normal cells do not accumulate lactose, which is split and metabolized as soon as it is taken up.

Cells of the second mutant type (cryptics or y^-), grown in presence of sufficiently high concentrations of thiomethylgalactoside, form normal amounts of galactosidase (as revealed by extraction), but no – or only traces – of permease. As long as they are physiologically intact, these cells hydrolyze galactosides at a much slower rate than normal cells possessing equal amounts of galactosidase. Moreover, the rate of hydrolysis is a linear function of galactoside concentration, instead of being hyperbolic, as in the wild type (fig. 5).

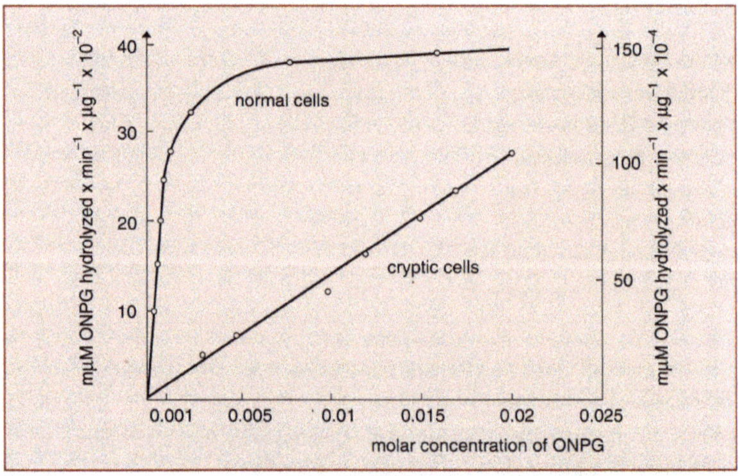

Figure 5. ***In vivo*** **hydrolysis of ortho-nitrophenyl-β-D-galactoside (ONPG) by whole cells of *E. coli*. Upper curve: wild type. Lower curve: cryptic (y^-) mutant. Ordinates on left apply to upper curve, on right to lower curve.**

This indicates that the rate of hydrolysis in these cells is limited by a diffusion process rather than by a catalyst. The organisms are, in particular, almost inert toward lactose, while their metabolic behavior towards other carbohydrates is normal. In other words, these organisms are specifically cryptic towards galactosides.

These properties of the cryptic mutants are immediately explained if β-galactosidase is effectively inside a highly permeable barrier which the galactosides can cross only by forming a complex with the permease. The existence of the two mutant types proves that permease and β-galactosidase are genetically and functionally distinct and that they form a metabolic sequence *in vivo*.

The study of the hydrolysis of galactosides *in vivo*, in wild-type organisms, also brings out the functional role of the permease. For instance, the hydrolysis *in vivo* of true galactosides is inhibited by thiogalactosides, allowing a determination of the affinities of the inhibitors for the total system. The pattern of affinities which is thus disclosed is quite different from that of β-galactosidase studied *in vitro*, but is very close to the pattern of affinities of the permease showing that, in normal induced wild-type, the permease, rather than the galactosidase, is the limiting factor for the hydrolysis of galactosides and that it controls *in vivo* the communications between the galactosidase and the external medium.

FUNCTIONAL RELATIONSHIPS OF PERMEASE: INDUCTION

Galactosides are not only substrates or specific inhibitors of permease and galactosidase; they are also inducers of the two systems. Therefore, by controlling the intracellular concentration of inducers, the permease should control the kinetics of induction of β-galactosidase and of its own induction.

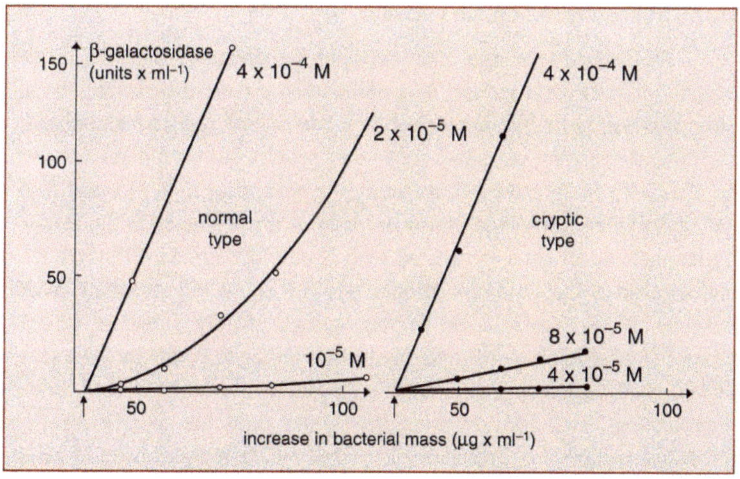

Figure 6. Kinetics of induced β-galactosidase synthesis in normal *E. coli* and in the cryptic mutant type.
Inducer: Isopropylthiogalactoside at the concentrations indicated.
Carbon source: succinate.

This is borne out by the experiments summarized in figure 6. The cryptic mutants, devoid of permease, show linear kinetics of induction at all concentrations of inducer. In contrast, with the normal cells, at low concentrations of inducer, the induction of the permease-galactosidase system is autocatalytic, since permease is a system which concentrates its own inducer.

GENETIC RELATIONSHIPS OF GALACTOSIDASE AND GALACTOSIDE PERMEASE

Although most mutations affect specifically either permease or galactosidase, giving rise to the "cryptic" and "absolute negatives" mentioned previously, some appear to suppress both simultaneously. They are deletions of the *lac* locus of the bacterial chromosome which comprises both the *z* and the *y* cistron. But another mutant type deserves special mention: this is the constitutive in which *both* the permease and β-galactosidase are formed in the absence of inducer. The mutation to constitutivity may occur in "cryptic" organisms devoid of permease. These "constitutive cryptics" which form very large amounts of β-galactosidase without external inducer but are unable to metabolize lactose, may in turn mutate to a permease-positive type able to grow on lactose. When this occurs, the permease is also constitutive. Therefore, a single-step mutation controls the constitutive *versus* inducible character of *both* galactosidase and permease. It should be added that this pleiotropic constitutive mutation is closely linked to the loci coding specifically for the structure of galactosidase and permease. We shall have to examine further the relationships between the inducible and constitutive states of this locus when we examine the mechanisms of induction.

GALACTOSIDE PERMEASE AS PROTEIN

At least *in vivo*, the functioning of the permease depends on the integrity of sulfhydryl groups; it is inhibited by *p*-mercuribenzoate, the inhibition being partially suppressed by cysteine and reduced glutathione. The thiogalactosides, permease substrates, protect against the inactivation, the ones with the greater affinity being the best protectors. Fox and Kennedy have taken advantage of this property of the permease in the following manner: intact cells are treated with N-ethylmaleimide (NEM), a reagent specific of the sulfhydryl groups which it covalently modifies, in the presence of thiodigalactoside (TDG), a high affinity substrate of permease. After several centrifugations and washings to eliminate the excess NEM and TDG, the cells are treated with radioactive NEM in the absence of protecting TDG. This, in principle, should allow the specific labeling of permease. One can indeed extract from the membrane fraction (with detergents) a radioactive protein which is the product of the *y* gene: the synthesis of this protein is inducible and it is distinct from β-galactosidase. The protein is firmly linked to the membrane, from which it cannot be extracted with the usual aqueous buffers, but is extractible with buffers containing ionic or non-ionic detergents.

An understanding of the mechanism of the permease has progressed through the cloning of the *y* gene and the examination of its protein sequence inferred from the DNA sequence. The expressed protein has been purified to homogeneity from the inner membrane, incorporated in proteoliposomes, and it has been shown that it is active as a monomer. It is a protein with 417 amino acid residues (fig. 7).

```
                  10                  20                  30
M Y Y L K N T N F W M F G L F F F F Y F F I M G A Y F P F F
                  40                  50                  60
P I W L H D I N H I S K S D T G I I F A A I S L F S L L F Q
                  70                  80                  90
P L F G L L S D K L G R K Y L L W I I T G M L V M F A P F F
                  100                 110                 120
I F I F G P L L Q Y N I L V G S I V G G I Y L G F C F N A G
                  130                 140                 150
A P A V E A F I E K V S R R S N F E F G R A R M F G C V G W
                  160                 170                 180
A L C A S I V G I M F T I N N Q F V F W L G S S G C A L I L
                  190                 200                 210
A V L L F F A K T D A P S S A T V A N A V G A N H S A F S L
                  220                 230                 240
K L A L E L F R Q P K L W F L S L Y V I G V S C T Y D V F D
                  250                 260                 270
Q Q F A N F F T S F F A T G E Q G T R V F G Y V T T M G E L
                  280                 290                 300
L N A S I M F F A P L I I N R I G G K N A L L L A G T I M S
                  310                 320                 330
V R I I G S S F A T S A L E V V I L K T L H M F E V P F L L
                  340                 350                 360
V G C F K Y I T S Q F E V R F S A T I Y L V C F C F F K Q L
                  370                 380                 390
A M I F M S V L A G N M Y E S I G F Q G A Y L V L G L V A L
                  400                 410           417
G F T L I S V F T L S G P G P L S L L R R Q V N E V A
```

Figure 7. Sequence of *E. coli* β-galactoside permease. One-letter code.

Circular dichroism studies of the pure protein and examination of the hydropathic profile of its sequence have allowed the proposal of a model of its secondary structure: the polypeptide possesses a short hydrophilic amino-terminal sequence on the internal face of the membrane, followed by 12 transmembrane hydrophobic domains, each being an α-helix, connected by more hydrophilic segments, and then a hydrophilic 17 amino acid C-terminal end, again on the internal face of the membrane (fig. 8). Evidence supporting the general features of the model has been obtained from laser Raman spectroscopy, limited proteolysis, immunological studies, and chemical modification. Exclusive support for the 12-helix motif has been obtained from an extensive series of lac permease-alkaline phosphatase *(lac Y-phoA)* fusions.

Mechanistic studies of the galactoside permease have greatly bene-

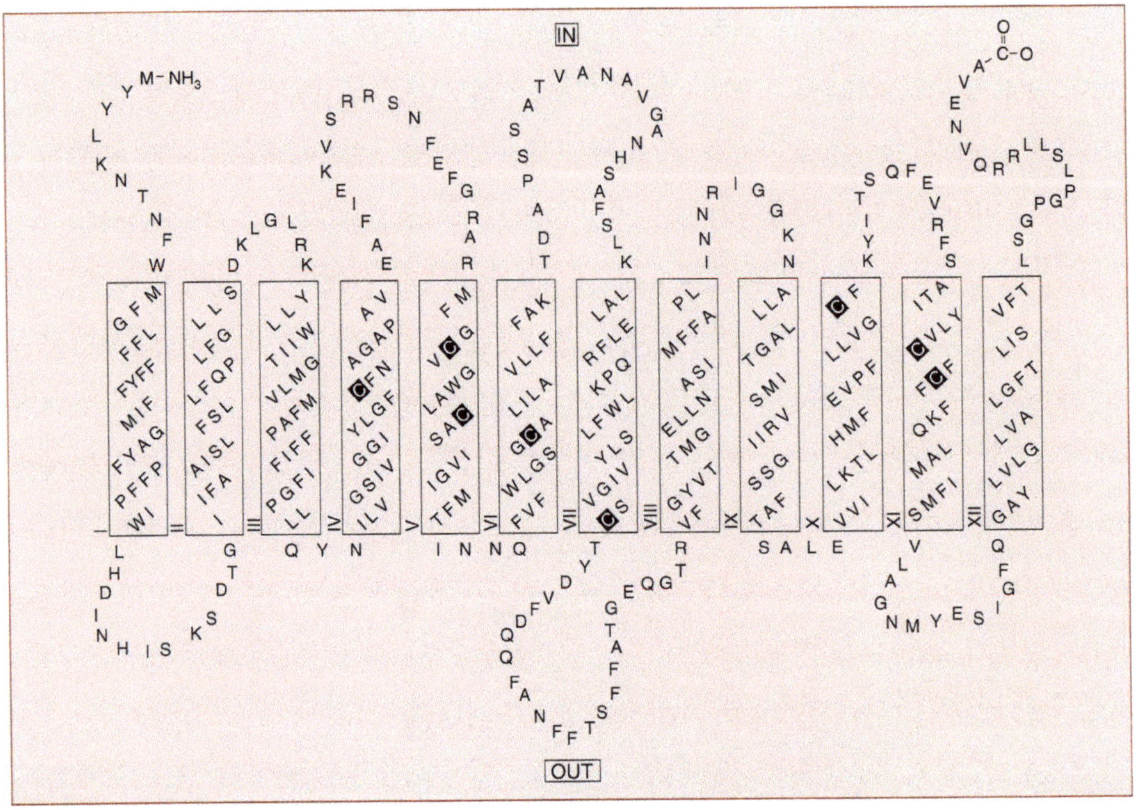

Figure 8. Model for the secondary structure of the β-galactoside permease based on a variety of data. The cysteine residues are highlighted on black backgrounds. One-letter code.
(From P. R. van Iwaarden, J. C. Pastore, W. N. Konings & H. R. Kaback, *Biochemistry* **30**, 9595-9600. With permission from the American Chemical Society.)

fited from the utilization by Kaback and his colleagues of membrane vesicles from the beginning of the 1970s to this day. The vesicles are prepared by osmotic lysis of *E. coli* spheroplasts under well-defined conditions; they are generally closed right-side out sacs of a diameter from 0.1 to 0.5 m, delimited by a double-layer membrane 70 Å thick. They are devoid of cytoplasmic proteins and of cofactors, and are unable to synthesize ATP by oxidative phosphorylation. However, in presence of a suitable energy source, they are able to accumulate sugars, amino acids and metals at rates comparable to those observed with intact cells. The accumulation obeys strictly the same genetic determinism as the bacteria from which they have been prepared. For example, vesicles prepared from a *y*- strain do not accumulate thiogalactosides. Respiration is indispensable for transport which is not accompanied by the phosphorylation of substrate.

If a pH gradient is imposed, the most acid part being outside the vesicle, accumulation of lactose takes place within the vesicle. A membrane potential, negative inside, generated by a gradient of potassium ions, also causes the accumulation of lactose. These results have revealed that the force leading to lactose accumulation is the protonmotive force through the membrane. Reconstituted vesicles containing only β-galactoside permease inserted in a phosphatidylcholine bilayer are perfectly able to transport lactose, the entry of each molecule of the disaccharide being coupled with the simultaneous movement of a proton from outside to the interior of the artificial vesicle; the permease actually contains, in addition to its galactoside binding site, a specific site for binding the protons. Under physiological conditions, the proton gradient necessary for active transport is generated by the electron flux originating from donors like $NADH^+$. We are thus in the presence of a further example of the unifying concept proposed by Peter Mitchell. Several sugars, amino acids, and succinate are also transported via this mechanism of proton symport.

Kaback and his colleagues have studied further the mechanism of β-galactoside/H^+ symport by site-directed mutagenesis of *y*, the lactose permease gene.

We shall try to summarize the main results of these studies. The studies of Fox and Kennedy, reported above, suggested the involvement of an essential sulfhydryl in or near the binding galactoside site. Subsequently, the substrate-protectable Cys residue was identified as Cys148, predicted to be in putative helix V. However, when this Cys is converted to a Gly or to a Ser, the permease is still active, although it is still inactivated by NEM; this inactivation is not protected by TDG. Therefore, although Cys148 is important for substrate protection against NEM in-activation, it does not play a direct role in the mechanism of lactose/H^+ symport. The seven other cysteines have been subjected individually to site-directed mutagenesis and only Cys154 is important for the symport although its mutants bind galactosides normally. Furthermore, recently (1991), Kaback has obtained a lactose permease where all the eight cysteines have been replaced, and which catalyzes active lactose transport, in a manner that it is resistant to inactivation by NEM, as well as efflux and equilibrium exchange. Then, individual putative intramembrane-charged residues have been replaced by neutral residues. Some of these replacements abolish lactose transport. The analysis of double replacements provides strong evidence that the putative transmembrane helix VII lies next to helices X and XI in the tertiary structure of the lactose permease.

Other mutants have been obtained where individual tyrosine, proline, tryptophan, arginine, histidine, or glutamic acid residues have been changed. The results show that disulfide-sulfhydryl interchange is not an integral part of the mechanism, that the pathways for H^+ and lactose translocation may share common residues, that the association of ligand probably involves large-scale environmental changes with small secondary structure alterations, and that the pathway for lactose may involve "binding sites" that include hydrophilic segments 2, 8 and 12 on the periplasmic surface, residues within putative helices IX and X and hydrophilic segments 5 and 11 on the cytoplasmic surface. Furthermore, the data in their entirety suggest possible arrangements of the 12 transmembrane helices in the permease, whereby key regions of the protein containing residues vital for various aspects of function are within reasonable proximity. One possible arrangement is shown in figure 9, although others can be postulated.

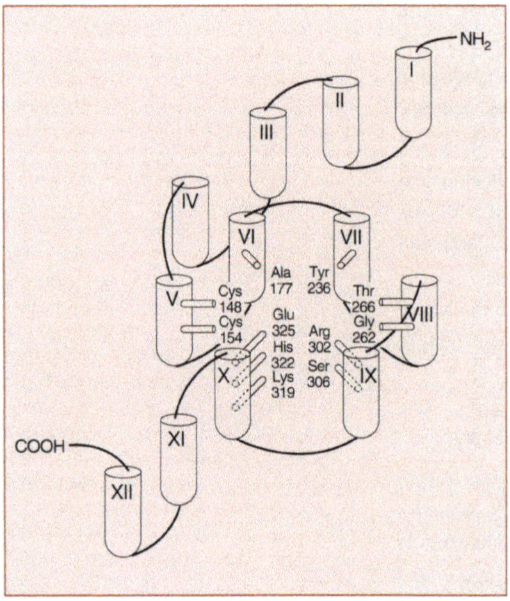

Figure 9. Hypothetical model for the tertiary structure of the β-galactoside permease. The model depicts 12 cylindrical transmembrane segments, in an α-helical conformation, connected by hydrophilic segments. (From. R. J. Brooker, *Res. Microbiol.* **141**, 309-315 (1990). With permission from *Research in Microbiology*.)

An interesting approach to model building deserves to be somewhat detailed: co-expression of the *lacY* gene encoding either the N-terminal or the C-terminal halves of the permease, or the first two putative helices and the remainder of the protein, leads to membrane insertion of the peptides and formation of functional complexes. No transport activity is observed when gene fragments are expressed individually. These experiments have been extended to new constructions based on the hypothetical model, and have provided evidence for a physical interaction between independently expressed pairs of complementing peptides, by co-immunoprecipitation with an anti-C terminal antibody. In addition, since helices I and II are independently inserted (they are separated in the model by an external loop) and form a stable complex, there is a strong argument against the hypothesis that the N-terminus inserts into the membrane as a helical hairpin.

Although the model is supported by biophysical, biochemical, immunological, and genetic studies, it is essential that it be confirmed by crystallography at an adequate resolution or by nuclear magnetic resonance studies.

Periplasmic binding proteins

Nossal and Heppel observed, 25 years ago, that a certain number of proteins are liberated from the bacterial cell envelope by an osmotic shock in the cold after treatment by a variety of agents. Among these proteins are some whose unique function seems to be the binding of small molecular weight substances, such as sugars, amino acids, or ions. The ease with which these proteins are released suggests that they are located near the cell surface, in the periplasm, a space between the cell wall and the the cytoplasmic membrane[2]. In many cases, it has been shown that cells having been subjected to an osmotic shock have a reduced capacity for accumulating various molecules. The property of the periplasmic proteins for restoring a normal accumulation, and the fact that they present association constants for their specific ligands of the order of the constants of intact cells, suggest that these proteins play an essential role in transport. Further arguments come from genetics: mutants of *E. coli* unable to accumulate sulfate are devoid of the corresponding periplasmic protein; the same is true of mutants unable to accumulate, respectively, arabinose or galactose.

In addition to those devoid of a sulfate-binding periplasmic protein, there are other mutants unable to accumulate sulfate which possess a normal binding protein; one has to admit that the mutation has affected another component of the system in those mutants; ten times more sulfate is needed to reach the half-saturation of the transport system than required for the periplasmic binding protein. Similar mutants, affected in their specific permeation, but with a normal periplasmic protein, have been described in the case of maltose and of arabinose.

The sulfate-, arabinose-, maltose- and galactose-binding proteins have been crystallized as well as the ones binding phosphate, valine-leucine-isoleucine, and lysine-arginine-ornithine[3] (LAO). The structure of the LAO binding protein has been solved at 2.7 Å resolution and shows that the molecule has a bilobal structure and that its topological structure, although different from other amino acid-binding proteins, is similar to that of the sulfate- and maltose-binding proteins. High sequence homology between LAO and the histidine-binding protein, which share the same membrane-bound protein complex (see below) permit the definition of functional regions responsible for the ligand binding and for the interaction with the membrane complex. Furthermore, the high degree of sequence similarity suggests that they have probably evolved from a tandem gene duplication event and function similarly.

The crystal structure of the arabinose-binding protein has been refined at 1.7 Å resolution, which allows a quantitative description of the interaction between the protein and arabinose. The refined structure of the complex of the protein with the sugar clearly shows an almost enclosed sugar molecule, the bound sugar being unable to leave without a confor-

2. The term "periplasmic protein" should be taken more as an operational definition than as an indication of a physical location, since a protein such as EF-Tu (elongation factor Tu), which must be functioning well inside the cell, actually is released by a treatment very similar to osmotic shock, while a number of phosphatases, which are typical periplasmic enzymes as far as their function is concerned, are poorly released by osmotic shock. The periplasmic space is almost isoosmotic with the cytoplasm. The osmotic pressure is thus maintained across the outer membrane and not across the inner membrane.
3. Many other periplasmic proteins have been sequenced.

mation change. This change allows essential residues of the protein to be properly oriented for binding. A hinge-bending motion between the two domains of the protein results in closure of the structure around the ligand (Venus flytrap hypothesis) and shields it from water. The ligand-induced conformational change is believed to be crucial to the function of the periplasmic proteins in transport. It generates the appropriate stereochemistry for the specific interaction of the liganded protein with membrane-bound protein components, thus initiating the translocation process.

This brings us to the actual passage of the ligand (which is still in the periplasm) across the cytoplasmic membrane. In the case of the histidine and maltose transport, it has been shown that the membrane potential is both insufficient and unnecessary for energizing periplasmic transport. Incontrovertible evidence that ATP is the direct energy source has been obtained by proteoliposome reconstitution using purified periplasmic proteins and bacterial phospholipids. The reconstitution of the histidine permease utilized pure histidine-binding periplasmic protein (coded by *hisJ*), about 80% pure complex comprising hisQ and hisM, two hydrophobic membrane-bound components and hisP, an ATP-binding membrane-bound protein, *E. coli* phospholipids and internally trapped ATP. The reconstituted system has an efficiency comparable to that in reconstituted right-side out vesicles or whole cells. It shows that the periplasmic protein alone is insufficient to explain the transport. Periplasmic proteins are subject, upon binding of their specific ligands, to conformation changes leading to an affinity change and to the delivery of the ligand to the external part of the cytoplasmic membrane where other hydrophobic membrane-bound components of the system would effect the actual translocation. Evidence has been obtained that these components (as shown for HisQ and HisM) are accessible on both the inner and outer membrane with the carboxyl termini to the inside surface of the cytoplasmic membrane and their amino termini in the periplasmic space. There are indications that both proteins span the membrane five times. ATP binds to one of these components and is hydrolyzed concomitantly with transport.

Maltose transport across the periplasmic membrane of *E. coli* is dependent upon the presence of the periplasmic binding protein MBP, the product of the *malE* gene. The products of the *malF*, *malG* and *malK* genes form a membrane associated complex that catalyzes the hydrolysis of ATP to provide energy for the transport event. Mutants have been isolated that have acquired the ability to grow on maltose in the absence of MBP. After reconstitution of the complex into proteoliposomes, measurements of the ATPase activity of mutant and wild-type complexes in the absence and presence of MBP have been carried out: they reveal that the wild type complex hydrolyzes ATP rapidly only when both maltose and MBP are present. In contrast, the mutant complexes have acquired the ability to hydrolyze ATP in the absence of maltose and MBP. The rate of hydrolysis in different complexes is directly proportional to the growth rate (on maltose as the carbon source) of the different mutants from which they were derived, indicating that the constitutive ATP hydrolysis, and presumably the cyclic conformational changes of the complex, produce maltose transport in the absence of MBP. These results suggest that the ATP hydrolysis is not directly coupled to ligand transport even in the wild-type cells and that one important function of MBP is to transmit a transmembrane signal through the transmembrane proteins MalF and MalG to the MalK protein on the other side of the membrane, so that ATP hydrolysis can occur. The authors of this work suggest that this function of MBP may be extended to periplasmic proteins in general (fig. 10).

MBP is the first binding protein involved in transport in which the structures of the liganded and the unliganded forms have been determined and refined at high resolution. The unliganded structure (open form) shows a rigid-body "hinge-bending" between two globular domains by approximately 35°, relative to the maltose-bound structure (closed form), opening the sugar binding groove located between the two domains.

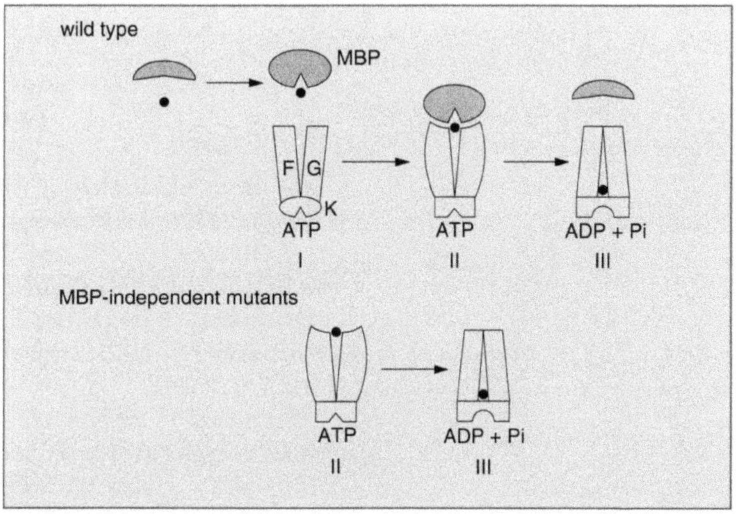

Figure 10. Model for maltose transport. In the wild type, MBP is in its open conformation, binds maltose, and is converted into a closed form. This form interacts with the MalFGK complex embedded in the membrane, in conformation I with the MalK ATPase being inactive. The binding of the liganded MBP affects the conformation of the transmembrane proteins F and G, which results in the activation of the MalK ATPase (II). Hydrolysis of ATP is then thought to push the ligand through the channel (III). In the MBP-independent mutants, the MalFGK complex is already in its activated conformation II.

(From A. L. Davidson, H. A. Shuman & H. Nikaido, *Proc. Natl. Acad. Sci. US* **89**, 2360-2364 (1992). With permission from Dr. H. Nikaido.)

Phosphotransferases. The PTS system

In 1962, it was shown that α-methylglucoside is partially accumulated within *E. coli* cells as a phosphate ester. Four years later, it was observed that the ester can be detected only during one minute, which suggested a specific mechanism of dephosphorylation, since in longer experiments the α-methylglucoside is found unchanged in the intracellular space.

Due to the efforts of Roseman and his colleagues, who analyzed this system for more than 25 years, a phosphotransferase was isolated in 1964 from several bacterial species which catalyzes the following sequence of reactions:

$$\text{phosphoenolpyruvate} + \text{HPr} \xrightarrow{\text{I, } Mg^{2+}} \text{HPr P} + \text{pyruvate} \qquad (1)$$

$$\text{HPr P} + \text{sugar} \xrightarrow{\text{II, III}} \text{sugar-phosphate} + \text{HPr} \qquad (2)$$

The sum of these two reactions leads to:

$$\text{phosphoenolpyruvate} + \text{sugar} \xrightarrow[Mg^{2+}, \text{ HPr}]{\text{I, II, III}} \text{sugar-phosphate} + \text{pyruvate} \qquad (1{+}2)$$

The unusual feature of this system is that phosphoenolpyruvate, rather than ATP, is the phosphoryl donor.

Crude extracts of *E. coli* contain a protein kinase that phosphorylates Enzyme I at a histidine residue. The activity of this ATP-dependent kinase depends on the presence of NAD^+ (or $NADP^+$). NADH reverses the activation by NAD^+, and thus the kinase level depends on the NAD^+/NADH ratio.

Enzyme I catalyzes the transfer of the phosphoryl group of phosphoenolpyruvate to a specific histidine residue (N5-atom of His15) of HPr which is a small thermostable protein[4]. The phosphohistidine is an energy-rich compound which transfers its phosphoryl group to "enzyme III" (a misnomer, enzyme III being actually a phosphocarrier protein like HPr), a peripheral membrane protein that interacts with enzyme II, the actual channel integral protein which is responsible for the final step, the

4. If *E. coli* is subjected to an osmotic shock, the cells lose between 50 and 80% of their HPr and, simultaneously, the capacity to concentrate α-methylglucoside by the same factor. If such cells are incubated with the purified thermostable protein, they recover their initial accumulation capacity. HPr has been sequenced. Its structure has been determined by X-ray crystallography and, in solution by three-dimensional and nuclear magnetic resonance, after labeling it heavily with ^{15}N and ^{13}C.
In *B. subtilis* and other Gram-positive bacteria, HPr is, in addition, phosphorylated at Ser46.

transfer of the phosphoryl group to the sugar that is being translocated. A functional enzyme complex has been obtained *in vitro* from the four purified proteins and an extensive study of the mechanism involving several critical amino acid residues is being carried on at present, involving, among others, the use of site-directed mutagenesis.

In *E. coli*, the phosphotransferase system is active on glucose (and α-methylglucoside), fructose, mannose and mannitol (in *Staphylococcus aureus* and *Bacillus subtilis*, similar systems are respectively responsible for lactose and sucrose transport). Whereas HPr and enzyme I are constitutive and soluble and participate in the transport of all sugars handled by the PTS, enzymes II and III are specific for particular sugars and inducible: there are different enzymes II and III for glucose, mannose and fructose. The three-dimensional structure of the enzyme III specific for glucose (III^{Glc}) has recently been solved. The molecule is a β-sheet sandwich with six antiparallel strands on either side. The active site is a shallow, extremely hydrophobic, depression near the center of one face, and is a good candidate for interaction with HPr. Mutants unable to synthesize a normal HPr or enzyme I have lost simultaneously the capacity to utilize numerous carbohydrates while mutants of enzymes II or III are unable to transport only a particular sugar.

Since the PTS is the mode of transport and of the introduction of many sugars into the metabolic network, the question arises as to whether it is general. The answer is negative. Among others, the accumulation of β-galactosides is inhibited by uncoupling agents such as azide and dinitrophenol which are without effect in the accumulation of α-methylglucoside by the PTS system. Moreover, membrane vesicles from a mutant devoid of enzyme I accumulate, normally, β-galactosides by the system linked to respiration.

In addition to its translocation function, the PTS has important regulatory functions. It regulates the uptake of its sugar substrates by controlling the adenylate cyclase and inhibiting the uptake of inducers by the non PTS-permeases. This aspect, which is in full development, is discussed at length in a 1990 review by Roseman and associates referred to in the references at the end of this chapter.

A few well-identified cases of specific cellular permeability

AMINO ACID PERMEASES

At the same time as the initial work on β-galactoside permease was being done, the same research team discovered specific permeases for amino acids in *E. coli.* A single permease catalyzes the intracellular concentration of the branched-chain amino acids valine, leucine and isoleucine; another is responsible for the entry into the intracellular space of methionine and of its structural analogue, norleucine. Later on, a third permease was described for the aromatic amino acids. Other systems were found responsible for the entry and accumulation of histidine, arginine, proline, etc.

Mutants devoid of one or the other amino acid permeases have been selected by layering the wild-type cells on plates containing a toxic structural analogue of the amino acid. These analogues are toxic because they are incorporated in proteins by replacing their natural counterparts at their normal positions in the protein sequences, thus leading, most of the time, to inactive enzymes. Some resistant colonies appear on the plates. Among these are found some which resist the toxic action of the analogue because they are devoid of the permease responsible both for the entry of the natural amino acid and the analogue. Thus, bacteria devoid of the arginine permease have been isolated among the resistants to canavanine (an arginine analogue); bacteria unable to concentrate glycine using D-serine as the analogue, and others lacking the histidine permease among the triazole alanine-resistant bacteria. Using this and other selection methods, practically all the specific permeases for all the amino acids have been identified and the genes governing their synthesis have been localized on the *E. coli* or the *S. typhimurium* chromosome. Membrane vesicles, identical to those used for the study of the galactoside permease, have been shown to concentrate proline, whereas membranes from bacteria without proline permease are unable to do so. Proline is thus concentrated by a system similar to the galactoside permease. In contrast, the protein responsible for the transport of valine, leucine and isoleucine, involves a periplasmic protein similar to the one involved in histidine transport.

PEPTIDE PERMEASES

The growth of *E. coli* auxotrophs requiring valine, leucine or isoleucine, is inhibited competitively by either of the two amino acids not required for growth. It has been shown conclusively that this competition is at the level of the common permease for the three amino acids, which has an association constant of the same order for the three compounds. However, if peptides containing the amino acids are used as competitors, they exert only a slight inhibitory effect when they are used at a concentration 100 times higher than the ones where the free amino acid exerts a total inhibition of growth. Thus, peptides escape the amino acid permease, and since they can be hydrolyzed at a slow rate and provide the free amino acids when they have entered the cells, there must be some specific mechanism for their entry. The isolation of mutants unable to transport a given amino acid, but growing on peptides containing this amino acid, has confirmed the independance of the accumulation mechanisms.

There exist an immense number of peptide structures, 400 for the dipeptides, 8000 for the tripeptides, etc. This precludes the existence of highly specific transport systems. The availability of radioactive peptides has provided a solution to this problem.

The existence of two independent peptide permeases has been shown in *E. coli*: one responsible for the transport of dipeptides, the other for oligopeptides. Dipeptides can be transported to a limited extent by the oligopeptide permease, but the opposite is not true. The two systems require a free amino terminal group, but only the dipeptide permease activity depends on a free carboxyl terminal group.

Mutants having lost the dipeptide permease (*dpp*, 14 mn) and retained the oligopeptide permease have not been thoroughly studied yet. Oligopeptide permease mutants are, however, available and have been isolated in the following way: triornithine is toxic for *E. coli* and mutants resistant to it can be isolated. The resistance is due to the lack of oligopeptide permease and therefore to the incapacity for bringing the toxic tripeptide inside the cells. Auxotrophs carrying this mutation cannot grow on oligopeptides, but grow perfectly on free amino acids or dipeptides (fig. 11).

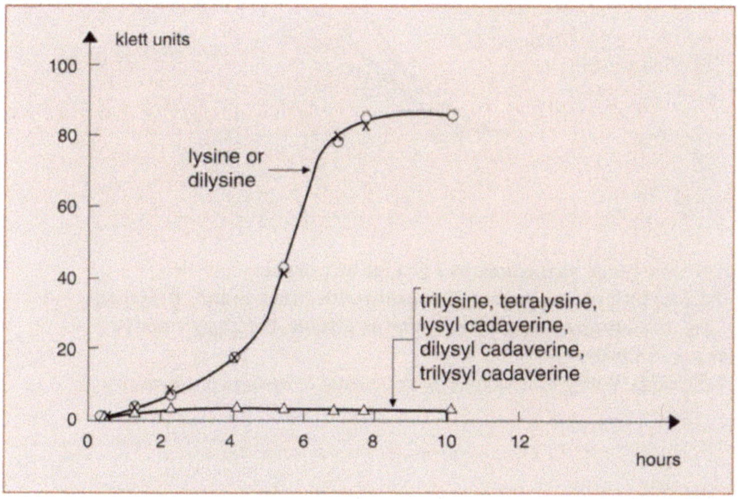

Figure 11.
Growth of an *E. coli* lysine auxotroph resistant to triornithine. Each culture contains either lysine or an equivalent concentration of lysine residues.
(From J. W. Payne & C. Gilvarg, *J. Biol. Chem.* **243**, 335-340 (1968). With permission from the *Journal of Biological Chemistry*.)

A single transport system exists for all oligopeptides, as shown by figure 12 which illustrates the competition, measured by the inhibition of the growth rate, of the utilization of trilysine or tetraglycine by trityrosine, in the corresponding lysine and glycine auxotrophs.

The oligopeptide permease (*opp*, 23 mn) belongs to the multicomponent periplasmic binding dependent transport, linked to ATP hydrolysis.

Porins

The cytoplasmic membrane of Gram-negative and most other bacteria is supported on the outside by the peptidoglycan layer, which gives the bacteria their characteristic shape and protects them from osmotic lysis. This peptidoglycan consists of a network of amino sugars (N-acetylglucosaminyl-N-acetylmuramyl dimers) forming long linear strands co-valently linked together through their muramyl residues by

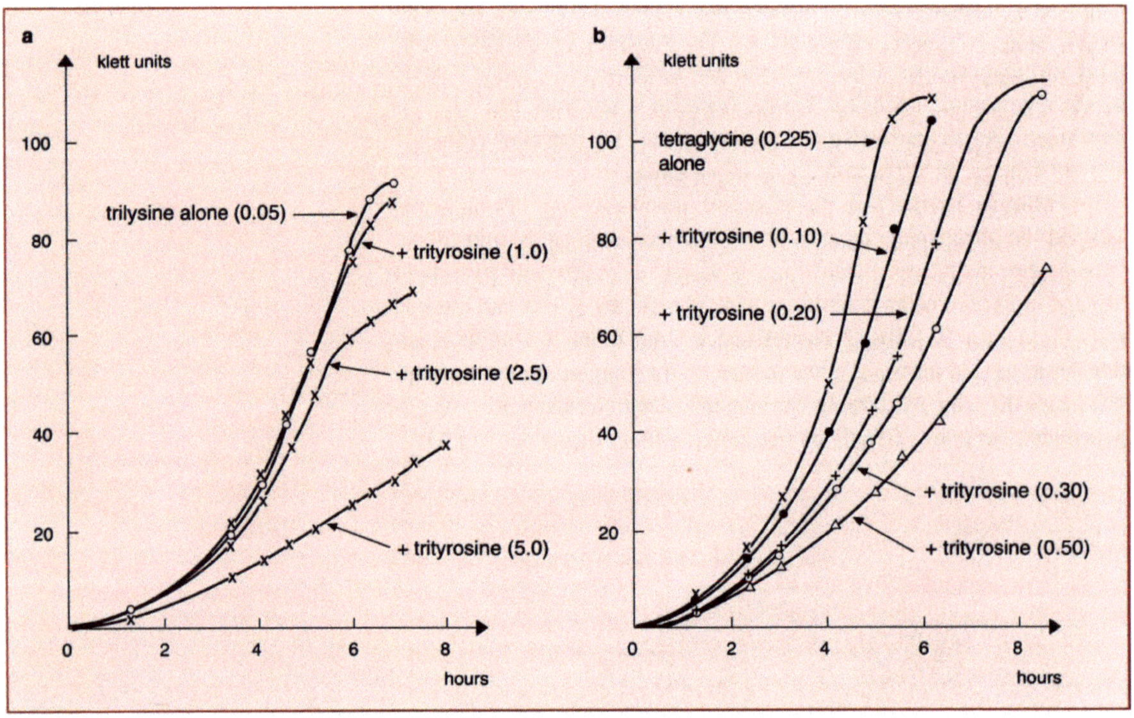

Figure 12. Effect of trityrosine on the growth of: a) a lysine auxotroph, b) a glycine auxotroph. Trityrosine has been selected because of its absolute lack of structural relationship with either lysine or glycine. Trityrosine does not affect the growth rate of the mutants growing on free lysine or glycine. Numbers refer to concentrations of oligopeptides in micromoles per milliliter.
(From J.W. Payne, *J. Biol. Chem.* **243**, 3395-3403 (1968). With permission from the *Journal of Biological Chemistry*.)

short peptides. The components of the outer membrane are bound to the peptidoglycan in such a way to form a tight network.

Porins, also called "peptidoglycan-associated proteins" or "matrix proteins", are very important for Gram-negative bacteria because all hydrophilic nutrients must pass through them in the outer membrane. They are diffusion channels often tightly bound to the peptidoglycan layer. Porins from many species of bacteria have been isolated and characterized. Each species contains a very limited number of porins; for example, according to the strain examined, *E. coli* contains only one (OmpF) or two (OmpF and OmpC) porins. The OmpF protein is produced at low osmolarity, while the OmpC protein is produced to a greater extent in high osmolarity. Their expression depends on two genes, *ompR* (75 mn), coding for a cytoplasmic repressor which binds to the promoter regions of both the *ompF* (21 mn) and the *ompC* (48 mn) genes, and *envZ* (75 mn), coding for an inner transmembrane protein, EnvZ, which functions as an osmolality sensor located in the cell envelope and can phosphorylate the OmpR protein. The phosphorylated OmpR exhibits an increased binding affinity to the *ompF* and *ompC* promoters. OmpF and OmpC exist as tight trimeric complexes which are not dissociated by sodium dodecylsulfate unless the protein is denatured by heating.

Most porins form wide channels filled with water in the outer membrane, which allow the diffusion of different types of solutes according to their molecular weight, the larger exclusion limit being of 5,000 daltons for the porin of *Pseudomonas aeruginosa.*

The porins fall into two classes: non-specific porins which form channels that permit general diffusion of hydrophilic molecules below a certain size, and thus are responsible for the exclusion limit of the outer membrane; and substrate-specific porins that act as facilitated diffusion channels for specific substrates by virtue of having a substrate-specific binding site in their channel. To date only six substrate-specific porins have been defined: growth of *E. coli* on maltose is accompanied by the synthesis of the LamB protein which is specific for the permeation of maltose and maltodextrins across the outer membrane; the nucleoside-specific porin Tsx of *E. coli (*coded by *tsx,* 9 mn); and, in *Pseudomonas aeruginosa*, a basic amino acid-specific porin (OprD), a phosphate-specific porin (OprP), a glucose-specific porin (OprB) and a polyphosphate-specific porin (OprO).

Some porins are inducible: phosphate limitation leads to the induction of the OprP protein of *Pseudomonas*, which allows the passage of phosphate. The induction of OprO requires, in addition, that cells be in the stationary growth phase.

The primary structure of many porins has been either determined by direct sequence of the protein or deduced from the DNA sequence of their genes. Electron diffraction studies, Fourier transform infrared studies, have shown that the polypeptide chain of porins traverses the membrane more than a dozen times. The β-structures are predominant. Whereas the function of the general diffusion porins is understood through the pore characteristics, chemical modification, and genetic mutations, our knowledge of their structure has suffered from the lack of good X-ray data. The selectivity of the specific porins could not be fully understood until the structure of the binding site inside the specific pore channels was available.

But recently the crystal structures of a porin from *Rhodobacter capsulatus,* of *E. coli* OmpF and PhoE have been determined by X-ray analysis. They reveal a homotrimeric structure, each of the three subunits consisting of a 16-stranded antiparallel β-barrel containing a pore. A long loop inside the barrel contributes to a constriction of the channel where the charge distribution affects the ion selectivity. The structures of OmpF and PhoE are compatible at the molecular level with the functional characteristics of the two porins and with their alterations by known mutations.

Functions of porins OmpF, OmpC and PhoE were analyzed at various pHs. Preliminary results from lipid bilayer membranes and liposomes swelling assays indicated that *in vitro* porin has at least two open-channel configurations with a small and a large size. The small channels were stabilized at low pH, while the larger channels were detected under basic conditions. The size switch occurs over a very narrow range near neutral pH, and the two major open-channel configurations responded differently to variations in voltage (applied in order to measure the electrical conductance across the lipid bilayer). The results obtained by modification of histidine 21 of OmpC and OmpF from *E. coli* with diethylpyrocarbonate (a specific histidine reagent) suggests that this residue is involved in the pH-induced change in channel size.

Porins have been extracted from the mitonchodrial outer membrane of organisms as diverse as *S. cerevisiae, N. crassa,* human B-lymphocytes, *Paramecium* and *Dictyostelium discoideum.* Their sequences show no particular homology to those of bacteria, although the function of porins as channels are similar to those of the mitochondrial porins.

Conclusion

At least four different systems playing a role in the passive or active transport of small solutes in bacteria have been identified: porins linked to the outer membrane, periplasmic proteins linked to cytoplasmic membrane proteins and where the transport has been shown in a limited number of cases to be linked to ATP hydrolysis, membrane carriers linked to respiration (*e.g.* β-galactoside permease), and the phosphotransferase (PTS) system where vectorial translocation is linked to phosphorylation of the substrate.

Comparison of amino acid sequences show strong conservation not only among bacterial transporters but also between them and many transporters of animal cells. A unifying design among membrane transporters is gradually emerging: common structural motifs and evolutionary origins among transporters with various energy-coupling mechanisms suggest that many transporters contain a module forming a transmembrane channel through which the solute may pass with size or/and steric constraints. Energy coupling mechanisms can then be viewed as secondary features added to the fundamental translocation units.

SELECTED REFERENCES

Coupling of energy to transmembrane solute translocation in bacteria, 6th Forum in Microbiology, *Res. Microbiol.* **141**, 281-395 (1990).

β-galactoside permease

G. N. Cohen & J. Monod, *Bacteriol. Revs.* 21, 169-194 (1957).

H. R. Kaback, *Ann. Rev. Biophys. Biophys. Chem.* **15**, 379-319 (1986).

Periplasmic proteins

F. A. Quiocho, *Ann. Rev. Biochem.* **55**, 287-315 (1986).

G. Ferro-Luzzi Ames, *Ann. Rev. Biochem.* **55**, 397-425 (1986).

A. L. Davidson, H. A. Shuman & H. Nikaido, *Proc. Natl. Acad. Sci.* US **89**, 2360-2364 (1992).

Phosphotransferase system

N. D. Meadow, D. K. Fox & S. Roseman, *Ann. Rev. Biochem.* **59**, 497-542 (1990).

Peptide permeases

J. W. Payne & C. Gilvarg, *Adv. Enzymol.* **35**, 187-244 (1971).

Porins

R. Benz, *Ann. Rev. Microbiol.* **42**, 359-393 (1988).

H. Nikaido, *Molecular Microbiology* **6**, 419-423 (1992).

H. Nikaido & M. H. Saier, Jr., *Science* **258**, 936-942 (1992).

CHAPTER III

Allosteric enzymes

One of the most remarkable properties of the cell is its capacity to coordinate efficiently its different biochemical activities. A steady state is thus maintained between the different catabolic processes and the thousands of synthetic reactions necessary for the reproduction of the cell.

The last thirty years have seen a striking advance in our knowledge of numerous regulatory processes, and in our understanding of their role in the cellular economy.

The present chapter will be confined to an examination of the properties common to many enzymes whose activity is subject to metabolic regulation. These controls will be examined in some detail when certain regulated functions are encountered.

Experiments carried out with *E. coli* some forty years ago at the Department of Terrestrial Magnetism (!) of the Carnegie Institution of Washington showed clearly that each end product of a biosynthetic sequence is part of the system responsible for its synthesis. Thus, the *de novo* synthesis of different amino acids from glucose is halted when these amino acids are added to the growth medium. The mechanisms which underlie the control by end products became clear when it was shown that each of these essential metabolites possesses the property of *repressing* the synthesis of one or several of the enzymes belonging to its biosynthetic pathway and also of *inhibiting* the activity of the early enzymes (often the first enzyme) in the pathway.

These two control mechanisms are distinct, as shown by the existence of mutants with one or the other controls affected. Nevertheless, the two mechanisms often occur in the regulation of the same metabolic pathway.

Which are the enzymes subject to control? Every highly complex metabolic system contains some reactions which produce intermediates occurring at branch points[5]. The branches may include biosynthetic pathways leading to different end products, or mixtures of degradative and biosynthetic pathways. In this entangled network of reactions, certain enzyme activities occur at strategic points[5], where their regulation becomes important for the preservation of the delicate equilibrium required to integrate the different metabolic functions. During evolution, natural selection has retained those enzymatic species with structures best suited as

5. To illustrate this point, it suffices to consult one of the many versions of the metabolic maps currently available.

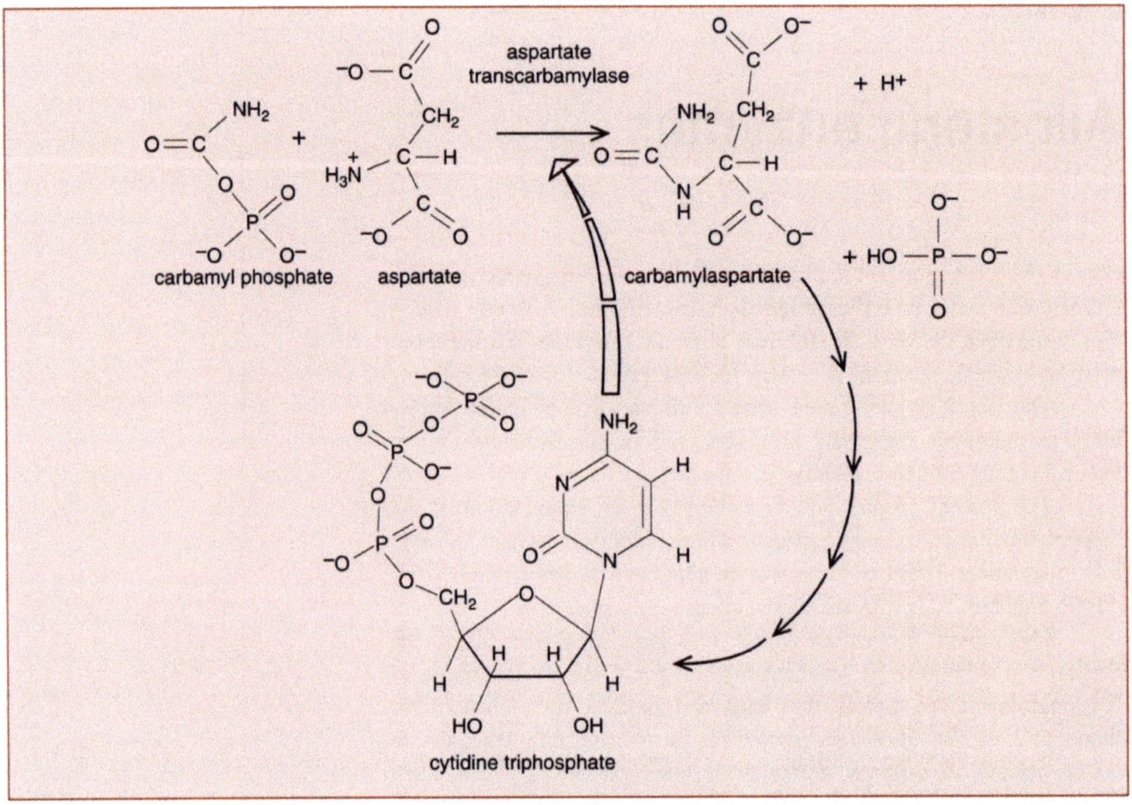

Figure 1. Feedback inhibition in the synthesis of pyrimidines by *E. coli*. Cytidine triphosphate is an allosteric inhibitor of aspartate transcarbamylase.

targets for activations or inhibitions endowed with a regulatory signification. The existence of control mechanisms is one of the principal bases on which most teleological arguments are built. The natural selection of these systems can make every natural phenomenon appear useful[6].

Since many metabolic functions are largely reversible, mass action might be thought to play a significant part in regulation. However, most pathways involve one or several irreversible steps which could not be controlled by mass action. Moreover, it is a general observation that the intracellular concentration of most intermediary metabolites is vanishingly small, indicating that mass action has only a limited role and suggesting that other mechanisms must intervene in metabolic regulation.

The examination of a large number of enzymes subject to control of their activity reveals certain common characteristics, structural as well as kinetic, which are not generally found for other enzymes. The most fundamental of these characteristics is the ability of this particular class of enzymes to be activated or inhibited by metabolites other than their

6. In this respect, in order to avoid any implicit reference to divine foresight and the raising of doubts about the theological bias of its user, C. S. Pittendrigh and B. D. Davis have suggested the word *teleonomy* to make explicit the shift in meaning of teleology as used by the modern biologist.

substrates or analogues. Often, there is no structural similarity between "effectors" (activitors or inhibitors) and substrates. A few examples will be sufficient to substantiate this:

L-isoleucine is a specific inhibitor and L-valine a specific activator of *E. coli* L-threonine dehydratase (also called L-threonine deaminase), the first enzyme specifically involved ("committed") in the synthesis of isoleucine. It could be argued that these compounds are analogues of L-threonine because they share a carboxyl and amino group in the same configuration. However, threonine dehydratase is not inhibited by any other naturally occurring amino acid other than L-isoleucine. Many more cases of strict specificity together with the absence of structural relationship between effector and substrate will be encountered throughout this book. We shall cite some examples here in order to make a general rule. Let us take the case of aspartate transcarbamylase, the first enzyme in the biosynthesis of pyrimidines; in *E. coli,* this enzyme is inhibited by cytidine triphosphate and activated by adenosine triphosphate. Examination of the structural formulas of the substrates, carbamyl phosphate and aspartate, and of the inhibitor (fig. 1) is enough to show that there is no similarity between them.

Likewise, in the bacterium *Rhodopseudomonas spheroides,* hemin inhibits the synthesis of δ-aminolevulinic acid from glycine and succinyl CoA. This acid is the compound from which the specific porphyrin synthesis diverges from the common catabolic path (fig. 2).

The very fact that the effector is not a steric analogue of the substrate has led to its being given the name *allosteric effector*, and the name *allosteric sites* to sites on the enzyme to which the effectors specifically

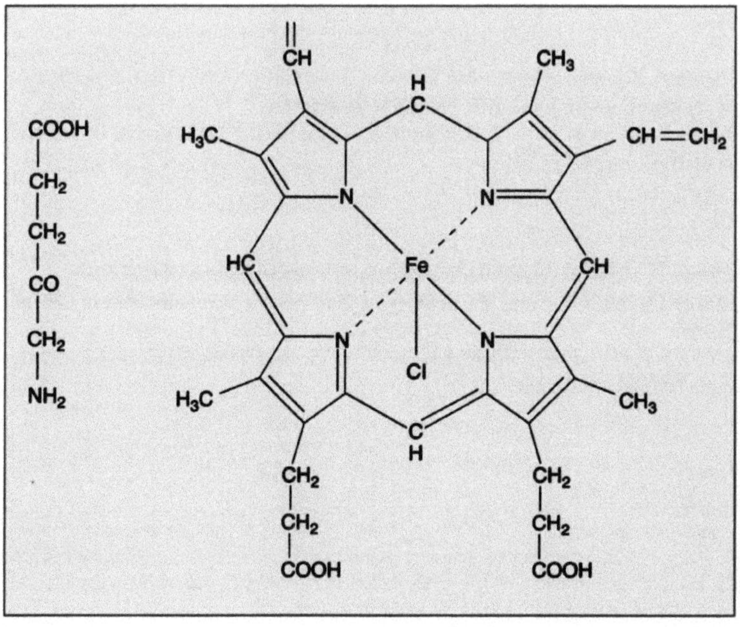

Figure 2. δ-aminolevulinic acid (ALA) is a precursor of hemin. Hemin inhibits the synthesis of ALA from simpler precursors. Examination of the structural formulas above shows that, even with the best will, no structural similarities can be found between the two compounds.

bind. By extension, the enzymes subject to control by these effectors are called *allosteric proteins or allosteric enzymes*[7].

One of the most frequent characteristics of the allosteric enzymes is the atypical relationship between their activity and the concentration of substrate and effector concentration. For the majority of enzymes, this relationship is described by a rectangular hyperbola, representing graphically the Henri-Michaelis equation (fig. 3).

In the case of many enzymes subject to allosteric control, this relationship is represented by a sigmoid curve (fig. 4).

Such a curve indicates that at least two molecules of substrate react with the enzyme and that the binding of one molecule of substrate in some way makes the binding of the second easier. Expressed differently, there is a cooperative effect in binding more than one substrate molecule to the enzyme. One often finds similar cooperative interactions in the binding of allosteric effectors, which suggests that allosteric enzymes contain more than one allosteric site per molecule (fig. 5). These interactions must reflect some fundamental property. Whatever the mechanism, they have a physiological significance: a sigmoidal response of the enzymatic activity to increasing concentrations of substrate or effector is in practice a "threshold" effect. At concentrations lower than the threshold, the enzymatic activity is essentially unaltered by changes in substrate or effector concentration, but on crossing the threshold, important activity changes are obtained by only slight changes in concentration. These threshold effects make the adjustment of enzymatic activity very sensitive to minimal changes in concentration within a narrow range.

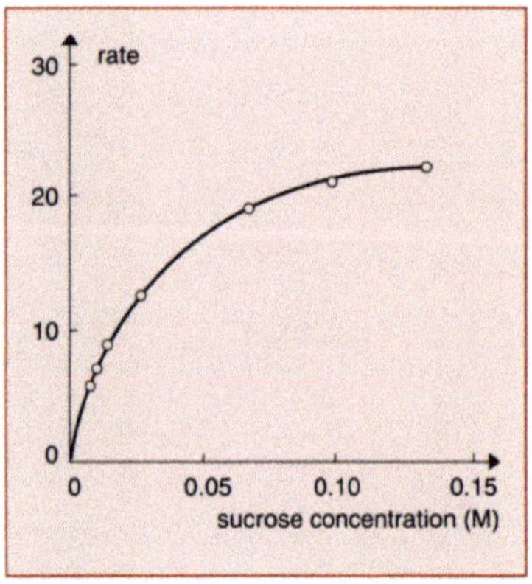

Figure 3. Hyperbolic relationship between the concentration of sucrose and its rate of hydrolysis by sucrase.

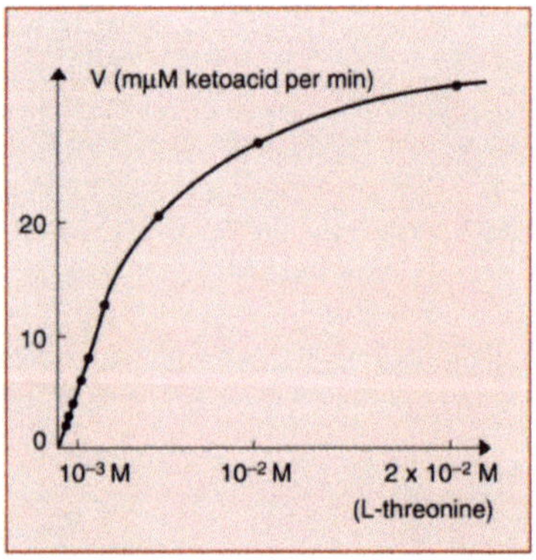

Figure 4. Sigmoid relationship between threonine concentration and the rate of its deamination by *E. coli* threonine deaminase.
(From J.-P. Changeux, with permission from the Société Française de Biochimie et de Biologie Moléculaire.)

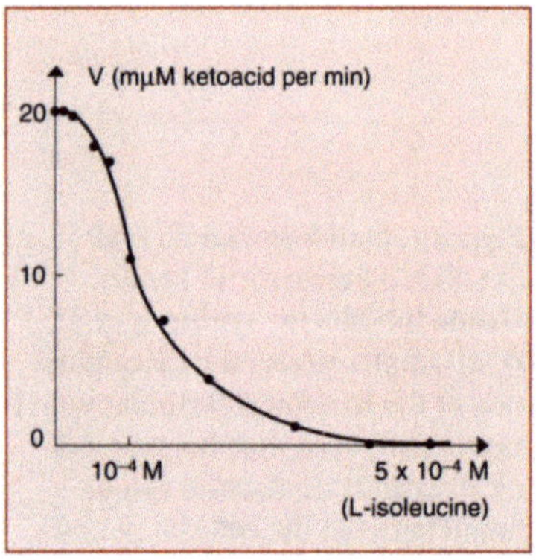

Figure 5. Sigmoid relationship between the concentration of isoleucine (allosteric inhibitor) and the activity of biosynthetic threonine deaminase of *E. coli*.
(From J.-P. Changeux, with permission from the Société Française de Biochimie et de Biologie Moléculaire.)

7. This terminology has been universally adopted and the term allosteric interaction is used to describe the interaction of any small molecule, including substrates, with binding sites other than the sites responsible for the catalytic activity of the protein.

Allosteric inhibition and activation

Classical kinetic analysis of the enzyme inhibitions caused by allosteric effectors show that they are reversible and can be competitive, non-competitive or of the mixed type, with kinetics quite often obeying very complex laws. When inhibition is non-competitive or mixed, the inhibitor binds to sites distinct from the catalytic site. Competitive inhibition is more difficult to explain. In the case of "classical" enzymes, competitive inhibitions are obtained with steric analogues of the substrate, and it is reasonable to assume that the substrate and the inhibitor compete with each other for occupation of the catalytic site. In some of the cases of interest here, the absence of structural similarity between the highly specific allosteric inhibitors and the substrates of the enzymes they inhibit seems to exclude the possibility of such a competition, since the catalytic sites are usually characterized by a strict degree of specificity. In fact, allosteric effectors have been found to bind to sites separate from the catalytic sites; numerous allosteric enzymes can be made by mutation or by various treatments insensitive (desensitized) to their allosteric effectors without their catalytic activity being affected[8]. In one case, at least, it has been possible to physically separate the different subunits of an allosteric enzyme subject to competitive inhibition and to show that one kind of subunit carries the catalytic site, and the other the binding site for the allosteric effector. This proves unequivocally that the two types of sites reside on different parts of the protein and that the apparent competitive inhibition does not result from a direct interaction with the catalytic site. Rather, the inevitable conclusion is that the binding of the inhibitor causes some conformational change in the enzyme which results in a diminished affinity for the substrate at the catalytic site.

Some allosteric effectors cause conformational changes which increase the affinity for the substrate. In the case of the enzymes showing a sigmoidal relationship between their activity and substrate concentration, the addition of these allosteric activators changes the sigmoid relationship to a hyperbolic one. In a number of cases, it is found that allosteric activators permit a modulation of the enzymatic activity to the advantage of the cell, as in the case of the threshold effects described above for allosteric inhibitors.

A number of allosteric enzymes are protected against thermal inactivation by their allosteric effectors (fig. 6). Such effects are attributable to an increased stability of the protein due to the conformational changes which accompany the reversible binding of allosteric ligands to their specific sites.

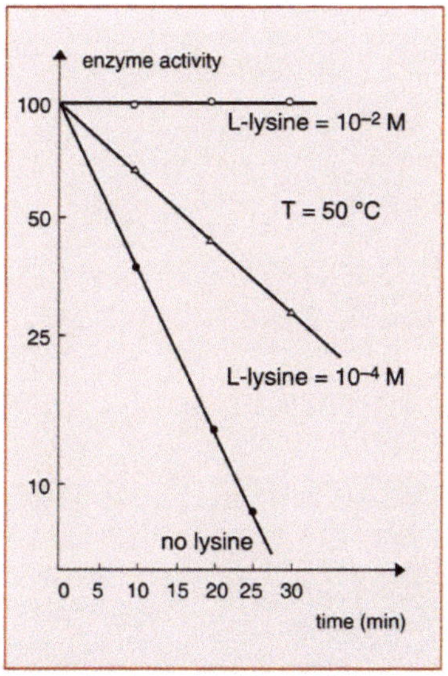

Figure 6. Protection of *E. coli* aspartokinase III against inactivation by heat. The inactivation can be totally prevented by L-lysine, a specific allosteric inhibitor of this aspartokinase.
(From P. Truffa-Bachi & G. N. Cohen, *Biochimica et Biophysica Acta* **113**, 531-541 (1966). With permission from Elsevier Science Publishers.)

8. The desensitizing agents can be as various as mercurials, cold treatment, dialysis, freezing, urea, limited proteolysis, high ionic strength, changes in pH.

Polymeric nature of allosteric enzymes. Monod, Wyman, and Changeux's model

All carefully studied allosteric enzymes are made up of subunits. Reversible associations and dissociations are found with some of them, which may or may not be accompanied by inactivation or a number of changes in susceptibility to allosteric effectors. Although this absence of a general rule would indicate that association and dissociation are not inherent in the allosteric effects, it is possible that such an absence indicates that the strength of the binding non-covalent interactions between subunits differs from enzyme to enzyme.

On the other hand, the polymeric nature of allosteric enzymes has been a decisive element in shaping the ideas about the mechanism of allosteric regulation. One of the models which has withstood the test of time was proposed by Monod, Wyman, and Changeux in 1965 and can be described by the following hypotheses:

1) Allosteric proteins are oligomers whose protomers are associated in such a way that they all occupy equivalent positions. This implies that the molecule possesses at least one axis of symmetry.

2) On each protomer, there exists one and only one site capable of forming a stereospecific complex with a given ligand. In other words, the symmetry of each group of stereospecific sites is the same as the symmetry of the molecule.

3) The conformation of each protomer is constrained by its association with the other protomers.

4) There are at least two states (R and T) in equilibrium, in which allosteric oligomers can exist. These states differ in the distribution and/or energy of the interprotomer bonds and, accordingly, in the conformational constraints imposed on the protomers.

5) As a result, the affinity of one or several of the stereospecific sites for the corresponding ligand is modified when there is a transition from one state to the other.

6) During these transitions, the molecular symmetry and the symmetry of the conformational constraints are conserved.

Let us analyze the consequences of this model: in the absence of substrates and allosteric effectors, the distribution of molecular species among the different states will be a function of the free energies of formation of these states, which are in turn determined by the strength of the interactions between protomers of the different states. In the case where a given ligand has a greater affinity for one of the conformational states, the presence of this ligand at low concentration will result in the preferential binding of a single ligand molecule to the protomer in the state for which it has the greater affinity. This will bring about a displacement of the equilibrium in favor of that state, and will thus facilitate the subsequent binding of additional ligand molecules, because of the simultaneous formation of more than one reactive site (the exact number will depend on the number of identical protomers in the oligomer). It is evident that the sigmoid relationship between activity and substrate (or effector) concentration is perfectly explained by this model. Monod, Wyman, and Changeux have elaborated a mathematical treatment of the model which accounts for a good many allosteric enzymes.

Let us analyze the interactions of an ideal allosteric enzyme with a single ligand F endowed with differential affinity toward the two accessible states. In the absence of ligand, the two states, symbolized as R_0 and T_0 are assumed to be in equilibrium.

Let L be the equilibrium constant for the $R_0 \rightleftharpoons T_0$ transition. In order to differentiate it from the dissociation constants of the ligand, the authors called it the "allosteric constant". Let K_R and K_T be the microscopic dissociation constants of a ligand F bound to a stereospecific site, in the R and T states, respectively. By reason of symmetry and because the binding of any one ligand molecule is assumed to be intrinsically independent of the binding of any other, these microscopic dissociation constants are the same for all homologous sites in each of the two states. Assuming n protomers (and therefore n homologous sites) and using the notation $R_0, R_1, R_2 ... R_n; T_0, T_1, T_2 ... T_n$ to designate the complexes involving $0, 1, 2 ... n$ molecules of ligand, the successive equilibria can be written as follows:

$$
\begin{array}{lcl}
R_0 & \rightarrow & T_0 \\
R_0 + F \rightarrow R_1 & & T_0 + F \rightarrow T_1 \\
R_1 + F \rightarrow R_2 & & T_1 + F \rightarrow T_2 \\
\cdots\cdots\cdots & & \cdots\cdots\cdots \\
R_{n-1} + F \rightarrow R_n & & T_{n-1} + F \rightarrow T_n
\end{array}
$$

Taking into account the probability factors for the dissociations of the $R_1, R_2, ... R_n$ and $T_1, T_2 ... T_n$ complexes, we may write the following equilibrium equations:

$$T_0 = L\,R_0$$

$$R_1 = R_0\, n\, \frac{F}{K_R} \qquad T_1 = T_0\, n\, \frac{F}{K_T}$$

$$R_2 = R_1\, \frac{n-1}{2}\, \frac{F}{K_R} \qquad T_2 = T_1\, \frac{n-1}{2}\, \frac{F}{K_T}$$

$$\cdots\cdots \qquad \cdots \qquad \cdots\cdots$$

$$R_n = R_{n-1}\, \frac{1}{n}\, \frac{F}{K_R} \qquad T_n = T_{n-1}\, \frac{1}{n}\, \frac{F}{K_T}$$

Let us now define two functions corresponding respectively to:

(a) the fraction of protein in the R state:

$$\bar{R} = \frac{R_0 + R_1 + R_2 + ... + R_n}{(R_0 + R_1 + R_2 + ... + R_n) + (T_0 + T_1 + T_2 + ... + T_n)}$$

(b) the fraction of sites actually bound by the ligand:

$$\bar{Y}_F = \frac{(R_1 + 2R_2 + ...+nR_n) + (T_1 + 2T_2 + ... + nT_n)}{n\ (R_0 + R_1 + R_2 +......... + R_n) + (T_0 + T_1 + T_2 +..........+T_n)}$$

Using the equilibrium equations, and setting

$$\frac{F}{K_R} = \alpha \quad \text{and} \quad \frac{K_R}{K_T} = c$$

we have, for the "function of state" $\bar{R}$:

$$\bar{R} = \frac{(1+\alpha)^n}{L(1+c\alpha)^n + (1+\alpha)^n}$$

and for the "saturation function" $\bar{Y}_F$:

$$\bar{Y}_F = \frac{Lc\alpha(1+c\alpha)^{n-1} + \alpha(1+\alpha)^{n-1}}{L(1+c\alpha)^n + (1+\alpha)^n}.$$

In figures 7(a) and 7(b), theoretical curves of the $\bar{Y}_F$ function have been drawn, corresponding to different values of the constants L and c. In such graphs the cooperative effect of the ligand, predicted by the symmetry properties of the model, is expressed by the curvature of the lower part of the curves. The graphs illustrate the fact that the "cooperativity"

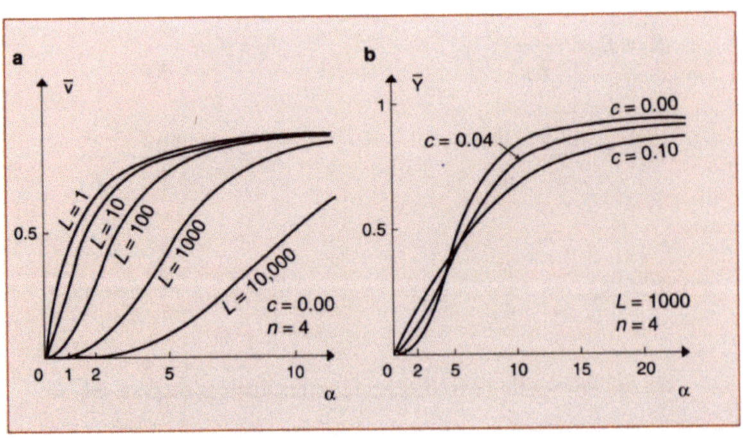

Figure 7. a & b, theoretical curves of the saturation function $\bar{Y}_F$ function drawn to various values of the constants L and c, with n =4, *i.e.*, for a tetramer. (From J. Monod, J. Wyman & J. P. Changeux, *J. Mol. Biol.* **12**, 88-118 (1965). With permission from the *Journal of Molecular Biology*.)

of the ligand depends on the values of L and c. The cooperativity is more marked when the allosteric constant is large (*i.e.*, when the $R_0 \rightleftharpoons T_0$ equi-librium is strongly in favor of T_0) and when the ratio of the micro-scopic dissociation constants ($c = K_R/K_T$) is small.

It should be noted that when $c = 1$ (*i.e.* when the affinity of the ligand towards both states is the same) and also when L is negligibly small, the Y_F function simplifies to:

$$\bar{Y}_F = \frac{\alpha}{1+\alpha} = \frac{F}{K_R + F}$$

that is, to the Henri-Michaelis equation.

The model therefore accounts for the homotropic cooperative ef-fects which are invariably found in allosteric proteins.

The properties of the model with respect to heterotropic inter-actions between different allosteric ligands has also been analyzed by Monod, Wyman, and Changeux and results in an equation expressing that the heterotropic action of an allosteric ligand upon the saturation function for another allosteric ligand should be to modify the homotropic inter-actions of the latter.

The main merit of the model is that it is suitable for experimental analysis almost thirty years after its inception, now that a number of al-losteric enzymes have been crystallized and that their tridimensional structure can be often analyzed in the liganded and unliganded state.

However, there is no *a priori* reason for believing that all enzymes subject to allosteric control have become so by the same mechanism. Each case must be examined individually.

It is useful to describe two classes of allosteric effects:

a) homotropic effects, or interactions between identical ligands;

b) heterotropic effects, or interactions between different ligands.

Homotropic effects are present in most allosteric proteins for at least one type of ligand (substrate, inhibitor or activator). While heterotro-pic effects may be either cooperative or antagonistic, homotropic effects are always cooperative. Experimental conditions which change heterotropic interactions often change homotropic ones (*e.g.* the desensiti-zation of threonine dehydratase or aspartate transcarbamylase is invariably accompanied by the disappearance of cooperative effects between sub-strate molecules); in other words, desensitization of these enzymes is con-nected with a normalization of the kinetics to the Michaelis type (fig. 8).

Two classes of allosteric enzymes can be defined:

1) Those in which the weaker substrate affinity predominates in the absence of substrate. Since the other state (*R* state) has the greater affinity for substrate, the equilibrium is displaced in its favor when substrate is added. The affinity of the whole population of enzyme molecules for sub-strate thus increases progressively as more molecules of substrate are bound. In contrast, an allosteric inhibitor displaces the equilibrium in the direction of the state (*T* state) with the weaker affinity for the substrate.

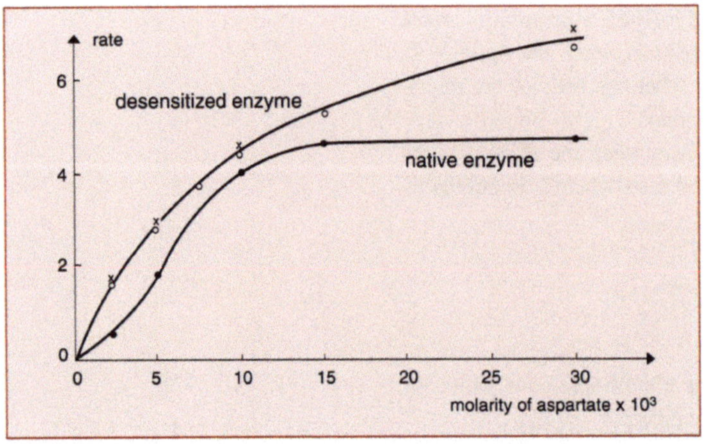

Figure 8. Kinetic differences between *E. coli* aspartate transcarbamylase and the same enzyme desensitized by heat or by a mercurial. Whereas the rate is a sigmoid function of the substrate concentration in the first case, the kinetics are "normalized" in both the cases of desensitization.

Enzymes in this category are called allosteric enzymes of type *K*. Both the substrate and the effector are allosteric ligands which affect the equilibrium between the two enzyme states, *R* and *T*. It is this change in the equilibrium which is defined as the allosteric transition. Cooperative homotropic effects are found for both the substrate and effector. The presence of the effector modifies K_m, the apparent affinity of the enzyme for its substrate and, conversely, the substrate concentration modifies the affinity of the enzyme for its effector.

2) In the second class, the two states of the allosteric enzyme have the same apparent affinity for the substrate. In this case, there is no effect of substrate or effector on the combination of the other ligand with the enzyme. Consequently, the effector is able to affect the enzymatic reaction only if the two states differ in their catalytic activity. Depending on whether the effector has a greater affinity for the active *R* state or inactive (or less active) *T* state, it will behave as an activator or an inhibitor. From the fact that it influences only V_{max} in the reaction and has no influence on the value of K_m, enzymes of this second class have been called enzymes of type *V*. Here the substrate is not an allosteric ligand, and cooperative interactions are not found between substrate molecules but only between effector molecules.

In practice, pure *K* systems will behave as enzymes inhibited competitively. The Henri-Michaelis curve will appear sigmoid rather than hyperbolic. Another useful representation is derived from an equation given by Hill in 1910. Letting V_{max} be the velocity of the enzyme when saturated with substrate, *v* the velocity at any non-saturating concentration and *S*, the substrate concentration, a plot of log $v/(V_{max} - v)$ as ordinate and log *S* as abscissa will give a curve with a linear portion of slope *n*. The number *n* (or Hill coefficient) is a measure of the interaction between the different substrate-binding stereospecific sites of the enzyme molecule. If the substrate causes an allosteric transition, *n* is always greater than unity.

In pure *V* systems, variation of enzyme activity with substrate concentration obeys classical Michaelis kinetics: *n* is therefore equal to 1.

But if the enzymatic activity is plotted against inhibitor concentration *I*, a sigmoid relationship is obtained. Plotted as log $v\ (V_o - v)$, where V_o is the non-inhibited activity and v the activity measured with a certain concentration of inhibitor, against log *I*, a curve is obtained which has a linear portion of slope n' which is a measure of the interaction between different inhibitor-specific sites of the enzyme molecule[9].

Conclusion

By their indirect nature, allosteric interactions do not depend on the structure or on the particular reactivity of ligands, but solely on the structural properties of the protein which acts as a relay. Growth, metabolism, and division require, in addition to the functioning of the main metabolic pathways responsible for the obtention of energy and of chemical building blocks, that the activity of the different metabolic pathways be precisely coordinated by a network of appropriate specific interactions. The creation and the development of such networks would have been impossible if they depended on strict direct interactions on the protein surface; such interactions would have been limited by the chemical structure, the reactivity or the absence of reactivity of the metabolites for which the existence of an interaction would have proven physiologically significant. The invention of indirect allosteric interactions depending on the structure of the protein, that is, in the last analysis, on the genetic code, would free molecular evolution from these limitations.

SELECTED REFERENCES

J. Monod, J. Wyman & J.-P.Changeux, *J. Mol. Biol.* **12**, 88-118 (1965).

M. F. Perutz, Mechanisms of cooperativity and allosteric regulation in proteins. *Quart. Rev. Biophys.* **22**, 139-236 (1989).

9. Obviously, n and n' can never exceed the number of protomers in the allosteric enzyme considered.

CHAPTER IV

Enzyme induction in catabolic systems

Enzymatic adaptation, as it had been studied by Karström can be defined as the induced appearance of specific enzymes after a preliminary "adaptation" of microorganisms on their substrates. The term *enzymatic adaptation* has been replaced by *enzyme induction*.

Specificity of induction

The relationships between the specificity of the induced enzymes and the specificity of induction are of capital importance for the interpretation of the phenomenon. Two levels of specificity can be defined for an enzyme: a) specificity of action, defined by the enzyme activity on different substrates; b) specificity of binding, defined by the affinity of the enzyme for substances forming a specific complex with it, whether these substances are substrates or not. These enzyme ligands comprise substrates and inhibitors.

What are the relationships between the inducing properties of a given substance and its substrate or ligand properties with regard to the enzyme it induces? There are three classes of hypotheses.

Functional hypothesis: The synthesis of the enzyme depends on its activity. The inducer acts as a substrate. The corollary of this hypothesis is that action and induction specificities are identical.

Equilibrium hypothesis: The synthesis of the enzyme is limited by the dynamic equilibrium:

$$\text{Precursors} \rightleftharpoons \text{Enzyme}$$

This equilibrium is perturbed when the enzyme is bound to a ligand:

$$\text{Enzyme} + \text{L} \rightarrow \text{Enzyme-L}$$

and more enzyme is produced. This hypothesis is based on the identity between in the induction and binding specificities.

Template hypothesis: The inducer plays an "organizing" role, not by combination with the enzyme itself but by its action on a precursor or on a synthesizing system. This hypothesis does not presuppose the identity between substrate, inducer or ligand. Monod, Cohen-Bazire, and Cohn have compared the properties of a series of sugar derivatives as inducers, substrates or ligands of β-galactosidase, an enzyme whose physiological function is the hydrolysis of the disaccharide lactose into glucose and galactose. Table I highlights their main results, which show that the specificity of combination of the purified enzyme is limited to the substances possessing the β-galactoside or the derived α-L-arabinoside configuration. Any substitution, inversion, or oxidation causes the complete or almost complete loss of affinity. α-galactosides are neither substrates nor ligands of the enzyme. Substitution of a sulfur atom in place of the natural oxygen atom of the galactoside link leaves the affinity almost intact whereas activity towards the analogue is entirely lost. As a result, for example, thiophenylgalactoside is a powerful competitive inhibitor of β-galactosidase. If one considers the inducing properties of these substances, the following conclusions can be drawn:

1) The inducing power is independent of the substrate property. This is best demonstrated by the inducing capacity of α-methylgalactoside and of melibiose. In addition, thiomethyl, thioethyl, and thioisopropylgalactosides which are among the best known inducers of β-galactosidase, are absolutely not hydrolyzed by this enzyme. Such non substrate inducer molecules have been called *gratuitous* inducers.

2) The inducing power is independent of the ligand property: there is no systematic correlation between the values of the affinity and the inducing power (Table I). For example, 2-deoxygalactose – which retains a significant affinity – has no inducing power. Phenyl-β-D-thiogalactoside (TPG), endowed with a considerable affinity, is not an inducer. However, it certainly penetrates the cells since it inhibits competitively β-galactosidase *in vivo*. In this respect, it is interesting to note that TPG, which inhibits the growth of *E. coli* on lactose, is devoid of any intrinsic toxic effect when maltose is the carbon source.

3) The inducing power is associated with the presence of an intact α- or β-galactoside link. The case of raffinose shows that the presence of this link is a necessary but not sufficient condition.

These results are incompatible with both the functional and the equilibrium hypotheses, and with any other hypothesis relying on a specific combination between the inducer and the enzyme as a *sine qua non* condition for induction to occur. We shall later reexamine the template hypothesis in its modern version.

De novo synthesis of β-galactosidase

Students of induction who were thinking in terms of mechanisms were confronted with the following problem: does enzyme induction correspond to the appearance of a new molecular species synthesized *de novo* from amino acids, or to a type of molecular reorganization which cannot be regarded as synthesis? For example, the conversion of trypsinogen to trypsin cannot be considered as the synthesis of a protein, although it implicates the appearance of a new protein species. Thus, the interpretation of induced biosynthesis must rely on an estimation of synthesis independent of activity measurements.

Table I. Comparison of the inducing, combination and substrate specificities of *E. coli* β-galactosidase.

Compound	Inducer[a]	Assayed as Ligand[b]	Substrate[c]
Galactose	420	30	+
β-D-galactosides			
methyl-	2800	10	+
n-butyl-	2800	400	+
o-nitrophenyl-(ONPG)	1060	1000	+
β-naphthyl-	42	200	+
4-glucose-*(lactose)*	2500	100	+
mannose-	2500	10	+
thiophenyl-	0	700	0
α-D-galactosides			
Methyl-	140	0	0
6-glucose-*(melibiose)*	2400	0	0
6-sucrose-*(raffinose)*	0	0	0
Carbon 2, 3, 4, and 6 substitutions of β-methyl-galactoside			
2-methyl-	20	0	0
2, 6-dimethyl-	0	0	0
3, 4-dimethyl-	0	0	0
2, 4, 6-trimethyl-	0	0	0
Reduced derivatives			
2-deoxygalactose	0	10	–
1, 2-dideoxygalactose	0	0	–
6-deoxygalactose*(fucose)*	0	0	–
Oxidation at carbon 6			
D-galacturonic acid	0	0	–
and its methylester	0	0	–
Suppression of carbon 6			
L-arabinose	0	0	0
methyl-β-L-arabinose	0	–	0
o-nitrophenyl L-α-arabinose	0	50	+
Sugars without galactoside cycle			
D-xylose, D-tagatose, D-mannose, D-glucose, methyl-β-D-glucoside, maltose, cellobiose	0	0	– or 0

a) activity determined after a 4 h growth in the presence of 10^{-3} M inducer, expressed as μmoles ONPG hydrolyzed/min/mg dry weight.
b) determined by measure of the competitive inhibition of the hydrolysis of ONPG, taken as reference substrate, and to which the value 1000 is arbitrarily attributed.
c) + and 0 mean respectively that the compound is hydrolyzed or not hydrolyzed; – = not tested.

Two research teams have established beyond doubt that β-galactosidase is synthesized *de novo* from amino acids. Hogness, Cohn, and Monod have grown *E. coli* in presence of radioactive sulfate ($^{35}SO_4^{2-}$) during several generations in order to obtain uniformly labeled bacteria. The cells are then washed to eliminate the excess radioactivity and transferred in a medium containing non radioactive sulfate and a gratuitous inducer of β-galactosidase. After several generations, the cells are collected and disrupted and the β-galactosidase purified from the extract, the last purification step consisting in a specific precipitation by anti-β galactosidase antibodies. The radioactivity of the isolated enzyme is about 4% of that of the enzyme obtained by induction in the presence of radioactive sulfate (of specific radioactivity identical to that used in the non-induced culture). This maximum of 4% is observed whether the mass increase during induction is of 10% or 100%. The authors conclude that if any precursors of β-galactosidase existed before induction, they must have been at a very low concentration. Spiegelman and Rotman have performed essentially the same experiment, using ^{14}C-lactate as a carbon source for growth. After having grown in this medium, the bacteria are induced in a non radioactive medium, the enzyme purified by electrophoresis on a starch gel and precipitated by the specific antibody. Here again β-galactosidase possesses less than 3% of the specific radioactivity of the other proteins present in the induced culture.

The results of these experiments schematized in figure 1 indicate that the contribution of preexisting proteins or of other preexisting carbon compounds is minimal, if any. They suggest, therefore, that the enzyme is synthesized from amino acids which in turn derive from the carbon source.

It is important to note that, had there been an important proteolysis of cell proteins, yielding free amino acids, β-galactosidase would have been radioactive. The fact that it does contain minimal amounts of radioactivity whatever the duration of induction suggests that there is no significant protein turnover during exponential growth.

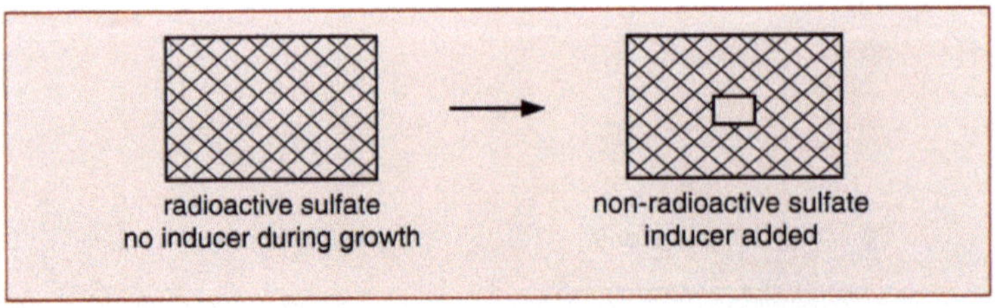

Figure 1. The left rectangle represents the total bacterial proteins, labeled by ^{35}S, synthesized in the absence of inducer. Right, after induction in a non radioactive growth medium, β-galactosidase appears as the sole non radioactive protein (recognized by specific antibodies).
(From D. S. Hogness, M. Cohn and J. Monod, *Biochimica et Biophysica Acta* **6**, 99-116 (1955). With permission from Elsevier Science Publishers.)

Constitutive mutants

If we consider Karström's separation of the sugar degradation enzymes into "constitutive" or "adaptative", we observe that it is purely empirical and does not implicate differences in the mechanism of synthesis of the enzymes belonging to the two classes. The main reason for refusing to consider different mechanisms has been that the same enzyme can be constitutive in one strain and inducible in another strain of the same species, or even partly constitutive; in the latter case, it is synthesized, albeit at a lower rate, even in the absence of inducer.

It has been unambigously demonstrated that the penicillinase constitutively produced by *Bacillus subtilis* is identical to the enzyme produced – in much larger amounts – when penicillin is present in the culture medium.

Mutations have been obtained which transform an inducible system into a constitutive one. These changes correspond to a single mutational event which does not affect the structure of the enzyme, but only its inducible or constitutive character: the inducible *E. coli* β-galactosidase has the same molecular weight, the same kinetic parameters, the same degree of activation by ions, the same pattern of heat inactivation and the same immunochemical specificity as the enzyme extracted from a constitutive mutant.

The isolation of constitutive mutants involves selection methods which vary from one case to the other and are based on a selective advantage conferred on the constitutive mutants; here are some examples of selection of mutants constitutive for the synthesis of β-galactosidase:

a) Thiophenyl-β D-galactoside (TPG) is an inhibitor of β-galactosidase, but mainly a powerful inhibitor of its induction. If a mutagenized population is layered on a solid medium containing lactose as the sole carbon source and TPG at a convenient concentration, the inducible bacteria will not be induced and will not grow; the only colonies which will appear arise from constitutive bacteria.

b) Phenyl-β D-galactoside is a substrate of β-galactosidase, but is devoid of inducing activity. If it is incorporated as the sole carbon source into a solid medium, only constitutive bacteria will be able to grow and to give rise to colonies.

c) Finally, a more general method is based on the difference in growth rates of the inducible wild type and the constitutive mutants in different media.

Pleiotropy of constitutive mutants

During the study of β-galactoside permease, two types of structural mutations had been observed: y^- mutations, which affect the structure of the permease; and z^- mutations, which occur within the structural gene of β-galactosidase.

When the permease activity of the β-galactosidase constitutive mutants was assayed, it was observed, with great surprise, that it was also constitutive, *i.e.,* it was present without a previous culture in the presence of an inducing galactoside.

One recalls that, during the same study, when the bacteria having accumulated radioactive thiomethylgalactoside, a minor spot was identified with 6-acetylthiomethylgalactoside. This compound is the product of an enzyme normally inducible in the wild type of *E. coli,* which catalyzes the following reaction:

$$\text{Acetyl CoA} + \text{thiogalactoside} \longrightarrow \text{6-acetylthiogalactoside} + \text{CoA}$$

The physiological function of thiogalactoside transacetylase is unknown to this day. What interests us here, however, is the fact that its synthesis is constitutive in the mutants constitutive for the galactosidase and the permease. We are thus confronted with a new type of mutation, a *pleiotropic* mutation, which affects several characters simultaneously. At this point, it is important to emphasize that it is not the structures of the three proteins which are affected by the pleiotropic mutation, which determines only the inducible or constitutive character of the synthesis of these catalysts. We shall call i^+ the wild type inducible allele, and i^- the mutated constitutive allele. We shall call the *i* gene a regulatory gene by contrast with the *z, y, ac* genes which are structural genes whose base sequence codes for the corresponding amino acid sequence.

Pleiotropic constitutive mutants have been found to affect the regulation of the synthesis of several enzymes involved in the degradation of arabinose, galactose, maltose, etc.

Genetic control and cytoplasmic expression of inducibility in the synthesis of β-galactosidase in *E. coli. Lac* repressor

Any hypothesis about the mechanism of induction must provide an interpretation of the differences between the inducible and constitutive systems. Since specific point mutations can transform a typically inducible system into a fully constitutive one, an analysis of the genetic structure and of the biochemical effects of such mutations should lead to such an interpretation.

If, for reasons of simplicity, one does not consider the transacetylase, the following eight genetic combinations can be obtained:

$i^+z^+y^+$, wild type, inducible for galactosidase and permease.

$i^+z^+y^-$, cryptic mutant (devoid of permease, but possessing a normally inducible galactosidase).

$i^+z^-y^+$, mutant inducible for permease, not producing active β-galactosidase. It has been shown in many z^- mutants, that an inactive protein is synthesized, with affinity for the *E. coli* anti-β-galactosidase antibodies (CRM protein, for *cross-reacting material).* In this class, the synthesis of CRM is inducible.

$i^+z^-y^-$, mutant devoid of the two activities, but where the synthesis of CRM is inducible.

$i^-z^+y^+$, constitutive mutant synthesizing permease and galactosidase without inducer added.

$i^-z^+y^-$, constitutive cryptic mutant, devoid of permease.

$i^-z^-y^+$, mutant constitutive for the syntheses of permease and CRM.

$i^-z^-y^-$, mutant without permease and constitutive for CRM.

Knowledge of the genetic structure of the *lac* region which commands the utilization of lactose is necessary for the comprehension of the capital experiment described later.

a) The recombination frequency between *z* and *y* mutations is very low, of the order of one hundredth of that observed between the *leu* (1 mn on the circular chromosome) and *gal* (17 mn) markers.

b) If, in the cross $y^+i^-z^-$ x $y^-i^+z^+$, the y^+z^+ recombinants are selected for their ability to grow on lactose-containing plates (on which none of the parents strains can grow), one observes that i^+ remains associated with z^+ in 85% of the cases.

c) the frequency of cotransduction of *i* and *z* (when z^+ is selected for) is higher than 90%, whereas the frequency of cotransduction of *i* with *y* is only of 70%.

These studies, and more detailed ones, show that the *z*, *y*, *ac*, and *i* genes are all grouped in the same region and that their topological relationships can be represented as shown in the scheme:

____/ - - - ______/________/_______/_______/_____ - - -

gal *i* *z* *y* *ac* *leu, thr*

It should be recalled that conjugation in *E. coli* involves the injection of the chromosome of a male Hfr (high frequency of recombination) into a receptor F⁻ (female) strain, which generally results in the formation of an incomplete zygote or merozygote. The recombination between the female chromosome and the segment of the injected male chromosome does not start before 60 to 90 minutes after injection; furthermore, segregation of recombinants from the merozygotes takes place only after several hours, which leaves ample time to study the interactions between the two genomes, their expression in the cytoplasm, and their dominance relationships. It is equally important for the interpretation of the results that the *contribution of the Hfr partner to the merozygotes is strictly chromosomic,* whereas the F^- bacteria have both a cytoplasmic and a chromosomal contribution. In order to be able to draw a conclusion from the experiments on the expression of the *z*, *y* and *i* factors, it is necessary to find conditions where the merozygotes, but not the parent strains, can synthesize β-galactosidase. The following cross provides such conditions:

$$\text{Hfr } z^+y^+i^+Sm^S \text{ x } F^- z^-y^+i^+ \ Sm^R$$

Sm^S and Sm^R being the alleles of the *Sm* gene determining streptomycin sensitivity and resistance, respectively.

The cross is made in the presence of the inducer (isopropylthio-galactoside, IPTG) and streptomycin. The F^- strain cannot synthesize the enzyme since its structural gene is inactivated. The Hfr strain cannot synthesize it either, because its growth is inhibited by streptomycin; the zygotes are not inhibited by streptomycin because they inherit the F^- cytoplasm and chromosome and the chosen Hfr strain transfers the *Sm* gene at a rather low frequency. Under these conditions, the enzyme is formed by the population with a time and amount dependency which is compatible only with a synthesis by the zygotes having received the z^+ gene (fig. 2).

The synthesis of enzyme begins a few minutes after the introduction of the z^+ gene (18 mn on the chromosome). If one takes into account that the number of zygotes having received z^+ is four to five times greater than the measured number of z^+Sm^R recombinants, and that each cell is, on the average, trinucleate (*i.e.* possesses three copies of z^+), one calculates that the rate of synthesis of the enzyme by the z^+ gene is normal.

Since all z^- mutants belong to the same cistron, and since the z^+ allele is obviously dominant over z^-, one can then make the following cross:

$$\text{Hfr } z^+i^+ \quad \text{x} \quad F^- z^-i^-$$

in the absence of inducer.

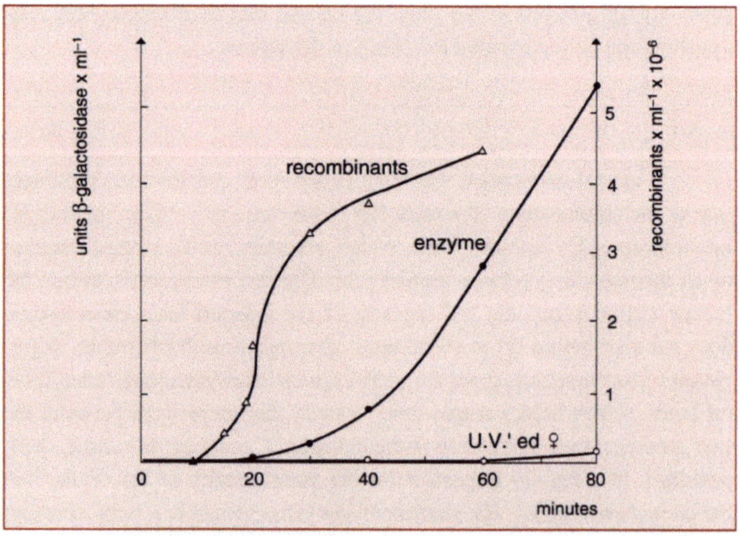

Figure 2. Synthesis of enzyme and appearance of recombinants during the cross described above, performed in the presence of inducer and streptomycin. A control where the F^- strain has been inactivated by UV irradiation is included. The z^+Sm^R recombinants are selected on plates containing lactose and streptomycin, after interruption of the cross by violent shaking.

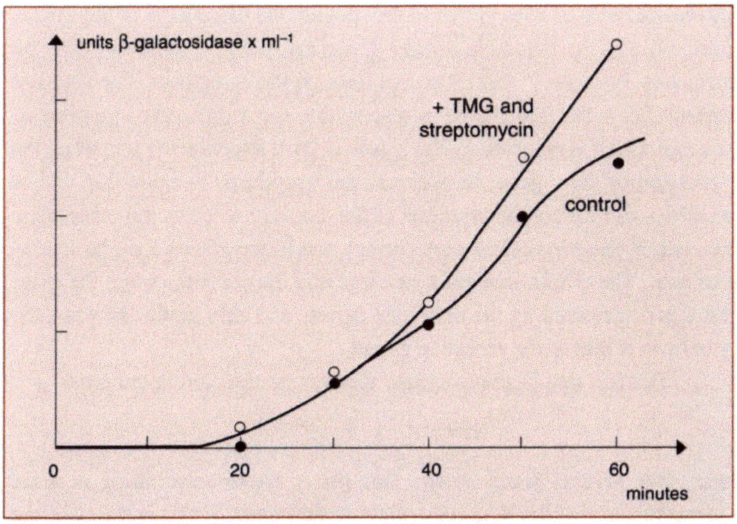

Figure 3. Synthesis of enzyme by the zygotes from the cross:
Hfr z^+i^+ x F$^-$ z^-i^-
At 20 mn and 25 mn respectively, streptomycin and TMG are added to an aliquot of the culture, in order to compare enzyme synthesis with and without inducer.
(From A. B. Pardee, F. Jacob & J. Monod, *J. Mol. Biol.* **1**, 165-178 (1964). With permission from the *Journal of Molecular Biology*.)

Under those conditions, neither of the parents can synthesize the enzyme. However, zygotes do synthesize β-galactosidase (fig. 3): during the first hour following the mixture, one observes a rapid constitutive synthesis. The cross has thus allowed an immediate and complete interaction of the male i^- and female z^+ genes. The possibility that this interaction is due to z^+i^- recombinants is excluded: the synthesis starts immediately after the injection of z^+ whereas genetic recombination takes place only 60 to 90 minutes after injection; on the other hand the *z* and *i* genes are so close that recombination between them is a rare event (such recombinants are 10,000 times less frequent than zygotes); the kinetics of enzyme synthesis is such that all zygotes must be synthesizing β-galactosidase.

There is the possibility that the interaction leading to enzyme synthesis is a consequence of the pairing of the homologous chromosome segments of the two parents. This possibility has been excluded by performing the following cross in the absence of inducer:

Hfr z^-i^- x F$^-$ z^+i^+

The zygotes obtained in this cross are identical to those of the preceding cross. However, no synthesis of enzyme is observed at any time, although adequate controls show that the conjugation and the injection of the male chromosome have occurred normally. The only difference between the two crosses is that in the second, the cytoplasm of the zygote is provided by the inducible partner. It can therefore be concluded that the interaction between i^- and z^+ occurs through the cytoplasm. The simplest explanation is that the i gene is responsible for a diffusible cytoplasmic message which is received by the z gene or by a structure commanding the expression of the z gene. In addition, the hypothesis is made that one of the alleles of i causes the synthesis of the message whereas the other allele does not. If these hypotheses are correct, one of the alleles must be strictly dominant. The dominance must be expressed immediately when the cytoplasm is contributed by the dominant parent, and only gradually when the cytoplasm is that of the recessive parent.

The fact that β-galactosidase is never synthesized in the cross

$$\text{Hfr } z^-i^- \text{ x } \text{F}^- z^+i^+$$

even after several hours means that the i^- constitutive allele is never expressed and that the dominant allele is the i^+ one. If this is the case, the i^+ allele should be expressed after a certain time in the cross

$$\text{Hfr } z^+i^+ \text{ x } \text{F}^- i^-z^-$$

which produces the enzyme constitutively.

In order to verify this assumption, the cross is repeated with strains containing additional markers:

$$\text{Hfr } z^+i^+Sm^sT_6^r \text{ x } \text{F}^-z^-i^-Sm^rT_6^r$$

and the synthesis of enzyme is followed, with or without inducer, in the presence of streptomycin and of bacteriophage T_6 (to prevent the induction of males and new crosses between recombinants). Figure 4 shows that, in the absence of inducer, the constitutive enzyme synthesis stops 90 minutes after the entry of z^+i^+. If, at this point, inducer is added, enzyme synthesis is resumed, showing that the zygotes z^+i^+/z^-i^- which were carrying a constitutive synthesis, have not been inactivated, but are now inducible. A necessary control shows that this transformation of constitutivity to inducibility is not due to the segregation of homozygous recombinants of the type $z^+i^+Sm^rT_6$, which happens much later.

From the above experiments, the following conclusions can be drawn:

– The i gene, under its active form (i^+), determines the synthesis of a diffusible cytoplasmic product which inhibits the expression of the structural genes z, y, and ac, unless an external inducer is added. When the product of the i gene is absent or inactive, there is no need for the addition of an external inducer to express the structural genes. The product of the i gene is highly specific and exerts no effect on other known systems. The above considerations do not inform us about the nature and the mode of action of the i gene. Two models are possible:

a) The β-galactosidase forming system requires the presence of an internal inducer, synthesized by both the i^+ and the i^- strains. The i^+

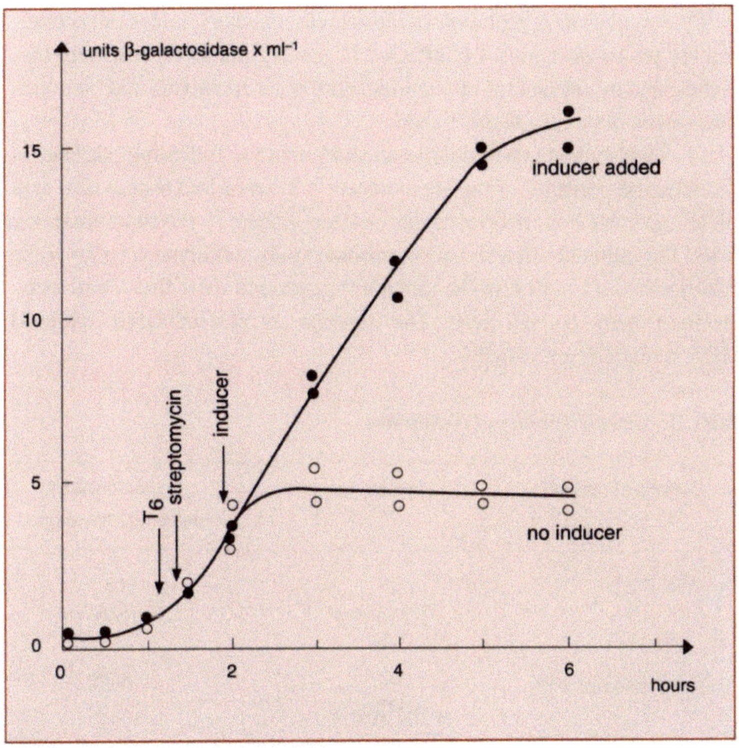

Figure 4. Cross performed in quadruplicate in the absence of inducer. At the times indicated by the arrows, a suspension of T_6 phage and streptomycin is added. Inducer is added to two of the cultures (•), omitted in the two others (○).

(From A. B. Pardee, F. Jacob & J. Monod, *J. Mol. Biol.* **1**, 165-178 (1964). With permission from the *Journal of Molecular Biology*.)

allele is responsible for the synthesis of an enzyme which destroys or inactivates this internal inducer. The i^- mutants do not produce this enzyme and accumulate the endogenous inducer. This model accounts for both the dominance relationships observed and for the kinetics of transformation of the constitutive zygotes into inducible ones.

b) The other model predicts that in the wild type (i^+), a specific *repressor* is synthesized. The inducer then acts as an antagonist of the repressor. The constitutive mutant i^- does not form an active repressor; therefore an exogenous inducer is not necessary. This model equally accounts for the observed facts.

As induction is highly stereospecific, a recognition site for the inducer must be carried either by the hypothetical repressor or by an element of the transcription or translation machinery recognized by the repressor. Whatever the case, one expects that mutations in the DNA segment coding for the repressor or for its target could abolish the recognition site for the inducer without affecting the interaction of the target with the repressor. A mutation of the repressor causing a loss of recognition for the inducer must be reflected in the zygotes as a permanent loss of the capacity for lactose utilization, and this incapacity should be dominant versus both the alleles i^+ and i^-.

Such mutants have been searched for in the following manner: in the immense majority of the *lac* mutants (798 out of 800 tested), the wild allele lac^+ was dominant over the mutated alleles. In the two remaining cases, the mutated alleles were dominant on the wild type $(i^+z^+y^+)$: the mutations could not be in the structural genes and their fine localization assigned them to the *i* gene. The diploids are not inducible (table II) These mutations are called i^s.

Table II. Non-inducibility of i^s mutants.

Genetic structure	Inducer	β-galactosidase units/mg dry weight
Wild: i^+z^+	0	< 5
	IPTG 10^{-4} M	7000
Non inducible: i^sz^+	0	10-20
	IPTG 10^{-2} M	10-20
Diploid: i^s/i^+	0	5
	IPTG 10^{-3} M	8
Diploid: i^s/i^-	0	8
	IPTG 10^{-3} M	8

Since an i^s mutation inactivating an enzyme destroying an endogenous inducer cannot be dominant over the wild allele i^+ allowing the synthesis of this enzyme, we are forced to accept the fact that i^s determines the synthesis of a super-repressor which cannot be antagonized by inducer and, as a consequence, to adopt the model of cytoplasmic repressors antagonized by exogenous inducers.

Operators and operons

That a given repressor specifically inhibits the synthesis of one or more proteins implies that at a given point of the expression of the gene into its encoded protein(s), this repressor must form a stereospecific combination with an element of the system. When this element is combined with the repressor, the expression is blocked. This hypothetical element which can switch on and off the expression of the system has been called *operator* by Jacob and Monod.

Whatever the nature of the operator and whatever the level at which it is involved in the regulation process, the specific configuration it must be endowed with to respond to the stereospecificity of the repressor should be altered by mutation. In particular, there should exist mutations of the operator rendering it unable to recognize the repressor.

In a diploid cell with a chromosome carrying the wild operator O^+ and the other the mutated operator O^c (unable to bind the repressor), the phenotype should be that of a dominant constitutive

$$\frac{i^+ \quad O^+ \qquad\qquad z^+}{i^- \quad O^c \qquad\qquad z^+} \qquad \begin{matrix} a \\ b \end{matrix}$$

because the repressor formed by i^+ will combine with the wild operator O^+ and the z^+ gene carried by chromosome a will not be expressed. But since O^c cannot combine with the diffusible repressor, the z^+ gene carried by chromosome b will be expressed freely.

Confirmation of the theory thus requires the isolation of dominant constitutive mutants, different from the i mutants, which are recessive.

An ingenious way of finding such mutants is to select constitutive lactose fermenters among inducible heterogenotes homozygous for the i^+ gene (i^+z^+/i^+z^+); only the very improbable event of two independent i mutations in the two i^+ genes can lead to the constitutivity due to the existence of an inactive repressor.

Constitutives have actually been isolated from these heterogenotes. Their genetic analysis shows that they possess the expected characteristics:

a) O^c is dominant over O^+ in O^+/O^c heterogenotes.

b) O^c possesses the same degree of constitutivity for galactosidase and permease. Many O^c mutants are partially constitutive and produce 10 to 40% of the inducible synthesis, which represents from 100 to 500 times more than the non-induced wild type.

c) O^c is insensitive to the super-repressor: i^sO^+/i^+O^c heterogenotes behave as constitutives.

In addition, O^c mutations display two additional properties which suggest the nature of the operator and allow one to guess at which level it intervenes in the mechanism of expression. First, all the O^c mutations are localized in a small segment situated at the extremity of the z gene, between i and z:

$$\overline{i \qquad O \qquad z \qquad y \qquad ac}$$

Second, when their behavior in heterogenotes is examined, one notes that only the genes situated on the same DNA segment as the O^c mutation are expressed constitutively.

In genetic language, one says that the O^c mutations act only in the *cis* position. For example, we know that an heterogenote z^+/z^- CRM which is homozygous for O^+i^+ synthesizes in the presence of inducer both active and inactive (CRM) galactosidase. If O^c is introduced on the chromosome or on an episome, one observes that the bacteria O^cz^+/F O^+z^- CRM produce active enzyme constitutively whereas the CRM protein is synthesized only in the presence of inducer. On the other hand the bacteria $O^+z^+/$ F O^cz^- CRM produce the CRM protein constitutively while galactosidase is normally inducible. Table III shows examples of diverse possible combinations.

Table III. Synthesis of β-galactosidase, of CRM and of thiogalactoside transacetylase by haploid or diploid mutants of the operator gene. Bacteria are grown in glycerol and induced, when necessary, by 10^{-4} M IPTG. The inactive protein (CRM) formed by certain mutants of the *z* structural gene, has the same antigenic properties as galactosidase, but is devoid of enzyme activity. Symbols *z* and *ac* refer to the genes coding respectively for β-galactosidase and thiogalactoside transacetylase. Symbols i^+ and i^- refer respectively to the inducible and constitutive alleles of the gene coding for the repressor. Symbols O^+ and O^c refer to the wild and constitutive dominant alleles of the operator gene.
***F* refers to the sexual factor of *E. coli*; when this factor incorporates the *lac* region of the bacterial genome, it forms a new genetic unit, F-*lac*, an episome, which can be added to the bacterial genome to form diploids for the *lac* region. Several F-*lac* episomes can coexist within the same cell, which explains why the organisms bearing the wild allele of the gene on the episome synthesize more enzyme than the wild type.**
The values are given in per cent of the values obtained with the induced wild type, except for the inactive CRM protein, which is estimated in β-galactosidase antigenic equivalents. In O^c mutants, values obtained vary somewhat with the culture conditions, especially in non induced cultures.

	Galactosidase		Transacetylase	
Genotype	Non induced	Induced	Non induced	Induced
$O^+z^+ac^+$	1	100	< 1	100
$O^cz^+ac^+$	25	95	15	110
$O^+z^+ac^-/F\,O^cz^+ac^+$	70	220	50	160
$O^+z^-ac^+/F\,O^cz^+ac^-$	180	440	< 1	220
$i^sO^+z^+ac^+/Fi^+O^cz^+ac^+$	190	210	150	200

	Galactosidase		CRM	
Genotype	Non induced	Induced	Non induced	Induced
O^+z^+	< 1	100	–	–
O^+z^+/FO^+z^-	< 1	105	< 1	310
O^cz^+	15	90	–	–
O^+z^+/FO^cz^-	< 1	90	30	180
O^+z^-/FO^cz^+	90	250	< 1	85

The β-galactosidase synthesized by the O^c mutants is identical to the inducible enzyme and to that produced by the constitutive i^- mutants, by all criteria available.

The preceding experiments lead to a rejection of a certain number of models for explaining the repression and the derepression (constitutivity). For example, each gene of a pathway regulated in a coordinated manner cannot have its own operator. One would expect, in this case, that a $O^+ \rightarrow O^c$ mutation would cause the constitutivity of one enzyme only.

The only model which has withstood analysis is that a single operator allows the expression of a sequence of genes and that this operator and the accompanying structural genes constitute an integrated topological and functional unit. We shall call such a unit an *operon.*

We know that the genetic information coded by DNA is transcribed as messenger RNA molecules; these molecules migrate toward the ribosomes where their information is translated into protein molecules. Messenger RNA is thus (in procaryotes) the primary product of the genes.

A remarkable and general property of the induction mechanisms is that all the genes whose expression depends on the same operator are contiguous on the chromosome; the operator itself being localized at the 5' extremity of the DNA sequence coding for the different genes. The most satisfactory explanation of all these facts is that the operator is at (or near) the genetic locus where the transcription of the information of a group of genes coding for the structure of contiguous genes involved in a given metabolic sequence occurs. We can then define the operon as a genetic unit of coordinated transcription. The repressor blocks transcription by specifically binding the operator, in a such way that no message is sent from the operon to the ribosomes, resulting in the arrest of the synthesis of the proteins coded by the structural genes. Figure 5 summarizes this theory.

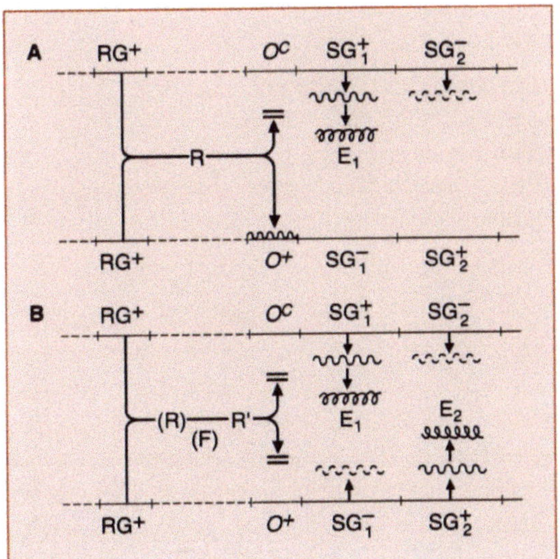

Figure 5. Schematized behavior of two diploids heterozygous for the operator and for the structural genes. Heterozygote for a constitutive operator *(O^c)* mutation/wild type *(O^+)*.
A. Non induced. The repressor made by the regulator genes (RG) blocks gene expression in the lower chromosome which carries a normal operator *(O^+)*. It does not act on the upper chromosome which carries a mutated operator *(O^c)* with a low affinity for the repressor. Only enzyme E_1 is produced constitutively since the upper chromosome carries a mutation in SG_2 (SG_2^-).
B. Induced. In the presence of inducer F, the repressor R is converted into the inactive form R'. Gene activity is allowed in both chromosomes. Both enzymes E_1 and E_2 are made.
(Adapted from F. Jacob & J. Monod.)

SELECTED REFERENCES

Specificity of induction. De novo *synthesis*

J. Monod, G. Cohen-Bazire & M. Cohn, *Biochim. Biophys. Acta* **7**, 585-599 (1951).

D. S. Hogness, M. Cohn & J. Monod, *Biochim. Biophys. Acta* **16**, 99-116 (1955).

Pleiotropy of the i *gene*

H. V. Rickenberg, G. N. Cohen, G. Buttin & J. Monod, *Ann. Inst. Pasteur* **91**, 829-857 (1956).

Thiogalactoside transacetylase

I. Zabin, A. Kepès & J. Monod, *J. Biol. Chem.* **237**, 253-257 (1962).

Repressors and operators

A. B. Pardee, F. Jacob & J. Monod, *J. Mol. Biol.* **1**, 165-178 (1964).

C. Willson, D. Perrin, M. Cohn, F. Jacob & J. Monod, *J. Mol. Biol.* **8**, 582-592 (1964).

F. Jacob & J. Monod, *Cold Spring Harbor Symp. Quant. Biol.* **16,** 193-211 (1961).

CHAPTER V

Transcription. RNA polymerase

At the end of the 1950s Hoagland and Zamecnik had succeeded in obtaining an *in vitro* protein synthesis from ribosomes and crude acellular extracts. Ribosomes cannot by themselves achieve a *specific* protein synthesis and require information originating, in the last analysis, in the genes.

A few minutes after the penetration of a virulent bacteriophage into *E. coli* cells, that is after the entry of the phage genes, the bacteria stop synthesizing their own proteins and produce only those proteins and enzymes necessary for manufacturing a few hundred copies of the infecting phage with its nucleic acid and its protein capsid, the new phages causing bacterial lysis in a few minutes. It was then shown that phage infection caused the immediate arrest of the synthesis of bacterial RNA while a new RNA species subject to a rapid turnover appears. Volkin and Astrachan confirmed this observation in 1956 and showed that this new RNA species corresponds to only 3% of the total RNA and that its base composition reflects that of phage DNA and not that of bacterial DNA. In 1961, Spiegelman showed that the RNA synthesized as the result of infection can form a complementary specific molecular hybrid with phage DNA, but not with bacterial DNA.

Everything became clear with the concept of *messenger RNA* formulated the same year by Jacob and Monod. This concept originated from their work on the regulation of enzyme synthesis which is treated elsewhere in this volume: they postulated the existence of an unstable species, messenger RNA, which transfers the genetic information from DNA to the site of protein synthesis, *i. e.,* the cytoplasmic ribosomes.

Three independent groups showed that a short-lived variety of RNA with a base composition analogous to that of DNA and different from that of ribosomal RNA existed also in uninfected bacteria.

Synthesis of messenger RNA. Bacterial RNA polymerase

The synthesis of the messenger (mRNA) is catalyzed by an enzyme, DNA-dependent RNA polymerase, present in the cell-free extracts of bacteria, of animal and of plant cells. In these preparations, it can be precipitated with DNA by the addition of protamine or streptomycin. In eukaryotic cells, it is localized in the nuclei.

In addition to DNA, the polymerization reaction requires the presence of the four nucleoside triphosphates, that of a divalent metal (Mg^{2+} or Mn^{2+}). The matrix DNA can be single or double stranded, or can be present as short oligonucleotide fragments. Finally, and above all, the reaction is characterized by the fact that the base composition of the single-stranded RNA product is defined by that of the template DNA. As in the case of DNA duplication, the new chain is complementary to that of the template DNA, which is of paramount importance for the transfer of genetic information. This process, catalyzed by RNA polymerase, has been called *transcription.*

In vitro, either of the two DNA strands can serve as matrix. *In vivo* however, only one of the DNA strands can serve as model. This was shown by using a microorganism whose two strands differ by their density in a cesium chloride gradient. After separating the two strands by column chromatography, it is found that only one of the chains hybridizes with that of the mRNA from the same organism.

The basic unit of transcription by RNA polymerase is the holoenzyme which can initiate and elongate the RNA chain. The holoenzyme contains five major species of subunits, α, β, β', σ and ω present in the molar ratios 2:1:1:1:1. The preparations contain strongly bound zinc. When the enzyme is deprived of the σ subunit, the enzyme is called "core polymerase", which can be obtained by a single chromatography of the holoenzyme on a phosphocellulose column under non-denaturing conditions.

The core polymerase is able to catalyze RNA synthesis, but the σ subunit is essential for transcription initiation from promoters. In addition, accessory factors, designated generally as "transcription factors", may be required for transcription initiation from some promoters. Some of these transcription factors associate with RNA polymerase in the absence of DNA. However, most factors interact only when both polymerase and factors are aligned along the same DNA strand, and thus the activation also requires a specific DNA-binding function.

The synthesis of RNA rests on the integrity of the β subunit, which is altered in rifampicin- and streptolydigin-resistant mutants. These antibiotics block RNA synthesis by interacting with the β-subunit of the wild-type polymerase. The σ and β' subunits are essential, respectively, for the selection of the promoter site and for the reconstitution of the holoenzyme from its isolated subunits. The genes corresponding to the β and β' polypeptides are adjacent on the *E. coli* chromosome and are synthesized in a coordinate manner.

The polymerase, when engaged in a ternary complex with its DNA template and a nascent RNA chain, is devoid of σ subunit, which favors the idea that the "core polymerase" is responsible for the elongation. The fact that the σ subunit can be lost during purification of the enzyme has suggested that other components might also have been lost. This approach has led to the isolation of ρ, a termination factor.

RNA synthesis can be decomposed into several steps, each one complex. It is generally accepted that for the transcription units needing only the holoenzyme for a selective transcription, four steps are necessary:

1) The attachment of the holoenzyme to the DNA matrix, during which the polymerase localizes the specific site where initiation can take place, and adopts an active conformation.

2) The initiation of the RNA chain, during which the enzyme catalyzes the coupling of ATP or GTP with a second ribonucleoside diphosphate to generate a dinucleoside tetraphosphate pppPupX with elimination of pyrophosphate; the dinucleoside tetraphosphate remains strongly bound to the RNA polymerase-DNA complex.

3) The elongation of the RNA chain, during which successive nucleoside monophosphate residues originating from the nucleoside triphosphates are added to the 3'-OH terminus of the growing chain.

4) The termination of the RNA chain, accompanied by the liberation of the free enzyme and of the newly synthesized RNA chain from the DNA matrix.

If one considers the selectivity of transcription, steps 1 and 4 (the attachment to DNA and the termination), are the most important: it has been shown unequivocally that the selection of a specific site for initiation does not require the presence of nucleoside triphosphates; the correct termination of a growing RNA chain occurs on a specific site of the DNA matrix with or without the accessory factor ρ.

The DNA region which controls a correct initiation is called the *promoter region.* This region may contain a variety of individual functions, which may or may not be physically separated. One can note, for example, the site(s) of recognition of the polymerase, the initiation sites proper, sites for the binding of positive regulators of the CAP type, and operator sites for the binding of negative regulators (repressors). Whereas the term *promoter* was initially meant, on the basis of genetic experiments, to designate "the site at which transcription is initiated", its meaning has been extended to designate the region of a transcription unit responsible both for the initiation of transcription and for its regulation.

An exhaustive study of the binding of the holoenzyme to bacteriophage T7 DNA was carried out by Chamberlin in 1972. The productive binary complexes bind at the initiation sites of the promoter regions and are able to selectively initiate the transcription. The kinetics of the selection suggests the holoenzyme separates the two DNA strands at the initiation site and that these *open complexes* are ready to initiate. Other authors have shown that the binding of RNA polymerase to phage λ DNA is accompanied by the opening of about 24 base pairs (bp) per DNA molecule. Four regions appear to be recognized by the enzyme, meaning that an average of 6 bp are opened per promoter site. At high ionic strength, the enzyme still binds, but there is no local opening and therefore no initiation. The correponding non-productive complexes are called *closed complexes.*

At least three processes are affected by the binding of the σ subunit to the polymerase: 1) the non-specific binding to the DNA helix is strongly reduced, due to a 10^6-fold increase in the dissociation rate of the non-specific complexes; 2) the σ subunit is required for the formation of the open complexes; 3) its binding to the core enzyme suppresses the spurious initiation at breaks occurring on a single strand and imposes a more strict initiation specificity.

The sequence of the promoter sites has been obtained by several methods, one of which is their isolation from the bulk of the DNA by nucleolytic digestion in the presence of the polymerase or of a repressor which are strongly bound to, and protect, the DNA region from the cleavage. The protected regions comprise the region recognizing the polymerase and the regions where the initiation takes place. By comparing a small number of promoter sequences, it wàs observed that each contained a sequence of seven base pairs, homologous to the TATAATG sequence, centered about 10 bp before the initiation site of messenger RNA. Later, Rosenberg and Court compared 46 promoter sequences and observed that the sixth position of this heptamer was invariably a thymine residue (TATAA**T**G). It appears that this base pair is essential to the promoter function and interacts in the same way with the polymerase in all promoters. The other positions of the heptamer are somewhat variable and one can extract from the comparison of all the promoters examined the consensus sequence

TAtaa**T**g (Pribnow box)

in which the base pairs represented by the capital bold letters are the less variable.

Upstream this "Pribnow box" centered at –10 bp from the initiation site, another region of homology has been detected and a consensus sequence for this region centered at –35 bp, has been extracted

tg**TTG**aca.

Initiation of transcription predominantly occurs 6 to 9 bp downstream the invariable **T** residue of the –10 region. The starting nucleotide is usually a purine (more often an adenine than a guanine). The mutations in the –10 sequence which most affect the promoter function are those of the best-conserved base pairs. Physical studies show that polymerase unwinds the double helix of DNA at this level.

Certain purine bases are protected by RNA polymerase from chemical modification by dimethylsulfate, whereas other bases are normally modified or even are more susceptible to methylation. The specific interference by the polymerase has allowed the detection of interactions between the enzyme and guanine residues in the large groove of DNA, the majority of contacts occurring in the –10 and –35 regions.

Termination of transcription in prokaryotes

The principal role of the termination of transcription is to maintain the different transcription units separate, so that there is no interference between the transcription of adjacent genes or operons. Comparing most procaryotic termination sites has revealed three common characteristics: a region of binary symmetry (that is, an inverted repeated sequence), GC-rich sequences of variable length (three to eleven contiguous base pairs) followed by a polyU sequence which terminates the transcript (fig. 1). In those cases, termination of transcription does not need additional factors. But there are other types of termination transcription sites which require the presence of the ρ factor. They are called ρ-dependent and do not present the structural characteristics of the above ρ-independent termination sites.

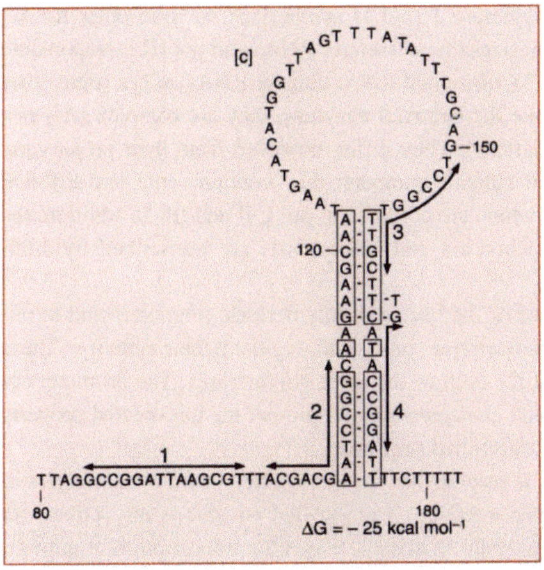

Figure 1. The ρ-independent terminator region of the metBL operon of *E. coli*. The free energy of the stem and loop structure is shown.
(From N. Duchange, M. M. Zakin, P. Ferrara, I. Saint-Girons, I. Park, S. V. Tran, M. C. Py & G. N. Cohen, *J. Biol. Chem.* **258**, 14868-14871 (1983). With permission from the *Journal of Biological Chemistry*.)

The active form of the ρ protein is a hexamer endowed with a nucleoside triphosphatase (NTPase) activity required for the liberation of the nascent transcript from the transcriptional complex. Its function depends on interactions with nascent RNA molecules that contain unpaired cytidylate residues. Results of a study where cytidine and cytidine analogues are shown to inhibit the binding of ρ to *cro* mRNA and its ATPase activity suggests that ρ has a specificity pocket in its polynucleotide binding site that can make hydrogen bond interactions with the side of the unpaired cytidine ring that normally faces away from the sugar ring. A few common characteristics of the ρ-dependent terminators are known, among them the presence upstream of the termination site of an unstructured guanine-poor region. The establishment of a consensus sequence is rendered difficult by the scarcity of ρ-dependent termination systems.

Some activators of transcription do not act by helping RNA polymerase to start transcription; rather, they prevent termination and as a consequence allow transcription of the downstream genes. Termination-antitermination as the means to switch genes on and off is employed in numerous systems in prokaryotes, and a few examples have been uncovered in eukaryotes as well. Two general mechanisms are responsible for antitermination. One, attenuation, will be examined in Chapter XI. The other was discovered while studying antitermination in phage λ: the N protein, coded by λ, interacts with RNA polymerase by masking its response to protein termination factors.

Eukaryotic RNA polymerases

The synthesis of cellular RNA which is accomplished in bacteria by a single polymerase is achieved in all eukaryotes by three distinct enzymes. RNA polymerase I (pol I) synthesizes the ribosomal RNAs, whereas pol II synthesizes the messenger RNAs and pol III is responsible for the synthesis of 5S ribosomal RNA, transfer RNAs and of some other small molecules. Like the bacterial enzymes, they are complex enzymes (MW ca. 500,000 daltons). They differ, however, from their prokaryotic counterparts in their subunit numbers: they contain some ten different subunits, certain of which are common to pol I, II and III. In addition, the genomes from mitochondria and chloroplasts are transcribed by additional polymerases.

As in the case of the bacterial polymerases, promoters and terminators limit the transcription units and regulate their activity. These signals are different for each of the three polymerases. The promoters of pol III have been well characterized and present the unexpected property of lying within the transcribed sequence.

Since pol II is responsible for the synthesis of the messengers, it has received the most attention. The purified enzyme is not active with natural promoters: in order to achieve a specific transcription, it requires other "general transcription factors" (TFIIA, TFIIB, TFIID and TFIIE), whose role is to recognize the TATA and the initiation sequences (see below) and to integrate the polymerase into the transcriptional complex. Five complexes have been identified by successive binding of TFIID, TFIIA, TFIIB, pol II and TFIIE. The addition of ribonucleoside triphosphates then generates new complexes which contain the correctly initiated complexes associated with the rest of the transcriptional machinery and with the DNA matrix.

Until recently, it was thought that pol I, II and III initiation factors were different. The polypeptide that binds directly to the TATA box (TBP or TFIID τ), first identified as a component of the class II initiation factor TFIID, is now known to be required for transcription by all three nuclear RNA polymerases. It thus resembles the essential subunits common to the three RNA polymerases in being a universal transcription factor. TBP is associated with other polypeptides, the TAFs, to form the TFIID multi-protein complex. The structure of TBP from the plant *Arabidopsis thaliana* has been determined by X-ray crystallography at 2.6 Å resolution. It is a highly symmetrical α/β structure which contains a heretofore undescribed DNA-binding fold, resembling a molecular saddle that sits astride the DNA. The DNA-binding surface is a curved, anti-parallel β-sheet. When bound to DNA, the convex surface of the saddle would be presented for interaction with other transcription initiation factors and regulatory proteins.

The recent finding that a homologue of the eukaryotic transcription factor TFIIB is found in the archaebacterium *Pyrococcus woesei* is of considerable interest. The degree of sequence similarity is very significant: this conservation implies a similar three-dimensional structure and is suggestive of a related function in a transcriptional initiation complex. It also is indicative that essential aspects of the eukaryotic transcriptional machinery were in place before the divergence of archaebacteria and eukaryotes; that is, much earlier in evolution than previously thought[10]. The analysis of mutants show that the DNA region which affects transcription is much longer than the bacterial promoters. Examination of several hundred eukaryotic genes has shown certain homologies on lengths extending over several hundred nucleotides upstream the initiation site by pol II. One of the most frequent homologies is the TATA sequence (so-called TATA box) centered around position –25. It appears in many genes to be required for the precise positioning of the transcription start, in a way like the –10 Pribnow box in prokaryotes. As in the case of the bacterial messengers, it is a purine that is first incorporated in the transcribed RNA. The 5' part of the transcript is then rapidly modified by "capping", which consists of a condensation of one molecule of GTP with the triphosphate located in 5' of the first messenger base to form a structure 3'-G-5'ppp5'-N-3' called "cap". The newly-added guanine is then methylated in the 7 position; positions 2' and 3' of the first and/or of the second adjacent nucleotide is (are) often also methylated.

Other homologies are located in variable locations between the TATA box and about the –200 position: in particular a sequence CCAAT has been an important element within the promoter of the eukaryotic genes transcribed by pol II. This sequence appears on either strand of the gene. In many constitutive genes, a GC-rich sequence is often found. These conserved sequences appear to be essential to the binding of the polymerase and to the initiation, as shown by deletion or by point mutations. It seems that at least two sequences upstream the gene are necessary for an efficient transcription of all genes, in addition to the TATA box. Sometimes, the relative position and orientation of these elements

10. In the same vein, the comparison between the ATP-dependent DNA ligase of the archaebacterium *Desulfolobus ambivalens* shows a very significant homology with eukaryotic ligases.

and their distance from the initiation point appear to be important; in some cases, it does not. From time to time, a gene is identified which lacks all of these sequences in its 5' flanking region.

It soon became clear that the polymerase alone could not bind to the promoter and initiate a selective transcription. Other protein factors, acting in *trans,* confer specificity to the purified pol II and allow the initiation of transcription at the sites that are utilized *in vivo.* Some of these factors are required for the initiation of all promoters, whereas others are necessary only for certain promoters. A current technique used for a first characterization of these factors is the DNAase I footprinting (fig. 2) or the gel retardation technique (fig. 3).

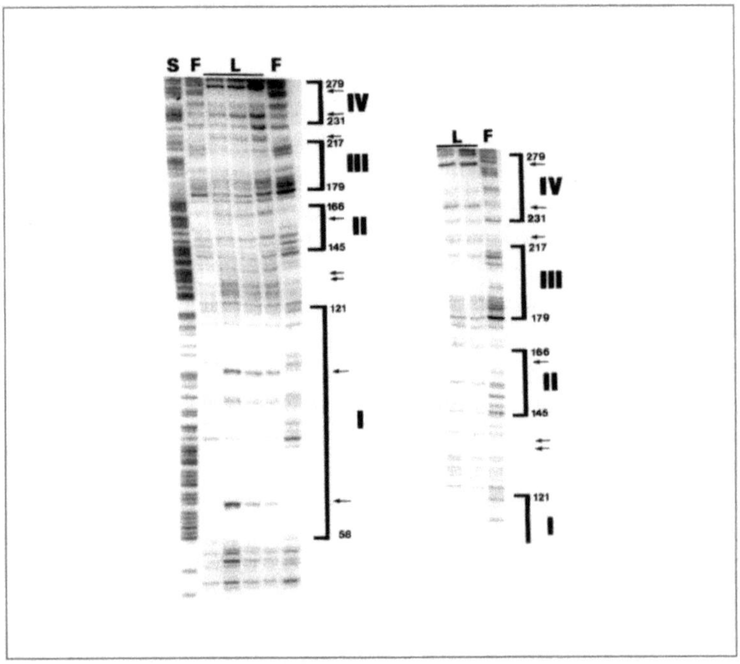

Figure 2. Four DNA motifs from the human transferrin gene enhancer interact with *trans*-acting factors present in liver nuclear extracts. Deoxyribonuclease I protection mapping, or footprinting, is a valuable technique for locating the specific binding sites of proteins on DNA. The basis of this method is that bound protein protects the phosphodiester backbone of DNA from DNAase I-catalyzed hydrolysis. Binding sites are visualized as "windows" or "footprints", by autoradiography of the DNA fragments that result from hydrolysis, following separation by electrophoresis on denaturing DNA sequencing gels.
S, sequence of the DNA fragment; F, digestion in the absence of nuclear extracts; L, digestion in presence of liver nuclear extracts. Four protected regions are clearly visible on each strand.
(From F. Boissier, C. Augé-Gouillou, E. Schaeffer & M. M. Zakin, *J. Biol. Chem.* **266**, 9822-9828 (1991). With permission from the *Journal of Biological Chemistry.*)

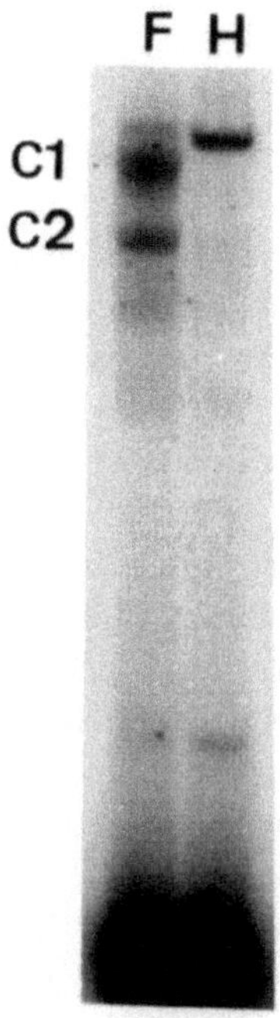

56 86
GCTCTTTGTTTGCTTTGCTTCTGTGTCAAC----
----AAACAAACGAAACGAAGACACAGTTGCGAG

Figure 3. Mobility shift DNA binding assay using polyacrylamide gel electrophoresis; a simple, rapid, and extremely sensitive method for the detection of sequence-specific DNA binding proteins in crude extracts. Proteins that bind specifically to an end-labeled DNA fragment retard the mobility of the fragment during electrophoresis, resulting in one or more discrete bands corresponding to the DNA-protein complexes. The fragment whose sequence is shown corresponds to part of motif I of the preceding figure.
Lane F: liver nuclear extract: two proteins bind (see the two arrows in motif I of fig. 2). H, HeLa cells nuclear extracts.
(From F. Boissier, C. Augé-Gouillou, E. Schaeffer & M. M. Zakin, *J. Biol. Chem.* **266**, 9822-9828 (1991). With permission from the *Journal of Biological Chemistry.*)

A factor named SP1 is necessary for the selective transcription of the SV40 virus genome. It binds the GC-rich regions not only of SV40, but also of those upstream the thymidine kinase gene from the herpes virus, the mouse dihydrofolate reductase gene, and of many others. The SP1 gene has been cloned and the protein expressed: it is glycosylated and contains in its C-terminal region three contiguous zinc finger motifs (see the chapter on protein-DNA interactions). It has been found that Zn^{2+} is required for the specific binding of SP1 to DNA. Other families of proteins acting in *trans* on proximal regions of the promoter have been identified as CTF/NF1 and *jun*-AP1, which are not zinc finger proteins (*jun*-AP1 is a leucine zipper protein: see the chapter on DNA-protein interactions).

When yeast cells are starved for amino acids, the synthesis of a protein called GCN4 is induced; it is a leucine zipper protein which binds upstream many amino acid biosynthetic genes and activates their transcription. GCN4 interacts directly *in vitro* with RNA polymerase, which strongly suggests that this interaction could be important in the formation of the transcriptional complex.

Analyses carried out by deletions or by mutagenesis show that the region of DNA which affects transcription is much longer than the promoter region: enhanced transcription is linked to DNA regions several kilobases upstream the transcription start. These *enhancer* regions possess fundamentally different properties from the *cis*-elements of promoters: 1) they are relatively long, sometimes covering several hundred base pairs; 2) they exert their action at considerable distances from the initiation point, up to several thousand base pairs; 3) they act in the two possible orientations; 4) their action does not depend on their localization: they can be found in 5' from the gene, reside within an intron, or even be in 3' from the gene, but they must reside on the same DNA molecule as the gene they activate; 5) finally, a given enhancer can function preferentially or exclusively in certain cell types. This specificity may be linked to the differential expression of genes during development, or in differentiated tissues of an adult organism.

Enhancers, like promoters, are the target of *trans*-acting proteins. This has been vindicated directly in many cases: in elements of murine mammary tumor virus, in a human gene coding for metallothionein, in the human gene transferrin, etc. Enhancers, like promoters, are sites where several proteins are assembled, certain of which bind directly to DNA while others are involved in more indirect aspects of the interaction. Certain of these proteins are identical or similar to the ones binding to the promoter regions. Little is known as to how the enhancers transmit their effects on long distances. If we accept that they act at the level of transcription initiation, we must also accept that they facilitate a limiting step; in addition, since they are sites where specific proteins bind, they must obey the rules governing the DNA-protein interactions. The formation of bends in the DNA could bring into contact the proteins interacting with enhancers and promoters and stabilize the transcriptional complex.

Other regions 5' of the gene contain positive or negative elements which stimulate or repress genetic expression. They are, generally, isolated and not clustered as in promoters and enhancers, and found within the first upstream kilobase; among these elements, one notes those which respond to hormones and metal ions. It has been shown that the protein binding to the hormone-responding element is the very receptor of steroid hormones which activate the corresponding gene. Chambon has shown that three discrete regions of the receptor correspond respectively to the hormone-binding domain, to the DNA-binding domain, and to the transcription-activation domain. The activation of the first domain upon hormone binding activates the second domain and the two activated domains induce in turn the activation of transcription by protein-protein interactions.

Transcription termination in eukaryotes

At the 3'-extremity of most eukaryotic messengers, one finds long sequences of adenylic acid residues. These polyA sequences and their nuclear precursors are not coded by DNA, but are added by an enzyme called polyA polymerase, after the gene transcription to pre-mRNA by pol II.

PolyA polymerase catalyzes the reaction

$$\text{RNA} + n\,\text{ATP} \rightarrow \text{RNA}\,(\text{A})_n + n\,\text{PP}_i$$

Ten to thirty nucleotides upstream the resulting polyA sequence, in a section of RNA coded by DNA, a sequence AAUAAA (or a similar one) is always found. However, this sequence could not by itself be the signal for polyadenylation, since it is also found in other parts of mRNA or pre-mRNA not followed by "polyA tails". In fact, pol II continues to elongate the RNA chain further from the polyadenylation site, sometimes by several thousand nucleotides. The nascent pre-mRNA is then cleaved by an endonuclease which creates a correct 3'-terminus which is then polyadenylated; the endonuclease degrades rapidly the downstream sequences. The upstream AAUAAA hexanucleotide is part of the cleavage signal. There is evidence that complexes between proteins and small nuclear RNAs are also involved in termination.

The terminated pre-mRNA molecule still contains the transcripts of the non-coding introns present in the genes. The elimination of introns is realized through the molecule of pre-mRNA itself. by excision and circularization of the intron; this involves a series of transfer reactions: in the presence of monovalent ions and Mg^{2+}, a guanine nucleoside attacks the phosphodiester bond on the 5'- end of the intron and is added to the intron by a classical 3'-5' bond; the 3'-OH bond thus formed attacks the phosphodiester bond at the other extremity of the intron thus creating a spliced joint: the intron is liberated and undergoes an autocyclization yielding a molecule which possesses at one extremity a linear sequence of 15 nucleotides ended by the G residue which was added in the first step and, at the other extremity, the remainder of the cyclized intron. This entirely novel reaction catalyzed by RNA can be reproduced *in vitro* in the absence of any protein.

In eukaryotes, polyA polymerase is essential for the maturation of mRNA. However, neither the specific role of the polyA tail nor its mechanism of formation are fully understood, despite the cloning of the bovine, yeast and vaccinia virus genes coding for the enzyme.

Substantial amounts of polyA sequences have been found in prokaryotic mRNAs from *Bacillus subtilis*, *Bacillus brevis* and *E. coli.* Two specific mRNAs in *E. coli*, those of the outer membrane lipoprotein and tryptophan synthetase α subunit, were found to be extensively polyadenylated. The gene coding for polyA polymerase in *E. coli* has been identified (*pcnB*, 3 mn) and overexpressed after insertion into an expression vector, opening the way for studying the biosynthesis and function of bacterial polyadenylated mRNA.

SELECTED REFERENCES

Trans-acting transcription factors
R. G. Roeder, *Trends Biochem. Sci.* **16**, 402-408 (1991).

Eukaryotic RNA polymerases
A. Sentenac, *CRC Curr. Rev. Biochem.* **18**, 31-90 (1985).

Termination of transcription
J. W. Roberts, *Nature* **224**, 1168-1174 (1968).
S. Adhya & M. Gottesman, *Ann. Rev. Biochem.* **47**, 967-996 (1978).
D. I. Friedman, in *The Bacteriophage*, vol. II. Plenum, New York (1988).

CHAPTER VI

Negative regulation

Each of the hypotheses implicitly or explicitly put forward in chapter IV has been borne out by relevant experiments: enzyme induction is correlated with the synthesis of a specific messenger RNA; the cytoplasmic repressor coded by the *i* gene has been isolated and characterized; its affinities for the inducer and for the operator have been determined; the repressor extracted from an i^s strain has lost its affinity for IPTG; the operator is a DNA segment which specifically binds the repressor; finally, the inducer displaces the repressor from its complex with the operator. We shall now briefly describe these experiments.

Induction correlated with the synthesis of a specific messenger

The study of early induction kinetics and the effects of very short inducer pulses have led to the conclusion that an unstable polynucleotide precursor is synthesized before the appearance of the enzyme. In addition, when the inducer is removed, the half-life of the residual β-galactosidase synthesis is of the same order of magnitude as that inferred for this unstable RNA.

However, the hypothesis that induction (and repression) act at the genetic level can only be demonstrated by the identification of a messenger RNA species *specific* for a given inducible system, and the establishment of a correlation between the rate of its synthesis and that of the corresponding induced enzymes.

The detection and the estimation of a specific mRNA corresponding to a gene or to a group of genes are based on the property of mRNAs to form specific molecular hybrids with the homologous single-stranded DNA when a heated mixture of the two polymers is cooled slowly.

In order to apply this technique to a single or to a few mRNA species, it was necessary to bring the sensitivity of hybrid detection to 0.01-0.02% of the *E. coli* genome, and to considerably reduce the noise due to the hybrids not pertaining to the specific genes studied. This can be achieved by obtaining an mRNA of high specific radioactivity and by increasing the relative concentration of the DNA segment carrying the genes under study. Such an increase in concentration can be obtained by using as DNA source the DNA from the defective phage λdg or from the F-*lac* episome in which specific bacterial genes responsible for galactose or lactose have been specifically inserted by recombination. The non-specific mRNAs can be eliminated by hybridization with the appropriate DNAs (that of phage λ or of *E. coli* F^+) devoid of these specific genes. We shall give here as an example the case of the galactose genes.

Four of the structural genes governing galactose degradation are closely linked on a small segment of the *E. coli* chromosome (the *gal* region), situated near the attachment site of prophage λ. Their coordinate expression is under the control of the same operator. The *gal* segment can be incorporated in the DNA of the transducing phage λdg. This DNA can then be used as acceptor in the hybridization experiments. As the DNA content of λdg is of the order of 1% of that of *E. coli,* and since the two DNAs contain the same amount of *gal* genes, the latter are 100 times more concentrated in the λdg DNA. In addition, the noise is greatly reduced by the replacement of *E. coli* DNA by the heterologous λ DNA.

The mRNA labeled with ^{32}P is purified from a non-lysogenic strain of *E. coli* whose galactose degrading enzymes are induced by fucose, a gratuitous inducer. The bacteria are grown for several generations in presence or absence of inducer and exposed to ^{32}P for a period not exceeding 0.7% of their doubling time. Under these conditions, ribosomal RNAs are essentially unlabeled and most of the radioactivity sediments between 8 and 10S. The radioactive material is mixed with DNA (from λ or λdg) previously denatured by heating. The mixture is incubated for one hour at 50 °C, for 14 hours at 40 °C; and then slowly cooled. The mRNA-DNA hybrids are separated from unhybridized mRNA by centrifugation in a cesium chloride gradient and treated by ribonuclease; the amount of radioactivity associated to DNA in the ribonuclease-resistant species is then determined.

The radioactivity associated to DNA (measured by optical density) is not significantly different in the λdg-mRNA hybrids obtained from non-induced cells and in λ-mRNA hybrids obtained from induced cells. This shows that the noise is essentially due to interactions between *E. coli* mRNA and λDNA. On the other hand, the amount of radioactivity associated to λdg is much higher when the mRNA has been obtained from induced than from non-induced bacteria. After subtraction of "non-specific" radioactivity, seven times more radioactivity is found associated to λdg DNA when cells are induced by fucose than when the cells were grown on glycerol alone. Note that the mRNA from glucose-grown cells is unlabeled (table 1).

There is a good correlation between the relative amounts of mRNA specifically hybridizable with the genes of the galactose operon and the rate of synthesis of the corresponding enzymes. A constitutive dominant strain *(O^c)* synthesizes more specific mRNA (table II)

Table I. Specific complexes between λdg or λ DNA and the ^{32}P labeled mRNAs originating from either non induced or fully induced bacteria.

	O.D. (260 nm)	cpm	cpm/O.D.	cpm/O.D. corrected for λ DNA hybridization
λdg DNA + mRNA from induced cells (glycerol + fucose)	2.03	1222	602	353
λdg DNA + mRNA from non-induced cells (glycerol)	2.07	627	303	54
λDNA + mRNA from induced cells (glycerol + fucose)	1.44	359	249	–
λdg DNA + mRNA from non-induced cells (glucose)	2.16	490	227	0

Table II. Synthesis of the enzymes of the galactose operon.

Growth conditions	Galactokinase	G-1-P-uridyltransferase
	mmoles/h/mg bacteria	
Non induced wild type (glycerol)	0.87	0.83
Induced wild type (glycerol + fucose)	13.6	12.2
O^c strain (glycerol)	24.5	23

Isolation of *lac* repressor

A central question of molecular biology concerns the nature of regulatory substances. The model of negative regulation proposed by Jacob and Monod implies that certain genes, called regulatory genes, are responsible for diffusible repressors which prevent the expression of other genes, called structural genes. The latter are organized as operons comprising *cis*-dominant operators, which are the targets of the repressors. Small appropriate molecules act either as inducers, by preventing repression, or as corepressors (leading to active repressors, as we shall see later). The simplest hypothesis for explaining the behavior of inducible systems is that the repressor is the direct product of the regulatory gene which binds to the operator site of a DNA molecule in such a way as to prevent the transcription of the operon by RNA polymerase. This model implies that the inducer combines with the repressor to form a complex which can no longer bind to the operator, thus allowing the free transcription of the messenger and the synthesis of the enzymes specific to the operon. However, other models could also account for the experimental results. The isolation of several repressors has led to the understanding of the regulation mechanisms at the molecular level. We will describe here the isolation and some of the properties of the Lac repressor

The detection method used involves the affinity of the hypothetical repressor with a radioactive inducer, isopropylthiogalactoside (IPTG), measured by equilibrium dialysis. In order to detect the binding, it is necessary to obtain a concentration of the receptor of the same order of magnitude as that of the dissociation constant of the repressor-inducer complex. In a y^- strain, semi-maximal induction requires a concentration of IPTG of 2 x 10^{-4} M; however, this value does not lead directly to an estimation of the equilibrium constant: since the enzyme level increases by a factor 1000 during induction, and since the rate of enzyme synthesis is inversely proportional to the repressor concentration, it follows that the concentration of free repressor must have fallen by a factor 1000 during induction. From the shape of the induction curve, one can deduce that the repressor extracted from the wild type must have an affinity of about 6 x 10^{-6} M for IPTG. If the binding is to be detected, one needs a repressor concentration of the same order. Since one expects very few repressor molecules per operator site, it appears probable that the repressor should be concentrated at least 100-fold before even a weak binding is detectable. Fortunately, a superinducible mutant of the *i* gene, i^t, has been isolated which synthesizes β-galactosidase with much lower concentrations of IPTG. With a diploid strain carrying two copies of i^t, the detection is still marginal (1000 cpm/0.1 ml outside the dialysis bag, 1040 inside the bag!). Further purification by classical techniques leads rapidly to satisfactory results (50-100% excess radioactivity inside the bag) for a protein concentration of 10 mg/ml.

Radioactive IPTG is reversibly bound and can dialyze in and out of the bag. The material which binds IPTG is indeed the product of the *i* gene: an i^s strain does not yield a material binding reversibly IPTG (fig. 1); the same is true for a suppressible i^- amber strain, which synthe-

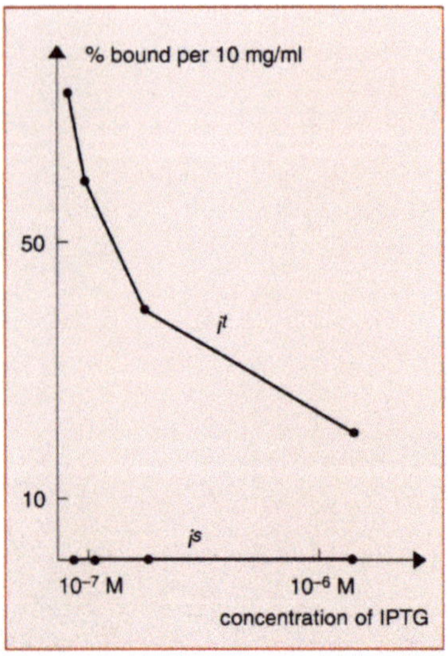

Figure 1. IPTG is not bound by an i^s strain.
The repressor fraction is isolated in parallel from 50 g batches of otherwise isogenic F' i^s/i^s and F' i^t/i^t strains. Aliquots of each extract are dialyzed against various concentrations of ^{14}C-IPTG during 8 h in an appropriate buffer. The excess radioactivity inside each dialysis bag is normalized for a protein concentration of 10 mg/ml.
(From W. Gilbert & B. Müller-Hill, *Proc. Natl. Acad. Sci. US* **56**, 1891-1898 (1966). With permission from Dr. W. Gilbert.)

sizes only an inactive fragment of the *i* gene product (fig. 2) or for a strain carrying a deletion of the *i* gene All these negative controls are significant, since extracts from these strains do not inhibit the binding from the i^t extracts.

If one plots the amount of bound IPTG against the ratio free/bound IPTG, one obtains a straight line, indicating a single type of binding site. The intercept on the ordinate gives the molarity of the binding sites and the slope is the inverse of the association constant. Figure 3 gives such straight lines for the i^t strain and for the wild type. One notices that, as expected, the repressor from the i^t strain has a higher affinity.

The equilibrium dialysis method allows an estimation of the affinities of IPTG competitors for the repressor binding site for the repressor: TMG is a less good inducer than IPTG. Galactose is a weak inducer of the i^t strain *in vivo*. Glucose has practically no affinity for the site. The values of the association constants are consistent with the identification of the purified material with the repressor (table III).

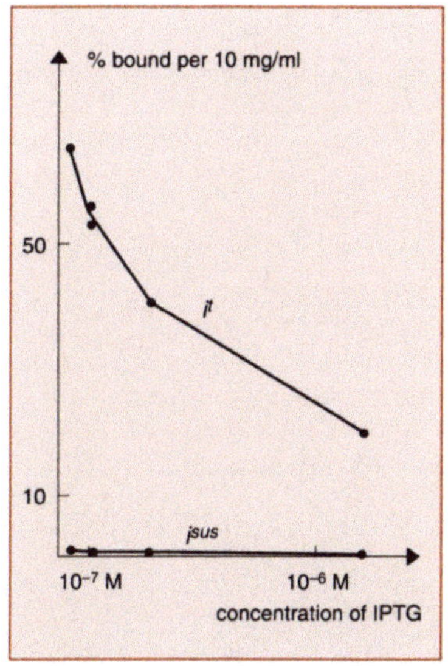

Figure 2. A suppressible i^- strain does not bind IPTG.
The repressor fraction is isolated from two otherwise isogenic strains F' i^{sus}/i^{sus} and F' i^t/i^t; the fractions are tested as mentioned in the legend of figure 1.
(From W. Gilbert & B. Müller-Hill, *Proc. Natl. Acad. Sci. US* **56**, 1891-1898 (1966). With permission from Dr. W. Gilbert.)

Table III. Dissociation constants of potential ligands of the Lac repressor.

Competitors	K_i for the i^t gene product
Non radioactive IPTG	5×10^{-7} M
Thiomethylgalactoside (TMG)	2×10^{-6} M
o-nitrophenylfucoside (ONPF)	2×10^{-5} M
Galactose	5×10^{-4} M
Glucose	3×10^{-2} M

The capacity to bind IPTG is insensitive to DNAase and to RNAase, but is destroyed by heating at 50 °C and by pronase. It sediments at 7.8S on a glycerol gradient. It is a non-inducible protein and calculations show that there are only 10-20 molecules of it per gene in the wild type, which amounts to about one ten-thousandth of the total weight of the *E. coli* proteins.

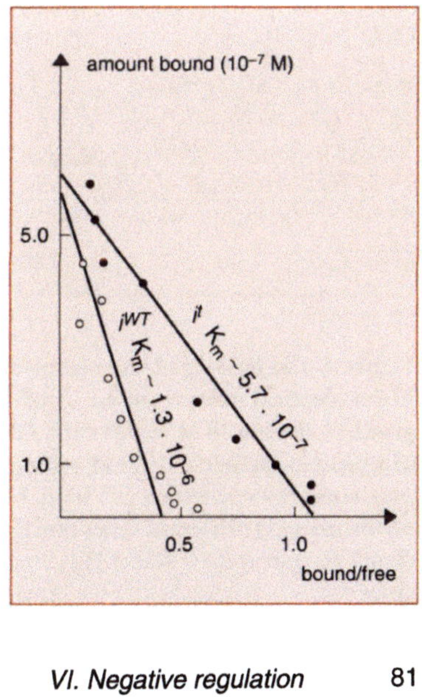

Figure 3. The association constants of the repressors isolated from the wild type and from the i^t mutant strain.
The repressor fractions from otherwise isogenic wild type or i^t strains are dialyzed against different concentrations of ^{14}C-IPTG, as in figure 1.
Ordinates: amount of IPTG bound for each IPTG external concentration, normalized for a protein concentration of 10 mg/ml.
Abscissas: ratio bound IPTG/free IPTG.
If there is only one type of binding site, Km is deduced from the relation:

$$\text{IPTG bound} = \text{number of sites} - \text{Km}\,\frac{\text{bound}}{\text{free}}.$$

(From W. Gilbert & B. Müller-Hill, *Proc. Natl. Acad. Sci. US* **56**, 1891-1898 (1966). With permission from Dr. W. Gilbert.)

Lac operator is a DNA sequence

Radioactive repressor can be obtained by blindly following the techniques used for the purification of non radioactive repressor. It is not necessary that the repressor be pure, but that it represent a reasonable proportion of the labeled material and that other proteins bound to DNA be eliminated, so that the observed results are really specific. One starts from a strain triploid for the *lac* segment (on the chromosome, on a defective ϕ80*dlac* and on an F*lac* episome), the proteins of which have been labeled by growth on ^{35}S. The proteins are fractionated by ammonium sulfate and on a DEAE-Sephadex column. After concentration of the material, it is deposited onto a glycerol gradient and the fraction sedimenting at 7.6S is retained. This material is mixed with phage DNA containing the *lac* region, incubated and sedimented on a glycerol gradient: a small radioactive peak is displaced from the 8S region and migrates with DNA, at 35-40S. Figure 4a shows a centrifugation in the presence of 10^{-2} M Mg^{2+}: a radioactive peak migrates with DNA, representing only 1 % of the total radioactivity present in the mixture, al-

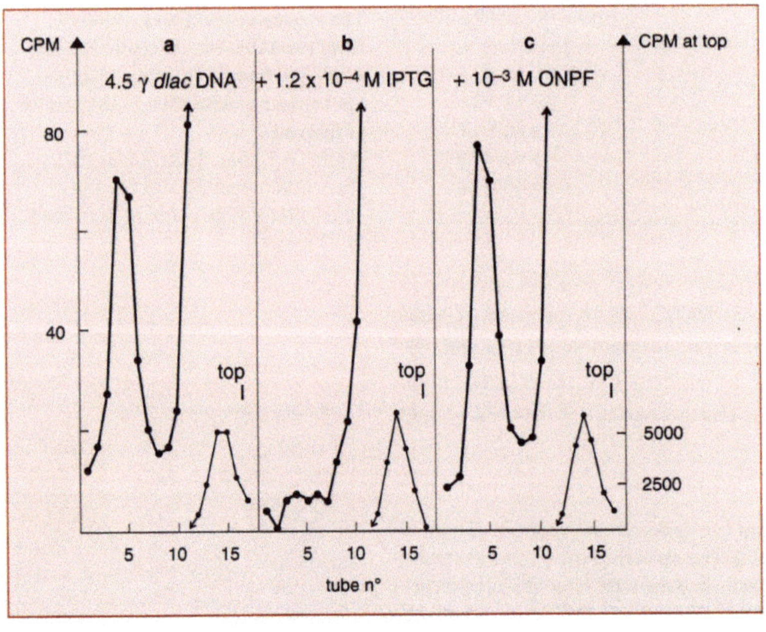

Figure 4. The binding of Lac repressor to the DNA of phage *dlac* and its displacement by the inducer.
Three identical mixtures of 4.5 µg of *dlac* DNA and of radioactive protein (7.8S) are centrifuged on a glycerol gradient, during 2h at 35,000 rpm. DNA alone would have generated an optical density peak in tubes 4 and 5. a) a distinct radioactivity peak sediments with DNA. The latter is in a 10-fold excess and sediments with the maximum radioactivity it can bind. b) the gradient solution contains IPTG. The binding of the repressor to DNA is abolished. c) ONPF eight times more concentrated than IPTG has no effect on the binding.
(From W. Gilbert & B. Müller-Hill, *Proc. Natl. Acad. Sci. US* **58**, 2415-2421 (1967). With permission from Dr. Walter Gilbert.)

though DNA is in great excess. Figure 4b proves that the material bound to DNA is actually the Lac repressor: no radioactivity migrates with DNA if 10^{-4} M IPTG is present. This effect is specific: ONPF, which binds to the repressor but does not induce the synthesis of β-galactosidase, is without effect on the binding (fig. 4c).

Ribonuclease does not affect the association. The purified, non-radioactive Lac repressor competes with the binding. There is no, or insignificant, binding of the repressor with phage DNA not carrying the *lac* genes (fig. 5).

A more significant test is provided by the DNA from phages carrying a mutated O^c operator gene in the *lac* segment: figures 6a, b and c

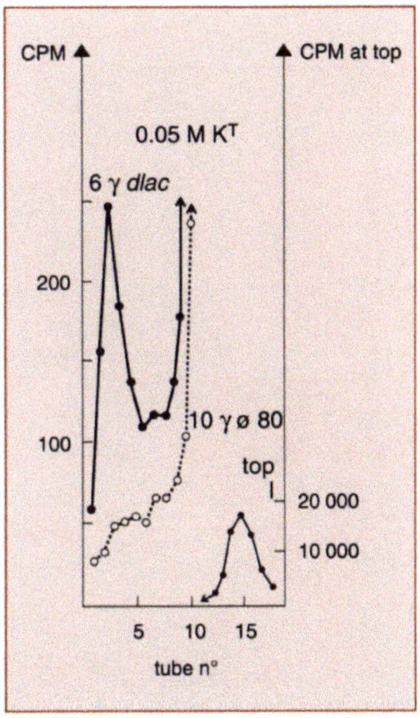

Figure 5. Binding specificity.
The same repressor preparation is centrifuged on two parallel gradients with two different DNAs. A mixture contains 6mg of purified DNA from *dlac* phage: the other contains 10mg of DNA of the parental phage (which does not carry the *lac* region).
(From W. Gilbert & B. Müller-Hill, *Proc. Natl. Acad. Sci. US* **58**, 2415-2421 (1967). With permission from Dr. Walter Gilbert.)

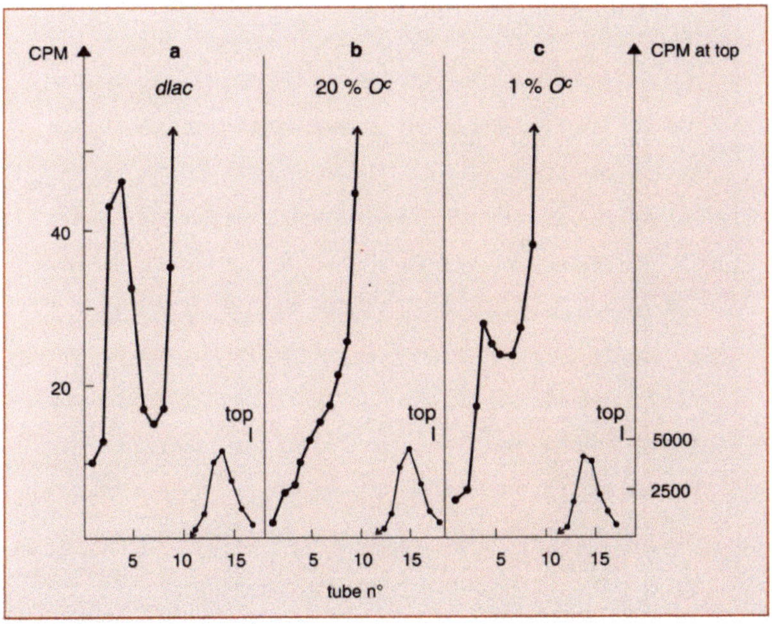

Figure 6. The Lac repressor binds specifically the *lac* operator.
The same preparation of Lac repressor is centrifuged in three parallel gradients with three purified DNAs originating from three different phages. a) 4.5 mg of DNA from a phage *dlacO*+ b) same amount of DNA carrying an O^c mutation leading to a 20% constitutivity: the repressor is not bound c) same amount of DNA carrying an O^c mutation leading only to a 1% constitutivity (10 times the basal level of an uninduced wild type strain): one can still detect a weak binding.
(From W. Gilbert & B. Müller-Hill, *Proc. Natl. Acad. Sci. US* **58**, 2415-2421 (1967). With permission from Dr. Walter Gilbert.)

show the compared binding on the DNA bearing the O^+ allele and the DNA from two different O^c alleles; one being a low-level constitutive (1% of the maximum specific activity after induction), the other possessing a higher degree (20%) of constitutivity. Whereas the binding of the repressor to the operator is still detectable in the first case, in the second, the radioactive characteristic peak is absent.

These experiments demonstrate that the repressor binds to a well-determined DNA sequence, the operator region. It binds only to double-stranded DNA.

Met Lys Pro Val Thr Leu Tyr Asp Val Ala Glu Tyr Ala Gly Val

Ser Tyr Gln Thr Val Ser Arg Val Val Asn Gln Ala Ser His Val 30

Ser Ala Lys Thr Arg Glu Lys Val Glu Ala Ala Met Ala Glu Leu

Asn Tyr Ile Pro Asn Arg Val Ala Gln Gln leu Ala Gly Lys Gln 60

Ser Leu Leu Ile Gly Val Ala Thr Ser Ser Leu Ala Leu His Ala

Pro Ser Gln Ile Val Ala Ala Ile Lys Ser Arg Ala Asp Gln Leu 90

Gly Ala Ser Val Val Val Ser Met Val Glu Arg Ser Gly Val Glu

Ala Cys Lys Ala Ala Val His Asn Leu Leu Ala Gln Arg Val Ser 120

Gly Leu Ile Ile Asn Tyr Pro Leu Asp Asp Gln Asp Ala Ile Ala

Val Glu Ala Ala Cys Thr Asn Val Pro Ala Leu Phe Leu Asp Val 150

Ser Asp Gln Thr Pro Ile Asn Ser Ile Ile Phe Ser His Glu Asp

Gly Thr Arg Leu Gly Val Glu His Leu Val Ala Leu Gly His Gln 180

Gln Ile Ala Leu Leu Ala Gly Pro Leu Ser Ser Val Ser Ala Arg

Leu Arg Leu Ala Gly Trp His Lys Tyr Leu Thr Arg Asn Gln Ile 210

Gln Pro Ile Ala Glu Arg Glu Gly Asp Trp Ser Ala Met Ser Gly

Phe Gln Gln Thr Met Gln Met Leu Asn Glu Gly Ile Val Pro Thr 240

Ala Met Leu Val Ala Asp Asn Gln Met Ala Leu Gly Ala Met Arg

Ala Ile Thr Glu Ser Gly Leu Arg Val Gly Ala Asp Ile Ser Val 270

Val Gly Tyr Asp Asp Thr Glu Asp Ser Ser Cys Tyr Ile Pro Pro

Leu Thr Thr Ile Lys Gln Asp Phe Arg Leu Leu Gly Gln Thr Ser 300

Val Asp Arg Leu Leu Gln Leu Ser Gln Gly Gln Ala Val Lys Gly

Asn Gln Leu Leu Pro Val Ser **Leu** Val Lys Arg Lys Thr Thr **Leu** 330

Ala Pro Asn Thr Gln Thr Ala Ser Pro Arg Ala **Leu** Ala Asp Ser

Leu Met Gln **Leu** Ala Arg Gln Val Ser Arg **Leu** Glu Ser Gly Gln 360

Figure 7. The sequence of the *E. coli* Lac repressor.

Specific protein-DNA interaction is essential for gene control in both prokaryotes and eukaryotes. Extensive biochemical and genetic studies with the Lac repressor indicate that the protein is a tetramer of identical subunits. Each subunit, 360 amino acid residues long (fig. 7), is composed of three major domains:

1) a small N-terminal domain, the "headpiece" (59 amino acid residues) recognizes the operator; mutations in the corresponding part of the *i* gene diminish or abolish operator binding without affecting inducer binding. If the repressor is treated by trypsin or chymotrypsin, it loses 1-79 as well as about 20 amino acids at the C-terminal end;

2) the core-protein, resistant to trypsin hydrolysis, (about 270 amino acids) binds the inducer and leads to the aggregation of dimers;

3) the C-terminal 30 amino acids, a leucine mini-zipper, mediate tetramer assembly.

How does Lac repressor, a tetrameric protein, recognize its target ? It is necessary to know the operator structure to try to answer this question.

The *lac* operator has been isolated as a double-stranded DNA segment protected by the Lac repressor against pancreatic DNAase digestion. The starting materials are bacteria lysogenic for an inducible defective and thermoinducible prophage carrying part of the *lac* genome including the operator region; the bacteria are induced at 42 °C and their DNA is fractionated by sonic disruption into fragments having an average length of 1000 nucleotides. The Lac repressor is added to these fragments under conditions optimal for its binding, and the mixture is digested by pancreatic DNAase, which hydrolyzes double-stranded DNA, but is without action on the protected segment. The mixture is filtered on a cellulose nitrate membrane which retains only the repressor-operator complex. The operator is eluted from the complex by isopropylthiogalactoside. The protected fragment is able to combine again with repressor and is a 24 bp long double-stranded segment. The DNA is denatured and each strand is used as a template for transcription by DNA-dependent RNA polymerase. The RNA copies of DNA are then sequenced[11]. If they are transcribed on paper as DNA sequences, one obtains

TGG AATTGT G A G C G G A T A AGAATT
ACC TTAACA C T C G C C T A T TGTTAA.

This sequence shows obvious symmetries. Sixteen bases are arranged on both sides of a binary symmetry axis which is not at the center of the total fragment: on the left, there are six bp symmetrically oriented to six other bp situated at the extreme right of the fragment; there are other base pairs symmetrically arranged along nine bases situated between the two principal regions.

11. At the time this experiment was done, methods for DNA sequencing were not yet available.

This symmetry could allow the Lac repressor to interact with DNA along its binary symmetry axis. Note that the two longer sequences are separated by one turn of the double helix (B form).

The operator sequence does not immediately lead to an understanding of the recognition between the two molecules. Although crystals of the repressor have been obtained, they are too small to lend themselves to an analysis by X-ray diffraction. Nuclear magnetic resonance studies of the Lac repressor operator complex have been carried out and it was found that the sequence from amino acid 17 to 25 is a DNA-recognizing α-helix. Genetic analyses of amino acid exchanges in this recognition helix and symmetric base-pair substitutions of the ideal *lac* operator revealed specific contacts between amino acids and base pairs and determined the orientation of the recognition helix within the major groove of the DNA which is opposite to that of λ and *cro* repressors (see the chapter on protein-DNA interactions).

Bound Lac repressor inhibits the unwinding reaction catalyzed by some *E. coli* DNA helicases. This inhibition is substantially weaker in the presence of IPTG. It is likely that replication forks encounter transcription complexes. The replication fork must also replicate regions of the chromosome that are complexed with other site-specific DNA-binding proteins like the various repressors of transcription. Such potential barriers to replication must be removed to allow DNA replication to proceed at a normal rate. Further experiments are needed to clarify this interesting problem.

SELECTED REFERENCES

G. Attardi, S. Naono, J. Rouvière, F. Jacob & F. Gros, *Cold Spring Harbor Symp. Quant. Biol.* **28**, 363-372 (1963).

W. Gilbert & B. Müller-Hill, *Proc. Natl. Acad. Sci. US* **56**, 1891-1898 (1966).

W. Gilbert & B. Müller-Hill, *Proc. Natl. Acad. Sci. US* **58**, 2415-2421 (1967).

J. E. Yancey-Wrona & S. W. Matson, *Nucleic Acids Research* **20**, 6713-6721 (1992).

CHAPTER VII

Enzyme repression in anabolic pathways

Description of the phenomenon

Forty years ago, the following observation was made at the Pasteur Institute: if wild-type *E. coli* was grown in the presence of an exogenous amino acid, the content of an enzyme involved in the biosynthesis of this amino acid was significantly lower in the cell suspensions or the extracts from these bacteria. Table I illustrates this observation in the case of methionine biosynthesis. The experiment was carried out with a strain constitutive for β-galactosidase, in order to ascertain the specificity of the effect observed. All the amino acids reduce somehow the galactosidase specific activity; there is no correlation between this reduction and the effects on the levels of methionine synthase. Whereas the addition of methionine does not affect β-galactosidase, it has a spectacular effect on methionine synthase levels. This effect can be obtained with rather low methionine concentrations: 500 μM of the L-isomer (table I).

Table I. Methionine synthase and β-galactosidase activities of cell suspensions of *E. coli* obtained after an overnight growth in presence of various amino acids.
(From M. Cohn, G. N. Cohen & J. Monod 1953.)

Amino acid during growth (6 mM)	% effect on β-galactosidase	% effect on methionine synthase
0	0	0
DL-isoleucine	– 7	– 36
DL-valine	– 24	– 18
DL-phenylalanine	– 15	– 27
DL-threonine	– 6	– 18
L-histidine	– 29	+ 23
DL-serine	– 30	+ 36
L-arginine	– 7	+ 36
L-proline	– 26	+ 36
L-tryptophan	– 10	+ 41
DL-methionine 500 μM)	*– 4*	*– 99*
DL-methionine (5 mM)	*0*	*– 99*

Table II. Tryptophan synthase activity of *Aerobacter aerogenes* having grown with various supplements.
The reaction followed is indole + serine → tryptophan and is performed in presence of pyridoxal phosphate. Activity expressed as nmoles of indole consumed/h/mg bacterial N.
(From J. Monod & G. Cohen-Bazire 1953.)

Supplement	Specific activity	% effect on tryptophan synthase
0	1300	0
L-histidine 2.5 mM	1500	+ 15
L-arginine 5 mM	1600	+ 20
DL-phenylalanine 5 mM	2100	+ 60
DL-methionine 2 mM	2300	+ 75
L-glutamate 2 mM	1200	− 8
L-proline 2 mM	1100	− 15
DL-serine 4 mM	1500	+ 16
L-tryptophan 2 mM	*360*	*− 72*

Since, in 1953, methionine synthase activity was only measurable in intact cells, there was no possibility of ascertaining that the observed inhibition was due to the absence of synthesis of a protein component of the system (since then, it has been found that this is indeed the case). The following control was made, however: cell suspensions do not lose their enzyme activity in presence of methionine, the effect being observed only after *growth* on this amino acid. In the case of tryptophan synthase (table II), the activity of which could be measured in cell extracts, there is no doubt possible: only growth in presence of tryptophan leads to a significant decrease in the level of tryptophan synthase. Mixtures of cell extracts from cultures grown in presence and in absence of tryptophan yield purely additive activities, eliminating the hypothesis that the observed effect could be due to an inhibitor. There is no indication of enzyme inactivation or of a direct inhibition: non-proliferating cell suspensions shaken in the presence of tryptophan retain a normal tryptophan synthase activity. The only alternative is that the presence of tryptophan inhibits the synthesis of the enzyme.

The observations made at the Pasteur Institute were extended to acetylornithinase, whose synthesis is inhibited by arginine, to the pyrimidine, purine, histidine, valine, isoleucine, biosynthetic enzymes, etc. So the phenomenon of inhibition of enzyme synthesis is largely observed in biosynthetic systems. Vogel chose the word *repression* to describe it. Repression is not limited to microorganisms.

A remarkable advance in the study of repression is due to an elegant experiment of Gorini and Maas on the regulation of the synthesis of ornithine transcarbamylase (OTCase), one of the enzymes involved in arginine biosynthesis. When arginine is added to an exponentially growing culture of *E. coli*, the synthesis of ornithine transcarbamylase stops immediately and the enzyme present in the cells is exponentially diluted during further growth. Thus, after about ten divisions, the intracellular concentration of the enzyme is about one hundredth that found in a control culture grown without added arginine. If the repressed bacteria are washed and resuspended in a medium without arginine, synthesis of OTCase resumes. This system provides an experimental study of the kinetics of enzyme formation, analogous to that consisting of adding an inducer to non-induced cells in order to study the kinetics of synthesis in an inducible system.

In the case of β-galactosidase, the addition of inducer causes the quasi-immediate synthesis of the enzyme at a constant rate. The increase of enzyme is described by

$$\% \; E_{max} = 100\,(1 - e^{-\mu t})$$

where E_{max} is the final enzyme concentration per cell and μ the rate constant of the equation for exponential growth, $dN/dt = \mu N$ (N = number of bacteria per unit volume). This means that the maximal enzyme concentration is reached only after several generations. However, the kinetics of OTCase synthesis is paradoxical: the level of enzyme reaches a maximum very rapidly (after about half a division) and then decreases during the rest of the exponential phase. Gorini and Maas, in order to explain this unexpected result, made the hypothesis that the synthesis of the enzyme is repressed not only by exogenous arginine, but also by the arginine endogenously synthesized. In the first part of the experiment, after growth on arginine and the washing, the cells contain very little enzyme and very little arginine; the enzyme is therefore synthesized very rapidly and arginine accumulates, repressing further enzyme synthesis (fig. 1).

In order to verify this hypothesis, it was necessary to find conditions where a level of intracellular arginine lower than that of the wild type growing on a minimal medium could be achieved. This was done by using an auxotrophic mutant which required arginine and was blocked before OTCase

$$\text{N-acetylornithine} \rightarrow \text{ornithine} \xrightarrow{\text{OTCase}} \text{citrulline} \rightarrow \text{arginine}$$

The growth of this mutant can be limited in a chemostat (Chapter I) by the rate V_1 at which arginine is delivered. If the pool size of the

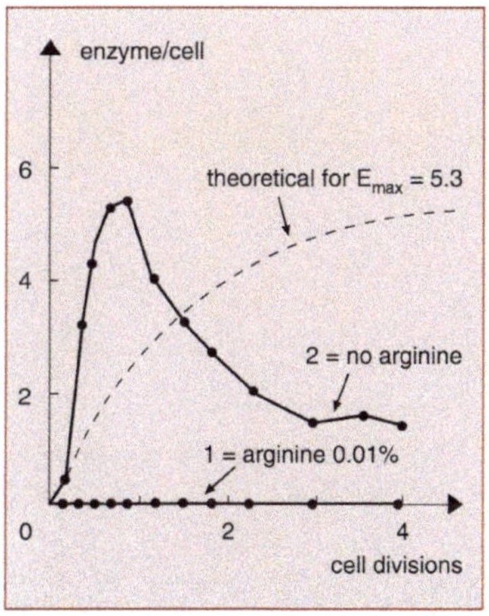

Figure 1. Synthesis of ornithine transcarbamylase in discontinuous culture. A preculture is made in the presence of excess arginine, and then transferred in the same medium (curve 1) or in a medium devoid of arginine. The enzyme activity is measured (arbitrary units) at different times during growth.
(From L. Gorini & W. K. Maas (1958). With permission from The Johns Hopkins University Press.)

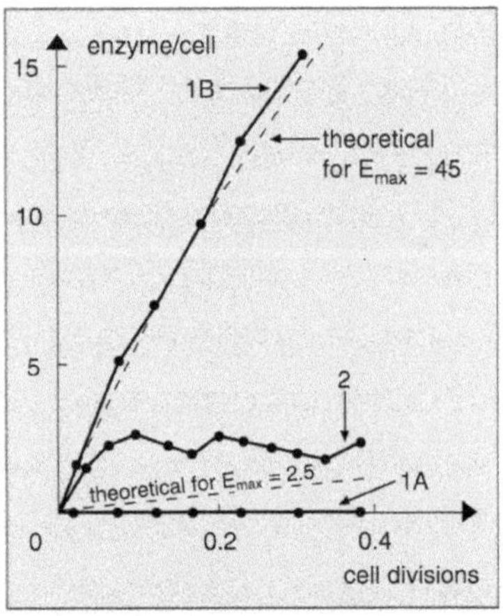

Figure 2. Synthesis of ornithine transcarbamylase in continuous culture (chemostat), using a *his⁻ arg⁻* double auxotroph.
Curve A: histidine is the limiting factor.
Curve B: arginine is the limiting factor.
(From L. Gorini & W. K. Maas (1958). With permission from The Johns Hopkins University Press.)

intracellular arginine is due to the difference between this rate V_1 and the rate V_2 of incorporation of arginine into proteins ($V_1 > V_2$), conditions are realized in the chemostat where $V_1 = V_2$ and the pool of free arginine is equal to zero. The mutant utilized was also auxotrophic for histidine, which allowed one to study the specificity of the observed effects. Figure 2 describes the results of the experiment. After a preculture with arginine and washing, the bacteria are transferred into the chemostat. When arginine limits growth, enzyme synthesis starts at the same rate as in the previous experiment, but the synthesis continues at the same rate since the intracellular conentration of arginine remains, at best, at a negligible level. If histidine limits growth and if arginine is delivered at a concentration exceeding by 1 μg/ml that necessary for keeping the growth rate at the imposed rate, the synthesis of OTCase is totally repressed.

One should note that, under these conditions of physiological derepression, the enzyme reaches an intracellular concentration 25 times higher than that obtained in the wild type growing in minimal medium. The important conclusion of this work is that the intracellular concentration of arginine (or of one of its derivatives) controls the rate of synthesis of ornithine transcarbamylase; this type of regulation had only been suggested by the experiments where it was shown that the final product of a biosynthetic pathway, provided exogenously, inhibited the synthesis of the enzymes involved in its own synthesis.

This inhibition of synthesis *(repression)* thus appears to be a general physiological mechanism of regulation of enzyme systems. One of the striking aspects of these studies is the high potential of the cells to synthesize a given enzyme, at a level far higher than that required for growth. This high potential and repression allow the cell to adapt rapidly to a change in the environment, from a medium in which a given metabolite is scarce to one where it is in excess. The capacity of small molecular weight metabolites to control the cellular enzyme level, observed in the above experiments, is reminiscent of enzyme induction. We shall see that this analogy is more than formal.

Isolation of derepressed (constitutive) mutants in biosynthetic pathways. Use of structural analogues

Any attempt to formulate a unitary model to acount for the mechanisms of enzyme induction of catabolic enzymes and of enzyme repression of biosynthetic enzymes rests on the isolation in the latter systems of mutants of the type we have called i^- and O^c.

In biosynthetic pathways, such mutants can be isolated with the aid of structural analogues.

An unexpected observation was made during the study of β-galactoside permease. One will recall that, in order to study its kinetics,

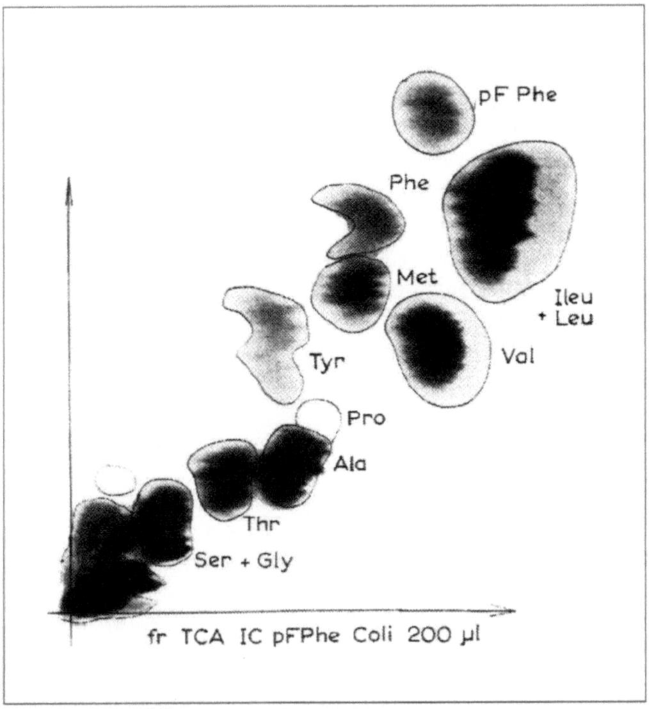

Figure 3. Incorporation of p-fluorophenylalanine in *E. coli* proteins. Two-dimensional chromatography of an acid hydrolyzate of the total proteins of *E. coli* grown in the presence of 1 mM DL-*p*-fluorophenylalanine.
(From R. Munier & G. N. Cohen, *Biochimica et Biophysica Acta* **31**, 378-391 (1959). With permission from Elsevier Science Publishers.)

chloramphenicol was added to stop the further synthesis of the permease during the experiment. One day, the chloramphenicol solution could not be readily found and *p*-fluorophenylalanine, which was reported to stop protein synthesis in yeast, was used instead. With great surprise, it was found that the addition of the analogue to an exponential culture of *E. coli* did not stop the increase in mass, but rather turned the exponential growth into a linear one.

A detailed analysis of the phenomenon revealed that *p*-fluorophenylalanine is incorporated instead of phenylalanine in proteins synthesized after its addition (fig. 3), and that the average degree of substitution in the overall proteins depends on the ratio endogenous phenylalanine/exogenous analogue. Similar results were obtained with norleucine, an analogue of methionine where the sulfur atom is replaced by an isosteric methylene group:

$CH_3–S–(CH_2)_2–CH(NH_2)–COOH$	$CH_3–CH_2–(CH_2)_2–CH(NH_2)–COOH$
methionine	norleucine

It has been possible to show that, within a given protein of *E. coli* (namely adenylate kinase which contains seven methionine residues per polypeptide chain), all proteins containing from one up to seven substituted residues are obtained by including norleucine in the growth medium and even that the fully substituted enzyme is perfectly active!

The probable reason for the obtention of linear growth has been discussed in the chapter on bacterial growth. Practically, the phenomenon of linear growth has a limit imposed by the cellular dilution of the indispensable component present before the addition of the analogue: the phenomenon will be observable if the analogue is added to an exponentially growing culture but no cell will give rise to a colony on a solid medium containing the analogue. However, if the plated population contains analogue-resistant mutants – for whatever reason –, they will give rise to colonies. This method has been used to obtain all sorts of interesting mutants, in particular derepressed (constitutive) mutants. An example will be detailed when we study the mutants derepressed for tryptophan synthesis which led to the isolation of the Trp repressor (chapter XXI).

SELECTED REFERENCES

Repression of biosynthesis of anabolic enzymes

J. Monod & G. Cohen-Bazire, *Comptes rendus* **236**, 530-532 (1953).

M. Cohn, G. N. Cohen & J. Monod, *Comptes rendus* **236**, 746-749 (1953).

H. J. Vogel, in *The Chemical Basis of Heredity*, W. D. McElroy & B. Glass eds., pp. 267-289. The Johns Hopkins University Press, Baltimore (1957).

L. Gorini & W. K. Maas, *Biochim. Biophys. Acta* **25**, 208-209 (1957).

L. Gorini & W. K. Maas, in *The Chemical Basis of Development*, W.D. McElroy & B. Glass eds., pp. 469-478. The Johns Hopkins University Press, Baltimore (1958).

Incorporation of amino acid analogues into proteins

G. N. Cohen & R. Munier, *Biochim. Biophys. Acta* **31**, 347-356 (1959).

R. Munier & G. N. Cohen, *Biochim. Biophys. Acta* **31**, 378-391 (1959).

D. B. Cowie, G. N. Cohen, E. T. Bolton & H. de Robichon-Szulmajster, *Biochim. Biophys. Acta* **34**, 39-46 (1959).

A. M. Gilles, P. Marlière, T. Rose, R. Sarfati, R. Longin, A. Meier, S. Fermandjian, M. Monnot, G. N. Cohen & O. Bârzu, *J. Biol. Chem.* **263**, 8204-8209 (1988).

CHAPTER VIII

Positive regulation

It has long been known that the presence of glucose (or of another easily degradable carbon source) inhibits the synthesis of certain bacterial enzymes such as β-galactosidase or tryptophanase from *E. coli,* or inositol dehydrogenase and histidase from *Bacillus subtilis*. This phenomenon was called glucose effect. In 1961, Magasanik gave it the more general name *catabolic repression*. Since many catabolic enzymes are inducible, the rate of their synthesis must then depend on the intracellular concentrations of the inducer and of the catabolite exerting its repressive effect.

The synthesis of catabolic enzymes is regulated in a manner such that their intracellular content be adequate for growth in a given environment. Inducers are signals indicating that the medium contains substances capable of being degraded by the inducible enzymes to yield ATP, pyruvate, etc. Catabolic repressors provide the information that these compounds exist in a sufficient amount. The synthesis of inducible enzymes will take place rapidly only when these compounds are at low level.

However, this is only a description of the physiological aspects of catabolic repression and says nothing about its genetic basis or the molecular basis of this type of regulation. In addition to the nature and specificity of the catabolic repressors, it is important to define their targets and their relation to the specific repressors we have encountered in the preceding chapters and whose action is counteracted by specific inducers.

An observation made by Makman and Sutherland in 1965 helped to clarify the problem: the addition of glucose to a culture of *E. coli* growing on glycerol causes a rapid fall in the intracellular concentration of 3'-5'-cyclic adenosine monophosphate (cyclic AMP or cAMP).

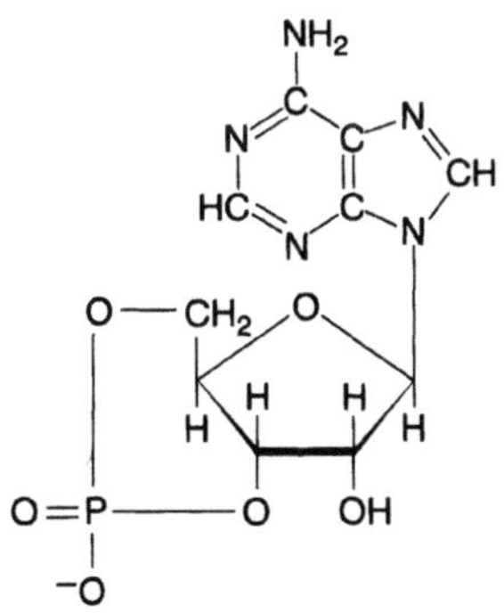

cyclic AMP

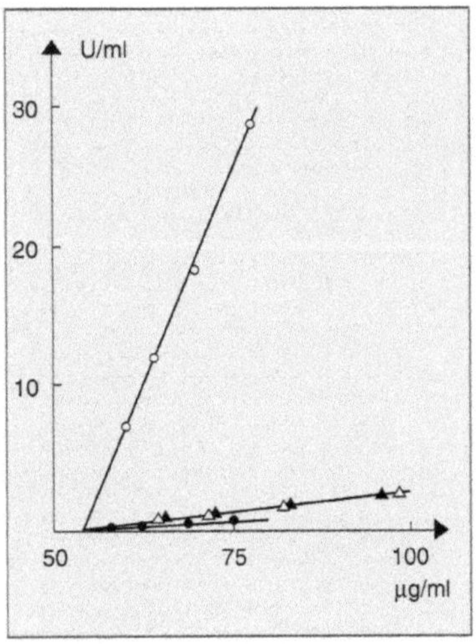

Figure 1. Effects of glucose and cyclic AMP on the synthesis of β-galactosidase by a culture of *E. coli*.
The slopes of the straight lines represent the differential rate of synthesis of β-galactosidase. The cultures were induced with 10^{-3} M IPTG at 55 mg dry weight/ml.cAMP (8. 10^{-3} M) was added five minutes before the inducer; •—•, control; ○—○, + cAMP; Δ—Δ, ▲—▲, controls irrelevant to this experiment.
(From A. Ullmann & J. Monod, *FEBS Letters* **2**, 57-60 (1968). With permission from *FEBS Letters*.)

It is now clear that the glucose effect is somehow linked to this depletion: Perlman and Pastan, and Ullmann and Monod observed, independently, in 1968, that the repressing effect of glucose on the synthesis of catabolic enzymes can be readily reverted by the addition of cyclic AMP to *E. coli* cultures (fig. 1).

How does cyclic AMP control enzyme synthesis?

Let us look again at the structure of the lactose operon, taking as a model the inhibition by glucose of β-galactosidase synthesis.

Promoter region (a new aspect)

Constitutive strains respond normally to catabolic repression, whether they are recessive (i^-) or dominant (O^c). On the other hand, the initial constitutive synthesis of β-galactosidase by the zygotes from the cross Hfr (i^+O^+) and an F^- strain with a deletion encompassing the *i* gene is subjected also to catabolic repression. Since these zygotes do not contain any product of the *i* gene, a role of the Lac repressor in the mechanism of catabolic repression is excluded.

A study of mutants sensitive to repression by the product of *i*, but whose β-galactosidase synthesis is no more sensitive to the presence of glucose in the medium, has led to the localization of a new region, *lacP*, which has been called promoter region. The promoter is characterized as an essential site for the formation of the products of the structural genes *z, y* and *a*. Promoter deletions lead to a considerable reduction of the rate of synthesis of these proteins when they are coded by structural genes in *cis* position (on the same DNA strand as the deletion), but do not alter the response of the *lac* operon to the Lac repressor when they are in *trans* position.

The order of the genes in the *lac* region is:

i p o z y a

Strains p^- with a point mutation in the promoter can be induced by IPTG: they synthesize β-galactosidase at a rate of the order of 5 to 10% of the rate of the parent p^+ strain. Such strains are sensitive to catabolic repression, but the effect of glucose is suppressed if cyclic AMP is added to the culture medium.

Mutations of the promoter of the *lac* region are specific: the synthesis of tryptophanase by a strain which does not show a glucose effect on β-galactosidase synthesis is normally subjected to catabolic repression by glucose.

Since the non-sensitivity to glucose is *cis*-dominant, this excludes the intervention in the process of a diffusible product whose synthesis would be under the control of *p*.

Role of cyclic AMP and of CAP protein in the binding of RNA polymerase to the promoter region

Experiments from numerous laboratories strongly suggested that catabolic repression affects gene expression at the transcription level (that of messenger RNA synthesis) rather than at the translation level (protein synthesis). Actually, the promoter region is the DNA site where RNA polymerase binds to initiate DNA transcription. In fact, it was during the study of catabolic repression that the promoter was first characterized, and the groups led by Beckwith, Miller, and Reznikoff established, from 1967 to 1975, that the promoter is a region of *lac* DNA necessary for the initiation of transcription of the *lac* operon.

A detailed analysis of the promoter region and of its role in transcription has been given in chapter V.

The sequence of the *lac* promoter region contains the information for binding another protein in addition to RNA polymerase. In 1969, Chambers and Zubay oberved that cyclic AMP is required for the DNA-dependent synthesis of the enzymes of the *lac* operon in the cell-free system they had developed (fig. 2).

A year later, Zubay, Schwartz, and Beckwith, theorizing that it was highly improbable that a mononucleotide intervenes directly and with the required specificity on the synthesis of catabolic enzymes, hypothesized that an intermediary protein existed which they called CAP (for catabolite gene activator protein). This hypothetical protein would have affinity for cyclic AMP, the complex CAP-cAMP once bound to DNA allowing the initiation of the messenger RNA at the promoter locus.

The detection of the CAP protein was facilitated by the existence of a mutant which synthesizes a defective CAP: this mutant was found during a systematic screening of *E. coli* variants unable to grow on lac-

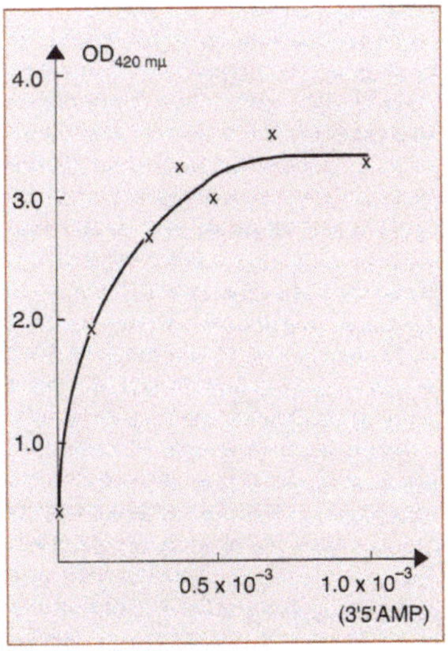

>

Figure 2. Ordinate: β-galactosidase synthesized *in vitro* (arbitrary units). Abscissa: cAMP concentration. The incubation mixture in Tris buffer pH 8.2 contains, in addition to the energy source and the inorganic ions required for optimal synthesis, all the amino acids, the four ribonucleoside triphosphates, the tRNAs, DNA and proteins from an *E. coli* strain not producing the Lac repressor.

(From D. A. Chambers & G. Zubay, *Proc. Natl. Acad. Sci. US* **69**, 1100-1103 (1972). With permission from Dr. G. Zubay.)

tose *and* arabinose; they all possess very low levels of the enzymes known to be subjected to catabolic repression. Certain are devoid of adenylate cyclase, the enzyme responsible for the synthesis of cAMP from ATP: they revert (phenotypically) in the presence of cAMP. Other produce normal quantities of cAMP; one of them is devoid of the CAP protein necessary for the "activation" of the lactose operon. If the cell-free system from this mutant is used to synthesize β-galactosidase, it needs the addition of CAP and thus allows the purification of the protein. The stimulation by cAMP requires the presence of CAP (table I).

Table I. Requirements for the cell-free synthesis of β-galactosidase by the mutant unable to synthesize an active CAP protein.
cAMP: 5 mM. Partially purified extract containing CAP: 10 μg/ml.

Cyclic AMP	CAP	β-galactosidase synthesized (arbitrary units)
–	–	1
+	–	1
–	+	1
+	+	5

The synthesis of β-galactosidase is proportional to the amount of CAP (fig. 3).

The same group of workers has repeated the experiment under conditions where the product detected during the cell-free synthesis is not β-galactosidase, but the messenger RNA corresponding to the transcription of the DNA segment coding for the lactose operon, by specific hybridization of mRNA to a defective bacteriophage carrying the operon. The results show that both cAMP and the CAP protein are required for the *in vitro* transcription of the *lac* operon. This transcription is inhibited by the Lac repressor.

The CAP protein was purified and obtained in the homogeneous state by Pastan in 1971. It binds *lac* DNA, and the association constant is increased in the presence of cAMP. The binding to DNA is strongly inhibited by cyclic guanosine 3'-5'-monophosphate (cGMP), an effect reproducing an *in vivo* observation.

The CAP protein, coded by the *crp* gene (74 mn), is a homodimer

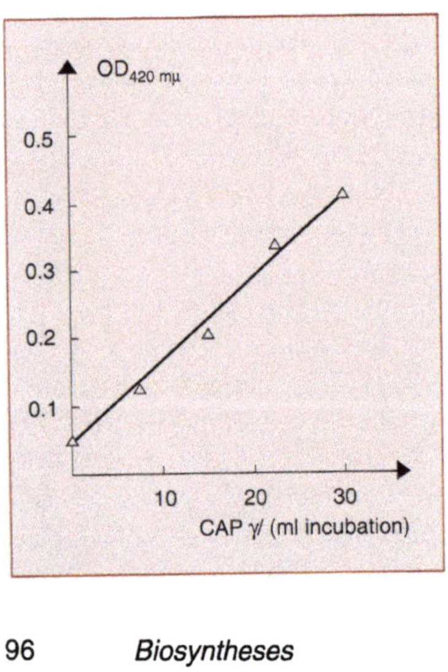

Figure 3. Synthesis of β-galactosidase as a function of the concentration of partially purified CAP.
The cell-free system is similar to that of figure 1, but the extract is from a strain which does not synthesize CAP. Ordinate: β-galactosidase synthesized, arbitrary units.
(From G. Zubay, D. Schwartz & J. Beckwith, *Proc. Natl. Acad. Sci. US* **66**, 104-110 (1970). With permission from Dr. G. Zubay.)

whose sequence has been established independently by Aiba and by Cossart and Gicquel-Sanzey in 1982. Each chain is made of 209 amino acid residues. Its mode of action will be examined in Chapter XXIX.

Synthesis and degradation of cyclic AMP

Cyclic AMP is formed from ATP in a reaction catalyzed by adenylate cyclase (*cyaA*, 85 mn):

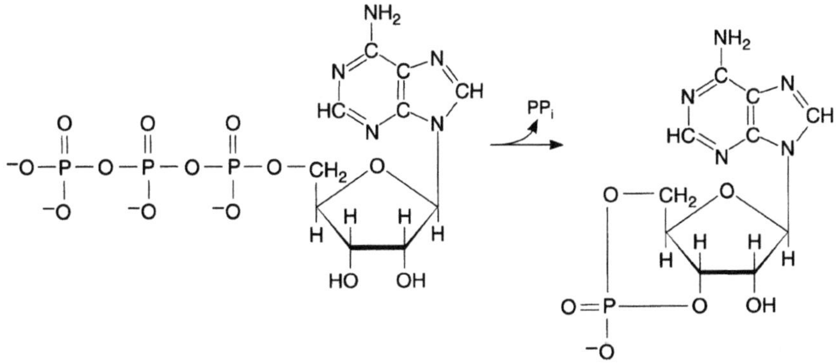

This reaction is driven toward the synthesis of cAMP by the subsequent action of a pyrophosphatase.

The cyclase is an integral membrane protein in eukaryotes; whereas early studies had reported this to be the case in *E. coli*, this result has been challenged by workers who found the cyclase in the cytoplasmic fraction of disrupted cells. The partially purified enzyme is very labile. The *cya* gene has been cloned and sequenced; it codes for a polypeptide of 848 residues.

Cyclic AMP is a very stable molecule. It takes a specific phosphodiesterase to hydrolyze it to AMP:

$$\text{cyclic AMP} + H_2O \xrightarrow{Mg^{2+}} \text{AMP} + H^+$$

How does glucose exert its inhibitory effect on *E. coli* β-galactosidase synthesis?

We must go back to the description of diauxie and of the PTS system. It will be noted that sugars A, defined in Monod's monograph, all exert a "glucose effect", in fact a catabolic repression. They are also all substrates of permeation through the phosphoenolpyruvate phosphotransferase system (PTS).

The underlying mechanisms for PTS-mediated repression occur first at the metabolic level. PTS controls the concentrations of both cAMP (and therefore the cAMP/CAP adduct) and the inducer of the operon. Progress in the understanding of the molecular mechanisms underlying these controls resided in the isolation of mutations (*crr*, 52 mn), which relieved simultaneously four non-PTS operons from PTS-mediated regulation. The *crr* gene was found to code for III^{Glc}. Both *in vivo* and *in vitro* experiments have shown that III^{Glc} interacts with a variety of PTS and non-PTS proteins. For example, when III^{Glc} is enclosed within natural membrane vesicles (see Chapter II), it inhibits directly, but incompletely, the activity of β-galactoside permease, while extravesicular III^{Glc} is inactive. III^{Glc} might thus participate in the glucose effect by excluding the inducer of β-galactosidase. On the other hand, a phosphorylated form of III^{Glc} activates strongly adenylate cyclase and increases the level of cAMP (and of cAMP-CAP), thus increasing the expression of the *lac* genes. The ratio of phosphorylated to unphosphorylated PTS proteins appears to regulate the activities of non-PTS proteins. The situation is far from being clear and it will require more work to correlate catabolite repression and diauxic curves at the molecular level.

SELECTED REFERENCES

Catabolic repression

B. Magasanik, *Cold Spring Harbor Symp. Quant. Biol.* **26**, 249-256 (1961).

Effects of cyclic AMP on glucose effect

R. S. Makman & E. W. Sutherland, *J. Biol. Chem.* **240**, 1309-1314 (1965).

R. Perlman & I. Pastan, *Biochem. Biophys. Res. Comm.* **30**, 656-664 (1968) & **37**, 151-157 (1969).

A. Ullman & J. Monod, *FEBS Lett.* **2**, 57-60 (1968).

The promoter

F. Jacob, A. Ullman & J. Monod, *Comptes rendus* **258**, 3125-3128 (1964).

The CAP protein

D. A. Chambers & G. Zubay, *Proc. Natl. Acad. Sci. US* **63**, 118-122 (1969).

G. Zubay, D. Schwartz & J. Beckwith, *Proc. Natl. Acad. Sci. US* **66**, 104-110 (1970).

Mode of action of cyclic AMP

A. Ullmann & A. Danchin, *Adv. Cyclic Nucleotide Research* **15**, 1-53 (1983).

N. D. Meadow, D. K. Fox & S. Roseman, *Ann. Rev. Biochem.* **59**, 497-542 (1990).

CHAPTER IX

Ribosomes

In the 1930s, cell biologists had surmised that the greatest part of RNA was localized in the cytoplasm and that there was a correlation between the amount of RNA and the protein synthesizing capacity of the cell. In the 1940s, Brachet and Caspersson established unequivocally this correlation with materials as diverse as the root tips of *Allium cepa,* imaginal disks of *Drosophila* larvae, secreting pancreatic cells and silk glands from the silkworm and with various methods such as specific dyeing techniques, specific nucleases and spectrophotometric methods on thin tissue slices.

By differential centrifugation, Claude identified a cellular fraction composed of small granules he called microsomes. Jeener and Brachet found that the microsomes contained RNA and made the hypothesis that they were involved in protein synthesis. The presence of Claude's microsomes in cytoplasm was confirmed by electron microscopy by Palade and Siekevitz who showed that these ribonucleoprotein particles, 200 Å in diameter, existed in all tissues. These particles, today called ribosomes, can be either attached to the endoplasmic reticulum in cells where this system is well developed, or exist as free particles.

From 1955 to 1957, studies in Zamecnik's laboratory showed with whole animals injected with radioactive amino acids and with cell-free preparations that the amino acids are first incorporated into the ribosomes from which they can be chased in the soluble fraction. It was shown that only 1% of the amino acids from the ribosomal fraction turn rapidly over, which suggested a steady state between the rates of formation and release of the polypeptide chains synthesized.

A great advance in the understanding of these particles was due to studies on bacterial ribosomes. Luria, Delbrück, and Anderson had observed as early as 1943, by electron microscopy of *E. coli* cells lyzed by bacteriophage, in addition to the virus particles, large amounts of granules of homogeneous size. Schachman, Pardee, and Stanier (1952), by ultracentrifugation of *E. coli* extracts, identified three well-separable constituents of respective sedimentation coefficients (uncorrected) 40S, 29S, and 5S. These constituents contain the essential of cellular RNA, the 40S constituent being composed of quasi-spherical particles. The sedimentation coefficients were corrected later by centrifuging in presence of Mg^{2+} ions, which stabilize the ribosomes.

Thus, by the middle of the 1950s, it had been established that quasi-spherical particles of a diameter 200-300 Å, formed in approximately equal proportion of RNA and proteins, were the site of protein synthesis in plants, animals, yeasts and bacteria. In 1956, Chao showed that the particles were composed of two unequal components, which could be separated in absence of magnesium, the metal being a bridge between the two parts. Tissières and Watson started to study the physicochemical properties of ribosomes at Harvard; there were at that time very few laboratories working on the subject, the most active being the Biophysics group of the Department of Terrestrial Magnetism of the Carnegie Institution of Washington, headed by Richard Roberts. A special meeting of the recently created Biophysical Society was held in Boston in 1958. It is during this meeting that Roberts first introduced the term ribosomes. In 1959, McQuillen, Britten, and Roberts showed that ^{35}S-amino acids originating from radioactive sulfate were first incorporated in the ribosomes, before they could appear in, or be chased to, other parts of the cell. Their elegant experiments were using ultracentrifugation in sucrose gradients. It became evident that with *E. coli,* experiments similar to those of Zamecnik in eukaryotes could make the understanding of the mechanism of protein synthesis easier. The names Watson, Kurland, Tissières, Schlesinger, Gros, and Zillig are associated with this early period (1959-1960).

Components of *E. coli* ribosomes

E. coli ribosomes have a sedimentation coefficient of 70S. They are composed of two subunits of 30S and 50S, containing respectively RNA molecules of 16S and 23S. They also contain a 5S RNA component, discovered by Monier and Rosset in 1963, and which was sequenced by Brownlee, Sanger, and Barrell in 1968.

The situation of ribosomal proteins is much more complex, as shown by Waller and Harris in 1961. Waller showed in 1964, by electrophoresis on starch gels and on carboxymethylcellulose, that the proteins of the 30S subunit were different from those of the 50S subunit. Two more years were necessary before the individual proteins could be identified with the techniques of protein chemistry.

We know today that the 50S subunit contains 31 polypeptides numbered from L1 to L34 (the discrepancy comes from errors in the early work) and the 30S subunit contains 21 proteins numbered from S1 to S21. All the proteins and the 16 and 23S RNAs have been sequenced through the efforts of two groups (Ebel in Strasbourg and Wittmann in Berlin). Secondary structures for the RNAs have been proposed and the binding site of many of the ribosomal proteins on the RNA have been determined.

Nomura and his colleagues have devised techniques of partial or total disassembly of the two subunits and have succeeded in reconstituting them by adding to the RNA of each subunit the proteins that had been eliminated. The order in which the proteins are added is not indifferent and Nomura has established a very detailed reassembly chart. The information necessary for assembly is entirely contained in the structures of the proteins and the RNA and there is no need for auxiliary proteins or enzymes. The *in vitro* auto-assembly differs only slightly from the *in vivo* assembly process and reflects the physiological process.

The predictions of the ribosomal proteins secondary structure have been calculated from their primary structure (a hazardous method), but also by circular dichroism and small angle X-ray diffraction. Mutants of practically all the ribosomal proteins have been obtained and have greatly helped the functional studies.

Different approaches have been used to unravel the topology of the ribosomes which is of paramount importance for the understanding of structure-function relationships.

a) Antibodies acting on individual proteins form dimers between two subunits which can then be examined by electron microscopy; the antibody can be localized on the surface of the subunit and the position of the antigenic determinant of a given protein can be determined. This method rests on the divalent nature of the antibodies and on the facts that antibodies react with their antigen within the intact ribosome, and that the two subunits possess distinguishable morphological features. This study was principally carried out in Wittmann's laboratory from 1973 to 1980 and the localization within the particle of the proteins belonging to one or the other subunit have been determined.

b) The proximity relationships of the different proteins within a subunit have been also determined by cross-linking bifunctional reagents which create a covalent attachment between two neighboring proteins.

c) Finally, pairs of proteins have been labeled with fluorescent markers and the measurement of the energy transfer has provided a measure of the relative distance between the centers of the proteins tested.

For example, experiments have disclosed that proteins S4, S7, S8, S13, S15, S17, S20 bind to the 16S RNA, S16 being in contact with S4 and S20. Other proteins (S3, S10, S12, S14 and S19), although not necessary for ribosome reconstitution, are necessary for their function. In the large subunit, L5, L18 and L25 bind specifically the 5S RNA. Furthermore, the complex between these three proteins and 5S RNA bind the oligonucleotide TpypCpGp, and suggest an interaction between 5SRNA and a molecule of transfer RNA bound to the ribosome. One or several proteins of the complex are necessary for the binding of 5S RNA to the 23S RNA. Bifunctional reagents have cross-linked the following pairs of the 30S subunit: S2-S3, S4-S6, S4-S8, S4-S12, S5-S8, S5-S9, S7-S8, S7-S9, S11-S18, S13-S19, and S18-S21. Other results indicate that proteins S2, S12, S15, and S18 are separated by a distance of 8 to 19 nm and that they must have an elongated shape. Protein S4 must extend over at least 17 nm, in keeping with the great number of reticulations of this protein with other ribosomal proteins. From all the preceding facts, it appears that the ribosome constitutes an extraordinarily complex machinery.

Ribosomes of eukaryotes

The eukaryotic ribosomes are composed of two subunits of 60S and 40S sedimentation coefficients (*vs* 50S and 30S in prokaryotes).The association of the subunits form a ribosome of 80S (*vs* 70S). The small subunit contains a molecule of 18S RNA and about 30 proteins, whereas the large subunit comprises three RNA molecules (5S, 5.8S and 28S) and 45 to 50 proteins. In spite of the difference in size, electron micrographs of the eukaryotic ribosomes reveal a morphology very similar to that of the prokaryotic ones

We shall not examine here the mechanistic aspects of translation of messenger RNA to protein by this machinery, these aspects being treated at length in many books, including the exhaustive one referred to below.

SELECTED REFERENCES

The Ribosome: Structure, Function and Evolution, W. E. Hill, A. Dahlberg *et al.* eds., 678 p. American Society for Microbiology (1990).

Eukaryotic ribosomes
I. G. Wool, *Ann. Rev. Biochem.* **48**, 719-754 (1979).

CHAPTER X

Genetic code, transfer RNAs and aminoacyl-tRNA-synthetases

Genetic code

The first ideas on an intimate connection between RNA and protein synthesis go back to Brachet and Caspersson. We know now that their observations did not bear on the messenger RNA, but on ribosomal RNA. Nevertheless, it is thanks to them that this connection became familiar. Then Beadle formulated the hypothesis called the "one gene-one enzyme correlation", but his ideas never reached the formulation of a linear code relating genes and proteins. In 1950, Caldwell and Hinshelwood published a theory according to which nucleic acid, by a process similar to crystallization, ensures the order of the amino acids in the protein. Their ideas were rather confusing, since they were considering 23 amino acids and five units in the nucleic acids, namely the four bases and ribose phosphate. The authors were not mentioning whether their doublet code was overlapping or not. This article, only occasionally quoted, exerted no influence on later theories. On the other hand, a paper published by Dounce in 1952 was well read: it treats mainly of the possible chemical mechanisms of protein synthesis, but it suggests a code where each nucleotide codes for an amino acid, which automatically leads to an overlapping code. He did not realize that his code could be rejected by the mere inspection of the few known polypeptide sequences then available.

The determination of the structure of the code was greatly helped by the discovery of the structure of DNA by Watson and Crick in 1953. Its simplicity impressed many scientists, including the cosmologist George Gamow who put forward in 1954 a theory according to which protein synthesis occurred on the surface of the DNA double helix, the sequence of bases within the structure forming a series of cavities, each of which was specific for a given amino acid. The theory does not mention how the amino acids recognize the cavities, but suggests that their side chains adapt stereochemically to them, without the aid of any enzyme. Gamow, like Dounce, was taking into account that the distance between residues of a stretched polypeptide chain is only 3.6 to 3.7 Å and that the code had therefore to be overlapping. Gamow's abstract model, historically important, was not taking into account useless chemical details: his ideas that protein synthesis did happen on a double-stranded DNA matrix turned out to be entirely wrong. He had the merit of having clearly understood that an overlapping code imposed restrictions on the amino acid sequences and that it should be possible to confirm or reject the several possible overlapping codes by inspecting the known sequences.

In 1955 Francis Crick wrote an article which was never published, but which circulated among specialists, called "On degenerate templates and the adaptor hypothesis". This paper lists the amino acids to be included in the code and demonstrates that Gamow's code is incompatible with the sequences of insulin and β-corticotropin and does not specify the direction of protein synthesis: he was already thinking that some mechanism should impose the direction in which DNA was read. This unpublished article contains the adaptor hypothesis for the first time, the adaptor being a small molecule able to form specific hydrogen bonds with a nucleic acid template, activation enzymes providing the specificity required for the combination between a given amino acid and its adaptor.

The next idea, formulated in 1957, was that of a code without punctuation (Crick, Griffith & Orgel). The same year, Brenner demonstrated that overlapping codes were impossible.

In 1959, the situation seemed hopeless. One expected that the sequences of a gene and of the protein it encoded should be colinear, but this was only shown independently in 1964 by Yanofsky and by Brenner.

The missing link was that of messenger RNA. When the experimental evidence for its existence became available, the code problem was solved. In 1961, Nirenberg and Matthaei discovered that polyU could serve as a messenger and that it coded for polyphenylalanine. Crick and Brenner showed unequivocally that the code was a triplet one, non overlapping and without punctuation. The groups of Nirenberg, of Ochoa, and Khorana prepared numerous artificial messengers, either by using the polynucleotide phosphorylase discovered by Grunberg-Manago and Ochoa in 1956 or by chemical synthesis. These polynucleotides had either random sequences or definite repeated sequences, and served for the preparation of polypeptides whose composition was analyzed. Finally, a binding method was devised by Leder and Nirenberg in 1964, which allowed the definition of a correspondance between each triplet (a codon) and an amino acid. This work led to the establishment of the genetic code: it is highly degenerated (leucine and arginine being coded by six different codons), non ambiguous, and it contains three nonsense codons, which do not code for an amino acid and cause a polypeptide chain termination when the message is translated by the ribosome (table I).

All we have learned since 1966 by the sequence of many proteins and nucleic acids has confirmed that the genetic code is practically universal. There are some deviations of the standard code in some systems, and, in many cases, variations occur at well-defined specific sites. However, these differences do not indicate a different origin of life. For example, in mitochondrial systems, the UGA termination codon codes for tryptophan and AUA codes for methionine instead of isoleucine.

Table I. The genetic code.
All triplet codons have a meaning. All the amino acids except methionine and tryptophan are represented by more than one codon. Synonymous codons form groups in which the third base is the less significant. Three codons lead to termination of translation (TER).
The order of bases is written in the direction 5' → 3'.

	Second base			
	U	C	A	G
U	UUU Phe	UCU Ser	UAU Tyr	UGU Cys
	UUC Phe	UCC Ser	UAC Tyr	UGC Cys
	UUA Leu	UCA Ser	UAA TER	UGA TER
	UUG Leu	UCG Ser	UAG TER	UGG Trp
C	CUU Leu	CCU Pro	CAU His	CGU Arg
	CUC Leu	CCC Pro	CAC His	CGC Arg
	CUA Leu	CCA Pro	CAA Gln	CGA Arg
	CUG Leu	CCG Pro	CAG Gln	CGG Arg
A	AUU Ile	ACU Thr	AAU Asn	AGU Ser
	AUC Ile	ACC Thr	AAC Asn	AGC Ser
	AUA Ile	ACA Thr	AAA Lys	AGA Arg
	AUG Met	ACG Thr	AAG Lys	AGG Arg
G	GUU Val	GCU Ala	GAU Asp	GGU Gly
	GUC Val	GCC Ala	GAC Asp	GGC Gly
	GUA Val	GCA Ala	GAA Glu	GGA Gly
	GUG Val	GCG Ala	GAG Glu	GGG Gly

AGA is a termination codon in the vertebrate mitochondrial RNA, but codes for an amino acid in the *Drosophila* mitochondria. There are also some deviations which have been found in the mitochondrial DNA of yeasts, of filamentous fungi, protozoa, and plants. Some differences have also been observed in the genome of some mycoplasmas and in the nuclear genes of at least four species of ciliated protozoa where standard termination codons code for glutamine.

Transfer RNAs

It appears useful to quote some lines from Crick's unpublished 1955 paper:

"... each amino acid would combine chemically, at a special enzyme, with a small molecule which, having a specific hydrogen-bonding surface, would combine specifically with the nucleic acid template. This combination would also supply the energy necessary for the polymerization. In its simplest form, there would be 20 different kinds of adaptor molecules, one for each amino acid and 20 different amino acids to their adaptors. Sidney Brenner, with whom I have discussed this idea, calls this the "adaptor hypothesis", since each amino acid is fitted to an adaptor to go on the template..."

At about the same period, Zamecnik's group was persuaded that an additional reaction should exist between the "activated" amino acid and its appearance in the microsome as a nascent polypeptide chain. The biochemists approached the problem, not in terms of hydrogen bonds and of translation of one language to another, but in terms of enzymes and energy. Holley (1956) followed the activation reaction of alanine in liver extracts and obtained indications of a reaction occurring after the activation, which was sensitive to ribonuclease, whereas the activation proper was not sensitive. The same year, Berg prepared, for the first time, an enzyme specific for the activation of a single amino acid, namely methionine, from a mixture obtained from yeast. There was a stoechiometric reaction between the enzyme and methionine, a result in contradiction with the catalytic nature of enzymes. The enzyme was not liberated and Berg thought that the reaction must be incomplete and started looking for the natural acceptor of the activated methionine, without ever having heard of Crick's adaptor hypothesis. He found it, and it was a small-size RNA, unlinked to microsomes and specific both for the activating enzyme and for methionine.

In 1956 also, Zamecnik and Hoagland added radioactive leucine to their cell-free extract and isolated a radioactive fraction of soluble RNA. When this fraction was added to a new cell-free extract, radioactive leucine was liberated from the soluble RNA in a few minutes and was found incorporated in the protein growing chain on the microsome.

The discoveries of the biochemists were then converging to a definition of a molecule that had all the characteristics of Crick's hypothetical adaptor. Soon, soluble RNAs and enzymes specific for each amino acid were found by Berg, Lipmann, Holley, and many others. Zamecnik found that all the 3' ends of soluble RNAs presented the CCA sequence to which the amino acids were linked by an energy-rich bond. The soluble RNA was named transfer RNA (tRNA) and the activation enzymes were named aminoacyl-tRNA transferases.

Holley could separate the different species of tRNA and sequenced the yeast alanyl-tRNA in 1966. His long task was rendered somewhat easier by the fact that nine of the 77 bases of the molecule were not the classical RNA bases A, G, C, and U, but belonged to different nucleosides (inosine, pseudouridine) or to modified nucleosides such as dimethylguanosine. This was the first sequence of a nucleic acid ever obtained. There were several G and C residues which could be paired so as to give the molecule the apperance of a cloverleaf. At the extremity of the central lobe of the cloverleaf, three nucleotides (anticodon) could pair with the triplet specific for the alanine codon of the template (by that time, messenger RNA and its properties were known). All tRNAs were subsequently found to have similar structures (fig. 1).

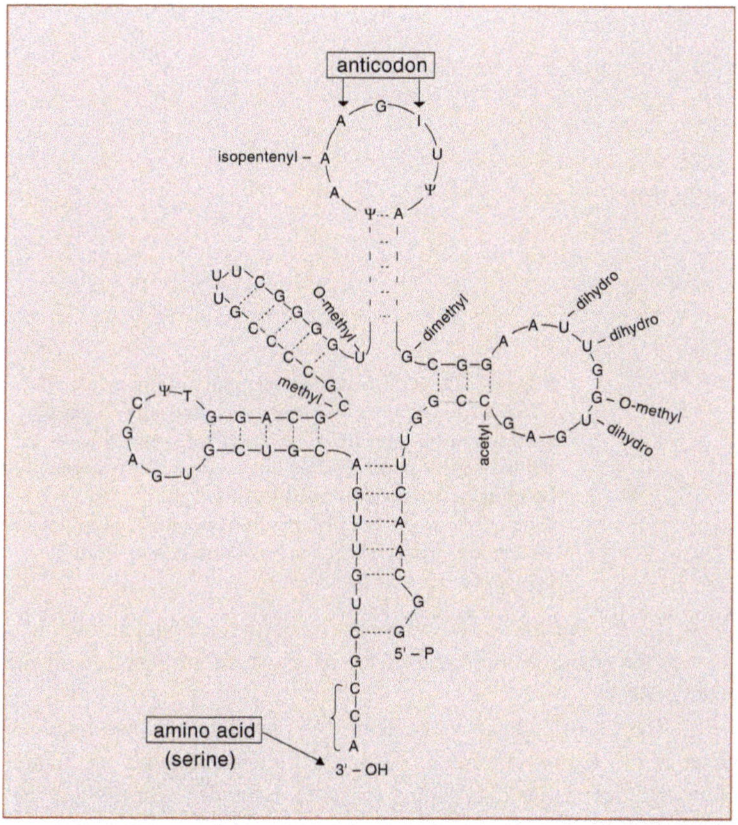

Figure 1. Cloverleaf model of a seryl-tRNA from yeast.
A, adenine; C, cytosine;
G, guanine; I, inosine; U, uracil;
T, thymine; Y, pseudouracil.
(From *Concise Encyclopedia of Biochemistry*, 2nd edition, Walter de Gruyter, Berlin and New York (1988). With permission from the publisher.)

In addition to the anticodon loop, there are two other loops, one always containing dihydrouracil, and one, called the pseudouridine loop, always containing the sequence GTΨC, which appears to play a role in the binding of aminoacyl-tRNA to the 50S subunit of the ribosome. A fourth loop may be present, but is not a constant feature of all tRNAs.

Very soon afterwards, attempts were made to crystallize the tRNAs in order to study their tridimensional structure. Rich and Klug determined the structure of phenylalanyl-tRNA from yeast in 1974-1975. Five years later, Rich and Moras determined respectively the tridimensional structure of N-formylmethionyl-tRNA and of aspartyl-tRNA which presented the same L structure as had been found for phenylalanyl-tRNA (fig. 2)

The tRNAs are not transcribed as such from their structural genes, but derive from precursors longer both in the 3' and the 5' side. These precursors are short-lived and are cleaved on the 5' side by a specific nuclease, ribonuclease P, which is a non-covalent complex of a small RNA molecule of 337 residues and a small protein of 20,000 Da; the cleavage reaction is catalyzed at almost the same rate by the RNA component in the absence of the protein. Another enzyme, ribonuclease D cleaves the precursors on the 3' side until it reaches the CCA extremity.

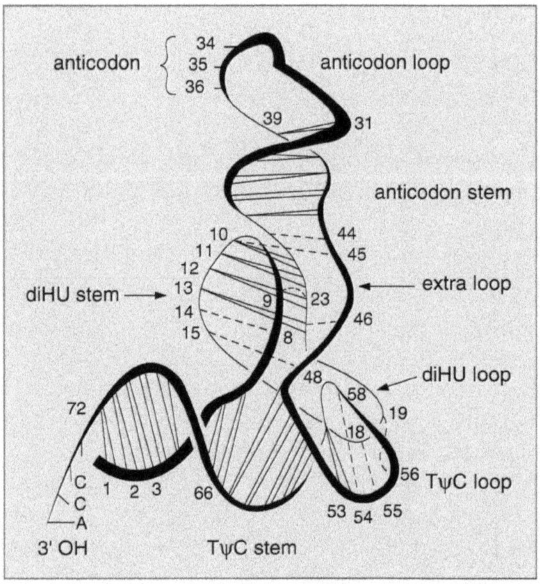

Figure 2. Three-dimensional structure of phenylalanyl-tRNA as determined by X-ray crystallography. Double solid lines represent hydrogen bonding between bases in double helical stems. Dotted lines represent hydrogen bonding between bases outside the helices.
(From *Concise Encyclopedia of Biochemistry*, 2nd edition, Walter de Gruyter, Berlin and New York (1988). With permission from the publisher.)

Other specific enzymes catalyze a host of posttranscriptional modifications of the ribose residues and of the bases which are found in all the mature tRNAs.

The genes coding for the pre-tRNAs are found in two main regions of the *E. coli* chromosome, in which clustered genes are found, often organized in operons coding for up to seven different tRNAs. However, genes coding for tRNAs are also found in the coding space between the genes coding for the16S and the 23S ribosomal RNAs and even 5S RNA. Their presence in the precursors of pre-ribosomal RNAs necessitates a coupling between the processing events of the two types of RNA. These tRNAs are obviously the most abundant ones since they are transcribed in a coordinate manner with the ribosomal RNAs.

Several isoaccepting tRNAs, the existence of which correlates with the degeneracy of the code, may have the same anticodon triplet; they are coded by different genes and are present at different concentrations; the codon usage to specify the same amino acid is not random: it is correlated with the occurrence of the corresponding codons in the mRNAs of the different organisms studied.

In contrast, there is no aminoacyl-tRNA synthetase specific to each tRNA corresponding to the same amino acid: the isoaccepting tRNAs are activated by the same synthetase.

The energy required for the synthesis of the aminoacyl bond originates from ATP. The enzyme first forms an amino acid adenylate:

amino acid + ATP $\longrightarrow$ aminoacyladenylate + pyrophosphate
(exchange reaction)

The intermediate remains strongly bound to the enzyme and, in a second step, transfers its amino acid to the specific tRNA:

aminoacyladenylate + tRNA $\longrightarrow$ aminoacyl-tRNA + AMP
(charge reaction)

The enzyme is thus able to recognize both a given amino acid and its cognate specific adaptor. It must have a binding site for each. Several of these transferases have been crystallized and their X-ray analysis has been carried out at high resolution. Co-crystals of the aspartyl-tRNA and its transferase have been obtained with yeast aspartyl-tRNA and their analysis should reveal how an aminoacyl-tRNA synthetase can achieve the discrimination between tRNAs corresponding to different amino acids and nonetheless bind the different isoaccepting tRNAs.

SELECTED REFERENCES

Colinearity of genes and proteins

C. Yanofsky, B. C. Carlton, J. R. Guest, D. R. Helinski & U. Henning, *Proc. Natl. Acad. Sci. US* **51**, 266-272 (1964).

A. S. Sarabhai, O. W. Stretton, S. Brenner & A. Bolle, *Nature* **201**, 13-17 (1964).

Genetic code

Cold Spring Harbor Symposia Quant. Biol. **31**, 1-762 (1966). The whole book is devoted to the unraveling of the code.

Transfer RNAs

S.-H. Kim, *Adv. Enzymology* **46**, 279-315 (1978).

Cold Spring Harbor Monograph Series. The Transfer RNAs. Part 1: Structure, Properties and Recog-nition. Part 2: Biological aspects. Cold Spring Harbor Laboratory (1980).

Aminoacyl-tRNA synthetases

P. R. Schimmel & D. Söll, *Ann. Rev. Biochem.* **48**, 601-648 (1979).

P. R. Schimmel, *Ann. Rev. Biochem.* **56**, 125-158 (1987).

CHAPTER XI

Attenuation

Ames and his colleagues had shown, in 1972, that the transcription of the histidine operon in *S. typhimurium* was not regulated by a repressor protein, but that the level of transcription depended on the intracellular level of $tRNA^{his}$ charged by histidine. These observations were confirmed and extended by Kasai in 1974. In the case of the *trp* operon, observations made from 1973 to 1976 indicated that repression could not be the only way by which *E. coli* regulates transcription. Mutants of tryptophanyl-tRNA synthetase presented anomalies in the regulation of the operon: addition of tryptophan to the cultures of these mutants did not repress maximally the expression of the operon; not only was transcription not initiated, but its progress from the initial segment was inhibited. Finally, *trpR* mutants devoid of a functional repressor did still respond to a tryptophan starvation by increasing the rate of synthesis of the *trp* messenger RNA.

When the transcription of the initial segment of the operon was analyzed in detail *in vitro* and *in vivo*, a ρ-independent transcription termination site was found upstream the structural genes of the operon. Tryptophan starvation reduces the termination at this site. This phenomenon has been called *transcription attenuation* by Yanofsky, and describes the early termination of transcription. In order to understand the mechanism of this regulation process, it is necessary to examine the anatomy of the regulatory and structural gene regions of the *trp* operon (fig. 1).

The results of the investigations carried out by Yanofsky on the tryptophan operon and by others on other amino acid biosynthetic operons can be summarized as follows:

Transcription is regulated by attenuation, the regulatory signal is charged tRNA, and the sensory event is the capacity to translate a short peptide coding region, rich in codons for the relevant amino acid and located in the initial segment of the transcript of the operon.

An RNA polymerase molecule which has escaped the regulation by the repressor initiates the transcription of the operon. Soon after the transcribed segment containing the ribosome binding site has been synthesized, a ribosome attaches to it. The polymerase which has started transcribing at the promoter can synthesize a small transcript ("leader transcript") and either stop before reaching the first structural gene, or synthesize a much longer transcript corresponding to the structural genes of the operon, depending on the physiological conditions.

The short transcript exhibits two very special characteristics:

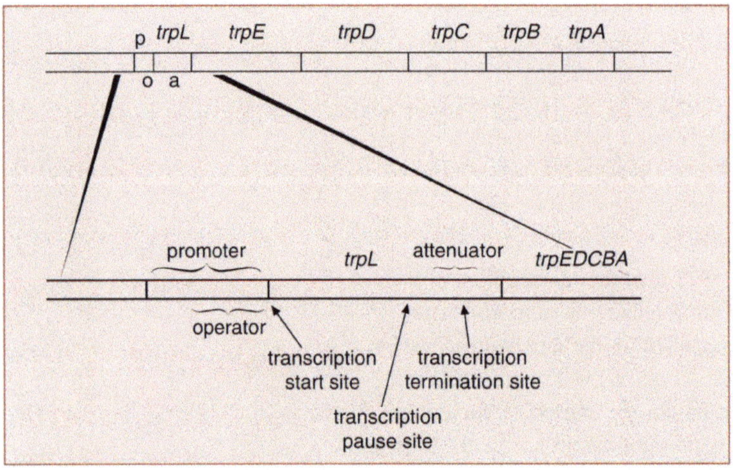

Figure 1. The regulatory and structural gene regions of the *trp* operon of *E. coli*. Transcription initiation occurs at the promoter-operator region. Transcription termination is regulated at an attenuator, in a transcribed 162 base-pair region, *trpL*, which encompasses the segment coding for the putative leader peptide. All RNA polymerase molecules transcribing the operon pause at the transcription pause site before proceeding further.
(From C. Yanofsky, with permission from *Nature* **289**, 751-758. Copyright (1981) Macmillan Magazines Limited.)

a) it comprises a small open reading frame coding for a peptide (*"leader peptide")* 15 to 32 amino acid residues long, according to the cases which have been analyzed, in which is found a high proportion of the codons corresponding to the amino acid (s) regulating the expression of the operon; these codons are called *regulator codons.* Depending on the levels of the appropriate tRNAs, translation of these regulator codons will or will not proceed, and this event determines whether transcription will terminate in the leader region. The actual synthesis of these peptides has been demonstrated;

b) certain regions of the transcript can adopt secondary "stem and loop" structures, mutually exclusive, one of these structures being recognized by RNA polymerase as a signal of transcription termination.

The secondary structure adopted by the RNA is determined by the position of the ribosome while it is translating the leader peptide. The progression of the ribosome, and thence its position on messenger RNA depends on the intracellular concentration of the amino acid regulating the operon through the charged tRNAs corresponding to the regulator codons. On the scheme (fig. 2), the regions noted 1, 2, 3 and 4 correspond to the regions of possible pairing of the messenger: the 3-4 pairing constitutes the *terminator* (signal of transcription termination); the 1-2 pairing is called the *protector* (structure which allows the formation of

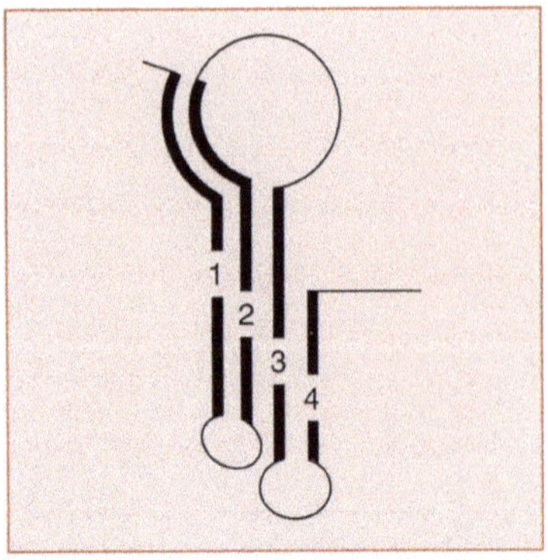

Figure 2. Scheme representing the four regions of possible pairing of the messenger RNA upstream the translation start of the structural genes.
(From C. Yanofsky, with permission from *Nature* **289**, 751-758. Copyright (1981) Macmillan Magazines Limited.)

the 3-4 structure) and the 2-3 pairing, which prevents the formation of the terminator is called the *preemptor*.

When the amino acid regulating the expression of the operon becomes limiting, the progression of the ribosome on the messenger ceases at the level of the regulator codons (since the corresponding tRNAs are not charged); the ribosome covering region 1 inhibits the formation of the protector (1-2 pairing). When region 3 is transcribed, it can therefore

Figure 3. Representation of the ribosome on the *trp* leader transcript in cells starved of Gly, Thr, Arg or Trp or not starved cells. A ribosome stalled over one of the codons of the transcript blocks about ten nucleotides in the downstream segment of the transcript, thereby determining which alternative secondary structure will form during transcription. The structure 2-3 will form only when cells are starved of Trp and Arg: the structure 3-4 (the ρ-dependent terminator) will not form and transcription will proceed into the structural genes.
(From C. Yanofsky, with permission from *Nature* **289**, 751-758. Copyright (1981) Macmillan Magazines Limited.)

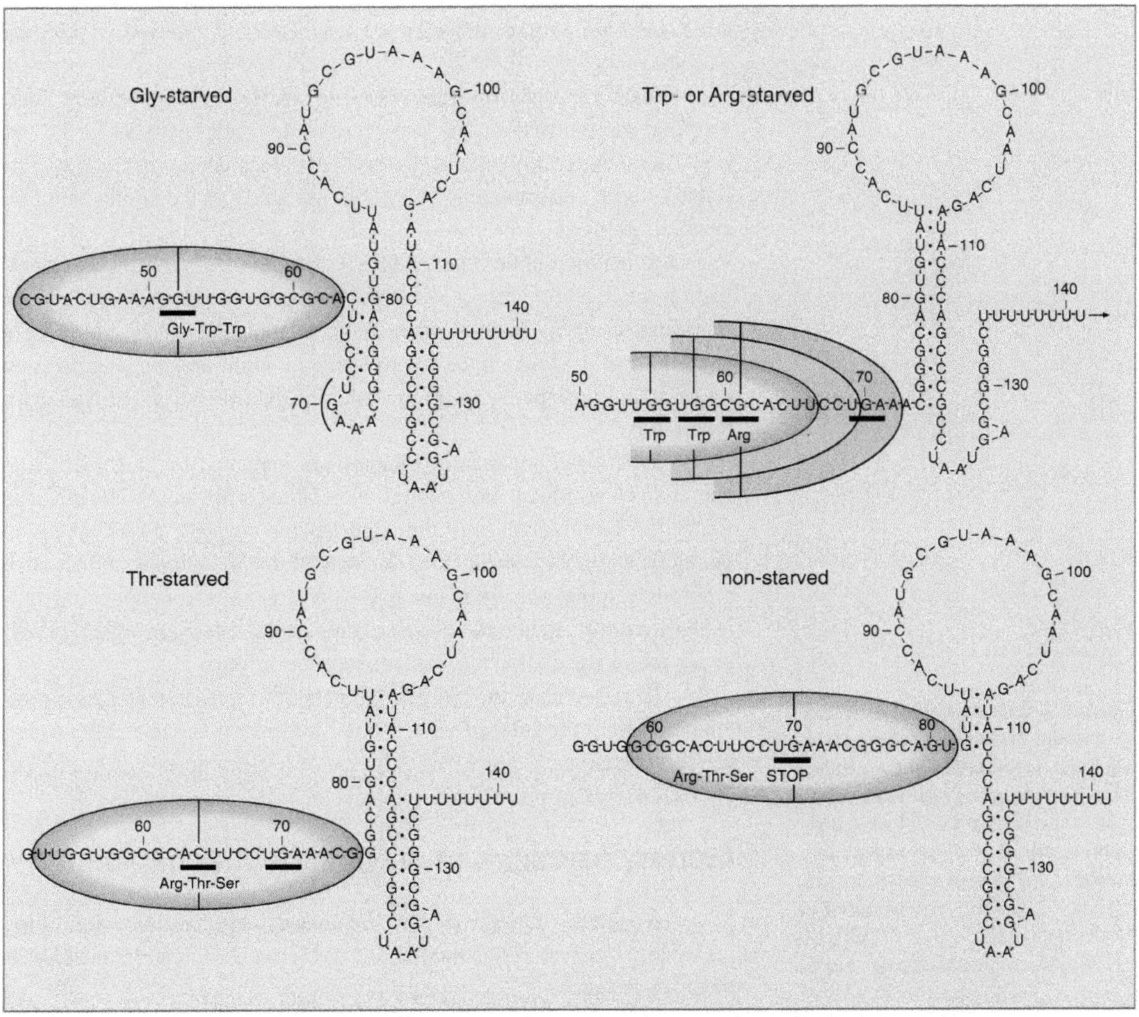

pair with region 2 and the premptor is formed. As a consequence, region 4 cannot pair with region 3 and in the absence of the terminator structure, polymerase transcribes the structural genes (fig. 3). It can be seen that not only tryptophan starvation, but also arginine starvation – which causes, in addition, a stalling of the ribosome because of the strategic position of the arginine codon in the leader transcript – results in the transcription of the tryptophan biosynthetic structural genes.

On the other hand, if the regulating amino acid is in excess or if the cell is starved for an amino acid whose codons are not topologically important in the leader transcript, the ribosome progresses normally on the messenger and translates region 2 as soon as it is transcribed. When region 3 is transcribed, it cannot pair with region 2, which is covered by the ribosome, and it is available for pairing with region 4 to form the terminator structure. The polymerase recognizes this termination signal and the structural genes are not transcribed (fig. 3).

If the stability of the 3-4 structure is decreased by mutations that affect the base pairing, transcription termination is relieved, in keeping with the model (fig. 4).

Such a mechanism implies a very precise coupling between transcription and translation. The sequences of the leader peptides of the *trp, pheA, his, leu, thr, ilvGMEDA* operons have been determined and will be found in each corresponding chapter of our study of the amino acid biosynthetic pathways.

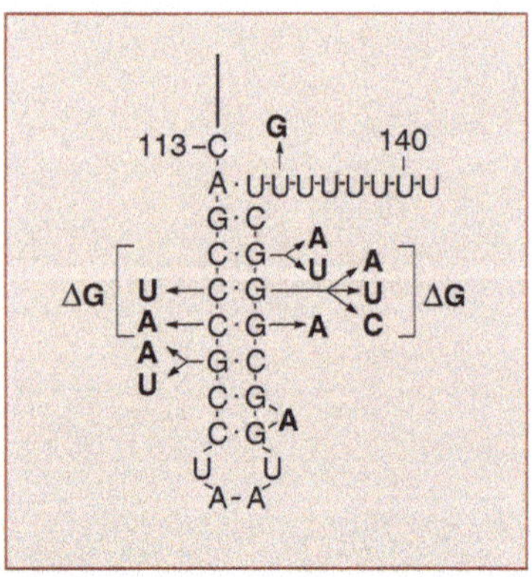

Figure 4. Base pair changes in the *trp* leader region that relieve transcription termination at the attenuator. The changes in the transcript are shown: all are in the 3-4 structure and reduce its stability except the single U, G change at position 135. (From C. Yanofsky, with permission from *Nature* **289**, 751-758. Copyright (1981) Macmillan Magazines Limited.)

Repression rather than attenuation exerts the largest regulatory effect on the *in vivo* transcription of the *trp* operon. Repression can reduce transcription up to 70-fold, whereas attenuation may reduce only by a maximum of 10-fold. Since tryptophan starvation augments expression by using both mechanisms, transcription of the *trp* operon can vary by a factor 700.

Whereas tryptophan biosynthesis is regulated by both repression and attenuation, this is not general. No evidence for repression has been found for the threonine, histidine and leucine operons, which appear to be regulated only by attenuation. In contrast, the methionine and arginine regulons respond to repression only: no structures corresponding to a leader transcript and to alternative pairing of the messenger RNA, or evidence for an attenuation mechanism, have been found.

If we compare the specific elements of repression to those of attenuation from the point of view of their information content, attenuation is more economical since its only specific element is a sequence of 50-100 base pairs of the leader sequence of each concerned operon.

General remarks on regulatory mechanisms

A simple analogy of the regulatory mechanisms controlling respectively enzyme regulation and enzyme synthesis can be made with the brick industry.

Allosteric inhibition or activation, as well as covalent enzyme modification, may be compared to a factory making less bricks or more bricks according to demand: if there are already too many bricks available, the factory will reduce its output or even close temporarily, but will be ready to function again as soon as there is a dearth of bricks.

Repression and attenuation correspond to a situation where less, or no more, *brick factories* will be built. As a result, when bricks are needed again, new factories will have to be built until the production corresponds to the demand.

It is impossible to conceive of a nuclear plant or an oil refinery without controls, captors, relays, and overflows. In the same vein, it is not surprising that an entangled network as complex as that of biosynthetic metabolism is endowed with microscopic feedback mechanisms: the contrary would be astonishing. The parallelism between a factory and metabolism is not a simple metaphor, and the underlying logic is similar. In most cases examined in this book, as in a thermostat or in Ktesibios clepsydra installed in Egypt three centuries before our era, or in the wine fountain of Heron of Alexandria, the logic is that of retroaction.

In the simplest case, that of feedback inhibition or of repression, the captor and the actuator are the same molecule, respectively the allosteric enzyme or the repressor, the final metabolite being an interdiction signal. We are in the presence of a true analogic device. In the case of the covalent modification of. *E. coli* glutamine synthetase, the metabolic signal is not perceived by the actuator, the modified enzyme, but by the modifying enzyme. In the case of attenuation, the signal, the aminoacyl-tRNA, is transmitted to the RNA polymerase actuator, through the relay of a site of transcription termination.

Another remarkable property of metabolic regulation concerns its hierarchy, its demultiplication. The double control of enzyme activity and of enzyme synthesis allows the establishment of a metabolic régime and a fine adjustment of the metabolic fluxes.

Finally, the economy of the setting up of the regulatory mechanisms during evolution should be emphasized. Despite their common logic, the great structural and functional diversity of allosteric enzymes suggests that in each case the control site was created independently from the catalytic site. Along the same line, any protein able to bind DNA is a candidate for becoming a repressor, through the elaboration of a site having affinity for a metabolite. Experiments of reconstruction of regulatory proteins and earlier experiments of selection of such proteins leave little doubt that the miracles of microscopic clockwork we detect in cells have been the result of molecular tinkering, that is of expediency.

SELECTED REFERENCES

T. Kasai, *Nature* **249**, 523-527 (1974).
C. Yanofsky, *Nature* **289**, 751-758 (1981).
C. Yanofsky, *J. Biol. Chem.* **263**, 609-612 (1988).

CHAPTER XII

How biosynthetic pathways have been established

In order to determine the pathways of biosynthesis, it is preferable to use actively growing organisms. Newly synthesized material can be recognized by the increase in cell numbers in a bacterial culture or the increase in weight in a growing animal such as a rat. Animal and plant cells grow slowly in general: the cells of the mammalian central nervous system grow only during the early life of the animal and then divide no further; muscle cells divide and grow slowly; liver cells divide about every three months; those in the intestinal mucosa have a division time easily measured in days. In contrast, bacterial cells can double their number every twenty minutes. Newly synthesized material, in a bacterium such as *E. coli*, arises solely from a single carbon source such as glucose or acetate, ammonia, sulfate, and inorganic phosphate. Such a bacterium is capable of intense chemical activity.

Nearly all the metabolic energy in the bacterium is used for biosyntheses. It is obvious that bacteria of this kind are the material of choice for the study of anabolic phenomena.

Several methods have been used to elucidate the pathways of biosynthesis for small molecules, amino acids, purines, pyrimidines, (which are the materials from which proteins and nucleic acids are synthesized), and vitamins (which are the cofactors or the substrates of important enzyme activities). We are going to survey and see examples of the most important methods: the use of isotopes, the use of auxotrophic mutants (unable to synthesize a given essential metabolite), and finally the examination of individual enzyme reactions.

Use of isotopes

Organisms can be grown on specific carbon sources such as glucose uniformly labeled with ^{14}C or labeled in the C1 or C6 position, acetate labeled with radiocarbon, or acetate doubly labeled in the methyl group with ^{14}C and in the carboxyl group with the non-radioactive heavy isotope ^{13}C. After growth, the organisms are collected, disrupted, and the protein fraction (for example) is isolated and subjected to acid or alkaline hydrolysis; the individual amino acids are isolated and the distribution of the individual carbon atoms is determined by laborious degradation methods. Such methods are generally very arduous and seldom provide unambiguous answers to the question concerning the detailed chemical reactions by which the carbon atoms of the nutrient are incorporated into an amino acid. More often, at best they allow one to exclude a pathway which might have seemed possible.

A much more useful and practical method is isotope competition, described below.

The principle is as follows: if uniformly-labeled glucose is provided as the sole carbon source for the bacterium, all the essential metabolites will be uniformly labeled. If a suspected intermediate is added to the growing culture as an unlabeled compound, it will dilute enormously the radioactivity of the labeled intermediate. As a result, the final metabolite will be non-radioactive, or at least its specific radioactivity will be greatly reduced[12]. Let us look as some applications of this method.

a) A culture of *E. coli* is grown in a synthetic medium containing uniformly labeled glucose and non-radioactive homoserine. After growth, the culture is centrifuged, washed, the cells are disrupted by sonication, and the protein fraction is isolated and hydrolyzed with 6N HCl. The resulting amino acids are separated by a bidimensional paper chromatography with different appropriate solvents so that all amino acids can be identified. The chromatograms are placed on a film sensitive to radiation; the spots corresponding to radioactive amino acids will blacken the film. In addition, non radioactive amino acids, if any, can be identified on the chromatogram by ninhydrin. The amino acids threonine, methionine, and isoleucine are found to have little or no radioactivity, while the specific activity of all the other amino acids is identical to that of the original glucose. The conclusion is reached that homoserine is probably an intermediate in the biosynthesis of threonine, isoleucine and methionine.

b) If the same type of experiment is carried out, but this time adding non-radioactive aspartate, the radioactivity of the protein aspartate is greatly reduced, and so is that of diaminopimelic acid (a cell-wall constituent in *E. coli*), lysine, and the same three amino acids (threonine, methionine, and isoleucine) of the previous experiment.

c) Upon addition of non-radioactive threonine as a competitor, the radioautograms of the protein hydrolyzate show that only threonine, and isoleucine from protein, are non radioactive (fig. 1 & 2).

d) Finally, when non-radioactive methionine and isoleucine are used as competitors, they affect only their own specific radioactivities.

Taken collectively, the results of these experiments suggest the following biosynthetic pathway:

aspartate → homoserine → threonine → isoleucine

aspartate ↓ diaminopimelate and lysine

homoserine ↓ methionine

12. In addition to its dilution effect, the exogenous addition often causes an allosteric inhibitory effect or a repression effect on the activity or synthesis of one or more enzyme of the pathway which result in a reduced or suppressed synthesis of the final metabolite.

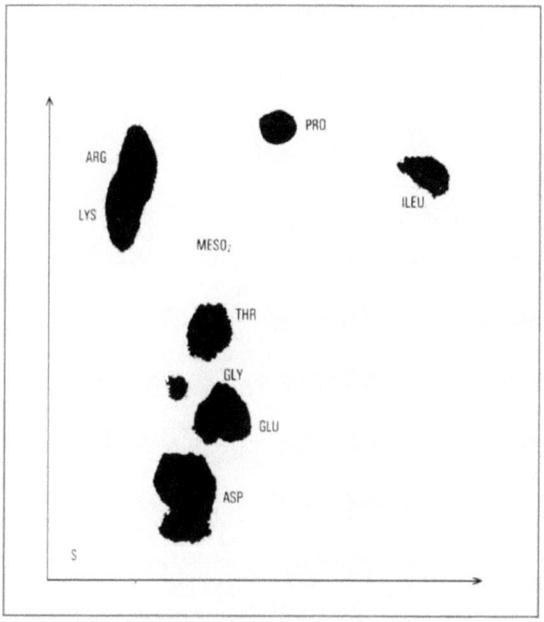

Figure 1. Autoradiogram of a bidimensional chromatogram of a protein hydrolyzate from *E. coli* cells grown in the presence of uniformly-labeled glucose.
(From R. B. Roberts, P. H. Abelson, D. B. Cowie, E. T. Bolton & R. J. Britten (1955). With permission from the Carnegie Institution of Washington.)

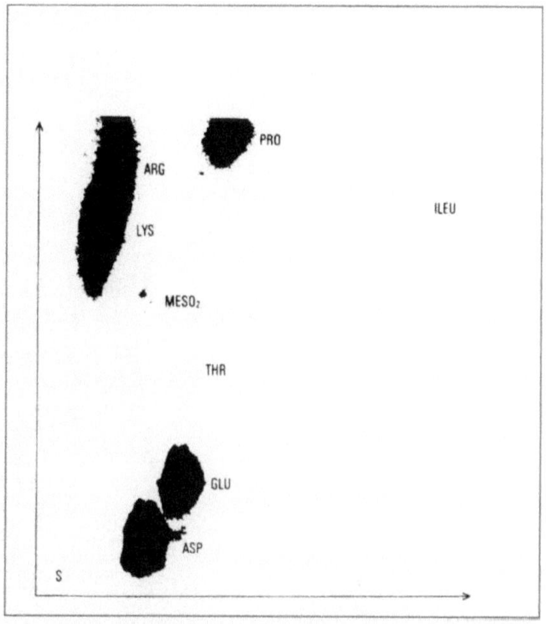

Figure 2. Same experiment as in figure 1, but *E. coli* was grown in the presence of non-radioactive threonine. Note that the radioactivity of threonine and isoleucine have disappeared.
(From R. B. Roberts, P. H. Abelson, D. B. Cowie, E. T. Bolton & R. J. Britten (1955). With permission from the Carnegie Institution of Washington.)

Many similar "isotopic competition" experiments have made it possible to establish clearly that the majority of the amino acids in the proteins and cell walls of *Escherichia coli* are derived from a small number of inter-mediates from the glycolytic or the tricarboxylic cycle: phosphoglycerate (serine, glycine, cysteine), pyruvate (alanine, valine, leucine), oxalo-acetate (aspartate, asparagine, lysine, diaminopimelate, threonine, methionine, isoleucine), α-ketoglutarate (glutamate, glutamine, proline, and arginine).

We shall examine these families in turn in later chapters. The aromatic amino acids phenylalanine, tyrosine, and tryptophan, and the heterocyclic amino acid histidine, are synthesized by a series of reactions which will be examined separately.

Use of auxotrophic mutants

In every wild-type bacterial population, there exist spontaneous mutants unable to synthesize one or another essential metabolite. They occur at a very low frequency, of the order of 10^{-8}. These mutants, whose growth depends on the exogenous addition of the metabolite they cannot synthesize, are called *auxotrophic* mutants, as against the wild-type *prototroph* from which they derive. The frequency of mutation can be increased by ultraviolet radiation or by treatment with chemical mutagens, some of which are highly effective. After mutagenesis, the problem is to select the desired mutant out of a whole population of prototrophs and mutants in which one has no interest.

One of the most frequently employed methods is to use selection by penicillin, developed independently by Davis, and by Lederberg and Zinder, and ameliorated by Adelberg and by Gorini.

The basis of the method is the fact that penicillin kills only actively dividing cells (the antibiotic acts by inhibiting the synthesis of the cell wall; as a result, the non-growing bacteria are protected). After mutagen treatment, the bacteria are allowed to divide a number of times in the presence of the growth factor for which auxotrophs are desired. The reasoning is as follows: bacteria have several nuclei, and therefore several identical chromosomes; the mutagen has acted on only one of the chromosomes at the specific locus coding for the synthesis of the required growth factor, for which the unaffected chromosomes still possess all the information. After a few divisions, the nuclei have become segregated among the different daughter cells, and the bacteria have become arranged into groups with a homogeneous make-up. Excess growth factor is eliminated by washing several times, and the culture is resuspended in a medium containing every growth factor (amino acids, purines, pyrimidines, vitamins) except the one for which auxotrophic mutants are being sought. The new medium also contains penicillin. Bacteria which have not been killed by the mutagen make up a population comprising prototrophic bacteria, mutants which are of no interest, and the mutants desired. The first two classes can divide and therefore will be killed by penicillin. The third class cannot grow and so will escape death. After removal of penicillin, plating on a suitable medium will, in principle, give colonies which are largely those of the mutants desired.

Let us consider a group of mutants obtained independently by this method, all of which are unable to synthesize a given amino acid, for example tryptophan. None of the mutants will grow in a synthetic medium which contains glucose as the only carbon source, but all will be able to grow if tryptophan is added. A class of mutants may be found which, under certain conditions, secretes into the growth medium a substance not found in the growth medium of the parental type. This substance has been identified as indole; it allows the growth of a second class of tryptophan-requiring mutants. This class, like the first, comprises individuals unable to synthesize tryptophan, but unlike the first class, also unable to synthesize indole. Mutants of this second kind can also accumulate a substance, anthranilic acid, which allows a third class to grow. Thus the third class is unable to carry out one or several reactions in the synthesis of tryptophan earlier than anthranilate production, and can only grow in the presence of anthranilate, indole or tryptophan. The biosynthesis of tryptophan can therefore be written schematically in the following provisional way:

$$\text{glucose} + NH_4^+ \xrightarrow{(3)} \text{anthranilate} \xrightarrow{(2)} \text{indole} \xrightarrow{(1)} \text{tryptophan}$$

The numbers in parentheses indicate "genetic blocks" of different kinds, that is, reactions that cannot be carried out.

Likewise, a mutant requiring both threonine and methionine for growth can see this double requirement satisfied by homoserine, in keeping with the results of isotopic competition described above.

Auxotrophic mutants have been obtained for practically every pathway, in particular for amino acid, purine, pyrimidine, and vitamin biosyntheses, and have proved to be immensely useful for the elucidation of the pathways. However, it remained to be shown, in each case, that the accumulated products are truly direct intermediates and not metabolic products of the actual intermediates.

Enzymatic analysis

The interpretation of the results obtained by the above methods is not always as easy and unequivocal as the concise summary just given would lead us to believe. The results must be considered as a point of departure for more detailed studies, and not as definite proof of the existence of the reactions postulated. In consequence, it is necessary to try to confirm them independently by using cell suspensions or better cell-free extracts to show that every postulated reaction does indeed take place and is missing in the mutants which are suspected of being unable to carry it out. For example, mutants requiring homoserine (or threonine plus methionine) for growth have been obtained in several bacterial and fungal species, and have been shown to lack the enzyme homoserine dehydrogenase which catalyzes the reduction of aspartic semialdehyde to homoserine. This result confirms the evidence obtained from the methods of isotopic competition and nutrition, and makes it certain that homoserine is an intermediate in the synthesis of homoserine and methionine.

SELECTED REFERENCES

Isotopic competition

R. B. Roberts, P. H. Abelson, D. B. Cowie, E. T. Bolton & R. J. Britten, *Studies in Biosyntheses in* Escherichia coli. Carnegie Institution of Washington, publ. n° 607 (1955).

Isolation of auxotrophic mutants

B. D. Davis, *J. Amer. Chem. Soc.* **70**, 4267 (1948).

J. Lederberg & N. Zinder, *J. Amer. Chem. Soc.* **70**, 4267-4268 (1948).

J. Lederberg, in *Methods in Medical Research*, J. H. Comrie Jr. ed., **3**, pp. 5-21. Year Book Publishers, Chicago (1950).

L. Gorini & H. Kaufman, *Science* **131**, 604-605 (1960).

CHAPTER XIII

Biosynthesis of polysaccharides

It has been known for about sixty years that the resynthesis of glucose from pyruvate and lactate is an important physiological process. However, progress has long been hindered by the well-entrenched belief that all the reactions of glycolysis were reversible and that gluconeogenesis was achieved by the same enzymes. This concept did not take into account that three reactions of the glycolytic pathway are highly exothermic and constitute important energy barriers to the reversal of glycolysis under physiological conditions. These three reactions are respectively catalyzed by hexokinase, phosphofructokinase, and pyruvate kinase.

We now know that gluconeogenesis makes use of all the reversible reactions of glycolysis except for three systems which substitute for the three impossible reactions which are strongly endergonic in the direction of glucose synthesis. These three reactions are: the synthesis of phosphoenolpyruvate by a reaction different from the one catalyzed by pyruvate kinase; the formation of fructose-6-phosphate by fructose 1,6-diphosphatase; and the hydrolysis of glucose-6-phosphate by glucose-6-phosphatase (fig. 1).

glucose ← glucose-6-phosphate → fructose-6-phosphate ←
fructose-1,6-diphosphate → glyceraldehyde-3-phosphate →
3-phosphoglycerate → 2-phosphoglycerate → phosphoenolpyruvate
← oxaloacetate ← pyruvate

Figure 1. Gluconeogenesis. All reactions except three are common to glycolysis and gluconeogenesis. The reactions specific to gluconeogenesis are indicated by reverse arrows.

Synthesis of phosphoenolpyruvate

Pyruvate phosphorylation can be achieved as follows:

a) By pyruvate kinase. This enzyme does not function in gluconeogenesis. In addition to the already-mentioned energetic obstacle, its concentration in liver and kidney – where the resynthesis of glucose is achieved – is insufficient to account for the efficiency of gluconeogenesis, and the affinity of pyruvate for it is too low for this enzyme to be efficient at intracellular substrate concentrations.

b) By a pathway involving the “malic enzymes”, which catalyze a reductive carboxylation of pyruvate to malate:

$CH_3–CO–COOH + CO_2 + NADPH + H^+ \rightarrow CH_3–CHOH–CH_2–COOH + NADP^+.$
pyruvate — malate

One can imagine that malate is oxidized to oxaloacetate by malate dehydrogenase and that oxaloacetate is transformed into phosphoenolpyruvate by phosphoenolpyruvate carboxykinase. This pathway is also eliminated due to the very weak concentration of "malic enzyme" in liver and kidney. Also, the low affinity of bicarbonate for the malic enzyme is incompatible, together with its weak intracellular concentration, with efficient functioning.

c) The pathway making use of two successive enzymes, pyruvate carboxylase and phosphoenolpyruvate carboxykinase:

1) pyruvate + ATP + CO_2 + H_2O → oxaloacetate + ADP + Pi
(pyruvate kinase)

2) oxaloacetate + GTP → phosphoenolpyruvate + CO_2 + GDP
(phosphoenolpyruvate carboxykinase)

The sum of these two reactions is:

3) pyruvate + ATP + GTP + H_2O → phosphoenolpyruvate + CO_2 + GDP

Two equivalents of nucleoside triphosphate are required for this overall reaction, which is therefore exothermic and does not present an energy barrier (the free energy of phosphoenolpyruvate hydrolysis is 12.8 kcal/mole, and that of two ATP equivalents is 14 kcal/mole).

Furthermore, pyruvate carboxylase and phosphoenolpyruvate carboxykinase are found in kidney and liver at high concentrations and the K_ms for their respective substrates are much lower than the intracellular concentrations of these substrates. Pyruvate carboxylase is virtually absent in the non-gluconeogenetic organs such as brain, heart, and skeletal muscle.

If pyruvate kinase were present in appreciable quantities in the same location as the two effective enzymes in gluconeogenesis, its reaction would be superimposed on theirs to give:

phosphoenolpyruvate + ADP → ATP + pyruvate
(pyruvate kinase)

pyruvate + ATP + GDP + H_2O → phosphoenolpyruvate + ADP + GDP + Pi
(pyruvate carboxylase + phosphoenolpyruvate carboxykinase)

Sum: GTP + H_2O → GDP + Pi.

Thus, the presence of the three enzymes in the same tissue or in the same subcellular fractions would lead to a futile cycle resulting in the useless hydrolysis of GTP!

Pyruvate carboxylase and its allosteric regulation

The enzyme has been purified to homogeneity from chicken liver mitochondria. It contains four moles of biotin per mole of molecular weight 655,000. It is sensitive to low temperatures which dissociate it into subunits with a sedimentation coefficient of 7S (the native enzyme has a sedimentation coefficient of 14.8S). The concomitant inactivation is partially reversed by restoring the original temperature: this phenomenon can be studied by ultracentrifugation and electron microscopy. Acetyl coenzyme A protects against the cold inactivation. The reverse reaction, the decarboxylation of oxaloacetate, is possible in principle, but the concentrations of inorganic phosphate required are so high that it is without physiological significance.

The reaction obeys the usual rules for carboxylation reactions catalyzed by biotin-containing enzymes:

$$\text{enzyme-biotin} + \text{ATP} + HCO_3^- \rightarrow \text{enzyme-biotin}{\sim}CO_2 + \text{ADP} + \text{Pi}$$

(enzyme carboxylation)

$$\text{enzyme-biotin}{\sim}CO_2 + \text{acceptor} \rightarrow \text{enzyme-biotin} + \text{carboxylated acceptor.}$$

(transcarboxylation)

Here, the acceptor is pyruvate and the product, oxaloacetate. The enzyme-biotin~CO_2 complex can be isolated and will carboxylate pyruvate in the absence of ATP. Acetyl CoA is necessary only for the carboxylation of the enzyme and is not involved in the transcarboxylation.

The enzyme is made up of four subunits of molecular weight about 150,000, each containing one equivalent of biotin. It should be noted that the activation of the enzyme by acetyl CoA at ordinary temperatures is not accompanied by a change in molecular weight; the activation does not involve an aggregation as has been described in other cases (*e.g.* effect of citrate on acetyl CoA carboxylase). A more likely explanation is that the binding of acetyl CoA to the enzyme strengthens the interprotomeric interactions preventing the inactivation by cold and causing the considerable increase in catalytic activity.

If the supply of oxaloacetate is low, citrate synthesis will diminish and acetyl CoA, the other substrate of citrate synthesis, will accumulate and will stimulate the synthesis of oxaloacetate through the activation of pyruvate carboxylase. This, in turn, will stimulate the production of citrate and other intermediates of the tricarboxylic acid cycle. Gluconeogenesis will similarly be facilitated, as will fatty acid synthesis, in which an essential step is allosterically activated by citrate.

Fructose diphosphatase and its allosteric regulation

The existence of this enzyme avoids the difficulty presented by the highly exergonic nature of the phosphofructokinase reaction. Fructose diphosphatase hydrolyzes fructose-1,6-diphosphate to fructose-6-phosphate and inorganic phosphate. Its importance in gluconeogenesis is attested by its exclusive presence in tissues which are gluconeogenetic or carry out an appreciable synthesis of glycogen. Some control of fructose diphosphatase activity is foreseeable for, in its absence, the simultaneous presence of phosphofructokinase and the diphosphatase would result in a futile hydrolysis of ATP by coupling of the two reactions:

	fructose-1,6-diphosphate	$\rightarrow$	fructose-6-phosphate + Pi
	fructose-6-phosphate + ATP	$\rightarrow$	fructose-1,6-diphosphate + ADP
Sum:	ATP	$\rightarrow$	ADP + Pi

Different laboratories have discovered that mammalian fructose diphosphatase is strongly inhibited by AMP and is reversibly activated by ATP and ADP. These effects are the opposite of those found with liver phosphofructokinase. This theoretically results in a fine control of the two enzymatic activities in a reciprocal fashion and avoids the wastage referred to above.

The two enzymes are also subjected to a reciprocal regulation by fructose-2,6-diphosphate, a signal molecule deriving from fructose-6-phosphate, independently discovered by Hers and by Pilkis in 1980. Fructose-2,6-diphosphate activates phosphofructokinase and inhibits the fructose-1,6-diphosphatase. When glucose is in excess, a hormone-controlled reaction cascade causes the synthesis of a great amount of fructose-2,6-diphosphate. As a result, glycolysis is accelerated and gluconeogenesis diminished. When glucose becomes scarce, the inverse effect is obtained and glucose synthesis is accelerated. Thus fructose-2,6-diphosphate concentration is important for determining whether glucose will be degraded or synthesized.

A bifunctional enzyme, 6-phosphofructo-2-kinase/fructose-2,6-diphosphatase catalyzes both the synthesis and the degradation of fructose-2,6-diphosphate. It is a homodimeric protein, each subunit being composed of 470 amino acids, and consisting of two catalytic domains.

Fructose diphosphatase in microorganisms

Mutants deficient in fructose diphosphatase have been obtained in *E. coli* (*fdp,* 96 mn). They cannot grow on acetate, glycerol, or succinate, demonstrating the importance of diphosphatase for neoglucogenesis; these mutants have an absolute requirement for hexoses, for they have no other means of synthesizing ribose-5-phosphate, for example, which is indispensable for the synthesis of nucleic acids, histidine, and tryptophan; or erythrose-4-phosphate required in the biosynthesis of the aromatic nucleus (see chapters dealing with these biosyntheses). Wild-type organisms possess higher levels of diphosphatase when they are grown on glycerol than when they are grown on glucose, for they must synthesize glucose (table I).

Table I. Levels of fructose diphosphate in organisms grown on glucose or glycerol.

Organism	Grown on Glucose	Grown on Glycerol
	Fructose diphosphatase specific activity (mmoles/mg/mn)	
Saccharomyces cerevisiae	1	25
Candida utilis	26	140

Glucose-6-phosphatase

This enzyme, membrane-bound inside the cytoplasmic reticulum in eukaryotic cells, catalyzes the final step of gluconeogenesis. Its regulation is compulsory and explains the constancy of the glucose concentration in blood, even under extreme conditions such as fasting. Its importance is underlined by the fact that patients with deficient liver glucose-6-phosphatase are hypoglycemic.

Glycogen synthase

The reaction catalyzed by glycogen phosphorylase is reversible and glycogen can be synthesized by it from glucose-1-phosphate in the presence of a branching enzyme. However, the phosphorylase reaction proceeds *in vivo* in the direction of glycogen degradation, and it was shown in 1957 by Leloir and his colleagues in Buenos Aires that glycogen is synthesized by a totally different mechanism, also used in the synthesis of starch and cellulose.

A glucosyl residue is transferred from a nucleoside diphosphate glucose to a primer whose length may vary, the primer being itself made up of a linear chain of glucosyl residues. The equation below shows the type of reaction common to the synthesis of the three polysaccharides.

$$\text{XDPG} + (\text{glucose})_n \rightarrow \text{XDP} + (\text{glucose})_{n+1}$$

In the case of glycogen, the glucose donor is uridine diphosphoglucose (UDPG). For starch and bacterial glycogen, the donor is adenosine diphosphoglucose (ADPG). The acceptor, $(\text{glucose})_n$, can be an oligosaccharide, but the best acceptor is glycogen itself.

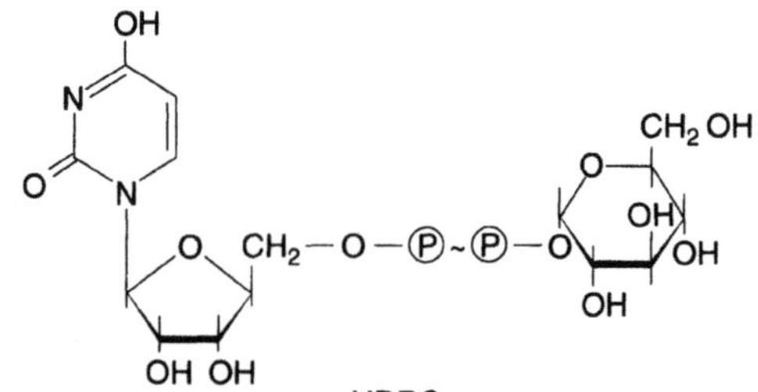
UDPG

UDPG is synthesized from glucose-1-phosphate and uridine triphosphate by a specific enzyme, UDPG pyrophosphorylase:

$$\text{UTP + glucose-1-phosphate} \rightarrow \text{UDPG + PPi}$$

Pyrophosphate, coming from the two outer phosphates of UTP, is rapidly hydrolyzed by a pyrophosphatase, which makes the synthesis of UDPG practically irreversible and drives the reaction in the direction of UDPG synthesis.

Glycogen synthase catalyzes only the synthesis of a linear polysaccharide made of α-1,4 linkages. A branching enzyme breaks an α-1,4 link and forms an α-1,6 link about every seven residues on the average.

In the last analysis, glucose-1-phosphate is the origin of glycogen; now in gluconeogenesis, the last intermediate before glucose is glucose-6-phosphate: we may remember that glucose-1-phosphate and glucose-6-phosphate are readily interconvertible by phosphoglucomutase.

Control of glycogen biosynthesis

Very soon after the discovery of glycogen synthase, it was observed that the enzyme functioned with maximum efficiency only in the presence of glucose-6-phosphate. However, this activation varied from preparation to preparation. The explanation of the variability was given by the existence of two interconvertible forms of the enzyme, a phosphorylated form depending on the presence of high levels of glucose-6-phosphate for its activity (D form), and a non-phosphorylated form independent of it (I form).

In skeletal muscles, adrenalin and glucagon bind to their receptors on the plasma membrane of target cells and activate adenylate cyclase which forms cyclic AMP (cAMP) from ATP. Intracellular cAMP activates in turn an inactive cAMP-dependent protein kinase by binding to its regulatory subunit which is then liberated from its (active) catalytic subunit. The active protein kinase can then phosphorylate glycogen synthase *and* phosphorylase kinase, which in turn phosphorylates phosphorylase. This is the basis of the coordinate regulation of glycogen biosynthesis and breakdown, since the phosphorylated phosphorylase is active and the phosphorylated glycogen synthase is inactive. In addition, phosphorylase kinase can directly phosphorylate glycogen synthase, which provides an additional safeguard, assuring that glycogen not be synthesized while it is being degraded.

Another reciprocal control is provided by a specific protein phosphatase which removes the phosphate group from phosphorylated phosphorylase, phosphorylase kinase, and glycogen synthase. This removal activates glycogen synthase and inactivates phosphorylase, directly and indirectly.

SELECTED REFERENCES

Glycogen synthesis and its regulation

L. F. Leloir, *Science* **172**, 1299-1302 (1971).

E. H. Fischer, *Bull. Institut Pasteur* **81**, 7-31 (1983).

H. G. Hers & L. Hue, *Ann. Rev. Biochem.* **52**, 617-653 (1983).

P. Cohen, *Control of Enzyme Activity, Outline Studies in Biology*, 96 p. Chapman and Hall, London and New York (1983).

CHAPTER XIV

Biosynthesis of lipids

Biosynthesis of short-chain fatty acids

During the late 1930s, Barker isolated an anaerobic bacterium which he called *Clostridium kluyveri*. This organism is able to grow on a mixture of acetate and ethanol and produces *n*-butyrate and *n*-caproate. From this organism, Barker and Stadtman obtained cell-free extracts which could carry the reactions leading to the C_4 and C_6 acids. Ethanol is oxidized to acetic acid in two steps, respectively catalyzed by alcohol dehydrogenase and an acetaldehyde dehydrogenase, depending on the presence of coenzyme A:

$$CH_3\text{–}CH_2OH + NAD^+ \rightarrow CH_3\text{–}CHO + NADH + H^+$$
$$CH_3\text{–}CHO + CoASH + NAD^+ \rightarrow CH_3\text{–}CO\text{–}SCoA + NADH + H^+$$

One notes that the reactive species acetyl CoA is formed, rather than acetic acid. A β-ketothiolase then condenses two molecules of acetyl CoA by joining the methyl group of one molecule with the carboxyl group of the other (head to tail condensation):

$$2\ CH_3\text{–}CO\text{–}SCoA \rightarrow \underset{\text{acetoacetyl CoA}}{CH_3\text{–}CO\text{–}CH_2\text{–}CO\text{–}SCoA} + CoASH$$

Acetoacetyl CoA is reduced by β-hydroxybutyryl CoA dehydrogenase:

$$CH_3\text{–}CO\text{–}CH_2\text{–}COSCoA + NADH + H^+ \rightarrow \underset{\beta\text{-hydroxybutyryl CoA}}{CH_3\text{–}CHOH\text{–}CH_2\text{–}COSCoA} + NAD^+$$

This compound is then dehydrated by a specific crotonase yielding crotonyl CoA $CH_3\text{–}CH{=}CH\text{–}COSCoA$, which is reduced to butyryl CoA $CH_3CH_2\text{–}CH_2\text{–}COSCoA$ by butyryl CoA dehydrogenase.

The synthesis of caproate occurs by an analogous series of reactions starting from the condensation of acetyl CoA with butyryl CoA giving rise to β-ketocaproyl CoA. If *Cl. kluyveri* is grown on ethanol and propionate, one obtains the odd-numbered acids, *n*-pentanoic and *n*-heptanoic acids by the same type of reactions.

Biosynthesis of long-chain fatty acids

SYNTHESIS OF ACETYL CoA

In addition to the synthesis generated by the pyruvate dehydrogenase complex, bacteria and plants can synthesize acetyl CoA from acetate and CoA by an ATP-driven reaction catalyzed by acetyl CoA synthetase.

$$CH_3COOH + CoA + ATP \rightarrow \text{acetyl CoA} + AMP + PPi$$ [13]

Animals synthesize acetyl CoA through the pyruvate dehydrogenase complex.

SYNTHESIS OF MALONYL CoA

After having observed, in 1957, that bicarbonate is required for this synthesis, Wakil identified malonyl CoA as the first intermediate in the synthesis of long-chain (C-10 to C-18) fatty acids. He discovered that its synthesis is catalyzed by a biotin-containing enzyme, the biotin being covalently linked to an ε-amino group belonging to a lysine residue. This enzyme, acetyl CoA carboxylase, catalyzes a carboxylation of acetyl CoA in the presence of ATP:

$$CH_3\text{–}COSCoA + CO_2 + ATP \rightarrow \underset{\text{malonyl CoA}}{HOOC\text{–}CH_2\text{–}COSCoA} + ADP + Pi$$

In yeast, acetyl CoA carboxylase is a tetramer of a multifunctional protein encoded by the *FAS3* gene.

E. coli acetyl CoA carboxylase has been resolved into three subunits, each of which catalyzes a partial reaction. Biotin is linked covalently to a small 16,688 Da protein (deduced from the DNA sequence), called "biotin carboxyl carrier protein" (BCCP). The carboxylation of the biotin is due to another subunit of the enzyme, biotin carboxylase, in close proximity to the biotin carboxyl carrier protein. The DNA sequence of the carboxylase is now available, and corresponds to a protein of a molecular weight of 49,000 (449 amino-acid residues). The gene encoding BCCP is located upstream of the biotin carboxylase gene and the two genes are transcribed as a single mRNA species that contains an unusually long untranslated leader preceding the *BCCP* gene. The last subunit is a transcarboxylase, responsible for the transfer of the activated CO_2 of carboxybiotin to acetyl CoA. The length and flexibility of the arm linking biotin to its carrier allow the activated carboxyl group to move from one subunit to the other.

13. *E. coli* contains in addition a single acyl CoA synthetase, catalyzing the same type of reaction on long-chain fatty acids (*fadD*, 39.5 mn), associated with the inner membrane, and that has been purified to homogeneity. It plays a pivotal role in long-chain fatty acid transport by catalyzing the thioesterification of exogenous fatty acid into metabolically active CoA thioesters prior to β-oxidation concomitant with transport.

The reaction catalyzed by acetyl CoA carboxylase is the rate-limiting step in fatty acid synthesis. The bacterial enzyme is allosterically inhibited by palmityl CoA, the end product of the synthesis. The mammalian enzyme is activated by citrate, which is not a substrate for the reaction. In the enzyme from adipose tissue and rat liver, this important activation is accompanied by a conformational change in the protein: its sedimentation coefficient, which is normally 18.8S, becomes 43S when it is incubated and centrifuged with citrate under the same conditions as those necessary for activation. The rate of activation of the enzyme is concentration-dependent, and there are grounds for thinking that activation and aggregation are two facets of the same phenomenon. Citrate accumulates when NAD-dependent isocitrate dehydrogenase is non-functional due to the lack of AMP; besides the effect of citrate on the carboxylase, the exclusive functioning of the other isocitrate dehydrogenase, the one requiring NADP, increases the amount of NADPH that becomes available for fatty acid synthesis. We should note that the addition of citrate to tissue slices causes an increase in fatty acid synthesis and a decrease in sterol synthesis, both of which start from acetyl CoA.

Mammalian acetyl CoA carboxylase is also regulated covalently by phosphorylation (which inactivates it), and dephosphorylation (which reactivates it).

FROM MALONYL CoA TO PALMITATE

It is known that the main compound synthesized from acetate by extracts purified from pigeon liver is palmitic acid, the saturated fatty acid with a linear chain of 16 carbon atoms. The overall equation representing the stoechiometry of its synthesis is the following:

$$\text{acetyl CoA} + 7\ \text{malonyl CoA} + 14\ \text{NADPH} + 14\ H^+ \rightarrow \underset{\text{palmitate}}{CH_3\text{-}(CH_2)_{14}\text{-}COOH} + 7\ CO_2 + 8\ \text{CoASH} + 14\ \text{NADP}^+ + 6\ H_2O$$

The carbon atoms of acetyl CoA are incorporated into the methyl group and the last methylene group of palmitic acid, whilst the carbon atoms of malonyl CoA appear as carbon atoms 1 to 14. Acetyl CoA can be replaced in this reaction by the butyryl, hexanoyl, or tetradecanoyl CoA derivatives, a fact that led to the belief that these compounds were probable intermediates in the synthesis of palmitate. We shall see that this is not so.

Work on the synthesis of palmitate has been mainly carried out with three systems, isolated respectively from *Escherichia coli*, the yeast *Saccharomyces cerevisiae*, and chicken liver. It has been shown in *E. coli* that the product of the condensation of acetyl CoA and malonyl CoA is an acetoacetyl derivative bound to a heat-stable protein fraction; this protein has been called *acyl carrier protein* (ACP).

ACP is a polypeptide of 77 residues in which a 4'-phosphopantetheine residue (see the section on pantothenic acid and CoA synthesis) is linked by a phosphodiester bond to a serine residue. ACP can be considered as a giant prosthetic group, a macro CoA. It is to the SH-group of phosphopantetheine that, during the synthesis, the acetyl, malonyl, and acyl groups of intermediate size are covalently linked.

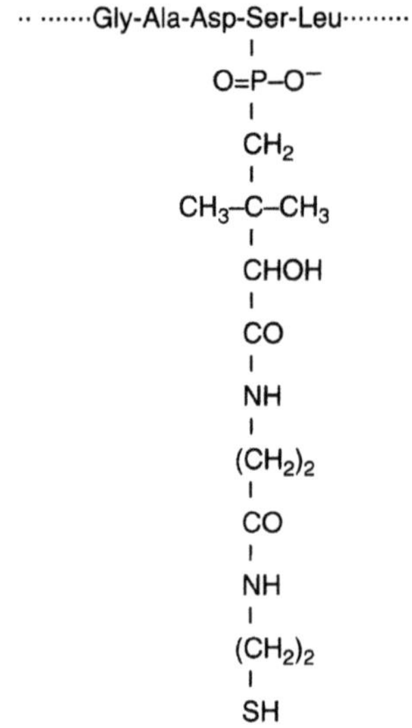

The acyl carrier protein

The gene that encodes the acyl carrier protein from *E. coli (acpS*, 43 mn) lies within a cluster of known fatty-acid biosynthetic genes.

The reactions that lead to butyryl-ACP are the following:

$$\text{malonyl CoA} + \text{ACP-SH} \rightleftharpoons \text{malonyl-S-ACP} + \text{CoASH}$$

$$\text{acetyl CoA} + \text{ACP-SH} \rightleftharpoons \text{acetyl-S-ACP} + \text{CoASH}$$

$$\text{acetyl-S-ACP} + \text{malonyl-S-ACP} \rightleftharpoons \text{acetoacetyl-S-ACP} + \text{ACP-SH} + CO_2$$

$$\text{acetoacetyl-S-ACP} + \text{NADPH} + H^+ \rightleftharpoons \beta\text{-hydroxybutyryl-S-ACP} + \text{NADP}^+$$

$$\beta\text{-hydroxybutyryl-S-ACP} \rightleftharpoons \text{crotonyl-S-ACP} + H_2O$$

$$\text{crotonyl-S-ACP} + \text{NADPH} + H^+ \rightleftharpoons \text{butyryl-S-ACP} + \text{NADP}^+$$

Once butyryl-ACP is formed, a secound round of synthesis takes place in which butyryl-ACP condenses with malonyl-ACP to form a β-ketoacyl-ACP with six carbon atoms; the successive reactions lead to caproyl-ACP, which is the substrate for a new elongation cycle. This continues until palmityl-ACP with 16 carbon atoms is obtained. This compound is not a substrate for the condensation enzyme: it is hydrolyzed and liberates ACP and free palmitate. Thus coenzyme A esters are not involved in this synthesis, after the first two steps which form the ACP derivatives of acetate and malonate. The next four reactions, previously shown to occur with CoA, reflect a lack of absolute specificity in the enzymes whose normal substrates are the ACP-bound derivatives.

The enzymes constituting the bacterial system dissociate when the cells are disrupted.

The fatty-acid synthetases of eucaryotes, in contrast to the bacterial ones, are well-defined multifunctional proteins. The one from yeast has a mass of 2200 kDa and appears under the electron microscope as an ellipsoid having axes of 250 and 210 Å. It consists of only two multifunctional polypeptide chains, α and β. The enzyme has an $\alpha_6\beta_6$ structure. The α chain (206 kD) carries the condensation enzyme, β-ketoacylreductase activities, and ACP, while the β chain (229 kDa) carries the the acyl transacylase, the malonyl transacylase, the β-hydroxyacyl dehydratase, and the enoyl reductase activities. The genetic study of yeast mutants has demonstrated that the genes coding for the seven proteins belong to only two complementation groups. This led to the reexamination of the previous belief that these proteins were distinct objects associated by non-covalent bonds. The yeast synthetase has been examined by negative-stain and cryoelectron microscopy, and a model has been proposed where the six fatty-acid synthesizing centers are composed of two complementary half-α subunits and a β subunit, an arrangement having all the partial activities of the multifunctional enzyme required for fatty-acid synthesis.

The fatty-acid synthetase from animals (230 kD) is a dimer of identical subunits organized in a head-to-tail manner, such that the cysteine of the active group of the β-ketoacylreductase (peripheral –SH group) of one subunit be very near the sulfhydryl group of the ACP of the other. Each subunit is organized into three globular domains linked by flexible hinges, as demonstrated by limited proteolysis experiments. Domain I plays the role of a condensation unit containing acetyl and malonyl transacylases and the condensation enzyme. Domain II plays the role of a reduction unit and contains ACP, β-ketoacylreductase, dehydratase, and enoylreductase. Domain III is responsible for the thioesterase activity which liberates the palmitic acid.

The elongation begins on subunit A with the condensation of the acetyl unit, fixed on a peripheral sulfhydryl with a malonyl residue fixed on ACP. A molecule of CO_2 is then liberated and an acetoacetyl-S-phosphopantetheinyl unit is formed on ACP, liberating the peripheral SH of the active site of the condensation enzyme. The acetoacetyl group is transferred on domain II of subunit B and is subject to the three-step reduction to butyryl-ACP. This C4 unit is then transferred from the SH of ACP to the SH of the condensation enzyme of subunit A. The enzyme is now ready for a second round. When the growing chain has reached a length of 16 carbon atoms, the palmitoyl residue still hooked upon ACP becomes the substrate of the esterase of the opposite chain. This remarkable translocation mechanism is reminiscent of that of the elongation of the nascent polypeptide chain on ribosomes. The flexibility and the maximal length (20 Å) of the articulated arm constituted by phosphopantetheine are essential for the functioning of this elaborated multifunctional enzyme. It is not necessary for the subunits to undergo important rearrangements to interact with the substrates: the latter reside on a long and

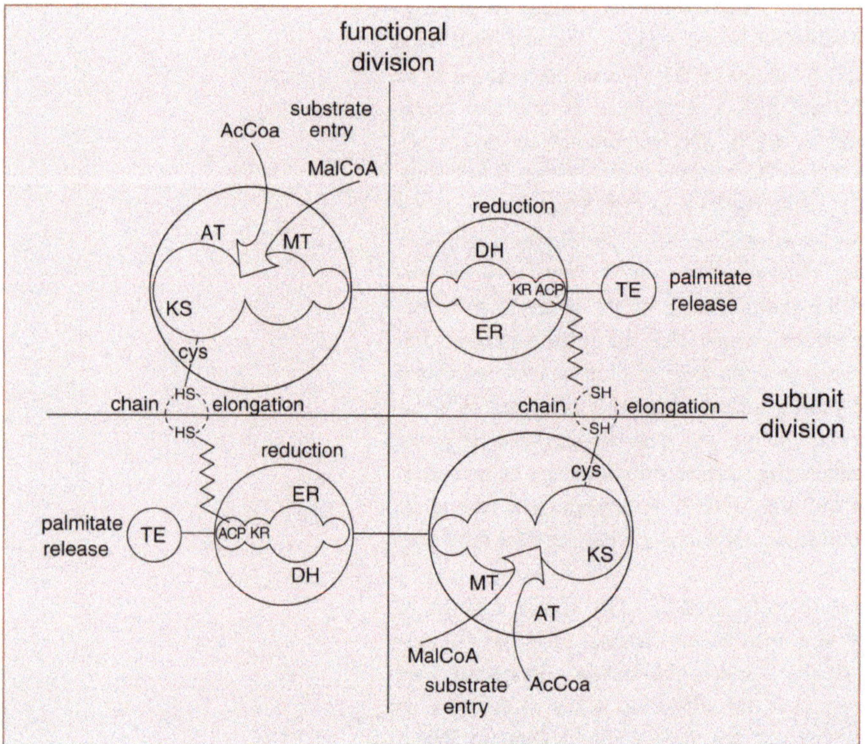

Figure 1. Map of the chicken fatty acid synthetase. The sketch is based on the results of limited proteolysis experiments and on evidence from assays of catalytic activities, and binding of substrates, of specific inhibitors or antibodies. Two subunits are drawn in head-to-tail arrangement (subunit division) so that two sites of palmitate synthesis are constructed (functional division). The abbreviations for enzyme activities are: AT, acetyl transacylase; MT, malonyl transacylase; KS, β-ketoacyl synthetase; KR, β-ketoacyl reductase; DH, dehydratase; ER, enoyl reductase; TE, thioesterase; and ACP, acyl carrier protein.
The wavy line represents the 4'-phosphopantetheine prosthetic group.
(From S. J. Wakil, J. K. Stoops & V. C. Joshi (1983). With permission from the *Annual Review of Biochemistry.*)

flexible arm which can reach the active sites of all the enzymes involved in the overall reaction, without dilution of the intermediates, which are sequestered and protected from competing reactions.

Fatty acids are synthesized in the cytoplasm whereas acetyl CoA is synthesized from pyruvate in the mitochondria. Acetyl CoA must therefore be transferred; this is done by condensation of acetyl CoA with oxaloacetate to yield citrate. Whereas mitochondria are impermeable to acetyl CoA, they are not to citrate. Once citrate is in the cytoplasm, it is cleaved by citrate lyase:

$$\text{citrate} + \text{ATP} + \text{COA} + H_2O \longrightarrow \text{acetylCoA} + \text{ADP} + \text{oxaloacetate} + \text{Pi}$$

Thus, the transfer is made at the expense of an ATP molecule.

Fatty acids longer than palmitate are synthesized by enzymes located at the internal side of the membrane of the endoplasmic reticulum, which add sequentially two-carbon atom units to the carboxyls of saturated and unsaturated acids. The same microsomal system introduces the double bonds in the long-chain fatty acids, at least until carbon atom 9. Mammals do not possess enzymes allowing the introduction of double bonds beyond that limit; so linoleic and linolenic acid, which possess double bonds respectively at C12, and C12 and C15, are essential fatty acids which must be present in the diet.

Regulation of yeast fatty acid synthesis at the genetic level

The two unlinked genes *FAS1* and *FAS2*, coding respectively for the β and α subunit of yeast fatty acid synthetase are mapped to yeast chromosomes XI and XVI, respectively. The synthesis of the multimeric $\alpha_6\beta_6$ heteromultimeric fatty acid synthetase requires the expression of *FAS1* and *FAS2* at comparable rates. This is achieved through the existence of an upstream activation site (analogous to the enhancers) having the same sequence in both cases: TYTTCACATGY (Y coding for any pyrimidine), functioning in any orientation. A protein *trans*-acting factor was identified which binds to this sequence. It is suspected that additional *cis*- and *trans*-acting elements are involved in the expression of the yeast fatty acid synthetase genes. The above consensus sequence is also present in the upstream regions of many other yeast genes involved in phospholipid biosynthesis (such as those coding for inositol-1-phosphate synthase and phosphatidylserine synthase) and a general control mechanism affecting most components of yeast membrane biogenesis has been suggested, implying the involvement of common *trans*-acting factors.

Each of the two yeast genes, *FAS1* and *FAS2*, encoding respectively the α and the β subunit, were individually deleted by one-step gene disruption. The deletion of *FAS2* does not influence the transcription of *FAS1*, whereas the transcription of *FAS2* is significantly reduced in the *FAS1*-deleted strain. This suggests an activating role of the β subunit on the expression of *FAS2*, or a repression of *FAS2* by an excess of its own product. On the other hand, differential proteolytic degradations remove the excess of either subunit, at least under starvation conditions. Thus, a combination of both regulation of expression and proteolysis of free *FAS* polypeptides may explain the equimolar amounts of α and β in yeast wild-type cells.

The gene *FAS3*, coding for yeast acetyl CoA carboxylase seems also to be coordinately regulated with *FAS1* and *FAS2*.

Regulation of fatty acid synthesis in bacteria

The multifunctional fatty acid synthetase from *Brevibacterium ammoniogenes* is composed of a single polypeptide of 3104 residues; the seven active centers of it have been identified at different locations of the polypeptide and the overall structure is formally consistent with a head-to-tail fusion of the two yeast genes *FAS1* and *FAS2*.

In *E. coli*, *fabA* (22 mn) codes for β-hydroxydecanoylthioester dehydrase, the introduction of the double bond of unsaturated fatty acids occurring at the C10 level. The FadR protein of *E. coli*, which acts as a repressor of fatty acid degradation, acts as an activator of the *fabA* gene: it binds at the – 40 region of the major *fabA* promoter. The transcription of *fabA* is markedly decreased when fatty acids are added to the cultures, and acyl CoA esters specifically release the FadR protein from its binding site.

Biosynthesis of triglycerides

A direct acylation of glycerol has never been observed. A particulate fraction and a soluble fraction from *Clostridium butyricum* have been obtained which, together, can acylate glycerol-3-phosphate in the presence of ACP and palmityl CoA. The same acylation is obtained directly, starting from palmityl ACP. The product of the reaction is lysophosphatidic acid:

$CH_2O–CO–R$	$CH_2–O–CO–R$
I	I
CHOH	CH–O–CO–R
I	I
$CH_2–O–PO_3H_2$	$CH_2–O–PO_3H_2$
lysophosphatidic acid	phosphatidic acid

In addition to lysophosphatidic acid, phosphatidic acid and monopalmitine resulting from the dephosphorylation of lysophosphatidic acid are obtained. On the other hand, the dephosphorylation of phosphatidic acid leads to an α-β diglyceride which is the common preecursor of triglycerides and choline; actually, an enzyme from chicken liver synthesizes triglycerides from diglycerides and acyl CoA:

$$R–CO–SCoA + \alpha,\beta\text{-diglyceride} \longrightarrow \text{triglyceride} + CoA–SH$$

Biosynthesis of phosphoglycerides

An intermediate compound, cytidine diacylglycerol, is formed by the reaction between cytidine triphosphate and phosphatidic acid, with the formation of pyrophosphate, the hydrolysis of which drives the reaction:

phosphatidic acid + CTP →

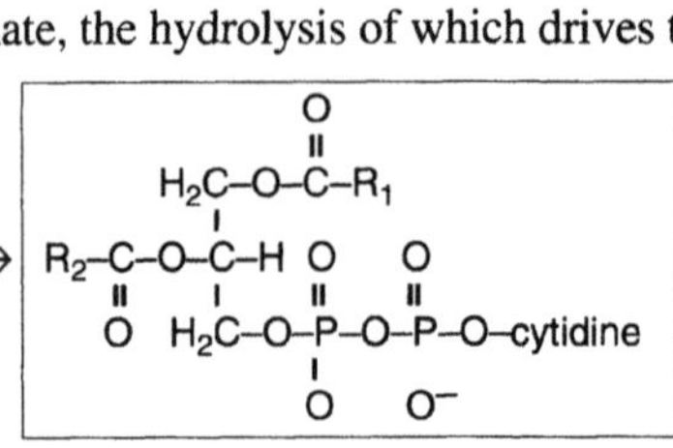

+ PPi

cytidine diacylglycerol

The activated phosphatidyl unit reacts with a polar alcohol while CMP is liberated. If the alcohol is inositol, phosphatidylinositol is formed.

Further phosphorylations by specific kinases lead to polyphosphorylated forms of phosphatidylinositol, whose hydrolysis yields two intracellular messengers, diacylglycerol and inositol-1,4,5-triphosphate, which elicit a great variety of physiological responses in many tissues. If the alcohol is serine, phosphatidylserine is formed.

In bacteria, phosphatidylserine is decarboxylated by a pyridoxal phosphate enzyme to phosphatidyl ethanolamine:

$H_2N{-}CH_2{-}CH_2{-}OR$ R = phosphatidyl residue

which is the substrate of three sucessive methylations to yield phosphatidylcholine, with S-adenosylmethionine as the methyl donor:

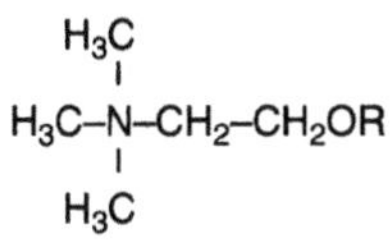

phosphatidyl choline

In mammals, who obtain it from their diet, choline is directly phosphorylated by ATP to phosphorylcholine, which then reacts with CTP to yield CDP-choline and pyrophosphate. CDP-choline is transferred to a diacylglycerol to finally give phosphatidylcholine.

Cyclopropane fatty acid synthase

Cyclopropane fatty acids have long been known to occur in the lipids of many eubacteria, although physiological and regulatory aspects of their formation have been almost exclusively been studied in *E. coli*. Their synthesis proceeds by the transfer of a methylene group from the activated methyl of S-adenosylmethionine to the *cis* double bond of an unsaturated fatty-acid chain, resulting in the replacement of the double bond with a methylene bridge. The enzyme catalyzing the reaction is found in the cytoplasm, the substrate being not the free unsaturated fatty acid but rather phospholipids containing it. The enzyme has thus one soluble substrate, S-adenosylmethionine, and an insoluble one, the phospholipid bilayer. The double bond has to be positioned 9 to 11 carbon atoms away from the ester bond linking the fatty acid to the phospholipid backbone, and is thus located within the hydrophobic core of the bilayer structures of the phospholipid vesicles or the membranes of intact bacteria. The recent cloning of the gene coding for the cyclopropane fatty-acid synthase *(cfa)*, its sequence, its overexpression, and the purification of the protein in milligram quantities, will certainly clarify the mechanisms of this fascinating enzyme.

SELECTED REFERENCES

Short-chain fatty-acid synthesis

H. A. Barker, *Bacterial fermentations*, Ciba Lectures in Microbial Biochemistry, 95 p. John Wiley & Sons, New York (1956).

Fatty-acid synthesis and its regulation

E. Schweitzer, in *Multifunctional proteins*, pp. 197-215, H. Bisswanger & E. Schmincke-Ott eds. John Wiley and Sons, New York (1980).

S. J. Wakil, J. K. Stoops, & V. C. Joshi, *Ann. Rev. Biochem.* **52**, 537-579 (1983).

N. Singh, S. J. Wakil & J. K. Stoops, *Biochemistry* **24**, 6598-6602 (1985).

CHAPTER XV

Biological fixation of nitrogen

Nitrogen fixation and photosynthesis are the two processes that ultimately feed the world.

Gaseous nitrogen (N_2) with a valence of 0 must be reduced to ammonia prior to incorporation into nitrogenous components of the cell. This process, called *nitrogen fixation,* is limited to certain procaryotic organisms called diazotrophs, able to synthesize the enzyme nitrogenase.

The contrast between the expensive chemical Haber-Bosch process for ammonia production from dinitrogen N≡N, which requires an elevated temperature and pressure, and the biological fixation of either the free-living nitrogen-fixing bacteria or the symbiotic *Rhizobia,* which occurs at ambient temperature and pressure, is striking. Each route accounts for approximately one hundred million tons of N_2 converted to ammonia per year.

Nitrogenase, the enzyme responsible for nitrogen fixation, uses approximately 34 iron and 2 molybdenum atoms to catalyze the reaction:

$$N_2 + 16\ MgATP + 8\ e^- + 8\ H^+ \longrightarrow 2\ NH_3 + 16\ MgADP + 16\ Pi + H_2$$

The substrate specificity of nitrogenase is relatively low; a number of other compounds, including N_3^-, N_2O, HCN, CH_3NC, CH_2CHCN, and C_2H_2 are also reduced by it. Some of these reactions involve the transfer of fewer electrons than the eight required to reduce gaseous nitrogen. The study of biological nitrogen fixation in whole cells and in extracts has been greatly aided by the introduction of an assay method using the substrate acetylene, which is reduced to ethylene

$$CH{\equiv}CH \xrightarrow[2\ H^+]{2\ e^-} CH_2{=}CH_2$$

The product can be easily quantitated by gas chromatography, and the reaction is a highly specific one since no enzyme system other than nitrogenase can reduce acetylene to ethylene[13].

13. Unlike wild type, certain Mo-dependent nitrogenases, which are expressed in non-nitrogen fixing mutants of *Azotobacter vinelandii* and have single amino-acid substitutions within a region of the MoFe protein α-subunit proposed to encompass the FeMoco binding domain (see below), are able to catalyze the further reduction of acetylene by two and four electrons to yield ethylene and ethane, respectively. These reductions are also catalyzed by the vanadium and "third" nitrogenases.

Nitrogenase is composed of two proteins: the Fe-protein (also called dinitrogenase reductase), and the MoFe-protein (also called dinitrogenase).

The Fe-protein is a homodimer (total molecular weight: 60,000) containing a single Fe4S4 unit per dimer with the atoms at the corner of a cube; it can be understood on the basis of the structure of ferredoxins, a class of low-potential Fe-S electron transfer proteins. The Fe-protein binds 2 MgATP per molecule, which are hydrolyzed with the concomitant transfer of an electron to the MoFe-protein. The Fe-protein possesses all – or at least part – of the apparatus necessary to transduce the hydrolysis generated by ATP hydrolysis into a very negative electron potential, which initiates the saturation of the triple bond in N_2. In the absence of the MoFe-protein and reductant, the Fe-protein exhibits no ATPase activity.

The amino-acid sequence of more than 20 different Fe-proteins has been determined and indicates that this protein is highly conserved (the most divergent sequences are 69% identical).The crystallographic structure of *Azotobacter vinelandii* Fe-protein has been resolved at 2.9 Å resolution. Many of the features uncovered are relevant to other systems in addition to nitrogenase: for example, the enzyme protobacteriochlorophyll reductase, which catalyzes the first committed step in photosynthetic growth, shows extensive similarity with the Fe-protein, especially for residues involved as cluster ligands, nucleotide binding, and interface regions, suggesting that it performs a similar mechanistic function.

The *A. vinelandii* MoFe-protein, which possesses a dinitrogen binding site, consists of four subunits in an $\alpha_2\beta_2$ arrangement (total molecular mass: 240 kD) the α and β-subunits being composed of 491 and 522 amino acids, respectively. Each molecule yields two equivalents of molybdenum as the MoFe cofactor, which is a complex of approximate composition $MoFe_7S_8$ (homocitrate) and has been detected through its ability to activate an Mo-deficient protein found in the extracts of mutant strains of bacteria. Several physical measurements have shown beyond doubt that the cofactor in solution is essentially identical to the Mo-containing center in the native protein, both being sulfur-bridged polynuclear clusters: the Mo environment in the protein includes coordination to oxygen or nitrogen containing ligands as well as sulfur bridged to iron. The cofactor, present in two discrete centers, accounts for 14 of the 30 iron atoms in a molecule of the MoFe protein. The remaining 16 atoms are present in four Fe4S4 clusters which transfer one electron per cluster.

A crystallographic structure at 2.7 Å resolution has been obtained for the MoFe-protein from *Azotobacter vinelandii*. Each center consists of two bridged clusters: the MoFe cofactor (FeMoco) has 4Fe:3S and 1Mo:3Fe:3S clusters bridged by three non-protein ligands; the four Fe4S4 clusters are bridged by two cysteine thiol ligands. Six of the seven Fe sites of the cofactor seem to have trigonal coordination symmetry, while the seventh has tetrahedral geometry and is liganded to the side chain of a cysteine. The molybdenum atom is liganded by three sulfurs in the cofactor, two oxygens from the homocitrate, and the imidazole side chain of a histidine. Neither the cofactor nor the Fe4S4 clusters are exposed to the surface (they are buried at least 10 Å below it), suggesting that substrate entry, electron transfer, and product release must involve a carefully-regulated sequence of interactions between the MoFe-protein and the Fe-protein in the native nitrogenase.

FeMoCo is probably the site of N_2 binding, although it is not clear whether the N_2 interacts primarily with the Mo, Fe, or S^{2-} components.

The overall mechanism is thought to be the following: the Fe-protein pumps one electron per two molecules of ATP hydrolyzed into the MoFe-protein. Each stroke advances the redox status of the MoFe-protein one step around an eight-step circular pathway; in fact, hydrazine N_2H_4 has been detected as an intermediate between N_2 and NH_3 upon denaturation of MoFe-protein during reduction of N_2.

Control of nitrogenase synthesis and activity

Nitrogen fixation is an intrinsically anaerobic process, due to the rapid irreversible inactivation of nitrogenase by even low partial pressures of oxygen. The mechanisms that protect nitrogenase from oxygen are varied. In many diazotrophs, more than one mechanism may be present. Obligate anaerobes such as *Clostridium pasteurianum* and *Desulfovibrio desulfuricans* are apparently devoid of any specific device for protecting nitrogenase from oxygen inactivation. Facultative aerobes *(Klebsiella pneumoniae, Bacillus polymyxa, Rhodospirillum rubrum)* are able to grow on combined nitrogen both aerobically and anaerobically, but can fix nitrogen only in anaerobiosis. Microaerophilic bacteria *(Azospirillum)* show a preference for low partial oxygen pressures when fixing nitrogen. Finally, *Azotobacter* can fix nitrogen in the presence of air, the nitrogenase being protected by a 2Fe-2S protein.

Cyanobacteria are the only organisms able to perform oxygenic photosynthesis that can also fix nitrogen. They have the simplest nutritional requirements and can grow with light in a mineral medium, using CO_2 as a carbon source and N_2 as a nitrogen source. This presents an obvious paradox. In most cyanobacteria, like *Anabaena variabilis*, nitrogenase is localized in the heterocysts, which are thick-walled cells lacking photosynthetic oxygen evolution. Heterocysts rely on carbohydrates, imported from the photosynthetically active vegetative cells. High respiration rates and a diffusion barrier for gases enables heterocysts to fix nitrogen aerobically (in *Anabaena flos-aquae*, the glycolipid area of the heterocyst envelope thickens as exogenous oxygen partial pressure increases, showing an O_2-induced mechanism for providing a greater diffusion barrier against inhibition of nitrogenase). Heterocystous cyanobacteria do not form heterocysts when grown photosynthetically with a combined nitrogen source (ammonia or nitrate), and nitrogenase synthesis is also repressed under these conditions; both heterocyst formation and nitrogenase synthesis resume as soon as the combined nitrogen source is removed: this occurs even in the absence of N_2, in an atmosphere of argon.

Expression of nitrogenase in heterocyst-containing cyanobacteria depends on the availability of combined nitrogen and oxygen, and upon cell differentiation. In addition to being under transcriptional control, enzyme activity is influenced by external factors which limit the carbon supply of heterocysts by vegetative cells and by excess ammonium chloride under certain pH conditions. These treatments cause inactivation of the enzyme correlated with a decrease of electrophoretic mobility of the Fe-protein, which suggests a posttranslational enzyme modification. This modification has been shown to occur in many different nitrogen-fixing microorganisms and in the case of *Rhodospirillum rubrum,* it has been shown to be due to an ADP-ribose covalently attached to an arginine of the Fe-protein, which interrupts the catalytic function of nitrogenase. The modification, catalyzed by dinitrogenase reductase ADP-ribosyltransferase, requires NAD^+ as a substrate (like ADP-ribosylation of the elongation factor catalyzed by diphtheria toxin), and ADP plus a divalent cation as cosubstrates. The modification is reversed under the action of dinitrogenase ADP-ribosyl glycohydrolase in the presence of MgATP and a divalent cation. The glycohydrolase is associated with the membrane and is extremely oxygen-labile. This reversible inactivation of nitrogenase upon a nutritional switch-off is reminiscent of the *E. coli* glutamine synthetase case which will be studied in detail in a later section.

Studies with nonheterocystous cyanobacteria have revealed no universal strategy that would allow nitrogenase to function in these organisms under ambient atmospheric conditions and concomitant with oxygen photosynthesis.

One of the most exciting discoveries in nitrogen fixation research in recent years was the finding that, in addition to the conventional molybdenum nitrogenase, *A. vinelandii* has two nitrogenases that do not contain molybdenum. The three nitrogenases are similar in structure and physico-chemical properties, and all include iron in three different types of cluster. One of the two "new" nitrogenases contains a vanadium analogue of FeMoco. No metals other than iron appear to be present in significant amounts in the third enzyme. Under normal conditions, synthesis of the two non-molybdenum nitrogenases is repressed in presence of nanomolar amounts of molybdenum. In addition, the synthesis of the third nitrogenase is also repressed by vanadium. The subunits of the three enzymes are encoded by three different genes. The non-molybdenum nitrogenases contain an additional subunit that has no equivalent in the molybdenum nitrogenase enzyme.

In the absence of fixed nitrogen, and aerobically at 30 °C, *A. vinelandii* synthesizes one of the three nitrogenases according to the metal content of the medium. Molybdenum prevents expression of the alternative nitrogenases and is required for expression of the genes (*nifHDK* operon) encoding molybdenum nitrogenase; vanadium prevents expression of the third nitrogenase and is required for expression of the *vnfDGK* operon. Although the detailed mechanism by which metals regulate nitrogenase synthesis is not known, some of the genes that affect this regulation have recently been identified.

SELECTED REFERENCES

R. H. Burris, *J. Biol. Chem.* **266**, 9339-9342 (1991).

Fe-protein

J. Kim & D. C. Rees, *Science* **257**, 1677-1682 (1992).

Mo-Fe protein

J. Kim & D. C. Rees, *Nature* **360**, 553-560 (1992).

Fe-MoCo

R. A. Ugalde, J. Imperial, V. K. Shah & W. J. Brill, *J. Bacteriol.* **158**, 888-893 (1984).

Oxygen relations of nitrogen fixation in cyanobacteria

P. Fay, *Microbiol. Revs.* **56**, 340-373 (1992).

CHAPTER XVI

Biosynthesis of the aspartic acid family of amino acids

Biosynthesis of aspartic acid and asparagine

Aspartic acid can be synthesized by two different reactions, both involving a substrate that participates in the tricarboxylic acid cycle. It can be obtained by transamination of oxaloacetic acid with glutamic acid. The biosynthesis of glutamic acid will be examined later:

$$COOH\text{–}CH_2\text{–}CO\text{–}COOH + COOH\text{–}CH_2\text{–}CH(NH_2)\text{–}COOH \rightleftharpoons COOH\text{–}CH_2\text{–}CH(NH_2)\text{–}COOH + COOH\text{–}CH_2\text{–}CH_2\text{–}CO\text{–}COOH$$

Of the different transaminases found in *E. coli* by Rudman and Meister, only two can synthesize aspartate. The major one has a high affinity for oxaloacetate and a low affinity for the keto analogues of phenylalanine and tyrosine. It is specified by the *aspC* gene (21 mn), and is a homodimer whose subunit has a molecular weight of 43,573. The other, coded by *tyrB* (92 mn), has a low affinity for oxaloacetate; furthermore, its synthesis is repressed by tyrosine.

A double mutation in *aspC* and *tyrB* causes a double requirement for aspartate and tyrosine (the theoretical phenylalanine requirement is met by another transaminase).

One would have thought that aspartate could also have been synthesized by aspartase, which catalyzes a direct amination of fumarate by ammonia:

$$COOH\text{–}CH\text{=}CH\text{–}COOH + NH_3 \rightleftharpoons COOH\text{–}CH_2\text{–}CH(NH_2)\text{–}COOH$$

However, an aspartate auxotroph lacking the *aspC* and *tyrB* gene products is not deficient in aspartase. It may be that the intracellular concentration of fumarate is too low for aspartate synthesis in a synthetic growth medium, unless fumarate is provided in the medium.

In *E. coli*, asparagine auxotrophy results from two mutations in two unlinked genes, *asnA* (84 mn) which encodes an ammonia-dependent asparagine synthetase A, and *asnB* (16 mn) which specifies an enzyme (asparagine synthetase B) in which the obligatory amino donor is glutamine. Whereas the presence of the ammonia-dependent amidation is restricted to procaryotes, the glutamine-dependent amidation is probably universally distributed.

Both reactions require ATP, the products being asparagine, AMP and pyrophosphate (and glutamate, in the case where glutamine is the amino donor):

$$\text{aspartate} + \text{ATP} + NH_3 \rightarrow \underset{\text{L-asparagine}}{H_2N\text{–}CO\text{–}CH_2\text{–}CH(NH_2)\text{–}COOH} + \text{AMP} + \text{PPi}$$

Whereas the ammonia-dependent enzyme is a homodimer of subunit molecular weight 36,688 (330 amino acid residues), the glutamine-dependent enzyme is a homotetramer of subunit molecular weight 57,000 (554 residues). The latter exhibits aspartate-independent glutaminase activity, and in that respect resembles many glutamine-dependent amidotransferases.

Both enzymes are inhibited by asparagine; the synthesis of both enzymes is repressed by asparagine, due to the presence of the product of the *asnC* gene, adjacent to *asnA* but transcribed in the opposite direction and coding for a 17,000 molecular weight modulator polypeptide. The activation of *asnA* transcription by this product is inhibited in the presence of asparagine.

A highly-conserved protein motif, characteristic of a class of aminoacyl tRNA synthetases and containing an invariant arginine residue, was found to align with a region of asparagine synthetase A. Every substitution of this arginine by site-directed mutagenesis, including its replacement by a lysine, results in a loss of asparagine synthetase activity. The results are consistent with the possibility that asparagine synthetase A evolved from an ancestral aminoacyl tRNA synthetase.

Biosynthesis of aspartate semialdehyde, common intermediate in biosynthesis of lysine, methionine, threonine, and isoleucine

A specific β-aspartokinase catalyzes the phosphorylation of aspartic acid in the β-position[14]:

$$COOH\text{–}CH_2\text{–}CH(NH_2)\text{–}COOH + \text{ATP} \rightleftharpoons \underset{\beta\text{-aspartyl phosphate}}{H_2O_3P\text{–}CH_2\text{–}CH(NH_2)\text{–}COOH} + \text{ADP}$$

The β-aspartyl phosphate thus formed is an unstable compound, and when aspartokinase is to be studied, the reaction is performed in the presence of hydroxylamine; the hydroxamate produced reacts with ferric ions to produce a colored compound for which a rather insensitive colorimetric assay exists. When purified enzyme is used, and ATPase activity has been eliminated, the reaction can be coupled with phosphoenolpyruvate and pyruvate kinase, the ATP being regenerated. The pyruvate

14. *E.coli* synthesizes three different aspartokinases which will be examined separately in the next chapter, because of their importance in the regulation of the biosynthetic pathway of the amino acids derived from aspartate.

formed is then reacted with lactate dehydrogenase and the oxidation of NADPH followed spectrophotometrically.

The substrate specificity of aspartokinase I, one of the three isofunctional aspartokinases, has been investigated. Even though the integrity of the α-amino group is required, the amide or esters of the α-carboxyl group are competent alternative substrates.

The next step is the reduction of β-aspartyl phosphate to aspartate semialdehyde, catalyzed by aspartate semialdehyde dehydrogenase (coded for by the *asd* gene, at 76 mn on the *E. coli* chromosome):

$$H_2O_3\,POOC{-}CH_2{-}CH(NH_2){-}COOH + NADPH + H^+ \rightarrow OHC{-}CH_2{-}CH(NH_2){-}COOH + Pi + NADP^+$$

aspartate-β-semialdehyde

This reaction is reminiscent of that catalyzed by triose phosphate dehydrogenase, which also involves the reduction of an acyl phosphate group to an aldehyde. In both cases, the suggested mechanism involves the formation of a thioester intermediate resulting from the attack of a cysteine thiolate on the substrate carbonyl group, followed by hydride transfer to $NADP^+$. Subsequent attack on the thioester intermediate by an oxygen anion of bound inorganic phosphate leads to an expulsion of the cysteine thiolate and formation of the phosphorylated product.

Aspartate semialdehyde dehydrogenase has been purified: it is a dimer of 77,000 daltons composed of two identical subunits. Alkylation of a cysteine at the active site by 2-amino-4-oxo-5-chloropentanoate, a substrate analogue, irreversibly inactivates the enzyme; aspartate semialdehyde protects against the inactivation. Substitution of this cysteine with an alanine by site-directed mutagenesis results in complete loss of enzyme activity. However, changing this cysteine to a serine yields a mutant enzyme with a maximum velocity that is only 0.3% that of the native enzyme, but which has retained the same affinity for substrates as the native enzyme, and the same overall conformation. The dehydrogenase is also inactivated by the coenzyme analogue chloroacetylpyridine-ADP. $NADP^+$ and NADPH, but not the substrate, protect against the alkylation.

Mutants with an inactivated *asd* gene require diaminopimelate, lysine, methionine, and threonine for growth. This requirement is met if only diaminopimelate and homoserine are provided, since homoserine is the common precursor of methionine and threonine (see below).

Biosynthesis of lysine from aspartate semialdehyde in bacteria

Aspartate semialdehyde condenses with one molecule of pyruvate to eliminate two molecules of water. The reaction product is dihydrodipicolinic acid:

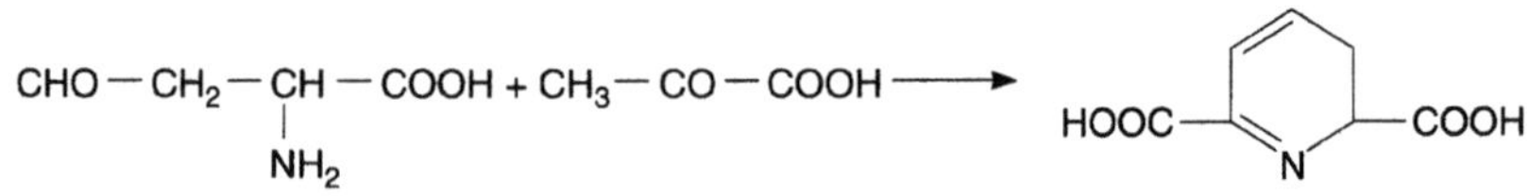

dihydrodipicolinate

The enzyme catalyzing this reaction, dihydropicolinate synthase, is coded by the *dapA* gene at 53 mn on the *E. coli* chromosome. It has been purified to homogeneity. The native protein has a molecular weight of 134,000; its subunit structure is not known.

The enzyme is inhibited by lysine; the lysine inhibition is cooperative and non-competitive toward both substrates.

Dihydrodipicolinate reductase (coded by *dapB* at 0.5 mn) has a native apparent molecular weight of 115,000. The *dapB* gene has been sequenced and codes for a 273 residues polypeptide of 28,798 daltons, suggesting that the enzyme is a tetramer. The product of the enzyme is tetrahydrodipicolinate. The reducing agent is NADPH.

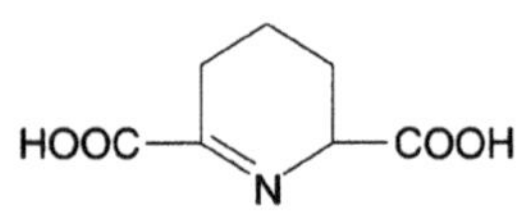

tetrahydrodipicolinate

The heterocyclic ring is then opened with a concomitant succinylation by tetrahydrodipicolinate succinylase (*dapD* 3.5 mn) in the presence of succinyl CoA to form N-succinyl-ε-keto-L-α-aminopimelic acid:

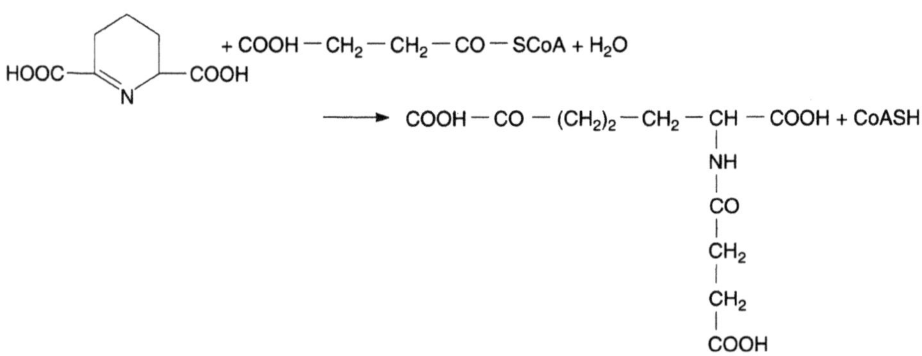

N-succinyl-ε-keto-L-α-aminopimelate

The nucleotide sequence of the gene has been determined: it encodes a polypeptide of 274 residues.

The succinylase has not been studied in detail.

The aminotransferase leading to succinyldiaminopimelate has been partially purified and characterized. Some auxotrophs, simultaneously requiring lysine and methionine, can grow on succinate in the absence of the amino acids. They have been identified as *suc* mutants with impaired synthesis of succinyl coenzyme A, which participates in lysine and methionine biosyntheses through succinylation reactions.

N-succinyldiaminopimelate desuccinylase has been partially purified. It requires Co^{2+} for activity. The enzyme is encoded by *dapE* at 52.5 mn on the *E. coli* chromosome, and its product is LL-diaminopimelate.

$$COOH{-}CH(NH_2){-}(CH_2)_3{-}CH(NH_2){-}COOH$$

diaminopimelate (DAP)

Diaminopimelate epimerase (*dapF*) catalyzes the interconversion of LL- and *meso*-diaminopimelate. Mutants of this gene can grow on *meso*-, but not on LL-diaminopimelate.

Finally, diaminopimelate decarboxylase (*lysA*, 61 mn) catalyzes the conversion of DAP to lysine.

$$\underset{meso\text{-DAP}}{COOH{-}CH(NH_2){-}(CH_2)_3{-}CH(NH_2){-}COOH} \rightarrow \underset{\text{L-lysine}}{NH_2{-}(CH_2)_4{-}CH(NH_2){-}COOH} + CO_2$$

The enzyme is specific for *meso*-DAP; the LL-isomer is not a substrate. The purified protein has a molecular weight of 200,000, as calculated from its sedimentation coefficient. Mutants devoid of the decarboxylase accumulate large amounts of DAP. Some *lysA* mutants alter the affinity of the decarboxylase for its coenzyme, pyridoxal phosphate, and require lysine or pyridoxine for growth. The nucleotide sequence of *lysA* indicates that it encodes a polypeptide of 420 residues (MW = 46,099) and suggests that the native decarboxylase is a tetramer.

Synthesis of dipicolinic acid, a substance present in the spores of Gram-positive bacilli

The structural events associated with spore formation have been elucidated by a combination of light and electron microscopy observations.

At the shift from exponential vegetative growth to the linear growth that just precedes the onset of sporulation, each cell contains two nuclear bodies which coalesce to form an axial chromatin thread. Then a transverse septum separates the cytoplasm and the DNA of the smaller cell (the future spore) from the rest of the cell. The smaller cell then becomes engulfed within the cytoplasm of the larger cell to produce a forespore. The cell is then committed to undergo sporulation and there is a rapid synthesis and deposition of new structures: the cortex and the spore coat that enclose the forespore. The maturing spore becomes refractile, but is still not heat-resistant. The development of heat-resistance closely follows the massive uptake of Ca^{2+} ions by the sporulating cell and the synthesis in large amounts of dipicolinic acid, a compound absent in vegetative cells.

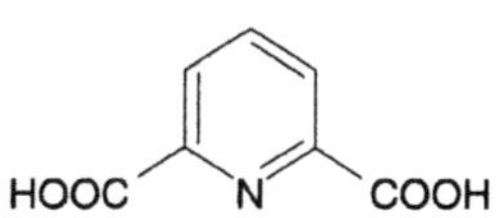

dipicolinic acid

In mature spores, the molar ratio dipicolinate/Ca^{2+} is close to unity and dipicolinate actually occurs as a calcium chelate. This compound represents 10 to 15% of the spore dry weight and is located within the spore. The synthesis of dipicolinic acid represents an offshoot of the lysine branch of the aspartate family pathway, at the dihydrodipicolinate level:

$$\begin{array}{c} \text{dihydrodipicolinic acid} \rightarrow \text{dipicolinic acid} \\ \downarrow \\ \text{lysine} \end{array}$$

Reduction of aspartate semialdehyde to homoserine, common precursor of methionine and threonine

An enzyme called homoserine dehydrogenase catalyzes the reduction of the aldehyde group of aspartate semialdehyde to a primary alcohol[15]. Whereas all the aspartic semialdehyde dehydrogenases are specific for NADPH, the homoserine dehydrogenases from different bacteria can use either NADH or NADPH. Depending on the source, one or the other of the pyridine nucleotides is the more active:

$$\text{COOH–CH(NH}_2\text{)–CH}_2\text{–CHO + NADPH + H}^+ \rightarrow \underset{\text{homoserine}}{\text{COOH–CH(NH}_2\text{)–CH}_2\text{–CH}_2\text{OH}} + \text{NADP}^+$$

The affinity label 2-amino-4-oxo-5-chloropentanoate inactivates homoserine dehydrogenase I. Homoserine and aspartate semialdehyde protect against this inactivation, caused by the covalent binding of one equivalent of the analogue per subunit.

Biosynthesis of methionine from homoserine

Apart from its role as protein constituent, methionine participates in the initiation of protein synthesis (as N-formylmethionine), as a universal methylation agent (as S-adenosylmethionine, SAM), and as a precursor of spermidine through SAM.

The methionine biosynthetic pathway is complex. The carbon skeleton derives from aspartate, the sulfur atom from cysteine, and the methyl group from the β-carbon of serine (fig. 1).

The first step of methionine biosynthesis is catalyzed by homoserine succinyltransferase, product of the *metA* gene (90.5 mn), which, in bacteria, transforms homoserine into O-succinylhomoserine in the presence of succinyl coenzyme A[16]:

15. In the next chapter, we shall see that there exist two distinct homoserine dehydrogenases in the *Enterobacteriaceae* and shall study their structure and regulation.
16. In yeast and *Neurospora crassa*, the acylation reaction is an acetylation, catalyzed by homoserine acetyltransferase.

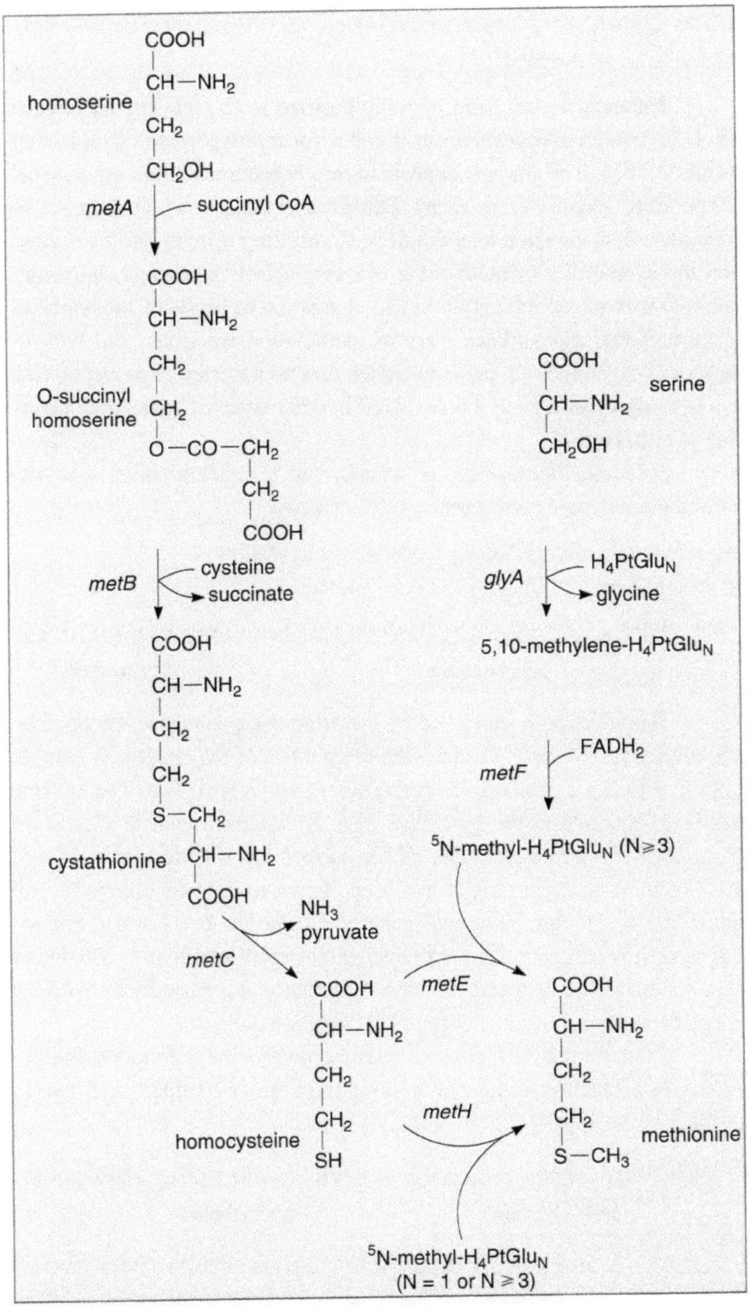

Figure 1. Biosynthesis of methionine from homoserine and serine in *Enterobacteriaceae*. Relevant enzymes and localization of the genes on the *E. coli* circular chromosome are given in the text. ^{5}N-methyl-H_4PtGlu stands for ^{5}N-methyl-tetrahydropteroylglutamate. The product of the *metH* gene can use the mono- or polyglutamate forms, whereas the product of the *metE* gene can only use the polyglutamate form of ^{5}N-methyl-H_4PtGlu as substrate.

$$COOH–CH(NH_2)–CH_2–CH_2OH + \text{succinyl CoA} \rightarrow \text{O-succinylhomoserine} + CoA$$

homoserine

The enzyme has been partially purified in *E. coli.* The *metA* gene has been cloned and sequenced; it codes for a polypeptide of molecular weight 35,673. The enzyme appears to be a homodimer, showing unusual temperature sensitivity in many *Enterobacteriaceae,* which appears to limit growth at elevated temperatures. Controlling growth rate by regulating the availability of methionine is a very effective strategy, since methionine is required for initiation and elongation of proteins, biosynthesis of purines and pyrimidines, various methylation reactions, and biosynthesis of polyamines. There is evidence that homoserine transsuccinylase is a heat-shock protein and is involved in other types of stress than elevation of temperature.

O-succinylhomoserine is transformed to cystathionine, a seven-carbon diaminoacid, in the presence of cysteine:

$$COOH–CH(NH_2)–CH_2–CH_2O–CO–(CH_2)_2–COOH + COOH–CH(NH_2)–CH_2SH \rightarrow$$

O-succinylhomoserine — cysteine

$$COOH–CH(NH_2)–CH_2–CH_2–S–CH_2–CH(NH_2)–COOH + HOOC–CH_2–CH_2–COOH$$

cystathionine[17] — succinic acid

This reaction is catalyzed by cystathionine-γ-synthase, encoded by the *metB* gene (89 mn). The reaction mechanism of the enzyme is thought to proceed via a pyridoxamine derivative of vinylglyoxylate. The enzyme has been obtained in the pure state in *S. typhimurium* and in *E. coli.* In both cases it is a homotetramer whose subunit has a molecular weight of 40,000; in *S. typhimurium,* it has been shown to contain four pyridoxal phosphate molecules. The *metB* gene of *E. coli* has been cloned and sequenced; it is composed of 1,158 nucleotides corresponding to 386 amino acid residues. The deduced N-terminal sequence agrees perfectly with the experimentally determined N-terminal protein sequence.

Cystathionine is cleaved to give homocysteine, ammonia, and pyruvate in a reaction catalyzed by cystathionase (cystathionine-β-lyase), product of the *metC* gene (89 mn).

$$COOH–CH(NH_2)–(CH_2)_2–S–CH_2–CH–COOH \rightarrow COOH–CH(NH_2)–(CH_2)_2–SH + \text{serine}$$

cystathionine — homocysteine

This is also a pyridoxal phosphate enzyme that has been purified in *E. coli* and *S. typhimurium.* The *E. coli* enzyme has been purified from a strain harboring a multicopy plasmid carrying the *metC* gene. Its pH optimum, substrate specificity, and kinetic parameters have been studied; 3,3,3-trifluoroalanine binds covalently to the enzyme and inhibits it irreversibly. The protein was, at first, reported to be a hexamer composed of

17. Note that cystathionine presents a second asymmetric center.

six identical subunits of MW = 45,000, each binding a molecule of pyridoxal phosphate. However, further studies suggest that the enzyme is a tetramer. The enzyme from *S. typhimurium* was not as well characterized. The *E. coli metC* gene was cloned and its nucleotide sequence determined; the deduced sequence of the protein (395 residues), supported by the experimentally determination of the first ten residues, shows strong homology with that of cystathionine-γ-synthase and thus points to a common ancestor for the two proteins (see section on protein evolution).

A mutation, *metQ* (39.5 mn), allows *metC* mutants to directly catalyze the formation of homocysteine from homoserine, thus bypassing the cystathionine normal intermediate.

The last step in methionine biosynthesis involves the methylation of homocysteine.

$$COOH\text{–}CH(NH_2)\text{–}(CH_2)_2\text{–}SH + \mathbf{R}\text{–}CH_3 \rightarrow COOH\text{–}CH(NH_2)\text{–}(CH_2)_2\text{–}CH\text{–}S\text{–}CH_3 + \mathbf{R}$$

The methyl donor **R** is ^{5}N-methyl tetrahydropteroylglutamate (N > 1), a member of the folate family, which is derived from ^{5}N-^{10}N-methylene tetrahydrofolate (generated by the conversion of serine to glycine)[18] by a reductase specified by the *metF* gene (89 mn). This reaction gives the methyl group specific for methionine biosynthesis. Two enzymes can catalyze this reaction: a transmethylase with an activity depending on the presence of vitamin B_{12} (product of the *metH* gene, 90.9 mn) which can use the mono- or polyglutamate form of ^{5}N-methyl tetrahydropteroylglutamate as methyl donor; and a vitamin B_{12}-independent transmethylase (product of *metE*, 85.5 mn) which can utilize as a substrate only the polyglutamate forms of ^{5}N-methyl tetrahydropteroylglutamate (N > 3). *E. coli* possesses both enzymes whereas most organisms possess only one. It does not synthesize vitamin B_{12} and uses one or the other transmethylase according to the availability of vitamin B_{12} in the medium. Strains with a mutation in *metE* require either methionine or vitamin B_{12} for their growth. A word of caution is necessary: vitamin B_{12} synthesis under anaerobiosis has been shown to occur. There is evidence that the vitamin B_{12}-independent enzyme, coded by *metE*, is a rather inefficient enzyme.

The mechanism of the vitamin B_{12}-dependent enzyme involves the methylation of the cobalt atom of the vitamin. Catalytic amounts of SAM are required for this priming step. After the methyl group is transferred to homocysteine, subsequent methyl groups come from ^{5}N-methyltetrahydrofolate. In addition to SAM, methyl B_{12}, and the folate derivative, the enzyme requires $FADH_2$ for activity.

The B_{12}-independent transmethylase involves neither a methylated coenzyme nor the other requirements of the vitamin-dependent en-

18. See chapter on the biosynthesis of amino acids deriving from phosphoglycerate and pyruvate.

zyme such as SAM or a reducing agent. Alkylation of Cys 726 results in complete loss of activity. It has been suggested that this thiol functions as an intermediate methyl acceptor in catalysis, analogous to the role of cobalamin in the reaction catalyzed by the vitamin B_{12}-dependent enzyme.

The sequence of *metE* has been determined, it has been cloned, and expressed as a polypeptide of 84654 Da. The sequence of the *metH* gene is known, it codes for an exceptionally long polypeptide of 132,628 Da showing no similarity with the *metE* product.

The methyl group transferred to homocysteine derives from the β-carbon of serine: the interconversion of serine and glycine involves a single enzyme, serine hydroxymethyltransferase, whose other substrate is tetrahydrofolate; this reaction also produces ^{5}N-^{10}N-methylenetetrahydrofolate, the substrate of the reductase coded by *metF*, which reduces it to ^{5}N-methyltetrahydrofolate which in turns transfers its methyl group to homocysteine by either of the two transmethylases.

S-adenosylmethionine (SAM) biosynthesis

S-adenosylmethionine synthetase catalyzes the only known route of biosynthesis of the intracellular alkylating agent S-adenosylmethionine and ATP:

$$\text{methionine} + \text{adenosine-}\underset{\alpha}{\text{P}}\text{–}\underset{\beta}{\text{P}}\text{–}\underset{\gamma}{\text{P}} \longrightarrow \text{S-adenosylmethionine} + \text{PPi} + \text{Pi}$$

This reaction is interesting because it is the phosphorus moieties in position α and β which are the source of the pyrophosphate produced. Before this discovery, it was believed that ATP reacted in three ways: 1) as phosphate donor to form ADP; 2) as pyrophosphate donor to form AMP; 3) as adenylate donor to form pyrophosphate. SAM synthetase synthesis showed, for the first time, that ATP can also react as an adenosine donor and that the ribose-phosphate bond is not as inert as had been thought.

The enzyme has been purified to homogeneity. It was found to have a molecular weight of 140 kDa and to be composed of four identical subunits. It is coded by the *metK* gene (64 mn), 1152 bp long, encoding a 384 residues polypeptide. In addition to the transferase reaction, the pure enzyme catalyzes a tripolyphosphatase reaction stimulated by SAM. The mechanism of the two reactions has been studied extensively, and the X-ray diffraction of the enzyme has been determined.

Several *E. coli metK* mutants have lower than normal levels of synthetase and increased levels of methionine biosynthetic enzymes. Since total inactivation of the *metK* gene results in residual enzyme activity, it seemed likely that *E. coli* possesses two different S-adenosylmethionine synthetases, as had been reported in *Saccharomyces cerevisiae*. Actually, a second enzyme has been found in *E. coli*, coded by the *metX* gene. Although antigenically related, the two enzymes can be distinguished. The *metX* gene is only expressed in rich media, while *metK* is only expressed in minimal medium.

The S-adenosylhomocysteine (SAH), formed as a by-product of SAM-dependent methylation reactions, is salvaged as the adenine base and homocysteine by a cyclical regenerative branch of the pathway, which has not been characterized genetically.

Biosynthesis of threonine from homoserine

Threonine is an isomer of homoserine in which the alcohol group is a secondary alcohol. The *thrB* gene (0 mn) encodes homoserine kinase, which catalyzes the phosphorylation of homoserine

$$\mathbf{R}\text{–}CH_2OH + ATP \rightarrow \mathbf{R}\text{–}CH_2O\text{–}PO_3H_2 + ADP + Pi$$

homoserine phosphate

where **R** = $COOH\text{–}CH(NH_2)\text{–}CH_2\text{–}$

The phosphate bond thus formed in homoserine phosphate is not energy-rich – unlike in aspartyl phosphate, where the phosphorylated group is a carboxyl. The native protein consists of two identical subunits of 29,000 daltons. The protein has been purified to homogeneity and requires Mg^{2+} for activity, which is stimulated by K^+. The entire nucleotide sequence of *thrB* is known and the deduced amino-acid sequence agrees with the first eight amino acids from the N-terminus determined by amino-acid sequencing.

The transformation of homoserine phosphate to L-threonine is catalyzed by threonine synthase, a pyridoxal phosphate enzyme:

$$CH_3\text{–}CHOH\text{–}CH(NH_2)\text{–}COOH$$

L-threonine[19]

The reaction mechanism of threonine synthase has been studied extensively with the enzyme from *Neurospora crassa.* It has been proposed that a water molecule is removed from homoserine, leading to vinylglycine, which is then rehydrated to threonine.

$$CH_2\text{=}CH\text{–}CH(NH_2)\text{–}COOH$$

vinylglycine

The positions where solvent hydrogen from water is introduced during the reaction have been determined. Threonine synthase is encoded by the *thrC* gene (0 mn), which has been cloned and sequenced. The first 40 N-terminal residues of the protein have also been determined by protein sequencing and are in perfect agreement with the sequence deduced from that of the gene. The molecular weight (47,000) of the polypeptide expressed in minicells, as well as the apparent molecular weight of the pure protein, (both determined by polyacrylamide gel electrophoresis under denaturing conditions) agree with the molecular weight deduced from the DNA sequence.

19. Threonine has two asymmetric centers.

Biosynthetic threonine dehydratase

Threonine is deaminated by a pyridoxal phosphate enzyme and yields α-ketobutyrate:

$$CH_3\text{–}CHOH\text{–}CH(NH_2)\text{–}COOH \rightarrow CH_3\text{–}CH_2\text{–}CO\text{–}COOH + NH_3$$

This enzyme (*ilvA*, 85 mn) is distinct from another enzyme, called degradative threonine dehydratase, which fulfills the same catalytic function but is not involved in isoleucine biosynthesis.

The *Salmonella typhimurium* enzyme has been purified; it consists of four apparently identical subunits, of a molecular weight of about 48,500. Only two of the subunits can bind pyridoxal phosphate. This half-of-the-sites reactivity has been encountered in several other cases.

Isoleucine biosynthesis

α-ketobutyrate condenses with a molecule of acetaldehyde derived from pyruvate to give α-aceto-α-hydroxybutyric acid, a reaction catalyzed by α-aceto-α-hydroxy acid synthase (*ilvB*, 82 mn):

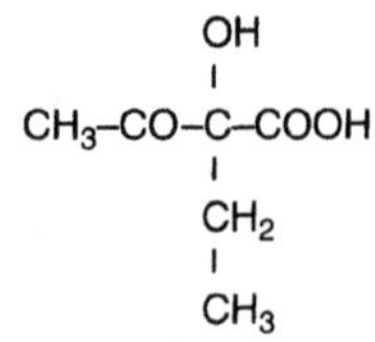

α-aceto-α-hydroxybutyric acid

This compound has the same number of carbon atoms as isoleucine but the carbon skeleton must be rearranged. This is brought about by an acetohydroxyacid isomeroreductase coded by *ilvC* (85 mn), which catalyzes a pinacol rearrangement accompanied by a reduction of the carbonyl group, leading to α,β-dihydroxymethylvaleric acid (trivial name, dihydroxyisoleucine):

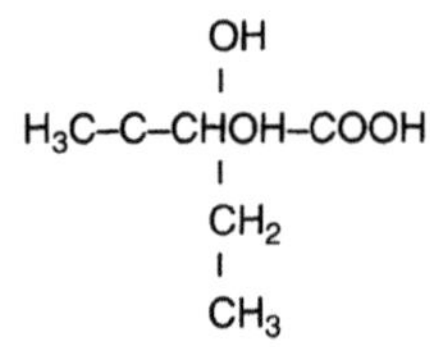

α,β-dihydroxy-β-methylvaleric acid

The *ilvD* gene (85 mn) specifies a dihydroxyacid dehydratase which catalyzes the loss of a molecule of water and the α-keto acid corresponding to isoleucine is produced:

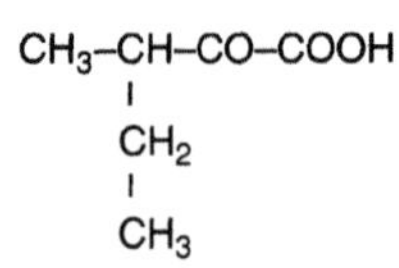

α-keto-α-methylvaleric acid (α-ketoisoleucine)

The last step of isoleucine synthesis is due to a transamination between α-ketoisoleucine and glutamic acid, catalyzed by a transaminase coded by the *ilvE* gene (85 mn):

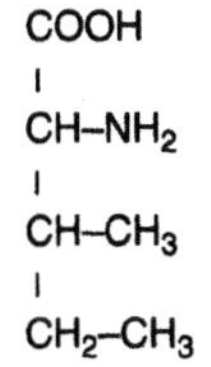

L-isoleucine

Summary of biosynthetic pathway of the aspartate family of amino acids

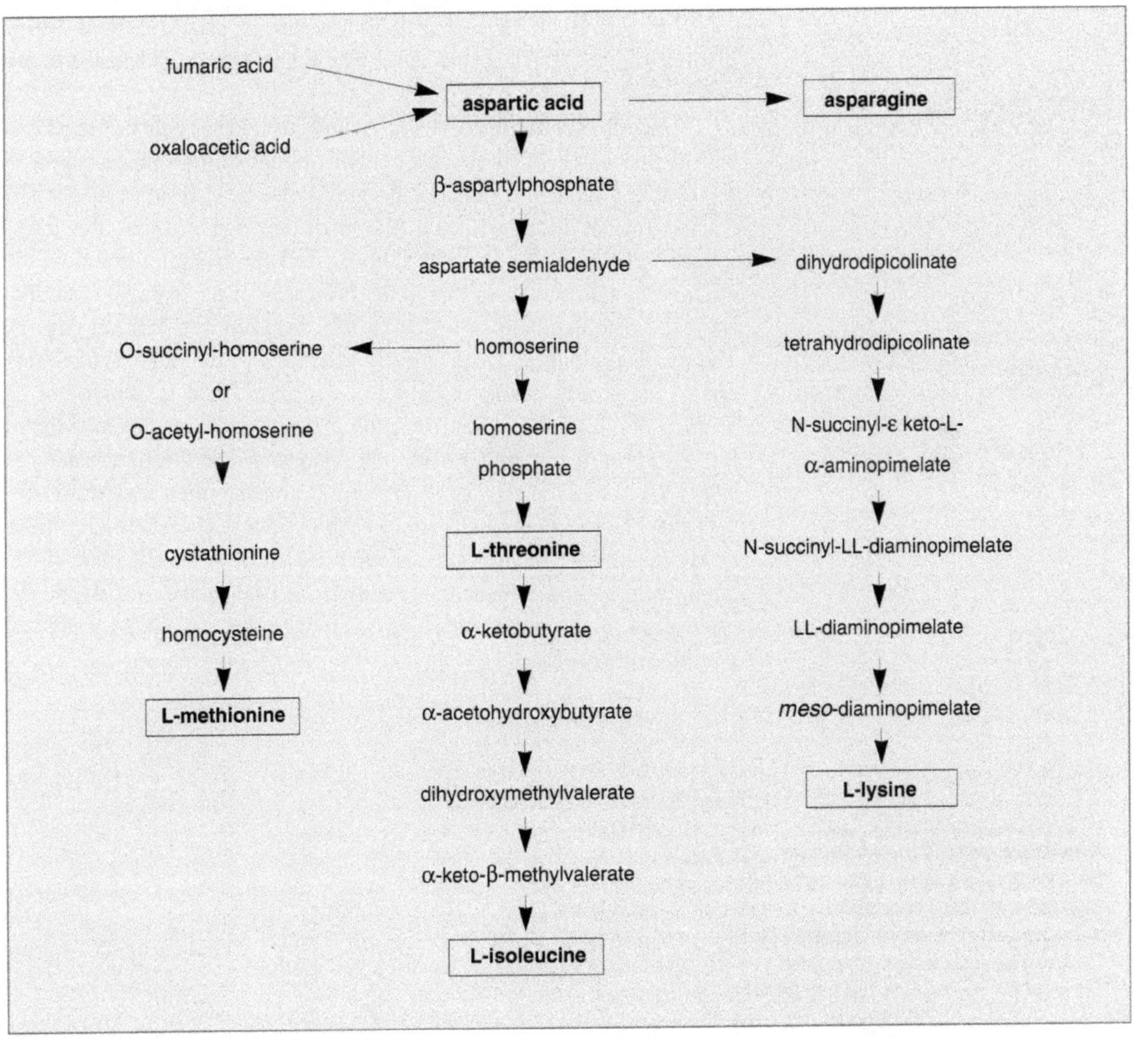

Remarks on some reactions encountered in this chapter

We have met with some interesting reactions during this study:

1) The phosphorylation of a carboxyl group, catalyzed by aspartokinase; this produces an energy-rich phosphate ester, whose reduction to an aldehyde is thus greatly facilitated.

2) The formation of intermediate cyclic compounds, the dihydro- and tetrahydrodipicolinic acids.

3) The protection of an amino group by succinylation to avoid an inopportune spontaneous cyclization.

4) Reactions requiring pyridoxal phosphate: transaminations (formation of N-succinyl-DAP and isoleucine), isomerization (formation of meso-DAP from the LL-isomer, threonine synthase), decarboxylation (DAP-decarboxylase), and deamination (threonine dehydratase).

It may be of interest to look at some of these reactions which involve pyridoxal phosphate as cofactor and whose mechanisms are understood.

Pyridoxal phosphate enzymes catalyze a wide variety of reactions with amino acids. Each enzyme contains pyridoxal phosphate bound as an internal aldimine to the ε-amino group of a lysyl residue. The initial mechanism for each pyridoxal phosphate enzyme involves the transimination between the internal aldimine and the α-amino group of the amino acid substrate, resulting in the formation of the external aldimine-enzyme-substrate complex which absorbs between 420-430 nm (fig. 2). The external aldimine is specifically oriented in each enzyme such that one of the groups on the α-carbon of the amino acid is released as an electrophile, resulting in the formation of a resonance-stabilized anion of the α-carbon of the amino acid. This resonance-stabilized carbanion is called the quinonoid complex and exhibits an absorption maximum near 500 nm. The reaction specificity of each pyridoxal phosphate enzyme is determined by the nature of the group released from the α-carbon of the amino acid substrate. Reaction specificity is determined, in part, by the orientation of the bound substrate relative to the conjugated pyridoxal phosphate ring system.

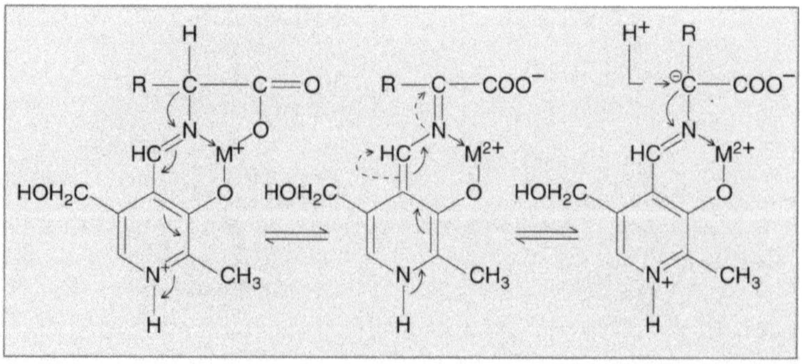

Figure 2. Proposed mechanism for the pyridoxal phosphate and metal ion (M) catalyzed racemization of amino acids. In the second formula, it can be seen that the α-carbon atom of the amino acid has lost its asymmetry.

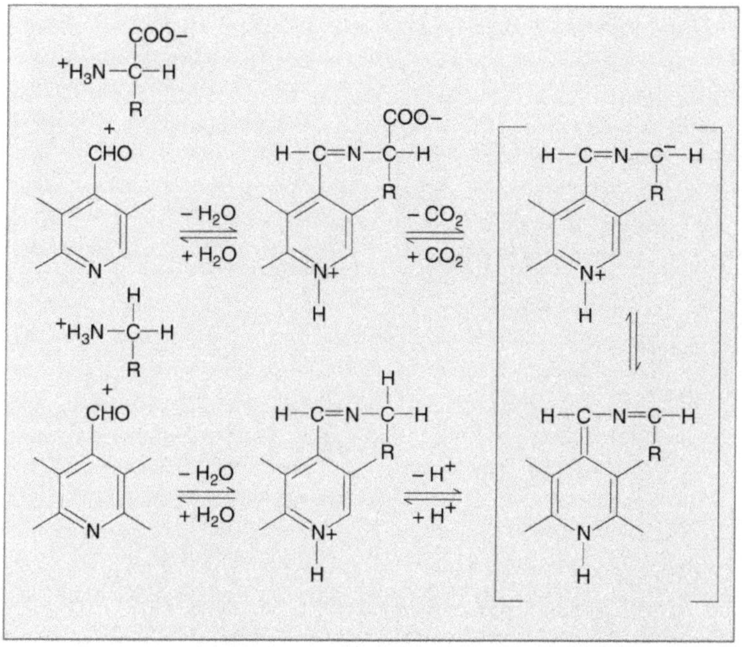

Figure 3. Proposed mechanism for amino acid decarboxylation catalyzed by pyridoxal.

Figure 3 indicates the mechanism postulated for amino acid racemization and decarboxylation. These mechanisms, and the other mechanisms of pyridoxal-phosphate catalyzed enzyme reactions, are examined in detail in the book by Meister referred to in the selected references below.

SELECTED REFERENCES

To be consulted for further information on the topics treated in this chapter and the following (up to, and including, chapter XXII):

A. Meister, *Biochemistry of the Amino Acids* (2nd ed.), vols 1 & 2. Academic Press, New York (1965).

Amino Acids: Biosynthesis and Genetic Regulation (K. M. Herrmann & R. Somerville eds.), 453 p. Addison-Wesley Publishing Company, Reading, Mass. (1983).

Escherichia coli and *Salmonella typhimurium. Cellular and Molecular Biology.* (J. L. Ingraham, K. B. Low, B. Magasanik, M. Schachter & H. E. Umbarger eds.) American Society for Microbiology, Washington, D.C. (1987).

CHAPTER XVII

Regulation of the biosynthesis of the amino acids of the aspartic acid family in *Enterobacteriaceae*

Cells have two main means at their disposal for regulating their metabolic reaction rates:

1) They can use small molecules for certain target enzymes, which, as a rule, belong to a certain class of proteins, the allosteric enzymes. These ligands bring about conformational changes which sometimes result in a lower enzyme *activity* (allosteric inhibition) and sometimes in a higher (allosteric activation).

2) The presence of essential metabolites in the cytoplasm at a certain concentration causes the repression, at the level of *transcription*, of the biosynthesis of enzymes leading to the biosynthesis of such metabolites.

I. A PARADIGM OF ISOFUNCTIONAL AND MULTIFUNCTIONALENZYMES AND OF ALLOSTERIC EQUILIBRIUM

We have seen in the preceding chapter that the biosynthetic pathway leading from aspartic acid to diaminopimelic acid and lysine on the one hand, and to methionine, threonine and isoleucine on the other, is not a linear sequence of reactions, but a highly branched chain. Let us look at the implications of this fact for growing bacterial cells.

a) Consider a linear sequence of reactions:

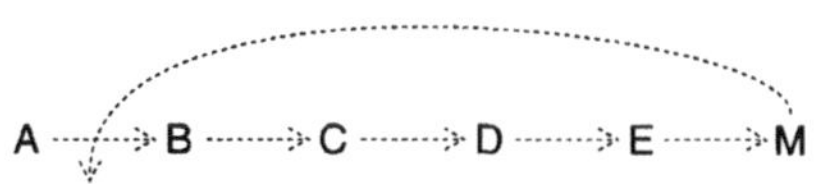

leading to the synthesis of the essential metabolite M. In the presence of excess M in the intracellular pool, the enzyme carrying out the conversion A $\rightarrow$ B will be inhibited by M if, as is generally the case, this enzyme is an allosteric protein. The presence of M may also repress the synthesis of the enzymes of the whole chain. These control phenomena, which are economical, do not cause any detrimental side effects, because the final product M is in excess; when production of M falls sufficiently, the activity of the first enzyme will increase and the synthesis of the enzymes of the chain will take place at a rate which is greater than under the conditions of repression. This new state will hold until M again reaches a concentration high enough to cause inhibition and repression. Oscillations in the size of the pool are therefore expected and have been observed experimentally in several instances.

b) Now consider a branched reaction sequence such as that of the aspartate family of amino acids:

aspartate
↓
aspartyl phosphate
↓
aspartate semialdehyde → lysine
↓
homoserine → threonine → isoleucine
↓
↓
↓
methionine

Examination of the diagram of the reaction sequence shows that accumulation of any one of the end products and control by either of the above regulatory processes would reduce the production of aspartyl phosphate consecutive to the decreased aspartokinase activity and consequently to a deficiency in the production of the other end products.

The solution of this potential dilemma differs according to the organism studied. Let us first consider the case of the *Enterobacteriaceae*, in particular that of *Escherichia coli.*

Two aspartokinases in *E. coli*

Of all the amino acids known to occur naturally as protein constituents, only two, L-lysine and L-threonine, were found to influence the aspartokinase activity of crude extracts. Neither of the two amino acids

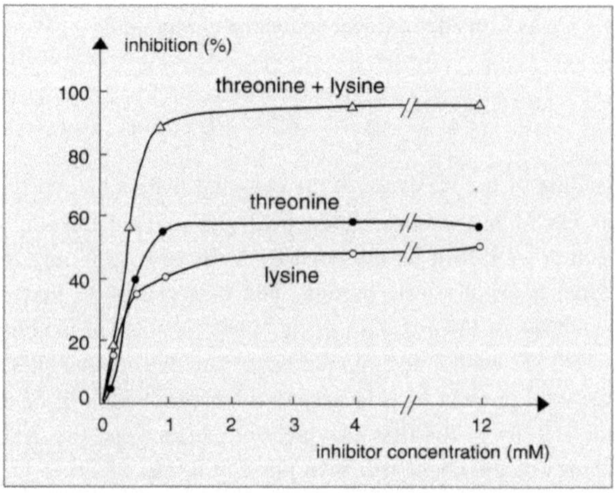

Figure 1. The inhibitory effects of L-lysine, of L-threonine, or both, on aspartokinase activity. (From E. R. Stadtman, G. N. Cohen, G. Le Bras & H. de Robichon-Szulmajster, *J. Biol. Chem.* **236**, 2033-2038 (1961). With permission from the *Journal of Biological Chemistry*.)

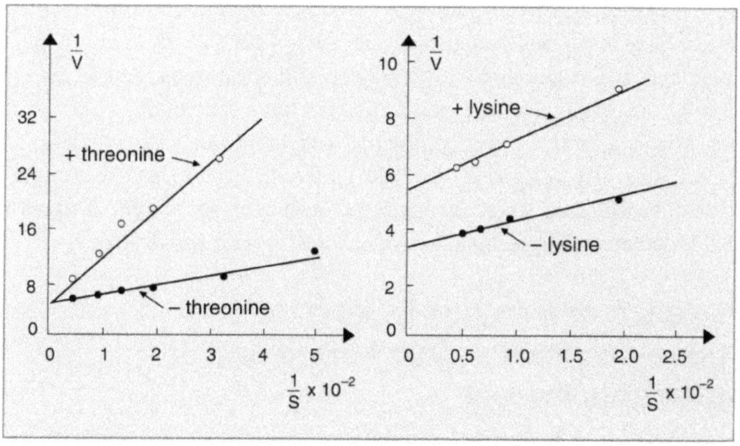

Figure 2. Inhibition analysis of the lysine- and threonine-sensitive aspartokinases.
(From E. R. Stadtman, G. N. Cohen, G. Le Bras & H. de Robichon-Szulmajster, *J. Biol. Chem.* **236**, 2033-2038 (1961). With permission from the *Journal of Biological Chemistry*.)

totally inhibit the activity but, as shown in figure 1, when they are added simultaneously the total inhibition is the sum of that observed for each independently. This suggests that crude extracts contain at least two aspartokinases and this interpretation is borne out by the physical separation of the two enzyme activities by ammonium sulfate fractionation.

A kinetic analysis of the inhibitions *vs.* aspartate concentrations shows that the threonine sensitive enzyme is competitively inhibited whereas the inhibition by lysine of the other kinase is non-competitive (fig. 2)

The synthesis of the lysine-sensitive enzyme is repressed when the organism is grown in the presence of lysine: only the threonine-sensitive enzyme is present in the extracts. On the other hand, the synthesis of the threonine-sensitive enzyme is subject to a multivalent repression by threonine and isoleucine.

The fact that a single organism may contain more than one enzyme catalyzing the same biochemical reaction was not an uncommon finding at the time (for example, the biosynthetic and degradative threonine dehydratases). The discovery, however, of the existence of two aspartokinases in *E. coli* was of special significance because it offered a reasonable explanation for the multiplicity of enzymes. The rational solution to the dilemma posed by the regulation of branched biosynthetic pathways was the synthesis of multiple, isofunctional enzymes which catalyze the synthesis of the common precursor, aspartyl phosphate, each enzyme independently subject to feedback inhibition and to repression by different end-product metabolites.

Threonine-sensitive homoserine dehydrogenase of *E. coli*

This enzyme, which catalyzes the third reaction between aspartate and homoserine, is inhibited very specifically by L-threonine. The inhibition is non-competitive versus both substrates, homoserine and NADPH. L-threonine is the best stabilizing agent, along with KCl. By using an inhibitor analogue, α-amino-β-hydroxyvaleric acid, mutants resistant to inhibition by threonine have been selected which have been useful in mapping the structural gene for this homoserine dehydrogenase. The synthesis of homoserine dehydrogenase, like that of the threonine-sensitive aspartokinase, is repressed when threonine and isoleucine are present in excess, and is derepressed upon limitation of either of the two amino acids.

Isolation of a mutant lacking the lysine-sensitive aspartokinase and of revertants thereof

The presence of several aspartokinases makes it awkward to apply the penicillin selection method for isolating directly mutants having lost one or the other aspartokinase. The principle of the selection is easy to understand: in the presence of threonine and isoleucine, homoserine synthesis is repressed and its activity is inhibited by threonine; as a result, growth is severely inhibited due to the artificial methionine limitation thus created[20]. Now if, among the mutagenized population, there are some individuals who have lost the lysine-sensitive aspartokinase activity, since the remaining kinase has its synthesis repressed and its activity inhibited, these individuals will have no means of synthesizing their diaminopimelate and lysine: they will not grow and will not be killed by penicillin. Such an individual has been isolated; its phenotype is that of a prototroph growing perfectly on minimal medium, but exhibiting a conditional requirement for diaminopimelate in the presence of threonine, isoleucine, and methionine. The defect causing this phenotype was localized on the *E. coli* chromosome.

Spontaneous revertants of this mutant were selected on the basis of their capacity to grow on threonine plus isoleucine plus methionine. Some of them were heavy threonine excretors. They had not reacquired the lysine-sensitive aspartokinase, but the other kinase was resistant to end-production inhibition by threonine, thus permitting a more rapid flow of aspartyl phosphate and aspartic semialdehyde, and a normal rate of synthesis of DAP and lysine. Because they overproduce threonine, the levels of aspartokinase and of homoserine dehydrogenase are at the repressed value, precluding the easy study of these enzymes. Therefore, the part of their genome responsible for the excretion character was transduced into an Ile^- strain, in order to enable derepressed synthesis in a chemostat limited by isoleucine.

It became immediately apparent (fig. 3) that the revertants possess a modified threonine-sensitive aspartokinase: a concentration of threonine which causes a complete inhibition in the wild-type extracts causes

20. If methionine is added, normal growth occurs.

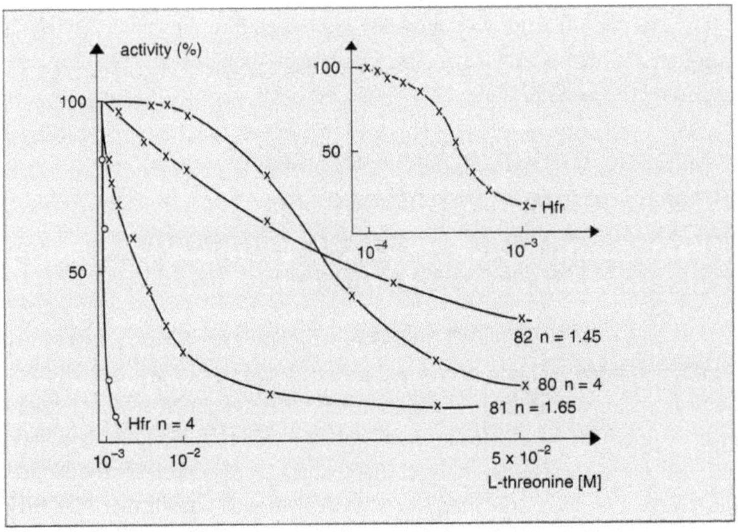

Figure 3. Inhibition by threonine of threonine-sensitive aspartokinases. The inset is an enlargement of the wild-type (Hfr) curve.
(From J. C. Patte, P. Truffa-Bachi & G. N. Cohen, *Biophysica and Biochimica Acta* **128**, 426-439 (1966). With permission from Elsevier Science Publishers.)

either no inhibition or an incomplete inhibition, according to the mutant extract tested.

A surprising fact was then uncovered: the homoserine dehydrogenase of the revertants was *also* modified with respect to its capacity for being inhibited by threonine (fig. 4). The apparent substrate affinity constants of the two enzymes obtained from the revertants do not differ significantly from those obtained in the wild type. The fact that the two activities are modified as a result of a single mutation prompted the reexamination of a mutant where feedback-resistance of the threonine-sensitive aspartokinase had been obtained by resistance to α-amino-β-hydroxyvalerate: its homoserine dehydrogenase was also resistant to inhibition by threonine.

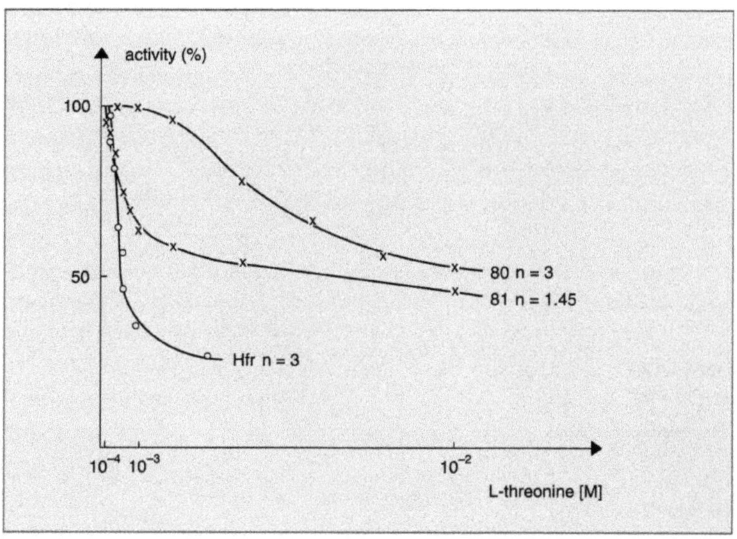

Figure 4. Inhibition by L-threonine of homoserine dehydrogenases from the wild type and from mutants.
(From J. C. Patte, P. Truffa-Bachi & G. N. Cohen, *Biophysica and Biochimica Acta* **128**, 426-439 (1966). With permission from Elsevier Science Publishers.)

The fact that the two enzymes are affected in the same way by a single mutational event led to the inescapable conclusion that they are carried by the same protein.

Evidence that threonine-sensitive aspartokinase and homoserine dehydrogenase of *E. coli* are carried by the same bifunctional protein

Observation provided indirect evidence that kinase activity is weakly inhibited by homoserine and $NADP^+$, and that dehydrogenase activity is notably inhibited by aspartate and ATP. A much stronger argument is provided by studies on heat stability: whereas NADPH is a very efficient protector against thermal inactivation, it also protects the kinase for which it is not a substrate. As a control, it was shown that NADPH does not protect the lysine-sensitive aspartokinase (efficiently protected by L-lysine) against thermal inactivation, although it catalyzes exactly the same reaction as the threonine-sensitive aspartokinase.

The single protein carrying the two activities was purified to homogeneity by standard procedures. It has a molecular weight of 360,000 as determined by ultracentrifugation and light-scattering studies. It is composed of four subunits as determined by cross linking. Gel filtration and isoelectric focusing showed that the subunits have an identical molecular weight and are identically charged. Finally, sequence studies showed that the subunits are identical. The sequence of the polypeptide chain was determined by protein sequencing and by deduction from the sequence of the cloned gene *thrA*. For reasons that will become evident later in this chapter, the *bifunctional* protein was called aspartokinase I-homoserine dehydrogenase I (AKI-HDHI).

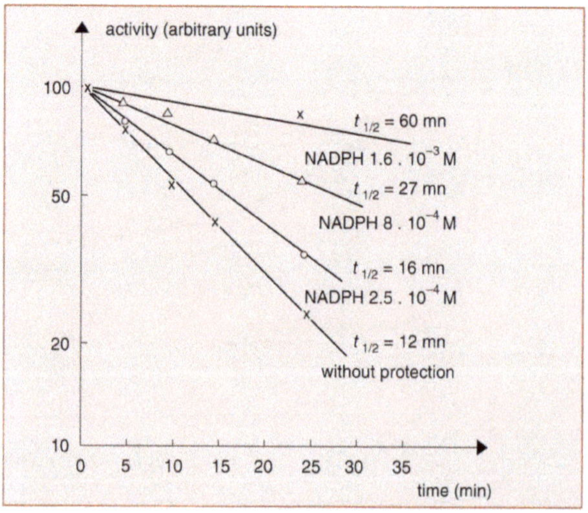

Figure 5. Protection against thermal inactivation of the threonine-sensitive aspartokinase activity by NADPH. Temperature: 44.8 °C.
(From J. C. Patte, P. Truffa-Bachi & G. N. Cohen, *Biophysica and Biochimica Acta* **128**, 426-439 (1966). With permission from Elsevier Science Publishers.)

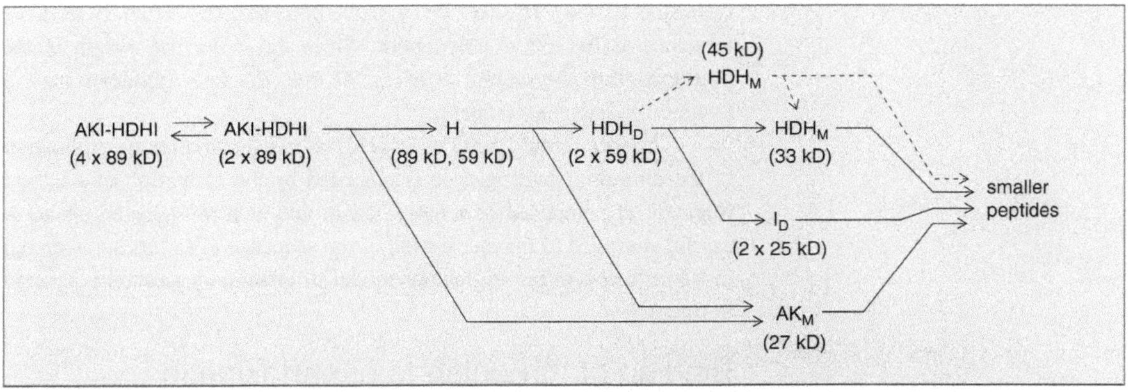

Figure 6. Pathway of the proteolysis of AKI-HDHI. The native tetrameric AKI-HDHI is dissociated into a dimer. The dimer is cleaved into the hybrid (H) and the monomeric aspartokinase fragment (AK_M). The intact chain of the dimer is then proteolyzed to a second AK_M fragment and the homodimeric fragment HDH_D is obtained. The next step in the proteolysis is the cleavage of HDH_D to produce monomeric homoserine dehydrogenase (HDH_M) and an inactive dimeric fragment (I_D). AK_M, HDH_M and I_D are further cleaved to give smaller fragments of about 10-12 kilodaltons.

(From A. Fazel, K. Müller, G. Le Bras, J. R. Garel, M. Véron & G. N. Cohen, *Biochemistry* **22**, 158-165 (1983). With permission from the American Chemical Society.)

When AKI-HDHI is subjected to mild proteolysis, the kinase is destroyed. A fragment carrying only the homoserine dehydrogenase activity resists proteolysis and was purified. It has the same carboxy-terminal sequence as the native polypeptide. This fragment is a homodimer whose subunit has a molecular weight of *ca.* 58,000 as opposed to 89,000 for the subunit of the native protein. The dehydrogenase activity of the dimer is totally insensitive to threonine. On the other hand, a tetrameric threonine-sensitive aspartokinase fragment was purified from a

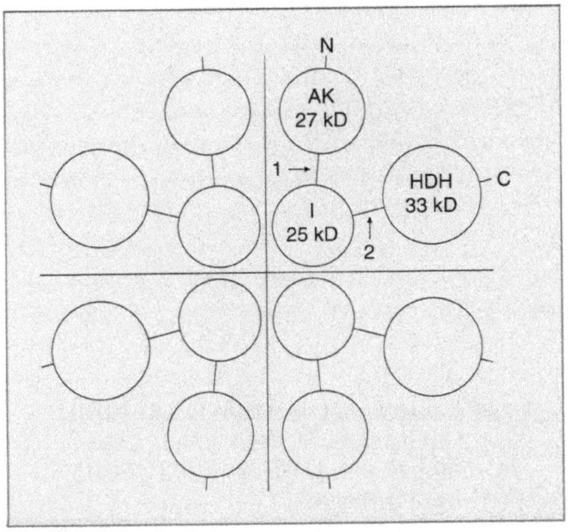

Figure 7. Domain I is depicted as responsible for the subunit contacts which generate the dimeric structure of H and HDH_D, and possibly the tetrameric structure of the native enzyme. The actual polymeric fragments obtained by limited proteolysis or by extraction from the *ochre* mutant (see text) can be visualized as the sum of I and of either the two catalytically active fragments, yielding respectively the corresponding dimer (HDH_D) or tetramer (AK from the *ochre* mutant). Arrows 1 and 2 indicate the sites of the corresponding proteolytic or mutational cleavages.

(From A. Fazel, K. Müller, G. Le Bras, J. R. Garel, M. Véron & G. N. Cohen, *Biochemistry* **22**, 158-165 (1983). With permission from the American Chemical Society.)

nonsense *(ochre)* mutant. As expected, it has the same N-terminal sequence as the native polypeptide. Since the molecular weight of the fragment from the mutant is of *ca.* 48,000, the two fragments have a common overlapping sequence.

A more refined study of proteolysis has shown that the production of the dimeric dehydrogenase is preceded by the formation of a hybrid fragment H composed of a native chain and of a proteolyzed chain. A careful study led to the elucidation of the sequence of events summarized in figure 6 and to the triglobular model of quaternary structure depicted in figure 7.

Binding of threonine to aspartokinase I-homoserine dehydrogenase I

The threonine binding experiments were performed using ^{14}C-L-threonine, either by equilibrium dialysis or by gel filtration on Sephadex G-25 preequilibrated with the ligand. Figure 8 shows that saturation of the binding sites is reached at free ligand concentrations in the 0.3-0.5 mM range, where eight moles of threonine are bound to one mole of enzyme. The same experiment, performed after treatment with –SH reagents – which desensitize the homoserine dehydrogenase activity (and destroy the aspartokinase activity) – shows the loss of binding capacity toward threonine. In the same vein, the homodimeric threonine-insensitive homoserine dehydrogenase proteolytic fragment does not bind threonine either.

We have seen that the inhibition of the two activities of the protein by threonine is strongly cooperative (see figs. 3 & 4). It is interesting to find that the binding of threonine also shows cooperativity. At non-saturating concentrations of threonine, the binding depends upon a number of factors, especially the concentration of KCl, which is found to dis-

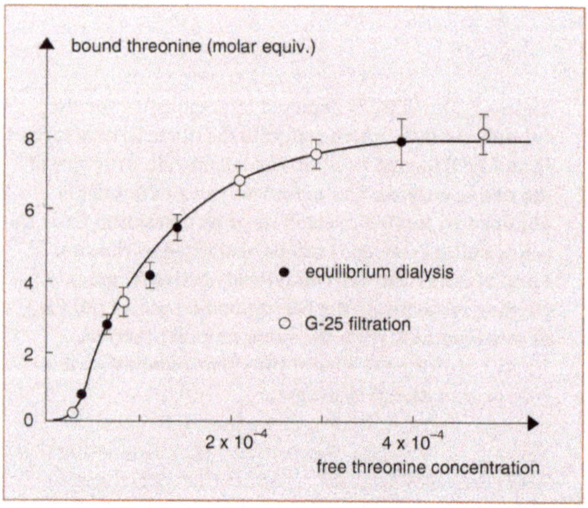

Figure 8. Binding of L-threonine to AKI-HDHI. (From P. Truffa-Bachi, M. Véron & G. N. Cohen, *CRC Critical Reviews in Biochemistry* **2**, 379-415 (1974), with permission.)

place in a very similar manner the inhibition curve of homoserine dehydrogenase and aspartokinase activities. In the presence of aspartate, the binding of threonine is also found to decrease, a result that could be expected since the inhibition of aspartokinase is competitive toward aspartate. The large cooperativity of inhibition by threonine (Hill number equal to four, near-complete inhibition of the aspartokinase activity) implies several allosteric sites for the protein. In this view, it is satisfactory to find that, at saturation, the protein composed of four identical subunits binds eight equivalents of threonine and it can be inferred that each native subunit carries two sites for binding the allosteric inhibitor (fig. 8).

Binding of pyridine nucleotides to aspartokinase I-homoserine dehydrogenase I

The binding of $NADP^+$ has been studied by the gel filtration technique using the ^{14}C-labeled coenzyme. The linear Scatchard plot indicates a simple Michaelian type of variation with ligand concentration. Extrapolation leads to a value of 4 moles of $NADP^+$ bound per mole of enzyme.

Effects of threonine on aspartokinase I-homoserine dehydrogenase I

Quantitative measures on the protein showed that, whereas many sulfhydryl groups could be titrated by specific reagents, essentially no –SH groups could be titrated in the presence of threonine (fig. 9). The magnitude of this effect makes it unlikely that it is due to a direct steric effect of the threonine molecules "covering" 18 cysteine residues. More probably, in the presence of the allosteric effector, the protein undergoes a conformation change such that its 18 cysteine residues become buried.

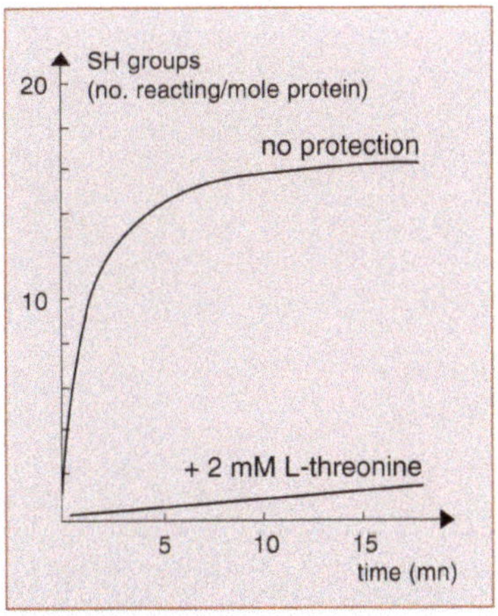

Figure 9. Titration of –SH groups of AKI-HDHI with dithio-bis-nitrobenzoic acid (DTNB). The reaction was performed at 20 °C, with 0.8 μM protein and 25 mM DTNB, with or without allosteric effector added. (From P. Truffa-Bachi, R. van Rapenbusch, J. Janin, C. Gros & G. N. Cohen, *Eur. J. Biochem.* **5**, 73-80 (1968), with permission.)

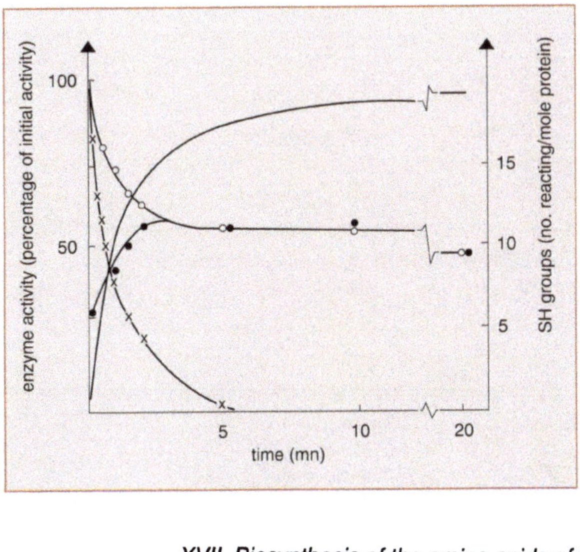

Figure 10. Effect of DTNB on the enzymatic activities of aspartokinaseI-homoserine dehydrogenaseI. –, titration of –SH groups; ○, homoserine dehydrogenase activity; •, homoserine dehydrogenase activity measured in the presence of 20 mM L-threonine; x , aspartokinase activity. No threonine was present in the mixture. Conditions essentially as in figure 9. Samples were diluted 50 times at various times; the activities and their inhibition by L-threonine were immediately tested. (From P. Truffa-Bachi, R. van Rapenbusch, J. Janin, C. Gros & G. N. Cohen, *Eur. J. Biochem.* **5**, 73-80 (1968), with permission.)

Aspartokinase inactivation and desensitization of the homoserine dehydrogenase activity were concomitant with the titration. At any given moment, the remaining aspartokinase activity is fully sensitive to threonine inhibition (fig. 10). When the reaction is performed in the presence of threonine, no effect on either enzymatic activity is found.

When a solution of the enzyme is prepared in a buffer containing threonine, slight differences in the absorption spectrum appear (fig. 11). The salient features are the peaks at 289 and 281 nm, and the complex trough around 269 nm. The analogy with the perturbation spectra of tyrosine and tryptophan residues dissolved in non-polar solvents suggests that both types of residues are involved and move to an environment less polar than water when threonine is added.

The fluorescence of most proteins is associated with their ultraviolet absorption with a major contribution of their tryptophan residues. When excited at 285 nm, the aspartokinase I-homoserine dehydrogenase I has a single emission band around 340 nm, entirely attributable to its aromatic residues (fig. 12). This figure also shows that the effect of 1 mM L-threonine on the excitation spectrum is comparable to the difference spectrum observed above: the emission maximum is shifted toward the longer wavelengths, and its intensity decreases by 12-15%. Some quenching – but no shift of the peaks – is also noted upon addition of other ligands.

Like other dehydrogenases, the complex formed with the reduced coenzyme has a characteristic fluorescence. When an excess of protein is added to a solution of NAPDH (fig. 13), the blue fluorescence of the coenzyme excited in the 340 nm absorption band is considerably increased; moreover, a new excitation band appears which corresponds to the ab-

Figure 11. Difference spectra in the presence of threonine. Baseline: reference and sample, no addition. Dashed line: reference, no addition; sample: 0.5 mM L-threonine. Aspartate spectrum: reference 15 mM L-aspartate; sample: 15 mM L-aspartate + 0.5 mM L-threonine. K spectrum: reference, 0.15 M KCl; sample, 0.15 M KCl + 0.5 mM L-threonine. (From J. Janin & G. N. Cohen, *Eur. J. Biochem.* **11**, 520-529 (1969), with permission.)

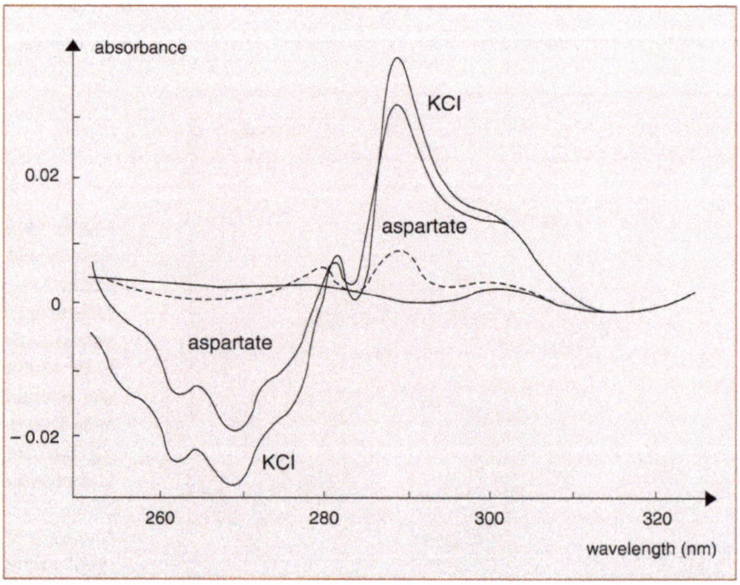

sorption of the protein, showing that excitation is transferred from an aromatic residue to the coenzyme. Finally, figure 13 also shows that the fluorescence of the protein-NADPH complex is strongly quenched upon addition of threonine. This considerable effect may be partly attributed to a perturbation of the same aromatic residues that leads to the difference spectrum, but the environment of the bound coenzyme is also modified, since both the fluorescence – upon excitation in the absorption band of the coenzyme – and the transfer fluorescence are quenched.

Taken collectively, these results concerning the effect of threonine on the microenvironments of cysteine and aromatic residues, can hardly be attributed to direct interactions of eight molecules of inhibitor with the groups involved in each case. They therefore indicate a conformational change of the protein from a "relaxed" form in absence of threonine, to a "tight" form in its presence.

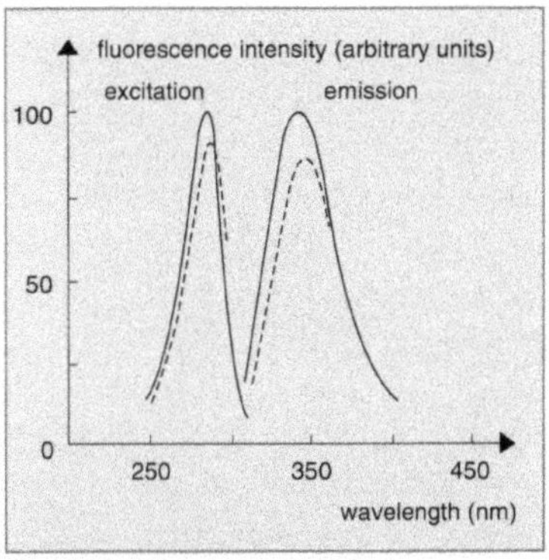

Figure 12. Fluorescence of aspartokinase I-homoserine dehydrogenase I. The excitation emission observed at 345 nm and emission (excitation at 282 nm) spectra were recorded (continuous line). Then 1 mM L-threonine was added and the same measurements repeated (broken line).
(From J. Janin, R. van Rapenbusch, P. Truffa-Bachi & G. N. Cohen, *Eur. J. Biochem.* **8**, 128-138 (1969), with permission.)

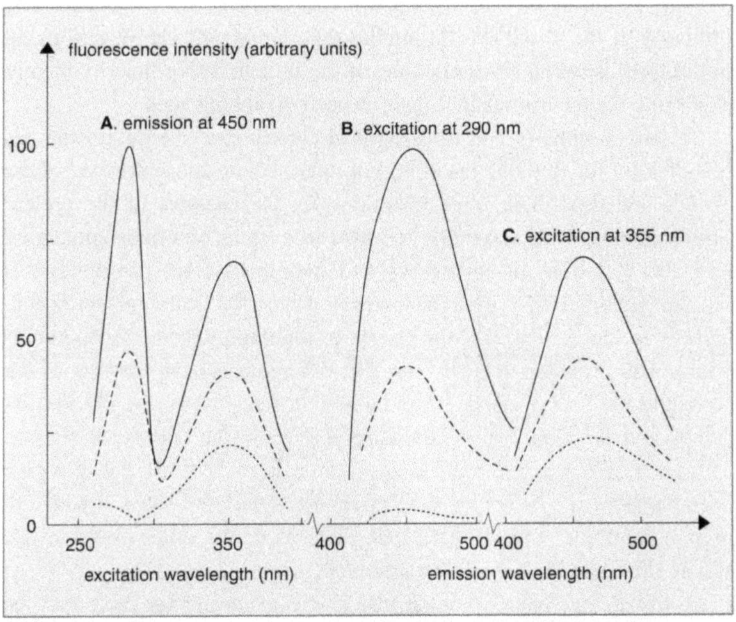

Figure 13. Fluorescence of the aspartokinase I-homoserine dehydrogenase I-NADPH complex. The fluorescence spectra of a solution of 1.5 μM protein and 3 μM NADPH were recorded before (continuous line) and after (broken line) addition of 1 mM L-threonine. Corrections were made for absorption. Emission was observed at 450 nm for the excitation spectrum (A); the excitation wavelength was 290 nm for the transfer emission spectrum (B) and 355 nm for the emission spectrum of NADPH (C). The dotted line represents the fluorescence of 3 mM NADPH under the same conditions.
(From J. Janin, R. van Rapenbusch, P. Truffa-Bachi & G. N. Cohen, *Eur. J. Biochem.* **8**, 128-138 (1969), with permission.)

Allosteric transition of aspartokinase I-dehydrogenase I

The difference spectra and the effects on the fluorescence of the protein can be used to define two conformations of the protein. Let T be the form with buried chromophores (low absorption at 269 nm and high absorption at 289 nm), and R the form with exposed chromophores. A straightforward interpretation of the data is that T is stable in the absence of added ligand or in the presence of threonine, and R is stable in the presence of potassium ions or of aspartate. As a consequence of this model, difference spectra similar to these should be obtained in the absence of added threonine when samples of protein in buffer containing KCl or aspartate are compared with similar solutions with no added ligands. This is, indeed, found to be the case. Aspartate and K^+, at least, are not structurally related, and it is *a priori* likely that the three ligands under study have different binding sites. Short-range perturbations of the aromatic residues of the protein would then be expected to have very different effects on the ultraviolet absorption when aspartate or K^+ are bound. The similarity of the actual spectra implies that the ligands act by shifting an equilibrium between conformations of the protein rather than by merely modifying the environment of their respective binding sites.

The simplicity and sensitivity of fluorimetric measurements has been helpful for studying the effects of intermediate concentrations of the ligands and describing more quantitatively the changes of the protein conformation. When threonine is added to enzyme solutions containing KCl, the variations in absorbance and fluorescence are coordinated. It can be assumed that both techniques detect the same phenomenon. Figures 14 and 15 describe the effects of the three ligands on the equilibrium. The ordinates R/R + T or T/R + T represent the fraction of the protein in the R or T states. It can be seen in figures 14a and 14b that aspartate and K^+ give positive homotropic effects (the curves are sigmoidal), and positive heterotropic efects (they help each other), but they give negative heterotropic effects with threonine. When threonine is added to protein solutions containing various amounts of the other ligands (figs. 15a & 15b), similar features are observed.

As a first approximation, the above experimental facts may be visualised by a two-state model.

$$R \underset{\text{aspartate, } K^+}{\overset{\text{threonine}}{\rightleftharpoons}} T$$

A strong correlation may be established between the measurements of absorption, of fluorescence, and of threonine binding, all of which are compatible for the existence of two states. Table I shows a similar correlation between the catalytic and regulatory properties of the protein. The competitive inhibition of aspartokinase by threonine

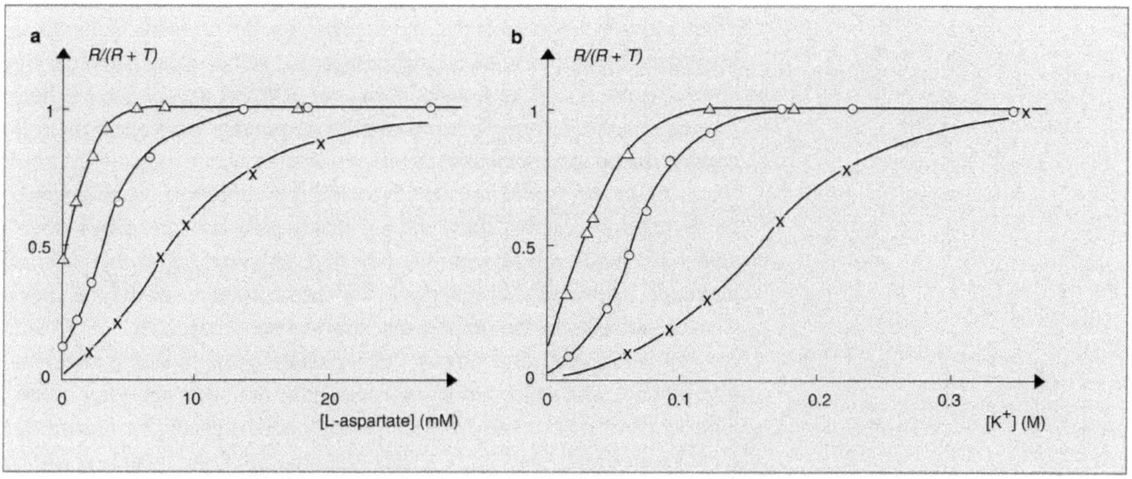

Figure 14. Effects of aspartate and K+ on the allosteric equilibrium. The values of R/R+T in the presence of various concentrations of ligands are derived from fluorescence intensity of the protein.
(a) effects of aspartate. Supplements: ○, none; △, 30 mM KCl; x, 40 μM L-threonine
(b) effects of K+. Supplements: ○, none; △, 2 mM L-aspartate; x, 40 μM L-threonine.
(From J. Janin & G. N. Cohen, *Eur. J. Biochem.* **11**, 520-529 (1969), with permission.)

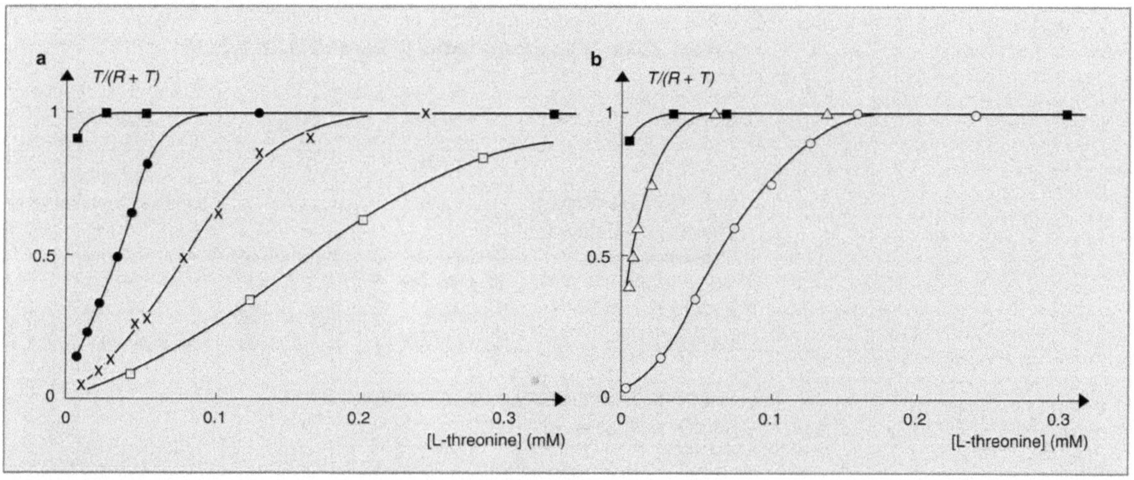

Figure 15. Effects of L-threonine on the allosteric equilibrium.
L-threonine is added to a solution of protein containing the following supplements:
(a) ■, none; ●, 60 mM KCl; x, 0.15 M KCl; □, 0,45 M KCl.
(b) ■, none; △, 5 mM L-aspartate; ○, 0,20 mM L-aspartate.
(From J. Janin & G. N. Cohen, *Eur. J. Biochem.* **11**, 520-529 (1969), with permission.)

(K type) can be accounted for by the preferential binding of aspartate to the R form. In contrast, the inhibition of the dehydrogenase activity and the analogous activation by potassium do not seem to be an effect of the R $\rightleftharpoons$ T transition on the binding of the substrates, for a noncompetitive and incomplete inhibition are observed (see text pp. 166-167 & fig. 4).

They may be due to an actual requirement for K^+ or, more likely, to an improper configuration of the dehydrogenase active sites. Although the binding constants of the R and T forms for NADPH are similar, the fluorescence of the T form is much smaller, indicating a different environment of the bound nucleotide.

Fast kinetic experiments were carried by stopped-flow and temperature jump in order to study the pre-steady state kinetics. When threonine was rapidly mixed with AK I-HDH I, an initial lag in the onset of threonine inhibition was observed. The rate constants for the lag times were dependent on threonine concentration but independent of protein concentration, thus reflecting an isomerization between active and inactive protein. The same result was essentially found in the temperature jump experiments, in the absence of any bound ligands, by raising the temperature from 20 to 28 °C. Thus, the binding of the ligands is irrelevant to the equilibrium which preexists in the absence of binding substrates or effectors. All the kinetic predictions of the allosteric model are fulfilled in the particular case of this protein.

Table I. General properties of the R and T conformations.

Property	T form	R form	Remarks
BINDING OF LIGANDS (dissociation constants)			shifts the equilibrium towards
Threonine	$K_T = 30\ \mu M$	$K_R > 1\ mM$ *	T form
Aspartate	$K_T > 20\ mM$	$K_R = 3\ mM$	R form
K^+ ions	low affinity	high affinity	R form
NADPH	$K_T = 0.3\ \mu M$	$K_R = 0.3\ \mu M$	no detectable effect
SPECTROPHOTOMETRIC PROPERTIES			
Relative fluorescence intensity	100	116	excitation at 280 nm, emission at 335 nm
Tryptophan and tyrosine residues	buried	exposed	
Enzymatic activities			
aspartokinase	none	active	T does not bind the substrate
Homoserine dehydrogenase	15-20%	100%	different configurations of the active sites

* Binding of threonine at saturation of KCl or aspartate.

Aspartokinase II-homoserine dehydrogenase II

If the homoserine dehydrogenase is carried by a single species, AK I-HDH I, it should be possible to obtain mutants that, following the loss of this activity, would require either homoserine or threonine plus methionine. Such mutants have been described in *Bacillus subtilis*, *Neurospora crassa* and *Saccharomyces cerevisiae*. Two such mutants have also been obtained in *E. coli*, strain B; despite separate efforts, it has, for some time, been impossible to obtain a similar phenotype in *E. coli* strain K12, the strain preferred (and most studied) by microbial geneticists. In consequence, a transduction has been carried out, introducing into K12 the mutated chromosomal piece corresponding to the inactivated homoserine dehydrogenase from *E. coli* B. The transductants can grow with threonine alone; in minimal medium, they eventually grow in 48 hours. They have no AK I-HDH I activity, but weak threonine-insensitive aspartokinase and dehydrogenase activities. These activities are greatly lowered by growth in the presence of methionine. Finally, methionine-requiring mutants of the transductants have been prepared which, when grown in a chemostat under methionine limitation, produce a high level of both threonine-insensitive activities. The synthesis of these two activities, regulated by the level of the intracellular methionine pool, have been called aspartokinase II (AK II) and homoserine dehydrogenase II (HDH II), the name aspartokinase III (AK III) being attributed to the lysine-sensitive enzyme.

During a systematic survey of spontaneous revertants of the transductants (that is, of mutants – devoid of AK I-HDH I activities – which can grow rapidly without added threonine), a mutant was isolated that synthesized constitutively AK II and HDH II. From this mutant, the purification of the two activities was achieved. The two activities could not be separated by the standard procedures of protein chemistry, and an homogeneous protein, aspartokinase II-homoserine dehydrogenase II, was isolated in the pure state.

It is amusing to note that in *S. typhimurium*, and other strains of *E. coli*, AK II-HDH II exists in such amounts that all the sophistication applied to its detection in the K12 strain of *E. coli* would have been unnecessary with these organisms.

The molecular weight of AK II-HDH II, determined by equilibrium sedimentation, is 185,000 ± 10,000. It is a dimer of identical subunits. In contrast to AK I-HDH I, it is not subject to regulation by allosteric control: neither of the two activities is inhibited by L-methionine, S-adenosylmethionine, L-threonine, or combinations, two by two, of methionine, threonine, isoleucine, and lysine. There is no cooperative binding of the substrates for either of the activities. ATP and aspartate do not inhibit homoserine dehydrogenase II; homoserine and NADPH do not inhibit aspartokinase II. NADPH, a good protector against thermal inactivation of homoserine dehydrogenase II, does not protect aspartokinase II against such treatment.

Upon limited proteolysis, as in the case of AK I-HDH I, aspartokinase activity gradually disappears and is finally completely destroyed, whereas homoserine dehydrogenase is essentially conserved. Two dimeric fragments are recovered, one being a dimer of 2 x 37,000 mol. wt., endowed with dehydrogenase activity, and one of 2 x 24,000 mol. wt., inactive. A careful study shows that, as in the case of the threonine-sensitive protein, a) the enzyme is triglobular; b) the kinase resides on the N-terminal part and the dehydrogenase in the C-terminal part, the inactive fragment being central.

The gene coding for AK II-HDH II, *metL* (89 mn), has been cloned and sequenced. The sequence homology between the two bifunctional enzymes will be discussed in chapter XXX.

Aspartokinase III

The enzyme is coded by *lysC* (91 mn). The gene has been cloned and sequenced. The enzyme has been purified to homogeneity. It is a dimeric protein of 2 x 50,000, which is in equilibrium with both a tetrameric species, favored by high ionic strength in the presence of lysine, and a monomeric species, favored by low ionic strength in the absence of lysine. Only the dimeric form is enzymatically active. However, lysine causes a cooperative allosteric transition leading to an inactive dimeric species, at protein concentrations below those required for tetramerization.

Certain so-called nonspecific hydrophobic amino acids (such as leucine, isoleucine, phenylalanine, and others) also inhibit aspartokinase III substantially at much higher concentrations. Synergism of inhibition by lysine, and low concentrations of the nonspecific amino acids, appears to be physiologically significant. The enzyme is also inhibited in an allosteric fashion by the lysine analogues β-aminoethylcysteine (thialysine) and selenalysine. Thialysine is a growth inhibitor to which resistant mutants have been isolated. Most excrete lysine in the growth medium. All the mutations analyzed lie within *lysC* and possess a desensitized aspartokinase III.

From homoserine to methionine

O-succinylhomoserine synthetase, the first committed step in methionine biosynthesis, is synergistically inhibited by methionine and S-adenosylmethionine (SAM). This is the only allosteric control in the specific methionine branch. The enzyme is inhibited by α-methylmethionine, and resistance to this analogue has been localized in the *metA* gene. Such mutants overproduce methionine and their homoserine succinyltransferase is now insensitive to methionine and SAM. The synthesis of the enzyme, as well as the synthesis of the rest of the enzymes leading to methionine and that of aspartokinase II-homoserine dehydrogenase II, is repressed by the presence of methionine in the culture medium.

From threonine to isoleucine

The biosynthetic threonine dehydratase is inhibited by isoleucine in a cooperative manner. It was during the course of his studies on this enzyme that Umbarger described the first case of feedback inhibition in a microorganism, and drew attention to the fact that its kinetics as a function of substrate concentration did not obey the Henri-Michaelis law. Umbarger, in his first paper on the subject, described the inhibition by isoleucine as competitive. To quote his own words: "when the double reciprocal plot of Lineweaver and Burk is employed, it is necessary to square the substrate (or inhibitor) concentration. This property of the data would be expected if the enzyme combined with two molecules of substrate or inhibitor" (figs. 4 & 5, Chapter III). Later it was confirmed that this enzyme neither follows the simple Henri-Michaelis law, nor obeys a purely bimolecular law with respect to substrate or inhibitor concentration. The results obtained have been interpreted as showing that several substrate or inhibitor molecules can react with a single molecule of enzyme and that, under these conditions, cooperative interactions exist between several molecules of ligand or several of the specific receptors on the enzyme. This research, together with work on aspartate transcarbamylase, is one of the foundations on which the theory of allosteric transitions has been built.

L-isoleucine, the allosteric inhibitor, confers considerable stability on threonine dehydratase. During inactivation (spontaneous or thermal), the enzyme becomes desensitized, *i.e.*, it is converted to a protein species insensitive to isoleucine inhibition (fig. 16).

This desensitization can also be obtained by treatment with mercurials, by incorporation of amino acid analogues into the enzyme molecule, or as the result of mutations. This was the first demonstrated case of desensitization. An important property of the desensitized preparations is that they no longer exhibit cooperative interactions between substrate molecules, the kinetics adhering closely to the Michaelis law. Whereas, in some

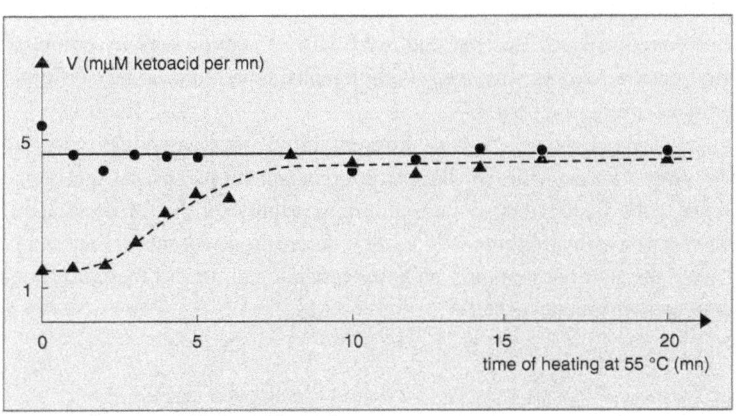

Figure 16. Heat inactivation of biosynthetic L-threonine deaminase. The solid line represents the enzyme activity in the absence of inhibitor; the dotted line, the activity measured with 10^{-2} M L-isoleucine.
(From J.-P. Changeux, with permission from the Société Française de Biochimie et Biologie Moléculaire.)

cases, desensitization and dissociation of an oligomeric protein occur together, in this particular case desensitization does not seem to be accompanied by a change in the apparent molecular weight of the enzyme.

Of all the other enzymes on the pathway, only acetohydroxyacid synthase is subject to feedback inhibition by valine in the K12 strain of *E. coli*. As a result, the addition of valine inhibits the growth of this organism by creating a phenotypic isoleucine requirement (see p. 233).

Multifunctional proteins

Only proteins that combine several autonomous functions on one polypeptide chain should be accepted as multifunctional proteins. Autonomy, in this sense, implies that each function is assigned to a different region – that is, a domain – on the polypeptide. The definition excludes enzymes that can catalyze different reactions using the same active site, such as asparaginase working as glutaminase, phosphoglyceromutase (which can carry three reactions using the same reaction center), or cystathionine-γ-synthase, which can catalyze a whole series of analogous reactions due to its relative lack of substrate specificity. As constituents of multifunctional proteins, we find mainly catalytic functions (AK-HDH I and II, yeast fatty acid synthetase, DNA polymerase I of *E. coli*, phosphoribosylanthranilate isomerase-anthranilate synthase of *E. coli*, tryptophan synthase of *Neurospora crassa,* flavocytochrome b_2-lactate dehydrogenase of yeast, DNA polymerase-5' exonuclease of bacteriophage T_4, to cite only a few). Certain polypeptide chains carry more than two catalytic functions. In addition to the already cited fatty acid synthetase, we can mention DNA polymerase I of *E. coli*, and a pentafunctional protein of the aromatic amino acid biosynthetic pathway in *N. crassa*. In certain cases, catalytic and non-catalytic domains are fused in a single polypeptide, like the microsomal cytochrome b_5 reductase and the mammalian myosin, two enzymes which have structural roles in addition to their enzyme activity; or the diphtheria toxin, which has a domain responsible for the binding to the cell-membrane and a domain responsible for the ADP-ribosylation of eukaryotic elongation factor EF_2. Finally, certain multifunctional proteins not endowed with a catalytic activity comprise independent binding sites for several ligands: repressors, some ribosomal proteins, immunoglobulins[21].

The existence of two different bifunctional enzymes carrying the same two activities in the same organism raises several questions: What is the selective advantage of bifunctionality for the aspartokinases-homoserine dehydrogenases? Do they derive from a common ancestor? And if so, was the common ancestor bifunctional, or did the corresponding gene fusion occur later?

21. See the book *Multifunctional Proteins* referred to at the end of this chapter.

We shall try to answer some of these questions when we review the evolution of proteins. Suffice it to say, at this point, that AK II-HDH II appears to be simpler: it is not inhibited by its end product, and it is a dimer, not a tetramer. Further comparative studies of these two proteins may lead to an explanation of some of their functional differences, especially as regards allosteric regulation, in terms of structural differences.

In the case of AK I-HDH I, one could argue that the end product, threonine, when in excess in the cell, inhibits the two activities simultaneously and that is a selective advantage that maintains a bifunctional enzyme. In the case of AK II-HDH II, no such argument holds. Since the reactions catalyzed by the two proteins are not sequential, but are the first and the third of the pathway, one could also think that one or two of the bifunctional enzymes form a multienzyme complex with aspartate semialdehyde dehydrogenase and maybe other enzymes of the threonine and methionine biosynthetic pathways. Despite many attempts, it has been impossible to demonstrate such protein asociations. Finally, as we shall see in the next chapter, in many bacterial genera, the aspartokinase and dehydrogenase activities are carried by different enzymes which are regulated in a variety of ways.

Thus, although AK I-HDH I was one of the first well-demonstrated multifunctional enzymes, it is still not clear why it is bifunctional.

II. REGULATIONS AT THE GENETIC LEVEL

Threonine operon

We have seen that there are five consecutive steps in threonine biosynthesis in bacteria. The first three are common to the biosynthesis of threonine and methionine; the first two are also the initial steps in lysine biosynthesis. The corresponding enzymes have been described above. Whereas the enzyme catalyzing the second step, aspartate semialdehyde dehydrogenase, is coded by the *asd* gene, localized at 75 mn, the bifunctional enzyme AK I-HDH I, homoserine kinase and threonine synthase, catalyzing respectively the first and third, fourth and fifth steps of threonine biosynthesis, are encoded by the genes *thrA*, *thrB* and *thrC,* which form a single transcription unit localized at 0 mn on the *E. coli* chromosome, the threonine operon.

The whole operon has been sequenced, including the 5' regulatory region and the termination region. The order of the operon, *thrA thrB thrC,* with transcription starting from *thrA* to *thrC,* has been established by analysis of phage Mu insertions and nonsense mutations in the *thr* cluster and by analysis of phage λ insertion into the *thr* regulatory region, upstream of, and adjacent to, the first structural gene of the operon, *thrA*. The λ insertion affects, in a pleiotropic manner, the expression of the three structural genes of the operon. Complementation analysis reveals that *thrB* and *thrC* consist of single cistrons, in keeping with the homopolymeric nature of their protein products, whereas *thrA* is composed of two cistrons, in accord with the bifunctional nature of aspartokinase I-homoserine dehydrogenase I. The intracistronic region between *thrA* and *thrB*, TGACATG, contains only one base pair while the boundary between *thrB* and *thr C* has no gap. The ribosome binding sites which mark the start of translation of *thrB* and *thrC* are thus within the structural genes of *thrA* and *thrB*, respectively.

Coordinate regulation of the threonine biosynthetic enzymes was demonstrated in both *E. coli* and *S. typhimurium* and it was observed that both threonine and isoleucine were required for repression[22], limitation in any one of the two causing an increased expression of the enzymes. This form of regulation was called *multivalent repression* (which is a misnomer, since the regulation occurs by attenuation). In fact, studies with a variety of regulatory mutations leading to constitutive pleiotropic expression of the structural genes causing a threonine excretion have shown that the mutations are localized upstream of *thrA*. They all lie in the threonine attenuator.

The sequence of the leader peptide involved in the attenuation mechanism has been established by Gardner:

Met Lys Arg **Ile** Ser **Thr Thr Ile Thr Thr Thr Ile Thr Ile Thr Thr**

and is seen to contain several threonine and isoleucine residues at which the uncharged specific t-RNAs can cause stalling of the ribosomes translating the RNA, favoring the transcription of the structural genes (See chapter XI on Attenuation).

Figure 17 describes the structure of the threonine operon of *Escherichia coli.*

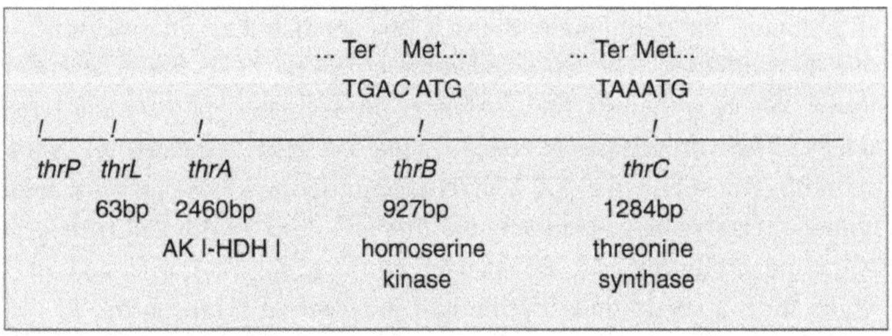

Figure 17. The threonine operon.

The abbreviations and symbols used are: Ter, termination codon; Met, initiation codon; *thrP*, promoter of the operon; *thrL*, sequence coding for the leader peptide.

22. We have seen that threonine is an intermediate in isoleucine biosynthesis, and both amino acids are involved in the regulation of the operon.

Regulation of the lysine regulon at the genetic level

One of the most striking features of the diaminopimelate-lysine pathway is that the structural genes are scattered over the *E. coli* chromosome. There are no multigenic operons.

The concept of the regulon arose from studies showing that a single transcriptional regulator can control several distinct operons. The first regulons described were similar to operons in that their members were involved in a specific physiological role, such as the uptake and utilization of maltose, or arginine biosynthesis.

Although the regulation of the synthesis of some of the lysine enzymes has not been demonstrated, and no common regulatory protein has been implicated in the repression of the *lys* and *dap* genes, the term *regulon* is used here to convey the notion of a physiological unit.

The specific activity of aspartokinase III, the lysine-sensitive species, varies in response to the intracellular lysine concentration. Approximately a 10-fold repression is observed when lysine is added to the minimal growth medium, whereas a 15-fold derepression is seen during lysine-limited growth in a chemostat. The number of active enzyme molecules may thus vary about 150-fold. But the molecular mechanism of this regulation remains obscure. The promoter and the start of transcription of *lysC* have been identified, and no signals similar to those described in attenuation have been found in the long sequence (308bp) existing between the transcription and the translation starts. No repressor protein has been identified.

Aspartate semialdehyde dehydrogenase, which is not inhibited by lysine, is repressed about 50% by it and lysine limitation causes a 20- to 30-fold derepression.

Variations of the lysine pool have no effect on dihydrodipicolinate synthetase, while the level of dihydrodipicolinate reductase may vary 15- to 20-fold from the repressed to the derepressed condition. The variation is minimal for the desuccinylase. The results reported for DAP-decarboxylase are the following: lysine represses the formation of the enzyme, which is induced by diaminopimelate.

In conclusion, the regulation of lysine biosynthesis at the genetic level is far from being completely understood.

Regulation of methionine biosynthesis at the genetic level

The *metA* gene has two transcriptional start sites located 74 nucleotides apart. The two promoters are expressed *in vivo*, but only one is regulated by intracellular methionine levels. The *metB*, *metC* and *met F* genes, coding respectively for cystathionine synthase, cystathionase and

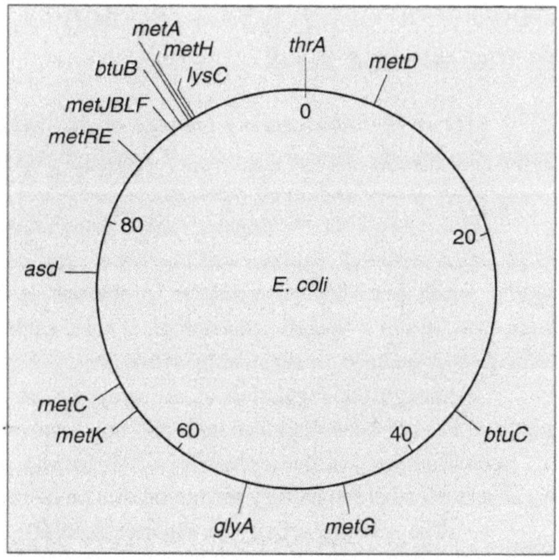

Figure 18. ***E. coli*** **chromosomal map showing the** ***met*** **genes. The** ***metR*** **gene was identified in** ***S. typhimurium.*** ***metG*** **is the methionyl-t-RNA synthetase,** ***btuC*** **is a gene coding for the transport of vitamin B_{12}. All the other genes are defined in the text.**

5,10-methylene tetrahydrofolate reductase, each have a single promoter. The *metA*, *metB*, *metC*, *metE* and *metF* genes are all repressed when excess methionine is added to the minimal growth medium, albeit to a different degree: the level of the enzymes varies from 300-fold for homoserine transsuccinylase, to 40-fold for cystathionine synthetase, 6- to 12-fold for cystathionase, 20-fold for the methylene tetrahydrofolate reductase and 60-fold for the vitamin B_{12}-independent transmethylase. There is thus no coordinate repression, reflecting the fact that the *met* genes do not form a single operon, but are scattered through the bacterial chromosome (fig. 18).

There is one notable exception: *metB* and *metL*, coding for cystathionine synthetase and AK II-HDH II, constitute an operon, *i.e.*, a single unit of transcription. These two genes belong to a cluster of genes, *metJBLF*, located at 89 mn on the chromosome.

Methionine also regulates the synthesis of aspartokinase II-homoserine dehydrogenase II *(metL)* and also that of S-adenosylmethionine synthetase *(metK)*. Although methionine affects the level of the vitamin B_{12}-dependent transmethylase, coded by *metH*, this effect appears to be the indirect result of repression of the synthesis of an activator protein (see below, Met R).

The expression of the *metE* gene is repressed by vitamin B_{12}; this is primarily due to a loss of MetR-mediated activation due to depletion of the coactivator homocysteine, rather than a direct repression by the metH-B_{12} holoenzyme, as it had been suggested (see below, MetR).

Mutants selected on the basis of their resistance to some methionine analogues were found to overproduce methionine, to be derepressed for the methionine biosynthetic enzymes, and to be resistant to repression by exogenous methionine. Many of them mapped to a chromosomal locus which was called *metJ*. The *metJ* gene product was thought to be a protein because some amber-suppressible *metJ* mutants could be isolated. Since *metJ*$^+$ was found to be dominant to *metJ* in partial diploids, it seemed likely that *metJ* was coding for a repressor molecule.

Not all the analogue-resistant mutants map to *metJ*. Some affect the *metK* gene, coding for S-adenosylmethionine synthetase (we have seen in the preceding chapter that this is an essential enzyme, and since S-adenosylmethionine cannot be transported by *E. coli*, the mutants must possess some residual enzyme activity). A class of *metK* mutants did excrete methionine, and were derepressed for the methionine biosynthetic enzymes which were not repressed by the presence of methionine in the growth medium. This suggested that S-adenosylmethionine, not methionine itself, was the corepressor of the *metJ* product.

Methionine repressor

The two members of the 89 mn cluster, *metJ* and *met F*, constitute separate units of transcription; *metJ* being transcribed divergently from *metB* and, consequently, a complex 276 base pair regulatory region is found between *metB* and *metJ*. There is a single promoter for *metB*, whereas *metJ* is transcribed from three separate points. Only two of the three *metJ* promoters are regulated by methionine. The repressor regulates its own synthesis.

The gene coding for *metJ* has been cloned and sequenced, and put under the control of a strong promoter in a multicopy plasmid. The amount of protein thus overexpressed represents 2% of the total soluble proteins, a 200-fold increase over the wild type.

The *E. coli metJ* gene is 312 bp long and codes for the Met repressor, a molecule of 104 amino acid residues (fig. 19). The repressor was found by sedimentation equilibrium to be a dimer.

The effect of the repressor on methionine biosynthetic enzymes has been assessed by measuring the β-galactosidase synthesis in a cell-free system derived from bacteria where the synthesis of this enzyme is

AEWSGEYISPYAEHGKKSEQVKKITVSIPLKVLKILTDERTRRQVNNLRH

* *

β_1 α_1

ATNSELLCEAFLHAFTGNPLPDDADLRKERSDEIPEAAKEIMREMGINPE

α_2

TWEY

Figure 19. Sequence of the *E. coli* methionine repressor.
The * signs indicate the amino acid residues making contacts with DNA inferred from the tridimensional structure of the repressor DNA complex. The underlined sections indicate the β-pleated strand and the α-helices of the protein respectively (see p. 362).

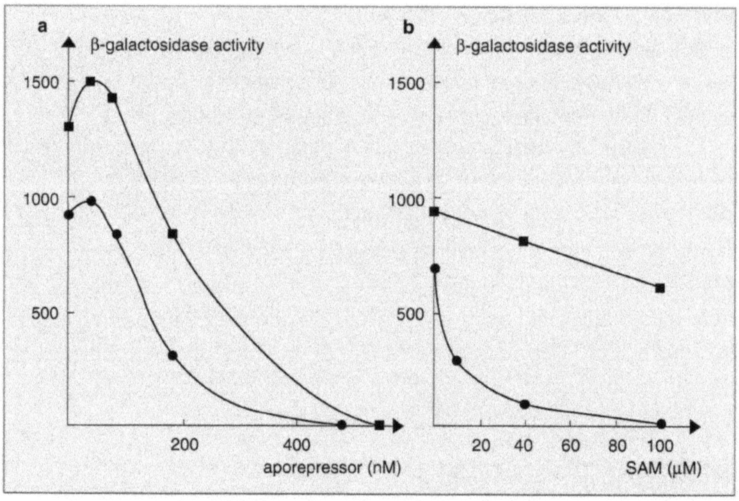

Figure 20. Inhibition of the synthesis of the galactosidase encoded by the *metF-lacZ* gene in a cell-free system by different concentrations of aporepressor (a) and S-adenosylmethionine (b). Template DNA was a plasmid carrying the gene fusion. (a): ● with or ■ without 50 μM SAM. (b): ● with or ■ without 400 nM repressor.
(From I. Saint Girons, J. Belfaiza, Y. Guillou, D. Perrin, N. Guiso, O. Bârzu & G. N. Cohen, *J. Biol. Chem.* **261**, 10936-10940 (1986), with permission from the *Journal of Biological Chemistry.*)

under the control of *metF (metF-lacZ* fusion). Figure 20a shows that at a constant concentration of DNA, the expression of the galactosidase is progressively repressed by increasing concentrations of the aporepressor, complete repression being attained at 600 nM of the protein. SAM lowers the effective concentration of the MetJ protein as a repressor. On the other hand, figure 20b shows that at a constant concentration of aporepressor, SAM enhances the repression in a concentration-dependent manner, half-maximum inhibition being reached at 10 μM and complete repression at 100 μM. Methionine had to be present in this experiment (0.16 mM in fig. 20a and 0.04 mM in fig. 20b) as a building block necessary for protein synthesis. Other assay systems have demonstrated *in vitro* that the aporepressor *plus* SAM have the same effect on the expression of the *metB* and *metL* genes,as they have on the *metJ* gene itself.

By equilibrium dialysis, it has been found that the methionine repressor binds two molecules of S-adenosylmethionine per dimer. The corresponding Scatchard plot is linear, showing that there is no cooperativity in the corepressor binding to the aporepressor. Also, equilibrium dialysis confirms that methionine does not bind at all to the aporepressor (fig. 21).

Since it was reasonable to assume that the repressor binding sites should be similar for all of the genes regulated by methionine, an exten-

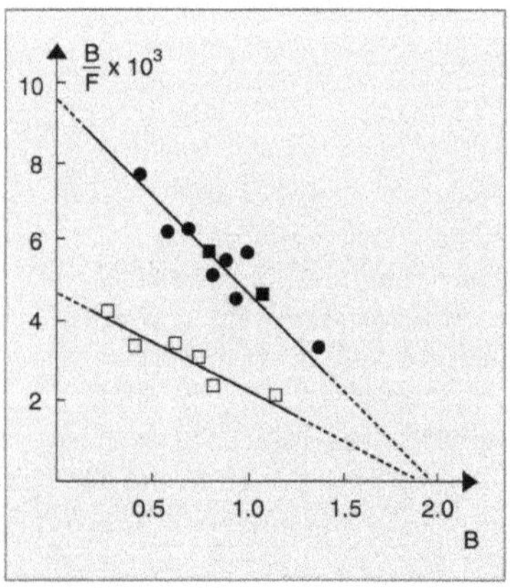

Figure 21. Binding of tritiated S-adenosylmethionine to the methionine repressor in the absence (□) or in the presence (●) of 80 mM KCl, as determined by equilibrium dialysis. Two experimental points (■) correspond to samples incubated with 2 mM L-methionine, in the absence of KCl.
(From I. Saint Girons, J. Belfaiza, Y. Guillou, D. Perrin, N. Guiso, O. Bârzu & G. N. Cohen, *J. Biol.Chem.* **261**, 10936-10940 (1986). With permission from the *Journal of Biological Chemistry.*)

sive survey of the regions upstream each structural gene was carried out. Comparison of the four 5' regions of *metC, metB, metA* and *metF* revealed the existence of a repetitive unit (R) eight nucleotides long which was named "Met box". In consequence, multiple alignments could be drawn by sliding each sequence with an eight-nucleotide periodicity. In the alignment presented in figure 22, out of the 128 positions considered, 89 matches and 21 transitions are found when the repetitive units are compared to their consensus sequence **R**. This consensus sequence is a perfect palindrome, AGACGTCT, which is present, under an altered

```
         - 50
metC :   ATATTCATGCTAGTTT A G A C A T C C
                                                -3
                          A G A C G T A T AAAAACAGGAATCCCG
         - 119
metB :   GGGATTTGCTCAATCT A T A C G C A A
                          A G A A G T T T
                          A G A T G T C C
                          A G A T G T A T
                                                -48
                          T G A C G T C C ATTAACACAATGTTTA
         - 97
metF :   CGCCCTTCGGCTTTTC C T T C A T C T
                          T T A C A T C T
                          G G A C G T C T
                          A A A C G G A T
                                                -26
                          A G A T G T G C ACAACACAACATATAA
         - 67
metA :   TTTTCTGGTTATCTTC A G C T A T C T
                          G G A T G T C T
                          A A A C G T A T
                                                -4
                          A A G C G T A T GTAGTGAGGTAATCAG

R    :                    A G A C G T C T
```

Figure 22. Comparison of the upstream regions of the *metC*, *metB*, *metF* and *metA* genes. The sequences 5' to the structural genes are presented discontinuously and have been aligned in order to focus on the presence of the underlying palindromic repetitive unit. Numbers indicate the positions relative to the adenine of the respective start codon taken as + 1. The – 10 promoter sequences are overlined and arrowheads indicate the transcription start signals. In the case of *metB*, the overlined hexamer is the – 35 box. The two underlined promoter sequences represent the – 35 and – 10 boxes of the first promoter of *metJ*.

(From J. Belfaiza, C. Parsot, A. Martel, C. Bouthier de la Tour, D. Margarita, G. N. Cohen & I. Saint Girons, Proc. *Natl. Acad. Sci. US* **83**, 867-871 (1986).)

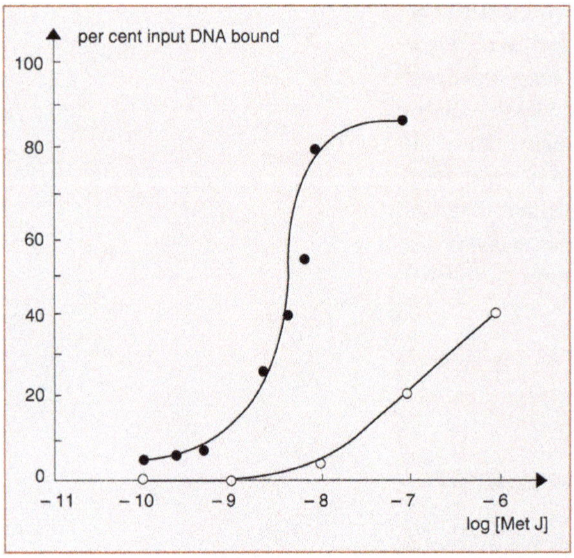

Figure 23. Filter binding of a fragment comprising the perfect consensus sequence 5'-dAGACGTCTAGACGTCT-3' 16-mer in the presence (●) or absence (○) of 100 mM SAM, and in presence of various concentrations of methionine repressor (Met J).
(From S. E. V. Phillips, I. Mainfield, I. Parsons, B. E. Davidson, J. B. Rafferty, W. S. Somers, D. Margarita, G. N. Cohen, I. Saint Girons & P. G. Stockley, with permission from *Nature* **341**, 711-715. Copyright (1989) Macmillan Magazines Limited.)

form, two to five times in the Met boxes. The *metB* and *metJ* genes, being transcribed divergently, share the same Met box.

It was then found that the *metE* and *metR* genes also possess those repetitive units (3 for *metE*, and 4 for *metR*). The two exceptions are *metK* and *metH* which, although regulated by methionine, do not seem to possess the consensus sequence.

It is possible that the difference between the number of boxes and/or their sequences are related to the different extents of repression elicited by the Met repressor; other explanations are, of course, possible.

In order to analyze the interaction between the Met repressor and its target, a nitrocellulose filter-binding assay using radioactive oligodeoxynucleotides has been developed. These experiments have shown that DNA fragments containing two consecutive consensus Met boxes are tightly bound by repressor in the presence of saturating SAM. Nucleotide sequences containing only one Met box are not bound. Scatchard analysis of the binding to the synthetic operator shows that binding is cooperative with respect to repressor concentration (fig. 23).

The repressor protects the operator against nuclease digestion and the results strongly indicate the binding of an array of repressor dimers, centered on the 16bp operator site, but extending into the neighboring non-operator DNA. The same is true for a fragment containing the *metF* regulatory region where the five boxes are protected, as shown by a footprint experiment (see chapter V). These observations are consistent with the following *in vivo* data: the level of repression of defined mutants of the *metC* operator, the smallest one known (two Met-boxes) is increased when the sequence comes closer to the consensus sequence; the same type of experiment was carried out with *metF*, and a similar series of point mutations and deletions in any of the five Met boxes was analyzed with the same result; furthermore, the arrangement of all five boxes in tandem repeats is important for effective repression, since there were some point insertions within the operators which lowered repression ratios 100-fold.

Parallel with the experiments in which the sequences of the operator Met boxes were altered by spontaneous or site-directed mutagenesis, a number of mutant repressor proteins were prepared, also by site-directed mutagenesis, to probe the roles of various amino acids in operator binding. Knowledge of the tridimensional structure of the repressor-operator complex[23] was taken advantage of by making the mutations at residues contacting the DNA directly, and at those involved in the protein-protein contacts between adjacent repressor molecules. Effective repression, as expected, depends on cooperativity arising from these contacts.

metR gene and its product

Some mutants with normal *metF*, *metE*, and *metH* genes were found to express the two transmethylases at a very low level, resulting in methionine auxotrophy. This auxotrophy was overcome in strains carrying multicopy number plasmids with either the *metE* or the *metH* gene. These results implied that the methionine auxotrophy was due to an independent mutation resulting in an inability to synthesize enough homocysteine transmethylase enzymes to allow growth. The mutations are linked to *metE*, but lie outside the *metE* structural gene. Their locus was called *metR*. The *metR* genes from both *E. coli* and *S. typhimurium* have been cloned and sequenced. The corresponding polypeptide chains are 80% identical. The *E. coli* chain is 317 amino acid residues long and codes for a polypeptide of 35,628 Da. The native protein is a homodimer and is supposed to contain a leucine zipper, a motif characteristic of many eukaryotic DNA binding proteins (and of the *E. coli* Lac repressor).

The expression of *metR* is repressed by the Met repressor, product of *metJ* (the 5' flanking region of *metR* contains four Met boxes) and by the product of *metR* itself. Conversely, the MetR protein plays no role in the regulation of the *metJ* gene.

MetR activates both *metE* and *metH* expression and homocysteine, the substrate of both transmethylases, is the coactivator in the case of *metE* but not of *metH*. It has been suggested that this positive regulatory mechanism may apply when vitamin B_{12} is absent, *i.e.*, when there is no B_{12}-dependent transmethylase activity (catalyzed by the *metH* product). The resulting methionine starvation would then cause the derepression of the biosynthetic enzymes and the accumulation of homocysteine. The latter could then act as a signal to activate the synthesis of the vitamin B_{12}-independent transmethylase, coded by *metE*.

Recently, it has been shown in *S. typhimurium* that the MetR pro-

23. See the chapter on DNA-protein interactions.

tein activates *metA*, the gene for homoserine succinyltransferase and that homocysteine plays an inhibitory role in this *metR*-mediated activation. Physiologically, this result makes sense.

Regulation of isoleucine synthesis at the genetic level

The regulation of carbon flow into and through the pathways to isoleucine (and valine) appears to rest upon end product inhibition of the two initial enzymes, threonine dehydratase (see the section on allosteric enzymes) and acetohydroxy acid synthase.

The regulation of the enzymes leading from α-ketobutyrate, the product of threonine dehydratase, to isoleucine occurs solely at the transcription level, the enzymes being subject to multivalent repression (see valine biosynthesis) except in the case of the acetohydroxy isomeroreductase, whose synthesis is induced by either of its substrates (α-aceto-β-hydroxybutyrate or α-acetolactate). A detailed account will be given when we study the regulation of valine biosynthesis.

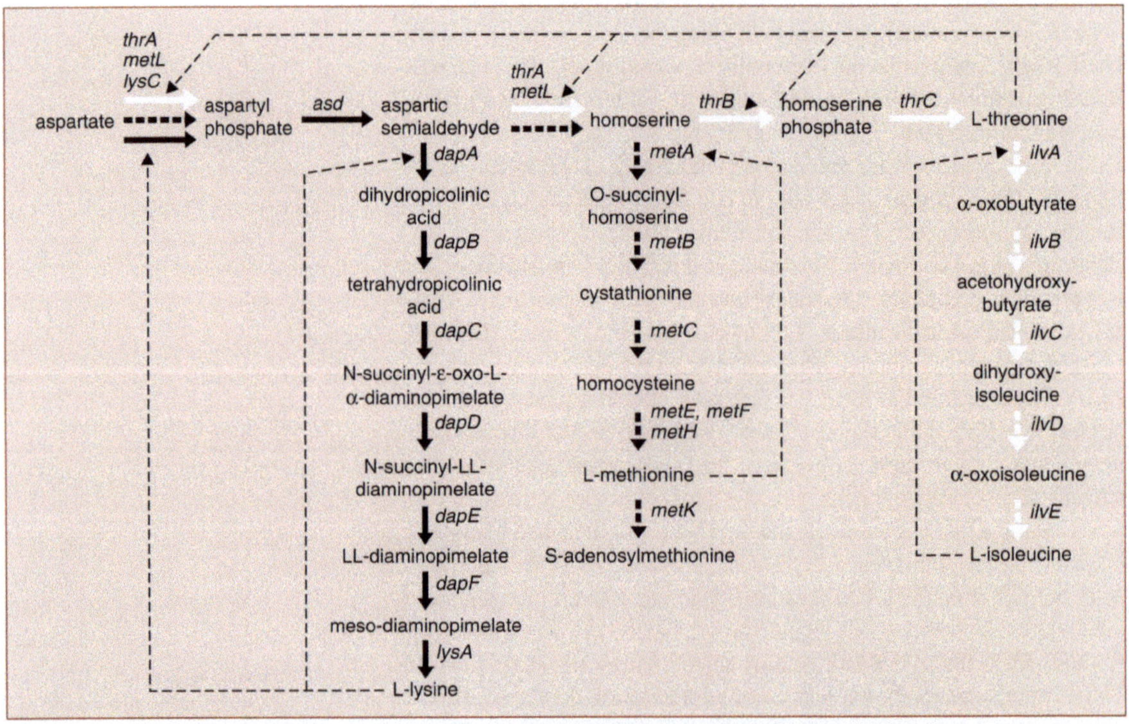

Figure 24. The pathway leading to lysine, threonine, isoleucine, methionine, and S-adenosylmethionine in *E. coli* and *S. typhimurium*. Details of the specific methionine pathway are given in figure 1, chapter XVI. The names of the relevant enzymes corresponding to the mentioned genes are given in the text. Long dashed arrows represent regulation at the level of enzyme activity. Expression of the genes is regulated specifically by the different end products of the pathway (white arrows, dotted arrows, and black arrows: regulation by threonine and isoleucine, methionine and lysine, respectively).

Appendix: More on regulons

The concept of regulons was generalized with the recognition of global responses, in which a number of functions are coordinately controlled in response to an environmental signal (temperature, pH, starvation, DNA damage, anaerobiosis, etc.). The factor mediating transcriptional regulation in a global response can be a repressor, an activator, an alternative sigma factor, or an effector of RNA polymerase.

The leucine-Lrp regulon (Lrp standing for leucine-responsive regulatory protein) was identified as having an activating or repressing effect on the transcription of many unrelated genes or operons. Some of these operons were found by chance, others were found by two-dimensional electrophoretic analysis of the proteins synthesized under different conditions, and others were found by analysis of mutants created by transpositions.

To cite only a few of the some 75 genes estimated to be regulated by Lrp:

a) *ilvIH*, *serA*, *gltD*, *leu*: activating effect of Lrp antagonized by leucine;

b) *sdaA* (serine deaminase), *tdh* (threonine dehydrogenase), *opp* (oligopeptide permease): repressing effect of Lrp antagonized by leucine;

c) *livJK* (uptake of isoleucine and valine): repressing effect of Lrp, leucine required;

d) *gcv*, *lrp* (glycine cleavage and synthesis of Lrp): respectively activating and repressing effect of Lrp, with no effect of leucine on either process.

One of the most striking aspects of the Lrp regulon is the number of different patterns of regulation caused by the interaction of Lrp and leucine. As in the case of some other important global regulators, like Crp, Lrp activates the expression of some operons and represses the expression of others. In some cases, activation requires, is counteracted by, or is independent of leucine. In the operons repressed by Lrp, the same categories are observed as far as the action of leucine is concerned. We do not yet know which aspect of the environment leucine represents.

Lrp is a 19 kDa DNA-binding protein, existing in solution as a dimer. In one case where an *in vitro* study has been conducted, that of the *ilvIH* gene, it has been shown that the purified Lrp protein stimulates transcription from the promoter at concentrations similar to those required for Lrp binding to *ilvIH* DNA.

In the case of the *ilvIH* operon, six Lrp binding sites are present with a several hundred base pair region upstream of the promoter region. Binding of Lrp induces an important DNA bending that can organize the assembly of a higher order nucleoprotein structure.

SELECTED REFERENCES

Isofunctional aspartokinases

E. R. Stadtman, G. N. Cohen, G. Le Bras & H. de Robichon-Szulmajster, *J. Biol. Chem.* **236**, 2033-2038 (1961).

Aspartokinases-homoserine dehydrogenases I and II. Structure and regulation of activity

G. N. Cohen & A. Dautry-Varsat, in *Multifunctional Proteins*, H. Bisswanger & E. Schmincke-Ott eds., pp. 49-121. John Wiley & Sons, New York (1980).

A. Fazel, K. Müller, G. le Bras, J. R. Garel, M. Véron & G. N. Cohen, *Biochemistry* **22**, 168-165 (1983).

Threonine operon and its regulation

M. Katinka, P. Cossart, L. Sibilli, I. Saint Girons, M. A. Chalvignac, G. Le Bras, G. N. Cohen & M. Yaniv, *Proc. Natl. Acad. Sci. US* **77**, 5730-5733 (1980).

C. Parsot, P. Cossart, I. Saint Girons & G. N. Cohen, *Nucleic Acid Research* **11**, 7331-7335 (1983).

J. F. Gardner, Proc. Natl. Acad. Sci. US 76, 1706-1710 (1979).

Regulation of methionine biosynthesis. Methionine repressor

I. Saint Girons, C. Parsot, M. M. Zakin, O. Barzu & G. N. Cohen, *CRC Critical Reviews in Biochemistry* **23** (suppl. 1), 1-42 (1988).

S. E. V. Phillips, I. Mainfield, I. Parsons, B. E. Davidson, J. B. Rafferty, W. S. Somers, D. Margarita, G. N. Cohen, I. Saint-Girons & P. G. Stockley, *Nature* **341**, 711-715 (1989).

Regulation of the synthesis of the branched-chain amino acid

Read the chapter by Umbarger in:

Escherichia coli and *Salmonella typhimurium. Cellular and Molecular Biology*, J. L. Ingraham, K. B. Low, B. Magasanik, M. Schachter & H. E. Umbarger eds., Vol. I, pp. 352-367. American Society for Microbiology, Washington, D.C. (1987).

Leucine-Lrp regulon

E. B. Newman, R. d'Ari & R. T. Lin, *Cell* **68**, 617-619 (1992).

Q. Wang & J. M. Calvo, *The EMBO Journal* **12**, 2495-2501 (1993).

CHAPTER XVIII

Other patterns of regulation of the synthesis of amino acids of the aspartate family

It appears that all *Enterobacteriaceae* make use of the three isofunctional enzymes to accomplish a differential control of the aspartokinase activity. The inhibition and repression patterns found in *Escherichia coli* and *Salmonella typhimurium* are also found in other coliform bacteria such as *Edwardsiella tarda, Aerobacter aerogenes* and *A. cloacae, Serratia marcescens, Erwinia carotovora* and *E. aroideae, Proteus vulgaris* and *Providencia* sp. The relative activity of the three aspartokinases (measured in crude extracts) differ greatly from strain to strain, even within a single species, and probably reflect a different degree of repression due to the different size of the individual free amino acids in the cytoplasm. Like *E. coli* and *S. typhimurium*, all strains examined contain a homoserine dehydrogenase partly inhibitable by threonine. The possible physical association between aspartokinase I and II with homoserine dehydrogenase in the other coliforms has not been examined, since this is a property determinable only after some degree of enzyme purification.

Concerted feedback inhibition of aspartokinase activity in *Rhodobacter capsulatus* (formerly *Rhodopseudomonas capsulata*)

This species possesses a single aspartokinase which is insensitive to feedback inhibition by any one of the essential metabolites, lysine, threonine, or isoleucine. However, if lysine and threonine are simultaneously present, considerable inhibition of the enzymatic activity is found. This absolute requirement for two or more of the end products for the accomplishment of inhibition has been called (by Howard Gest, who discovered it), *concerted feedback inhibition,* or multivalent feedback inhibition. This mechanism seems to be a less delicate control than the one using the isofunctional enzymes, since it does not allow the independent regulation of the first reaction of the branched pathway. Rather, it presents an alternative solution to the difficulty raised by the existence of such pathways. Figure 1 shows a typical experiment carried out with an extract of *Rh. capsulatus.* The concerted inhibition is never 100 %, a fact which would seem to be important for enabling the synthesis of methionine to proceed even in the presence of both allosteric effectors.

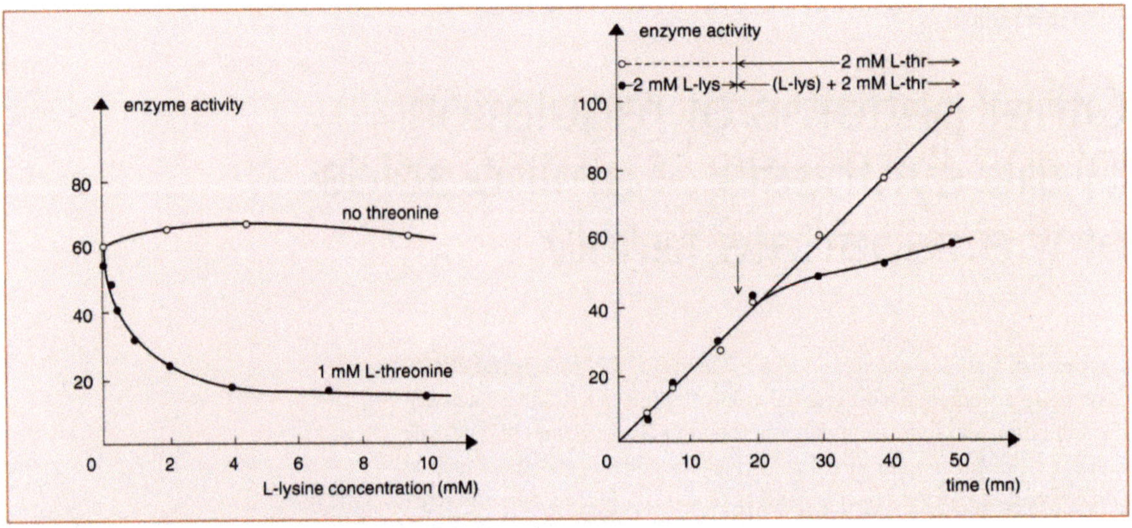

Figure 1. Two experiments showing the concerted feedback inhibition of the single aspartokinase from *Rh. capsulatus*. The experiment on the left shows the absence of inhibition by lysine alone, and the progressive inhibition on increasing the concentration in the presence of threonine. In the experiment on the right, empty circles represent the enzyme activity measured in the absence of allosteric effector, black circles the same activity in the presence of lysine. At time t = 18 mn, threonine is added to the two incubation mixtures. Inhibition occurs only when lysine is present.
(From P. Datta & H. Gest, *Proc. Natl. Acad. Sci. US* **52**, 1004-1009 (1964). With permission from Dr. Howard Gest.)

The synthesis of the single aspartokinase of *Rh. capsulatus* is repressed by the presence of methionine in the culture medium. As a result, the growth of this organism, a prototroph in the wild type, is partially inhibited by excess methionine, or by an excess of the combination of threonine plus lysine, growth being re-established in the presence of the three amino acids (fig. 2).

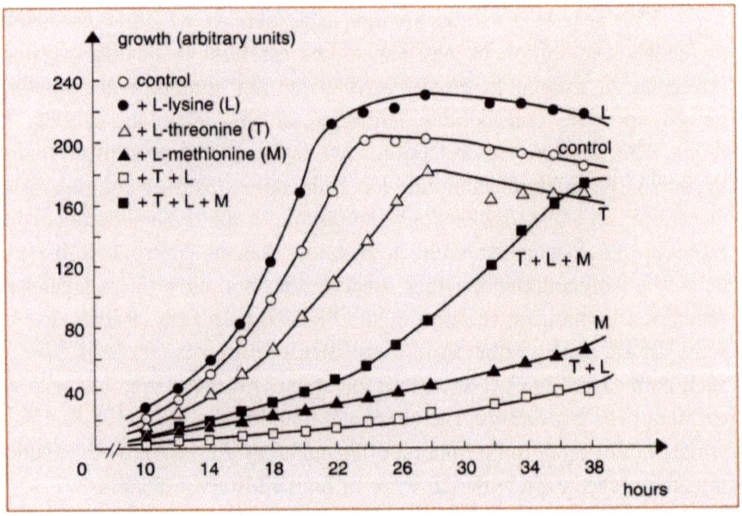

Figure 2. Effects of adding amino acids on the growth of *Rh. capsulatus*. L-lysine is added at a concentration of 10^{-3} M, L-threonine and L-methionine at 5×10^{-4} M. Only methionine and the combination of threonine + lysine have an appreciable inhibitory effect on growth. The mixture of the three amino acids restores growth to near normal.
(From L. Burlant, P. Datta & H. Gest, *Science* **148**, 1351-1353. Copyright (1965) by the AAAS.)

Pseudomonads

The fluorescent pseudomonads constitute the central taxonomic cluster of the genus *Pseudomonas*. Although they share many common phenotypic properties, a thorough analysis has shown that they can be subdivided into three main groups: *P. aeruginosa*, *P. putida* and *P. fluorescens*, of which the last two are further subdivisible by minor phenotypic traits into a number of different biotypes. It should be mentioned that this type of classification is an extremely conservative one, relative to the current formal classification of coliform bacteria; the phenotypic differences among the three species of fluorescent pseudomonads are no smaller than those which differentiate such genera as *Escherichia*, *Citrobacter*, *Salmonella* and *Shigella*.

Markedly different from the fluorescent pseudomonads in many phenotypic respects are the two *Pseudomonas* species of the acidovorans group, *P. acidovorans* and *P. testosteroni*, which likewise share many common properties. In the fluorescent pseudomonads, there is a common pattern of inhibition of a single aspartokinase by threonine and lysine, entirely different from that characteristic of coliform bacteria. The aspartokinase activity of these pseudomonads is subject to concerted inhibition by the two amino acids; each amino acid is, however, inhibitory when furnished at high concentration. Consequently, the concerted nature of the inhibition can be clearly shown only when relatively low levels of the effectors are used (table I).

Table I. Concerted feedback inhibition of the aspartokinase activity in *Pseudomonas aeruginosa*.
The values in parentheses represent the percent inhibition of the total activity.

Additions	Activity (nmoles/mn/mg)
None	26.8
L-lysine, 10^{-2} M	14.2 (47)
L-threonine, 10^{-2} M	6.5 (75)
Both effectors, 10^{-2} M	0 (100)
L-lysine, 10^{-3} M	18.8 (29)
L-threonine, 10^{-3} M	9.5 (65)
Both effectors, 10^{-3} M	2.6 (90)
L-lysine, 3×10^{-4} M	25.9 (3)
L-threonine, 3×10^{-4} M	28.7 (0)
Both effectors, 3×10^{-4} M	9 (66)

Methionine is without effect on the aspartokinase activity. This pattern has been confirmed for many strains belonging to all three species. It is different in several respects from the mode of concerted feedback inhibition originally discovered in *Rhodobacter capsulatus,* in which the single amino acids have little effect on aspartokinase activity, even at high concentrations.

In the strains representative of the *acidovorans-testosteroni* group, feedback inhibition is again concerted and resembles that characteristic of *Rh. capsulata* rather than that of fluorescent pseudomonads: the single amino acids do not inhibit appreciably even at 10^{-2} M.

A further difference between the fluorescent pseudomonads and those of the *acidovorans-testosteroni* group appears when their homoserine dehydrogenase activity is considered: whereas the enzyme is active with both pyridine nucleotides in the fluorescent group, the enzyme of the other group shows an absolute specificity for NADH.

The aeromonads have an uncertain taxonomic position, since they have properties that could permit assignment either to the pseudomonads or to the coliform group. The great weight traditionally assigned to the mode of flagellar insertion in bacterial taxonomy had led to their exclusion from the coliform group. In three species of aeromonads, *A. hydrophila, A. shigelloides, and A. formicans,* the three aspartokinases were found, however, with the inhibition and repression pattern characteristic of coliforms.

The conclusion, also reached earlier, through studies on the regulation of the aromatic biosynthetic pathway, is confirmed: specific regulation mechanisms are stable group characters, shared by the members of large taxa for which a common evolutionary origin can be postulated on other, independent, grounds.

Specific reversal of a particular feedback inhibition by other essential metabolites. Case of *Rhodospirillum rubrum*

This organism, like *R. capsulatus*, possesses only one aspartokinase and apparently a single homoserine dehydrogenase. The feedback inhibition by L-threonine has been examined in some detail. Both enzymes are inhibited by threonine, the inhibition being reversed by the addition to the reaction mixture of isoleucine in the case of aspartokinase, or of methionine or isoleucine in the case of the dehydrogenase. Table II shows that, in the absence of the inhibitor, isoleucine and methionine have an activating effect on aspartokinase.

Table II. Effect of certain amino acids on the aspartokinase activity of *Rhodospirillum rubrum*.

Additions (10^{-4} M)	Enzyme activity (arbitrary units)
None	41
L-isoleucine	74
L-methionine	70
L-threonine	0
L-threonine + L-isoleucine	47
L-threonine + L-methionine	4
L-isoleucine + L-methionine	69

The effect of different ligands on homoserine dehydrogenase has been studied with regard to the state of aggregation of the enzyme. Centrifugation experiments in sucrose gradients show that L-threonine stimulates an aggregation of the enzyme to an inactive form. This aggregation is reversed by the allosteric modifiers methionine or isoleucine. As can be seen in Figure 3, both forms of the enzyme can also be distinguished by gel filtration in a buffer containing the different effectors. These observations were interpreted as reflections of the importance in this orga-

>

Figure 3. Elution pattern of *Rhodospirillum rubrum* homoserine dehydrogenase on Sephadex G200 in the presence and absence of substrate and/or allosteric modifiers. A: enzyme filtered in buffer with no additions. B: enzyme filtered in the presence of threonine. C & D: enzyme filtered in the presence of various concentrations of threonine and homoserine. E: enzyme filtered in the presence of threonine. F & G: enzyme filtered in the presence of threonine and modifiers. The dotted curves come from different experiments and are included for comparison. Threonine evidently causes the enzyme to aggregate, the effect being reversed in the presence of sufficient substrate (C & D). In F & G, the high molecular weight characteristic of the threonine-treated enzyme is always present, but a considerable trailing is observed which extends into the region where species of lower molecular weight are expected.
(From P. Datta, H. Gest & H. J. Segal, *Proc. Natl. Acad. Sci. US* **51**, 125-130 (1964). With permission from Dr. Howard Gest.)

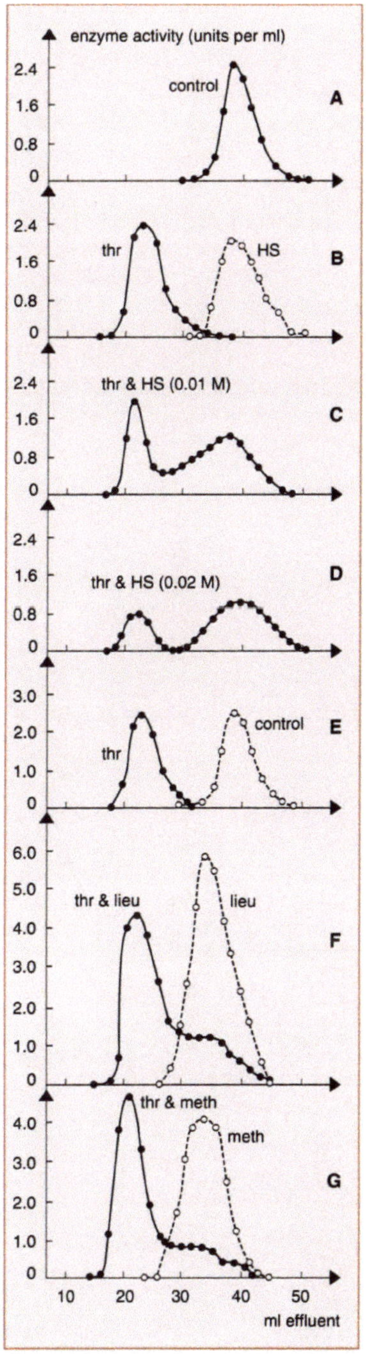

nism of monomer → polymer interconversions in controlling the activity of homoserine dehydrogenase.

In this organism, which possesses only one aspartokinase, an increase in the size of the intracellular pool of threonine above a critical level must produce a decrease in the concentrations of the common intermediates, due to allosteric inhibition of the kinase and the dehydrogenase. Now, it is found that the threonine dehydratase of *R. rubrum*, unlike that of *E. coli*, is practically insensitive to isoleucine inhibition; it is therefore probable that the synthesis of isoleucine proceeds freely from threonine in these circumstances, but the problem of achieving a normal synthesis of DAP, lysine, and methionine remains. The specific reversal by isoleucine of the allosteric inhibition (caused by threonine) of aspartokinase and homoserine dehydrogenase provides the answer: it appears that the increase in the isoleucine/threonine ratio is the signal for an accelerated production of the common intermediates required for the synthesis of DAP, lysine, and methionine.

Particular case of spore-forming bacilli

In the bacteria we have examined so far, the aspartate pathway has the relatively simple function of providing amino acids needed for growth; in non-growing cells, the pathway can be essentially quiescent. A more complex situation occurs in the genus *Bacillus* as a result of its characteristic life cycle. Under adverse nutrient conditions, members of this genus enter a resting state, the bacterial spore, which can effectively survive prolonged heat, dehydration, and mechanical stress. We have seen (chapter XVI) that the development of heat resistance is accompanied, among others, by the synthesis of large amounts of dipicolinic acid, a compound derived from dihydrodipicolinic acid, itself a precursor of diaminopimelic acid and lysine.

In addition, the process of sporulation involves the synthesis of large amounts of diaminopimelate for the synthesis of new structures such as the spore cortex mucopeptide. Aspartokinase has thus a dual function in *Bacillus* species: the synthesis of amino acids during growth, and the production of dipicolinate and diaminopimelate during sporulation.

Is this dual function reflected in the regulation of aspartokinase? The predominant pattern that emerges from a survey of several members of the genus is that of two distinct aspartokinases whose relative amounts vary during vegetative growth and sporulation. The major enzyme in growing cells (vegetative type or AK II)[24] is usually subject to feedback inhibition by lysine *(e.g., B. subtilis)* or by combination of lysine and threonine *(e.g., B. polymyxa)*, consistent with the idea that its primary function is to provide amino acids for protein synthesis. The predominant aspartokinase activity during sporulation (AK I)[24] is inhibited by the cell-wall constituent *meso*-diaminopimelate, allowing it to function even in the presence of high levels of protein amino acids to produce the structural components of the spore. To assure the efficient production of dipicolinate and diaminopimelate under these conditions, the lysine branch of the aspartate pathway in *Bacillus* species has two characteristic features. One is the absence of inhibition (by lysine) of dihydrodipicolinate synthase, the first committed step in the lysine branch, contrasting with the situation in *E. coli*, and allows the production of dipicolinate and diaminopimelate even in the presence of lysine. The other is the control of the conversion of diaminopimelate to lysine either by feedback inhibition of lysine decarboxylase by lysine, or by inactivation of the enzyme during sporulation, thereby preventing the needless diversion of diaminopimelate to lysine.

Whereas the structure of aspartokinase I has not been examined in detail, this is not the case of aspartokinase II, which has been thoroughly studied by the group of Henry Paulus in *B. polymyxa* and *B. subtilis.*

The enzyme is composed of two non-identical subunits of respective molecular weight 43,000 and 17,000, which are arranged in an $\alpha_2\beta_2$ structure. The immunological and chemical comparisons of the subunits of *B. subtilis* aspartokinase II revealed that the smaller β-subunit is identical with the COOH-terminal portion of the α-subunit. It was first believed that the two subunits were related by a proteolysis mechanism, β being a proteolytic fragment of α. This is not so and, unexpectedly, when the gene for aspartokinase II was cloned and sequenced, it became apparent that the two subunits are encoded by in-phase overlapping sequences and are the products of independent translation, the ribosome binding site for the β-subunit being part of the 3' sequence of the cistron coding for the α-subunit.

24. This numbering has nothing in common with the one used in *E. coli* and coliforms.

In *B. megaterium*, the existence of two aspartokinases are reported on the basis of their different patterns of repression and inhibition: the first, repressed by methionine, is inhibited in a concerted fashion by lysine and methionine, while the second is repressed and inhibited by threonine. Unfortunately, the two forms of aspartokinase have not been physically separated.

Some other cases

The presence of a single aspartokinase was demonstrated in *Rhodospirillum tenue*. The enzyme was partially purified. Inhibition by lysine, by threonine, and concerted inhibition by these two end-products are regulatory characters which have also been found in many other species. In *R. tenue,* aspartokinase is also subject to a hitherto not encountered type of concerted feedback inhibition, by L-threonine plus L-methionine. Methionine alone, even at high concentrations, is without effect.

The concerted feedback inhibition by threonine and methionine has been found in *R. tenue* only; the other photosynthetic bacteria examined, *Rhodospirillum molischianum*, *R. fulvum*, *R. rubrum* (see above), *Rhodobacter palustris*, *R. gelatinosa,* display concerted inhibition only by lysine and threonine, the pattern first discovered with *R. capsulatus*.

In *Rhodobacter spheroides,* there is no feedback inhibition by threonine, lysine, or methionine, single or in combination. The effector here is aspartate semialdehyde.

Although methionine does not participate in a concerted effect on the activity of the aspartokinase from *R. capsulata,* the synthesis of the enzyme is repressed by the presence of methionine in the culture medium. As a result, the growth of this organism is partially inhibited by excess methionine. In contrast with this behavior, the synthesis of *R. tenue* aspartokinase is not repressed by methionine, and this amino acid has no effect on the growth rate of the organism. In each organism, evolution has produced a different pattern leading to a similar result, whereby methionine can exert a regulatory function.

In *Micrococcus glutamicus* and in the yeast *Saccharomyces cerevisiae* (from which the single aspartokinase has been cloned and sequenced), the synthesis of the single homoserine dehydrogenase is specifically repressed by methionine. As in *E. coli*, it is threonine which is the allosteric inhibitor in *M. glutamicus*, while in *S. cerevisiae,* the activity is inhibited by threonine and methionine, the latter being by far the more effective.

Conclusion

The above examples have been chosen to illustrate the very marked variations which are met with in different organisms as means of reaching the same goal, *i.e.*, the harmonious integration of the rates of synthesis of the different cellular constituents.

SELECTED REFERENCES

Concerted feedback inhibition
P. Datta & H. Gest, *Proc. Natl. Acad. Sci. US* **52**, 1004-1009 (1964).
P. Data & H. Gest, *Nature* **203**, 1259-1261 (1964).
H. Paulus & E. Gray, *J. Biol. Chem.* 239, 4008-4009 (1964).

Pseudomonads
G. N. Cohen, R. Y. Stanier & G. J. Le Bras, *Bacteriol.* **99**, 791-801 (1969).

Rhodospirillum rubrum
P. Datta, H. Gest & H. J. Segal, *Proc. Natl. Acad. Sci. US* **51**, 125-130 (1964).

Spore-forming bacilli
H. Paulus, *J. Biosc.* **6**, 403-418 (1984).

CHAPTER XIX

Biosynthesis of the amino acids of the glutamic acid family and its regulation

I. BIOSYNTHESIS OF GLUTAMINE

Biosynthesis of glutamine. Cumulative feedback inhibition

This reaction is carried out by an enzyme, glutamine synthetase, whose occurrence is not restricted to microorganisms. It is essential not only for the synthesis of glutamine, a building block of proteins, but it also contributes, through its amide group, to the synthesis of many nitrogen containing substances. In addition, it is an obligatory intermediate in ammonia assimilation in ammonia-limited media.

Glutamine synthetase has been obtained in the pure state from different sources, in particular in *E. coli,* where it is encoded by the *glnA* gene, situated at 87 mn on the chromosome. It catalyzes the following reactions:

$$\text{L-glutamate} + NH_3 + \text{ATP} \rightarrow \text{L-glutamine} + \text{ADP} + \text{Pi}$$
$$\text{L-glutamate} + NH_2OH + \text{ATP} \rightarrow \text{L-glutamohydroxamate} + \text{ADP} + \text{Pi}$$

The second reaction is obviously non-physiological, but provides an easy colorimetric assay by reacting the hydroxamate with ferric ions.

In the absence of ammonia, the enzyme can also catalyze the following reactions, which are not believed to be physiologically significant:

$$\text{L-glutamate} + \text{ATP} \rightarrow \text{pyrrolidone carboxylate} + \text{ADP} + \text{Pi}$$
$$\text{L-glutamine} + H_2O \rightarrow \text{L-glutamate} + NH_3$$
$$N_1TP + N_2DP \rightarrow N_2TP + N_1DP$$

In this last reaction, N_1 and N_2 stand for different nucleosides.

In the presence of Mn^{2+} and AMP, the unadenylylated enzyme (see below) can also catalyze the formation of glutamate and pyrophosphate from glutamine and inorganic phosphate:

$$\text{L-glutamine} + 2\ \text{Pi} \rightarrow \text{L-glutamate} + \text{PPi} + NH_3$$

The allosteric control of glutamine synthetase deserves a detailed examination, for it differs from the schemes we have encountered so far for the branched biosynthetic pathways. First, it is important to recall that the glutamine amide group supplies the nitrogen atoms of the following molecules by means of reactions which we shall study in due course: tryptophan, adenylic acid, cytidylic acid, glucosamine-6-phosphate, histidine, and carbamyl phosphate. In addition, it has been shown in *E. coli* that glutamine is used as a substrate for transaminases which convert, respectively, pyruvate to alanine and glyoxylate to glycine.

It is very interesting to note that the eight compounds listed, plus serine, are inhibitors of the glutamine synthetase of *E. coli*, and also of organisms as different as *Salmonella typhimurium, Pseudomonas fluorescens, Neurospora crassa, Bacillus licheniformis, Chlorella pyrenoidosa* and *Anabaena* sp. Many other nitrogen-containing compounds not derived from glutamine are not inhibitors. The majority of the organisms studied do not possess isofunctional glutamine synthetases (as found for *E. coli* aspartokinase and homoserine dehydrogenase activities) and no glutamine synthetase shows the phenomenon of concerted feedback inhibition (as found, for example, for *Rh. capsulatus* aspartokinase).

Each of the inhibitors causes only partial inhibition, even at near saturation, when investigated individually; however, in combination, their effects are cumulative. Each inhibitor acts independently and as a result, when two or more end-products are present together at saturating concentrations, the residual activity is equal to the product of the residual activities observed with each of the inhibitiors at saturating concentration. The enzyme activity can be completely inhibited by the simultaneous presence of the nine inhibitors. Cumulative feedback inhibition provides an elegant regulatory mechanism for branched pathways which lead to the formation of numerous end products. Ideally, the fraction reduced by any given inhibitor is sufficient to curtail significantly the synthetic pathway to produce that end product while the other branches can still function more or less normally through the differential in binding constants. However, the inhibition of glutamine synthetase might prove disastrous for the economy of the cell if there did not exist individual control mechanisms for each branch.

The fundamental mechanism underlying the cumulative feedback inhibition is not yet clear; in fact, it is difficult to visualize a simple model which could account for the experimental observations; each model must explain the restricted inhibition by each of the effectors together with the absence of synergistic effects or antagonism at saturating concentrations. This last phenomenon implies that the enzyme must possess distinct allosteric binding sites for each of the inhibitors. Indeed, early studies indicated that this was the case: inhibition by glycine, cytidylic acid, or tryptophan is competitive with respect to glutamate, whereas inhibition by histidine or glucosamine-6-phosphate is competitive with respect to ammonia, and inhibition by alanine, adenylic acid, or carbamyl phosphate is not competitive with respect to either of the two substrates. This result implies at least three binding sites. Furthermore, 1 M urea (in the absence of chelating agents) desensitizes the enzyme to inhibition by tryptophan, histidine and glucosamine-6-phosphate; there must be therefore at least one additional site. Treatment of the enzyme with 30% acetone does not cause any loss in activity, but produces an increased inhibition by alanine and glycine: we have now six sites. Another difference arises from the observation that adenylic acid and histidine protect the enzyme against mercaptide formation with *p*-chloromethylsulfonate, while carbamyl phosphate and cytidylic acid accelerate the rate of inactivation by this reagent.

Taking all these results together, we reach the conclusion that each inhibitor must have its own binding site.

These observations were made on a mixture of adenylylated and unadenylylated glutamine synthetase (see below), which complicates the understanding of the mechanism of this cumulative feedback inhibition, since the adenylylated enzyme is highly sensitive to the above effectors, while the unadenylylated enzyme is much less susceptible to them.

As for all allosteric enzymes, the determination of the number of binding sites on a homogeneous preparation is absolutely necessary for the construction of a model which would account for the observed facts.

Recently, it has been unequivocally shown that fully unadenylylated glutamine synthetase binds glycine, alanine, and serine at the glutamate binding site. This result, obtained by X-ray cristallography, contradicts some of the above conclusions (Liauw, Pan & Eisenberg 1993).

Biosynthesis of glutamine. Covalent modification of glutamine synthetase

E. coli cells grown under nitrogen limitation (glutamate as the nitrogen source) contain active glutamine synthetase. In the presence of excess ammonia, the synthesis of the enzyme is repressed. Addition of NH_4^+ ions to derepressed cells not only causes a repression, but also an "inactivation", in a few seconds, of the enzyme already present. If cells containing the inactivated enzyme are transferred into a medium where ammonia is limiting, the enzyme is reactivated. The inactivation can be obtained *in vitro* in the presence of glutamine, ATP and Mg^{2+} ions, and of a specific inactivating enzyme. Only the glutamine synthetase activity proper is inactivated, whereas most of the glutamyl transferase activity (measured by glutamyl hydroxamate synthesis) remains unaffected. The conclusion is drawn that only part of the catalytic site of glutamine synthetase is altered by the inactivating enzyme.

The main difference between the two forms of the enzyme is in their ultraviolet spectrum: the difference between the two spectra has a maximum at 261 mm, which suggests that the ammonium chloride grown cells contain a bound nucleotide absent from the cells grown in the presence of glutamate. Since the inactive enzyme is reactivated by the action of snake venom phosphodiesterase, it is concluded that the modification caused by the inactivating enzyme is an incorporation of a bound nucleotide into glutamine synthetase. This is confirmed by the incorporation of radioactivity into the inactivated enzyme when the inactivation is performed in the presence of radioactive ATP labeled either in the adenine or with α ^{32}P. The bound nucleotide has been unequivocally identified as covalently attached 5'-AMP. It is presumed that the effect observed *in vivo* is due to a synthesis of glutamine from ammonia followed by an inactivation of glutamine synthetase by the inactivating enzyme.

Glutamine synthetase structure

The gene coding for the *E. coli* glutamine synthetase (*glnA*, 87 mn) has been cloned and sequenced. The deduced amino acid sequence is given in figure 1.

SAEHVLTMLNEHEVKFVDLRFTDTKGKEQHVTIPAHQVNA
EFFEEG**K**MFDGSSIGGWKGINESDMVLMPDASTAVIDPFF 80

ADSTLIIRCDILEPGTLQGTDRDPRSMSKRADEYLRSTGI
ADTVLFGPEPEFFLFDDIRFGSSISGSHVAIDDIEGAWNS 160

STQYEGGNKGHRPAVKGGYFPVPPVDSAQDIRSEMCLVME
QMGLVVEAHHHEVATAGQNEVATRFNTMTKKADEIDIYKY 240

VVHNVAHRFGKTATFMPKPMFGDNGSGMHCHMSLSKNGVN
LFAGDKYAGLSEQALYYIGGVIKHAKAINALANPTTNSYK 320

RLVPGWEAPVMLAYSARNRSASIRIPVVSSPKARRIEVRF
PDPAANPYLCFAALLMAGLDGIKNKIHPGEAMDKNL**Y**DLP 400

PEEAKEIPQVAGSLEEALNELDLDREFLKAGGVFTDEAID
AYIALRREEDDRVRMTPHPVEFELYYSV 468

Figure 1. The amino acid sequence of *E. coli* glutamine synthetase.

Each subunit (MW = 51,814 Da) has four cysteines, not accessible to –SH reagents when the enzyme is in its active conformation, two tryptophans, whose fluorescent quantum yield is very sensitive to the binding of substrates and effectors; 17 tyrosines, including Tyr 397* which is the adenylylation site; 16 histidines, one of which (His 269) whose oxidation by mixed-function oxidation reagents causes the irreversible inactivation of the enzyme. In addition, Lys 47* is covalently modified by an ATP analogue, which inactivates the enzyme in a competitive manner to ATP binding, suggesting that it is part of the ATP binding site.

* Indicated in bold letters in the sequence, figure 1.

>

Figure 2. A high-magnification picture of five superimposed images of unfixed molecules in the three characteristic orientations. The mean dimensions are indicated. When the molecule rests on a face, the subunits appear as a hexagonal ring (top). Molecules seen on edge show two layers of subunits, either as four spots (center) when viewed exactly down a diameter between subunits, or in general as two lines (bottom). The molecular structure is thus based on twelve identical subunits in two hexagonal layers, the units in one layer being directly above those in the other. Magnification: 3,160,000.
(From R. C. Valentine, B. M. Shapiro & E. R. Stadtman, *Biochemistry* **7**, 2143-2152 (1968). With permission from the American Chemical Society.)

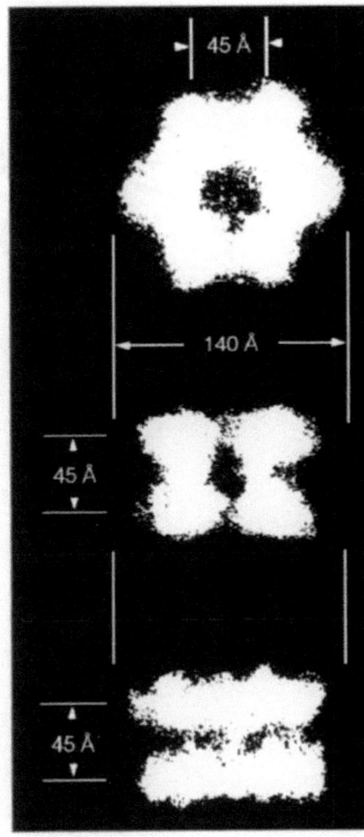

The sequence of the *Salmonella* enzyme is very similar, that of *Anabaena* is somewhat different, and the mammalian enzyme is distinctly different.

The enzyme is dissociated by guanidine hydrochloride into twelve identical subunits. Electron micrographs of the native enzyme show that it is made up of twelve subunits arranged in two layers of six. This double-decked hexagon has a 6-fold dihedral symmetry. The outer diameter of the hexagon is 140 Å and the subunit molecular dimensions are 45 x 45 x 53 Å (fig. 2).

The tridimensional structure of the completely unadenylylated enzyme from *Salmonella typhimurium*, which has a very similar primary sequence and which crystallizes isomorphously with the *E. coli* enzyme has been solved at 3.5 Å resolution. The 12 identical subunits are present in the asymmetic unit of the crystal. The model obtained is in general agreement with the electron microscopy. The outer diameter of the hexameric ring is 143 Å.

In each subunit, the small N-terminal domain (residues 1-103) is formed of six strands of antiparallel β-sheets, one large and one small α-helix. It is connected to a large C-terminal domain just before the long helix starting at residue 104. This domain contains seven long and six short α-helices and an extended β-sheet (six long and three short strands). At the top, six prominent strands of β-structure from the C-domain form a partial barrel around two metal ions. In addition, there are 25 turns and loops (fig. 3).

The view of glutamine synthetase down the 6-fold axis (fig. 4) reveals a petal-shaped molecule with an outer diameter of 143 Å. As anticipated from the electron micrographs taken some 20 years before the structure was elucidated, there is a central channel about 40 Å in diameter, but it is not completely open. The central loop referred to in the caption of figure 3 protrudes into the channel, so that the open inner diameter is only 24 Å. This loop is sensitive to various proteases. Six active sites, each marked by a pair of manganese ions, are roughly midway between the periphery of the glutamine synthetase molecule and the central channel. Each active site is at the interface between two subunits.

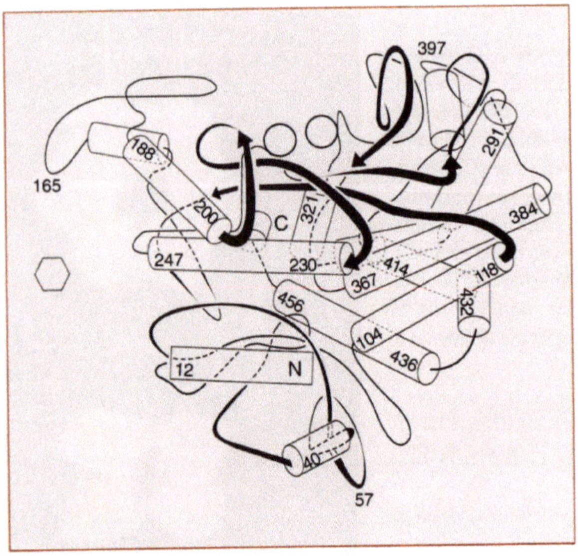

Figure 3. A schematic drawing of the glutamine synthetase polypeptide chain. The 6-fold axis is indicated by the hexagon at the left, and the metal ions by two circles in the upper center. Cylinders: α-helices; heavy arrows, the six strands surrounding the metals. Bottom left: the N-terminal domain (1-103). The C-terminal domain starts at residue 104. Note the central loop at the upper left.
(From R. J. Almassy, C. A. Janson, R. Hamlin, N. H. Xuong & D. Eisenberg, with permission from *Nature* **323**, 304-309. Copyright (1986) Macmillan Magazines Limited.)

Each of the two metal atoms necessary for catalytic activity is connected to the protein backbone by four ligands. The formation of an active site from the association of two subunits had already been found in several enzymes, among them aspartate aminotransferase (transaminase) and aspartate transcarbamylase.

This model, proposed by David Eisenberg and his colleagues, suggests a mechanism for the main functional role of glutamine synthetase: how the enzyme regulates the rate of synthesis of glutamine in response to covalent modification and feedback inhibition. The reader is encouraged to read their paper.

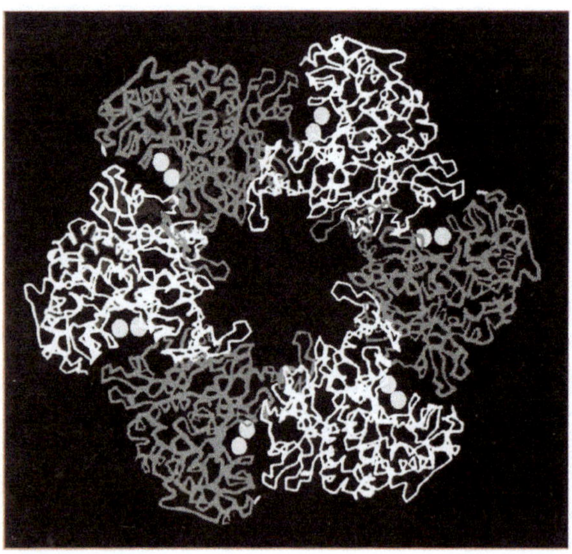

Figure 4. The upper half of the glutamine synthetase molecule, projected down the 6-fold axis of symmetry. The six subunits are shown as line segments connecting sequential α-carbon atoms. Six active sites are marked by pairs of Mn^{2+} ions (shown as spheres). The side-chain of Tyr 397, target of adenylylation, is between subunits at larger radius than the active site.
(From R. J. Almassy, C. A. Janson, R. Hamlin, N. H. Xuong & D. Eisenberg, with permission from *Nature* **323**, 304-309. Copyright (1986) Macmillan Magazines Limited.)

Reversible adenylylation of the *E. coli* glutamine synthetase

Glutamine synthetase from *E. coli* and other enteric bacteria was found to exist into two interconvertible forms: an adenylylated form, which is highly sensitive to allosteric inhibitors (the activity measured in this case is the glutamyltransferase), and an unadenylylated form, relatively insensitive to feedback inhibition. Adenylylation implies the specific transfer of the adenylyl moiety of ATP to Tyr 397 of each subunit of the dodecamer and is catalyzed by an enzyme which has been called glutamine synthetase adenylyl transferase. This tyrosine is known by the crystallographic model to be located near the interface, between two subunits, and to be close to the outer periphery of the hexameric ring. The linkage between the adenylyl group and the tyrosine hydroxyl is a phosphodiester bond.

The enzyme has been adenylylated by a spin-labeled analogue of ATP. The modified enzyme exhibits similar catalytic properties, pH profile, and inhibitor susceptibility as that of the glutamine synthetase adenylylated with normal ATP. The nitroxyl spin label allows the determination of the distance between the paramagnetic metal (Mn^{2+}) and the nitroxyl moiety of the probe. This permits a topographical analysis of the correlation between the structural metal binding site, the catalytic site, and the regulatory adenylylation site. The data show that the two latter sites are close to each other and that binding of substrates changes their distances. In addition, the data on the rotation correlation time confirm that the adenylyl site is located on the surface of the enzyme.

Each subunit can be adenylylated, so that a molecule of glutamine synthetase can have a maximum of 12 adenylyl groups covalently attached:

$$\text{glutamine synthetase} + n\,\text{ATP} \xrightarrow{Mg^{2+}} \text{glutamine synthetase (AMP)}n + n\,\text{PPi}$$

Adenylyltransferase can also catalyze the phosphate dependent removal of AMP from each subunit of glutamine synthetase.

The deadenylylation is not the reverse reaction of adenylylation:

$$\text{glutamine synthetase (AMP)}_n + n\,\text{Pi} \xrightarrow[\text{ATP}]{Mg^{2+}} \text{glutamine synthetase} + n\,\text{ADP}$$

Adenylyltransferase (coded by *glnE*, 66.7 mn) has been purified and is a monomer (MW = 115,000). It can catalyze the two reactions *in vitro* in the absence of the two regulatory proteins*, PII and uridylyltransferase/uridylyl removing enzyme. Biochemical and genetic evidence indicate that the two activities depend on separate sites on the same polypeptide.

* See below.

Regulation of glutamine synthetase activity by covalent adenylylation

The state of the cell with regard to nitrogen availability is defined by the intracellular ratio of glutamine to α-ketoglutarate. When ammonia is in excess, the rate of glutamine synthesis by glutamine synthetase increases, as does the conversion of α-ketoglutarate to glutamate by glutamate synthase and glutamate dehydrogenase. If ammonia becomes limiting, the rate of glutamine synthesis decreases. As a result, excess ammonia in the medium causes a high glutamine/α-ketoglutarate ratio and a low intracellular concentration of partially unadenylylated glutamine synthetase. When ammonia is limiting, this ratio becomes low and the intracellular concentration of active, unmodified, glutamine synthetase becomes high.

The regulation of glutamine synthetase activity requires the cooperation of three proteins: 1) the glutamine synthetase adenylyltransferase (ATase); 2) a protein called PII, which exists in two forms, a uridylylated form and a non-uridylylated form; 3) a bifunctional uridylyltransferase/uridylyl-removing enzyme (UTase/UR). A high concentration of glutamine activates the uridylyl-removing enzyme which, in turn, causes the deuridylylation of protein PII; PII interacts with the adenylyl transferase which causes the adenylylation of glutamine synthetase. The deadenylylation is carried out as follows: a high concentration of α-ketoglutarate activates the uridylyltransferase activity of the bifunctional protein and PII is uridylylated; the uridylylated form of PII interacts with the adenylyltransferase which causes the deadenylylation of glutamine synthetase.

Here again, the deuridylylation reaction is not the reversal of the uridylylation reaction. The two reactions can be written as follows:

$$\text{PII} + n\,\text{UTP} \xrightarrow[\text{Mg}^{2+},\ \alpha\text{-ketoglutarate}]{\text{uridylyltransferase}} \text{PII-(UMP}n) + n\,\text{PPi}$$

$$\text{PII-UMP}n + n\,\text{H}_2\text{O} \xrightarrow[\text{glutamine}]{\text{uridylyl removing enzyme}} \text{PII} + n\,\text{UMP}$$

PII (coded by *glnB*, 55 mn) is a tetramer consisting of identical subunits, each of a molecular weight 44,000. The uridylylation of PII occurs at one specific tyrosine residue in one or more subunits. The UTase/UR enzyme (coded by *glnD*, 4 mn) is a monomer of MW = 95,000.

Figure 5 summarizes the sequence of events.

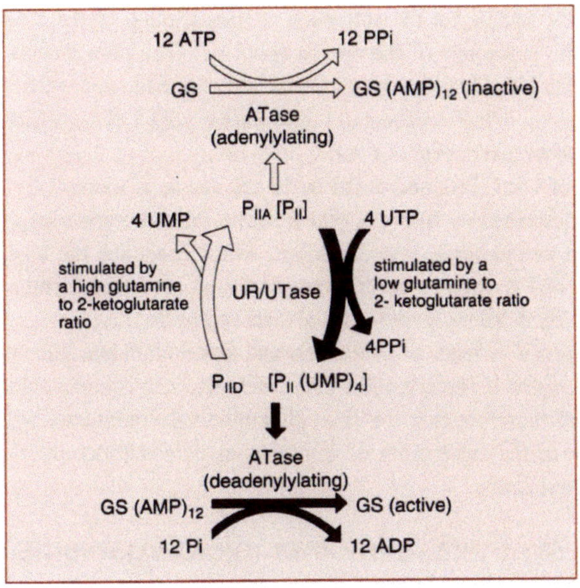

Figure 5. Covalent modification of glutamine synthetase (GS).
UR/UTase: uridylyl removing enzyme/uridylyl transferase.

Regulation of the synthesis of glutamine synthetase also involves the two forms of PII and UR/UTase

The *glnA* gene, coding for glutamine synthetase, belongs to an operon which comprises three structural genes, *glnA* itself and two genes, *glnG* and *glnL* (87 mn) coding respectively for two regulator genes, NR_I and NR_{II}. These proteins are homodimers whose subunits have a respective molecular weight of 110,000 and 68,000. NR_{II} has been shown to be a kinase which phosphorylates NR_I in the presence of ATP. The phospho-

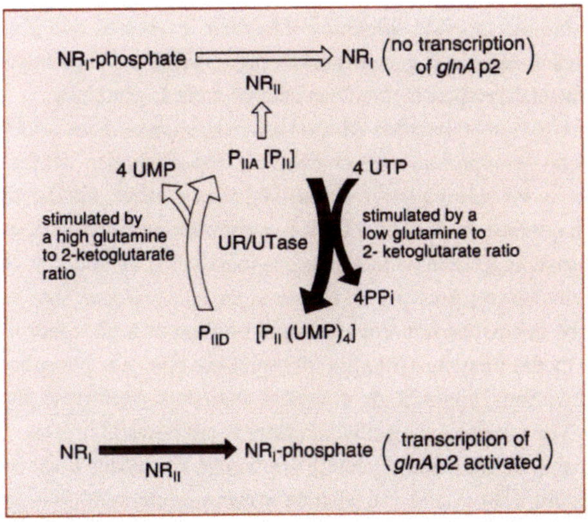

Figure 6. Covalent modification of NRI, the regulator of *glnA* transcription.
(From L. J. Reitzer & B. Magasanik, in *Escherichia coli and Salmonella typhimurium, Cellular and Molecular Biology* (J. L. Ingraham, K. B. Low, B. Magasanik, M. Schachter & H. E. Umbarger eds.), vol. I, pp. 302-320, American Society for Microbiology, Washington D.C., 1987. With permission from the American Society for Microbiology.

rylated NR_I is responsible for the activation of transcription of the *glnA* gene at one of the promoters of the operon (*glnA* p2). The phosphorylation is reversible, NR_{II} being a bifunctional enzyme endowed with a phosphatase activity which regenerates unphosphorylated NR_I, which cannot activate the transcription of *glnA*.

The role of UR/UTase and of PII in the regulation of transcription of *glnA* is best described by figure 6, which shows that the variations of glutamine and α-ketoglutarate concentrations, which modulate the level of uridylylation of PII, affect not only the activity of glutamine synthetase (by controlling its adenylylation) but also its synthesis.

The reciprocal effects of glutamine and α-ketoglutarate can be more fully appreciated if one considers that α-ketoglutarate is a precursor of glutamine and therefore that the ratio glutamine/α-ketoglutarate will vary in response to the availability of ammonia, as demonstrated by various *in vivo* experiments.

Glutamine synthetase in other microorganisms

Bacillus species do not have a global sophisticated system of regulation of nitrogen metabolism as demonstrated in *E. coli.* It is clear for example that *B. subtilis* lacks the wide range of adaptability with respect to the nitrogen source that is typical of *Enterobacteriaceae*. On the other hand, *B. subtilis* responds by sporulating when deprived of a rapidly metabolizable nitrogen source. In the same species, selective proteolysis may serve as an important regulatory mechanism: *in vitro* proteolysis of the enzyme is inhibited by the substrates metal-ATP and glutamate, and accelerated by the product, glutamine. Thus, the activity of glutamine synthetase might be controlled directly through the ratio of the concentrations of the substrates and/or the products in the cell. In *B. subtilis*, glutamine synthetase is also a dodecamer consisting of a two-layered structure of hexagonal rings.

The extreme thermophile *Bacillus caldolyticus* expresses two glutamine synthetases as separated gene products, regulated in a complementary fashion by the end-product metabolites derived from L-glutamine.

In the photosynthetic bacteria *Rhodobacter capsulatus*, *Rhodobacter spheroides,* and *Rhodopseudomonas palustris*, the glutamine synthetase is regulated by an adenylylation-deadenylylation system, similar to that of *E. coli*. In contrast, in *Synechocystis,* a cyanobacterium which does not possess this system, glutamine synthetase is inactivated by addition of ammonium ions to cultures growing on nitrate as the sole nitrogen source. This appears to be due to the non-covalent tight binding of a phosphorylated compound to the enzyme. Alkaline phosphatase (but not phosphodiesterase), and various physical or chemical tratments, reactivate the enzyme *in vitro.* The nature of the actual inhibitor is unknown at present.

The nitrogen-fixing actinomycete *Frankia* can fix nitrogen for its own growth in pure culture and can also be a root-nodule symbiont. In free culture, the main nitrogen assimilation occurs through glutamine synthetase. Two glutamine synthetases, GS1 and GS2 are synthesized under nitrogen-fixing conditions, but only GS1 is synthesized during growth on ammonia. Under N2-fixing symbiotic conditions, the synthesis of both GS1 and GS2 is repressed. Whereas GS1 consists of 12 subunits of about 59 kDa each and its synthesis is repressed by nitrogen starvation, the GS2 gene has high-sequence similarities to the genes of eukaryotic glutamine synthetases.

The existence of two glutamine synthetases in the bacterium *Bradyrhizobium japonicum* has been the subject of a very interesting debate. Whereas GSI from this organism is a dodecamer, as found in eubacteria and archaebacteria, GSII was found to have an octameric arangement and an amino acid sequence typical of eukaryotes. The obvious opportunity afforded by its symbiotic association with plants made the possibility of a horizontal gene transfer from the plant to the bacterium quite plausible.

II. BIOSYNTHESIS OF GLUTAMATE

In principle, glutamate synthase and glutamate dehydrogenase are both able to synthesize glutamate. Mutants lacking both enzymes have an absolute requirement for glutamate. But glutamate dehydrogenase is a completely dispensable enzyme, although the cell can use it if provided with a high enough concentration of ammonia.

Glutamate dehydrogenase

Glutamate dehydrogenase has been purified from *E. coli* (*gdhA*, 27 mn) and *S. typhimurium*. Its molecular weight is 300,000, with a relatively low apparent affinity for ammonia. It is a hexamer of identical subunits.

The reaction catalyzed,

$$\underset{\alpha\text{-ketoglutarate}}{COOH{-}(CH_2)_2{-}CO{-}COOH} + NH_3 + NADPH + H^+ \rightarrow \underset{\text{glutamate}}{COOH{-}(CH_2)_2{-}CH(NH_2){-}COOH} + H_2O + NADP^+$$

uses NADPH for the reductive amination, but not NADH.

Although there is no control of glutamate dehydrogenase synthesis by ammonia, repression is observed when glutamate is present in the growth medium of *E. coli*, but curiously this effect is not seen in *S. typhimurium*.

Glutamate synthase

When the ammonium ions concentration is sufficiently high, ammonia is incorporated directly into glutamine, glutamate, and asparagine. When ammonia is limiting, it is incorporated in glutamine only. Since glutamate is about eight times more abundant than glutamine in *E. coli* proteins, glutamine has to be somehow recycled in glutamate under these conditions. The enzyme responsible for this recycling is glutamate synthase, which catalyzes the following reaction:

glutamine + α-ketoglutarate + NADPH + H^+ → 2 glutamate + $NADP^+$

The enzyme has been purified from *E. coli* and *Klebsiella aerogenes*. It is a heterodimer whose two subunits which have molecular weights of 53,000 and 135,000 in *E. coli*. It contains both flavin adenine dinucleotide (FAD) and flavin mononucleotide (FMN), iron, and labile sulfide. Like many other amidotransferases, it can utilize a high level of ammonia instead of glutamine as the nitrogen donor. The ammonia-dependent activity is catalyzed by the isolated small subunit, which then performs essentially the same reaction as glutamate dehydrogenase.

A mechanism for glutamate synthase has been proposed, in which NADPH binds the small subunit and transfers its electrons to the large subunit, which reduces the flavin. α-ketoglutarate binds the small subunit and glutamine binds the large one. The amide group of glutamine is then transferred to ketoglutarate and the reduced flavin reduces a postulated iminoglutarate intermediate to glutamate. The loss of either subunit results in the loss of capacity to carry the overall reaction. This mechanism is supported by chemical and kinetic data.

The genes coding for the two subunits form an operon (*gltBD*, 69 mn), *gltB* and *gltD* coding respectively for the large and the small subunit.

Glutamate represses the synthesis of glutamate synthase, which is activated upon glutamate deprivation. The mechanism of this regulation is not yet clear.

III. BIOSYNTHESIS OF PROLINE

Biosynthesis of proline in *Enterobacteriaceae*

The radioactivity from ^{14}C-glutamate is incorporated into proline, and isotope competition experiments show that glutamate is a precursor of proline.

It was established during the 1950s that an intermediate of proline biosynthesis was glutamate γ-semialdehyde, arising from the reduction of the γ-carboxyl of glutamate.

$$OHC{-}CH_2{-}CH_2{-}CH(NH_2){-}COOH$$

glutamate-γ-semialdehyde

This compound is the higher homologue of aspartate-β-semialdehyde, the common precursor of lysine, methionine, and threonine, which we have encountered in a preceding chapter. However, it is expected that between glutamate and the semialdehyde, there should exist some activated intermediate. It was proposed in 1969 that such an intermediate could be γ-glutamyl phosphate. Unfortunately, this compound is extremely labile and cyclizes spontaneously to 5-oxopyrrolidine-2-carboxylate. However, if the postulated corresponding kinase reaction is carried out in the presence of hydroxylamine, a hydroxamate is obtained – which is to be expected if an energy-rich phosphate intermediate is formed. On the other hand, if the second step in the pathway, glutamate semialdehyde dehydrogenase, is studied in the reverse direction, the end product is 5-oxopyrrolidine-2-carboxylic acid, the cyclization product of glutamyl phosphate. It was concluded that the latter existed *in vivo* as en enzyme-bound intermediate.

In 1984, it was shown that the glutamyl kinase is active only in the presence of the enzyme catalyzing the second step, glutamate semialdehyde dehydrogenase. The complex is very labile *in vitro* and its dissociation leads to an inactive kinase.

The glutamate kinase has finally been purified from a strain containing a multicopy plasmid carrying the genes for the two enzymes, which are organized into a single operon, *proBA* (5.8 mn). It is assayed in the presence of an excess of the dehydrogenase. It is a homopolymer of six subunits of molecular weight of 38,952, as deduced from the DNA sequence of its coding gene *(proB)*. The reaction catalyzed

$$\text{glutamate} + \text{ATP} \rightarrow \gamma\text{-glutamylphosphate} + \text{ADP}$$

is inhibited by proline at a very low concentration.

Glutamic semialdehyde dehydrogenase catalyzes the second step of proline biosynthesis, the reduction of glutamylphosphate to glutamate semialdehyde. Whereas the subunit molecular weight is 43,503, as determined by the nucleotide sequence of the *proA* gene, there is a controversy concerning the molecular weight of the native enzyme, which could be either a tetramer or a hexamer.

Glutamic semialdehyde exists in equilibrium with Δ^1-pyrroline-carboxylic acid, which is the substrate of a specific reductase which reduces it to L-proline in the presence of NADPH:

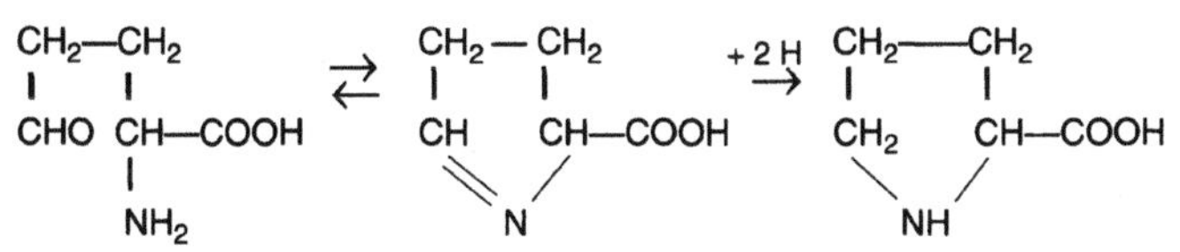

Δ^1-pyrroline carboxylate L-proline

From a strain harboring a plasmid carrying *proC* (8.9 mn), the gene coding for the reductase, a homogeneous preparation was obtained with a MW = *ca.* 300,000. The subunit, whose sequence was derived from the DNA sequence, has a molecular weight of 28,112.

The inhibition of glutamate kinase by proline is the major control mechanism of proline biosynthesis. Although it has been sought for in several laboratories, a negative control by end-product repression has not been found. It has even been excluded in a study performed with *Serratia marcescens,* by analyzing the expression of the *lacZ* structural gene fused with the *proBA* promoter. Other control mechanisms, such as a translational control at the level of the mRNA specified by the *proBA* operon (transcribed from *proB* to *proA*), have not been excluded.

Biosynthesis of proline in other genera

Bacteria have developed other ways to synthesize proline. Some – like the strict anaerobes *Clostridium sporogenes* or *Cl. botulinum* – use ornithine cyclase; others, like *Aspergillus nidulans,* rely on ornithine-δ-transaminase

$$\text{ornithine} + \alpha\text{-ketoglutaric acid} \rightarrow \text{glutamate semialdehyde}$$

In many plants, proline is accumulated in response to osmotic stress. From *Vigna aconitifolia,* a cDNA clone has been isolated which encodes a bifunctional enzyme, Δ^1-pyrroline-5-carboxylate synthetase, which corresponds to the two proteins encoded by *proA* and *proB* in *E. coli.*

Utilization of proline

Proline utilization by *E. coli* and *S. typhimurium* requires the expression of genes *putP* (encoding a proline transporter) and *putA*. Genetic evidence indicates that the PutA protein is both Put repressor and a respiratory chain-linked dehydrogenase. The *putA* gene from *E. coli* has been cloned and sequenced and the corresponding protein purified to homogeneity; it is a dimer and binds the *put* DNA *in vitro.*

IV. BIOSYNTHESIS OF ARGININE AND POLYAMINES

Biosynthesis of arginine

In the course of evolution, microorganisms have developed a mechanism analogous to the one we have already seen during the study of lysine biosynthesis to protect an amino group from involvement in spontaneous cyclization. But in this case, it is acetylation rather than succinylation that we find: a specific N-acetylglutamate synthase is both repressed and inhibited by arginine. In *E. coli,* where its structure is specified by *argA* (60 mn), it consists of a single type of subunit of molecular weight of *ca.* 50,000. Arginine is a potent feedback allosteric inhibitor of the enzyme. Desensitized mutants have been obtained.

$$\text{glutamate} + \text{acetyl CoA} \rightarrow \underset{\substack{|\\ NH\text{–}CO\text{–}CH_3}}{COOH\text{–}CH}\text{–}(CH_2)_2\text{–}COOH + \text{CoASH}$$

N-acetylglutamate

The *argB* gene (88 mn) codes for the next enzyme, N-acetylglutamokinase (MW of the primary polypeptide product, *ca.* 29,000), which phosphorylates the γ-carboxyl group of N-acetylglutamate, yielding an energy-rich compound, N-acetyl-γ-glutamylphosphate, which is labile and has not been isolated in the free state,

N-acetylglutamate + ATP → N-acetyl-γ-glutamylphosphate + ADP

and is hypothesized to be transferred onto the next enzyme without being liberated.

This enzyme, N-acetylglutamylphosphate dehydrogenase or reductase (*argC*, 88 mn), catalyzes the reduction of the phosphoryl ester to N-acetylglutamate semialdehyde. It has been somewhat purified and the molecular weight of the *argC* encoded peptide has been reported to be 47,000, although the gene coding for it encodes only 334 residues. Comparison of the sequences of the enzyme from *E. coli*, *Bacillus subtilis*, *Streptomyces clavuligerus*, *Saccharomyces cerevisiae*, and from the cyanobacterium *Anabaena*, highlight a cysteine residue present in a highly-conserved domain in the five proteins. This cysteine is probably located in the active site of the enzyme. The electron donor is NADPH.

The semialdehyde is the acetylated form of the intermediate occurring in proline biosynthesis, and were it not for the acetylation, all the semialdehyde would be converted to proline.

N-acetylglutamate semialdehyde undergoes a transamination reaction, catalyzed by N-acetylornithine transaminase (*argD*, 72.5 mn), which has been obtained in the crystalline state (MW of the native enzyme =119,000, of the subunit = 31,000)

$$\underset{\text{N-acetylglutamate semialdehyde}}{CHO\text{–}(CH_2)_2\text{–}\underset{\displaystyle NH\text{–}CO\text{–}CH_3}{\underset{|}{CH}}\text{–}COOH} + \text{glutamate} \rightarrow \underset{\text{N-acetylornithine}}{NH_2\text{–}(CH_2)_3\text{–}\underset{\displaystyle NH\text{–}CO\text{–}CH_3}{\underset{|}{CH}}\text{–}COOH} + \alpha\text{-ketoglutarate}$$

The acetyl group can now be removed. This reaction is catalyzed by a deacylase, N-acetylornithinase (*argE*, 88 mn), an enzyme depending for its activity on the presence of divalent cobalt ions and a thiol compound. The enzyme has been obtained in the homogeneous state and is a monomer of MW 62,000; it is rather unspecific and deacylates other substrates than N-acetylornithine.

Ornithine carbamoyltransferase, also known as ornithine transcarbamylase (*argI*, 97 mn) catalyzes the sixth step of arginine biosynthesis, the coupling of ornithine and carbamyl phosphate to yield citrulline:

$$\underset{\text{ornithine}}{NH_2\text{–}(CH_2)_3\text{–}\underset{\displaystyle NH_2}{\underset{|}{CH}}\text{–}COOH} + \underset{\text{carbamylphosphate}}{NH_2\text{–}COO\ PO_3H_2} \rightarrow \underset{\text{citrulline}}{\underset{\displaystyle CONH_2}{\underset{|}{NH}}\text{–}(CH_2)_3\text{–}\underset{\displaystyle NH_2}{\underset{|}{CH}}\text{–}COOH} + P_i$$

Ornithine transcarbamylase and carbamyl phosphate synthetase are by far the best known enzymes of the arginine regulon. Carbamyl phosphate synthetase (coded by the *carAB* genes), which provides the carbamyl phosphate necessary not only for arginine, but also for pyrimidine synthesis will be examined later in this book. Ornithine transcarbamylase is a homotrimer. In the K12 strain of *E. coli,* there is a second gene, *argF*, located at 7 mn on the circular chromosome, which codes for another ornithine transcarbamylase, also a homotrimer, which displays 86% amino acid homologies with the enzyme coded by *argI*. Other strains of *E. coli* and other *Enterobacteriaceae* do not possess this second gene. In K12, the products of each gene interact to form four trimeric isoenzymes. The same pattern of homotrimeric quaternary structure is found in the ornithine transcarbamylases of *S. typhimurium, Pseudomonas putida, P. aeruginosa, Aeromonas formicans,* and *S. cerevisiae.* The molecular weight of their identical subunits varies between 35,000 and 40,000. A notable exception is the enzyme of *B. subtilis,* which exists as an equilibrium mixture of dimeric, tetrameric and hexameric forms of a 44,000 MW subunit. Interestingly, the pattern of regulation is also different in *B. subtilis,* the enzyme being repressed by arginine in exponentially growing cells, induced by arginine at the end of the exponential phase, and inactivated during sporulation. The enzyme is the same under all conditions.

The kinetic properties of the *E. coli argI* enzyme have been thoroughly investigated: carbamylphosphate binds first, and Pi is released last.

The seventh step of this particular pathway is the addition to citrulline of an aspartic acid residue in the presence of ATP: ATP activates the ureido group of citrulline, a citrulline-adenylate intermediate being likely, and argininosuccinate, AMP and pyrophosphate are released:

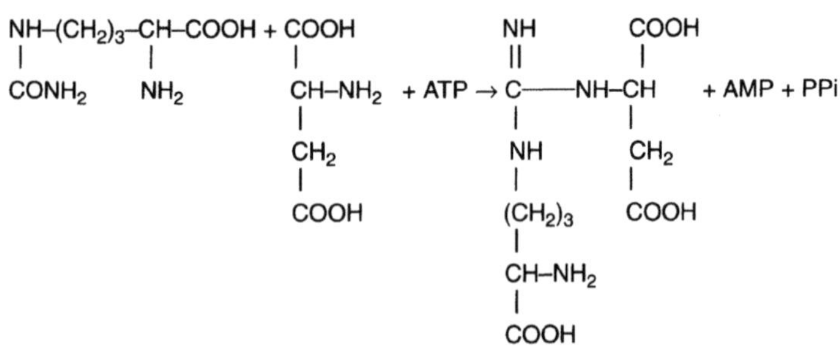

citrulline aspartate argininosuccinate

Argininosuccinate synthetase is coded by *argG* (68 mn on the *E. coli* map). The enzyme has not been studied in detail in bacteria, but is purified and well-known in mammals and in yeast. The yeast enzyme is a homotetramer (subunit MW = 48,000), while a basic polypeptide of the same molecular weight is predicted from SDS-gel electrophoresis of the *argG* product.

The last step is catalyzed by argininosuccinase (*argH,* 88 mn) which promotes a trans-elimination and hydrolyzes argininosuccinate to fumaric acid and arginine:

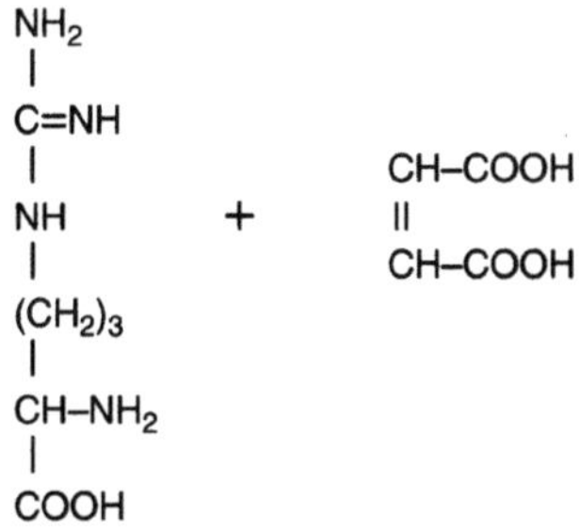

L-arginine

Here again, information on the bacterial enzyme is scanty. The purified mammalian counterpart is a tetramer of 50,000 Da subunits. Indirect evidence is available that the *E. coli* polypeptide coded by *argH* is of a similar size.

Regulation of arginine biosynthesis at the transcriptional level

If we examine the circular chromosome of *E. coli*, we notice that the structural genes of arginine biosynthetic enzymes are scattered all over the chromosome. They must therefore belong to several operons. The only genes to be clustered are *argECBH.* A fine analysis has revealed that the *argECBH* cluster is actually a divergent operon consisting of two arms, *argE* and *argCBH,* transcribed in opposite directions with an internal operator common region flanked by two convergent promoters.

The study of the 5'-flanking region of *argI*, *argF*, and of the *carAB* operon (specifying the structure of the two subunits of carbamylphosphate synthetase) has revealed the existence of operator regions which show a great similarity to that of the *argECBH* common operator and has permitted the derivation of a consensus sequence from the various operators, the Arg boxes, as we have seen in the case of the Met box operators of the methionine biosynthetic genes. Here, as in the case of methionine, the variation between the normally-found level and maximally derepressed levels of the various enzymes of the pathway is considerable: ornithine transcarbamylase levels can vary 400-fold, whereas N-acetylglutamate reductase, N-acetylglutamate kinase, and argininosuccinase, which belong to the same transcription unit (*argCBH*) vary coordinately only 50- to 70-fold[25]. The other individual gene products are also repressed, in a parallel but uncoordinated fashion, as are the products of the *carAB* genes and of *argR* coding for an arginine repressor.

25. N-acetylornithinase, although *argE* (transcribed divergently from *arg CBH*) belongs to the same operon, varies only 20-fold, due to the existence of a secondary weak promoter located outside of the control region.

Arginine repressor

In *E. coli,* the arginine repressor – in conjunction with arginine as the corepressor – controls the genes of the enzymes of arginine biosynthesis. The nucleotide sequence of the *argR* gene has been described. It corresponds to a monomeric subunit of molecular weight 16,500. The repressor is precipitable with arginine, which facilitates its purification. The native repressor is a hexamer (MW = 98,000). It binds to the Arg box of *argF* and to its own Arg box (autoregulation). In the presence of arginine, it inhibits the transcription of *argF* in an *in vitro* assay. In equilibrium dialysis experiments with arginine, the Arg repressor shows positive binding cooperativity, in contrast to the Met and Trp repressors.

An unexpected finding has been the implication of the arginine repressor as an auxiliary protein in site-specific recombination.

A similar arginine repressor, also a hexamer, plays a dual role in *B. subtilis.* It represses the arginine biosynthetic genes and activates the genes responsible for the enzymes of arginine catabolism in this organism. Although *E. coli* and *B. subtilis* are taxonomically distant, the *B. subtilis* repressor can functionnally complement *argR* for its two distinct functions in repression and recombination, probably by binding to similar regulatory sequences. Both repressors have been crystallized and should eventually permit a detailed comparison of the two systems.

Polyamine biosynthesis

We have seen how the acetylation of glutamate, the first reaction specific to the synthesis of arginine, is allosterically inhibited by this amino acid. In addition, in *E. coli*, the presence of excess arginine represses in a parallel, but not coordinate fashion, the synthesis of all the enzymes involved in its biosynthesis.

However, the synthesis of arginine cannot be considered as a perfectly linear sequence of reactions. In fact, one of the intermediates, ornithine, is the precursor of a diamine, putrescine:

$$\underset{\text{ornithine}}{H_2N{-}(CH_2)_3{-}CH(NH_2){-}COOH} \rightarrow \underset{\text{putrescine}}{H_2N{-}(CH_2)_4{-}NH_2} + CO_2$$

This reaction is catalyzed by ornithine decarboxylase (*speC,* 63.5 mn). The enzyme has been isolated as a dimer of MW = 160,000.

The physiological role of putrescine and of the two other polyamines known to occur in *E. coli*, spermidine and cadaverine, is still under discussion, although it is clear that they are required for optimal growth and that they must act through their polybasic nature.

The level of polyamines in *E. coli* is high, of the order of 30% of the arginine in the proteins: therefore, in the presence of excess arginine, the synthesis of ornithine would be reduced to a rate unsuitable for the synthesis of polyamines. Under these conditions, putrescine is synthesized *via* an alternative route: an arginine decarboxylase (*speA*, 63 mn) decarboxylates arginine to agmatine:

$$H_2N\text{–}C\text{–}NH\text{–}(CH_2)_3\text{–}CH\text{–}COOH \rightarrow H_2N\text{–}C\text{–}NH\text{–}(CH_2)_4\text{–}NH_2 + CO_2$$

NH NH_2 NH

arginine agmatine

Arginine decarboxylase is a tetramer of 296,000 Da, whose activity is inhibited by putrescine and spermidine.

Agmatine is then hydrolyzed by agmatine amidinohydrolase (also called agmatine ureohydrolase), specified by *speB*, localized at 63 mn on the *E. coli* chromosome.

$$H_2N\text{–}C\text{–}NH\text{–}(CH_2)_4\text{–}NH_2 + H_2O \rightarrow NH_2\text{–}(CH_2)_4\text{–}NH_2 + H_2N\text{–}CO\text{–}NH_2$$

NH

agmatine putrescine urea

Urea is not degraded by most strains of *E. coli* and its concentration is actually a measure of the agmatine pathway[26].

This alternative pathway is a new example of the diversity of the solutions arising in the course of evolution to the problem posed by the existence of branched pathways. As can be seen in the following scheme

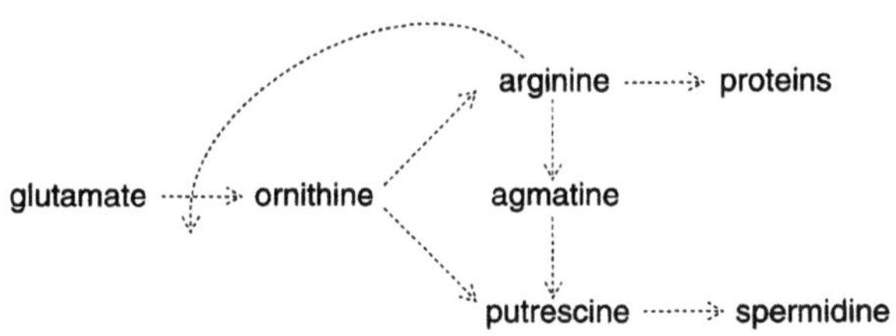

the synthesis of arginine is effectively part of a branched system, the branch point being situated at ornithine. The most economical way to produce putrescine is obviously to decarboxylate ornithine. However, when repressible strains of *E. coli* are in the presence of excess arginine, the activity of the first enzyme, N-acetylglutamate synthase, and the synthesis of the biosynthetic enzymes leading to arginine, are inhibited and the conversion of arginine to putrescine is the only alternative for the synthesis of polyamines.

The other important polyamine, spermidine, is a triamine and has the following structure:

$$H_2N\text{–}(CH_2)_3\text{–}NH\text{–}(CH_2)_4\text{–}NH_2$$

spermidine

which derives from putrescine by the addition of an aminopropyl group.

26. Other genera of bacteria degrade actively urea through a urease yielding ammonia and CO_2. Some clinical isolates of *E. coli* possess urease.

We shall see now how this addition occurs.

An enzyme, S-adenosylmethionine decarboxylase (*speD,* 3 mn), transforms SAM to S-adenosyl-5'-δ-methylmercaptopropylamine or decarboxylated S-adenosylmethionine:

$$SAM \rightarrow H_2N{-}CH_2{-}CH_2{-}CH_2{-}\overset{+}{\underset{\underset{CH_3}{|}}{S}}{-}Ado + CO_2$$

decarboxylated SAM

Spermidine synthetase catalyzes the reaction between decarboxylated SAM and putrescine to yield spermidine and 5'-thiomethyladenosine:

$$\text{decarboxylated SAM} + \text{putrescine} \rightarrow \text{spermidine} + \overset{+}{\underset{\underset{CH_3}{|}}{S}}{-}Ado$$

5'-methylthioadenosine

S-adenosylmethionine decarboxylase has been purified from *E. coli.* It has a molecular weight of 108,000 and is made of six identical subunits; as some other bacterial decarboxylases, it does not require pyridoxal phosphate, but covalently bound pyruvate. Some *speD* mutants require the addition of exogenous spermidine for optimal growth. Spermidine synthetase is a dimer of molecular weight 72,000. It can transfer the aminopropyl group not only to putrescine, and to the higher homologue cadaverine (the product of decarboxylation of lysine), but also to spermidine itself to form spermine, not normally found in *E. coli,* but among others in sperm cells where it stabilizes the tightly-packed DNA.

$$H_2N{-}(CH_2)_3{-}NH{-}(CH_2)_4{-}NH{-}(CH_2)_3{-}NH_2$$

spermine

The 5-methylthioadenosine formed during spermidine synthesis is not lost and serves to reconstruct methionine in the so-called methionine salvage pathway: methylthioadenosine phosphorylase cleaves it to adenine, itself salvaged, and 5'-methylthioribose-1-phosphate, further converted to α-keto-methylthiobutyrate, which is converted to methionine by a specific transaminase.

V. BIOSYNTHESIS OF LYSINE IN YEASTS AND MOLDS

It has long been known that lysine is synthesized via two mutually-exclusive pathways. Aspartate is a precursor of the diaminopimelate pathway, found in bacteria, in a small number of Phycomycetes and in green plants. Glutamate, or rather α-ketoglutarate, is the lysine precursor in Euglenids, some Phycomycetes, yeast and higher fungi, which use the so called aminoadipate pathway.

So far, in spite of urgent research, no organism has been found in which the two pathways exist together. It is tempting to think that the aminoadipic pathway may be more economical than the diaminopimelic acid pathway and has therefore been selected during evolution in yeasts and molds. Bacteria, in which diaminopimelic acid is a necessary compound for the cell wall construction, have only to acquire an additional decarboxylase to obtain their lysine, thus avoiding the necessity of synthesizing the enzymes of the aminoadipic pathway.

Aminoadipic acid pathway

The methyl group of acetate reacts with the carbonyl group of α-ketoglutarate to yield homocitric acid. This reaction is similar to the one leading to the synthesis of citrate from acetyl CoA and oxaloacetate or to the formation of isopropylmalate (β-carboxy-β-hydroxyisocaproate), which we shall meet in the study of leucine biosynthesis in a forthcoming chapter:

$$\underset{\alpha\text{-ketoglutarate}}{COOH{-}CO{-}(CH_2)_2{-}COOH} + \underset{\text{acetyl CoA}}{CH_3{-}CO{-}SCoA} \rightarrow \underset{\text{homocitrate}}{\begin{array}{c} COOH \\ | \\ HO{-}C{-}CH_2{-}COOH \\ | \\ (CH_2)_2 \\ | \\ COOH \end{array}}$$

Whereas only one homocitrate synthase is found in *N. crassa* and *Saccharomyces lipolytica,* two isoenzymes are found in *S. cerevisiae,* explaining the failure to isolate mutants blocked at this step in this organism.

Homocitrate dehydrase and homoaconitate hydratase catalyze two successive steps leading to the formation of homoisocitrate with the intermediate formation of homoaconitate, calling to mind the well-known catalysis by aconitase during the tricarboxylic acid cycle.

$$\underset{\text{homoaconitate}}{\begin{array}{l} CH{-}COOH \\ \| \\ C{-}COOH \\ | \\ (CH_2)_2{-}COOH \end{array}} \qquad\qquad \underset{\text{homoisocitrate}}{\begin{array}{c} HO{-}CH{-}COOH \\ | \\ CH{-}COOH \\ | \\ (CH_2)_2{-}COOH \end{array}}$$

Homoisocitrate dehydrogenase, reduces homoisocitrate in an NAD^+-dependent reaction with the presumed formation of enzyme-bound oxaloglutarate, which is then decarboxylated to α-ketoadipate, the higher homologue of α-ketoglutarate, formed in a similar fashion from isocitrate in the tricarboxylic acid cycle.

oxaloglutarate α-ketoadipate

A specific aminoadipate aminotransferase catalyzes a transamination with glutamate to yield α-aminoadipate and α-ketoglutarate:

$$COOH\text{–}(CH_2)_3\text{–}CH(NH_2)\text{–}COOH$$

α-aminoadipate

This diacid is reduced to the corresponding semialdehyde by a type of reaction we have already seen in connection with aspartic and glutamic acids. It has been reported however that it is catalyzed by a single enzyme depending on ATP, Mg^{2+}, and NADPH. The intermediate energy-rich phosphate has not been isolated.

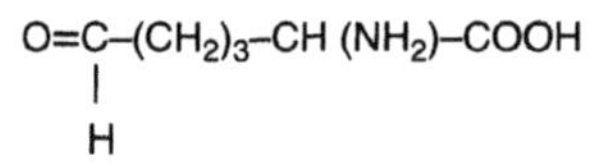

α-aminoadipate semialdehyde

Saccharopine dehydrogenase condenses the semialdehyde with a molecule of glutamic acid, with a concomitant reduction of the adduct by NADPH.

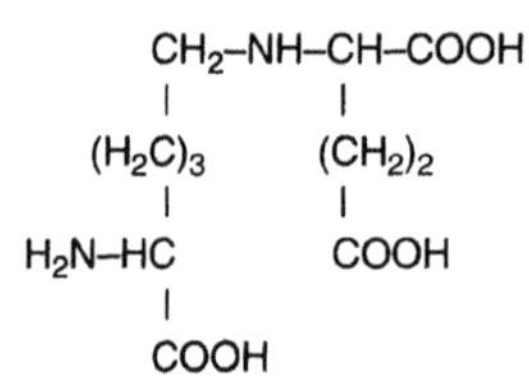

saccharopine

This saccharopine dehydrogenase is called glutamate-forming, because glutamate is formed when it is studied in the reverse direction.

The last reaction of this series is carried by a distinct saccharopine dehydrogenase, called lysine-forming, which in the presence of NAD^+ produces α-ketoglutarate and lysine.

Kinetic studies on the lysine-forming enzymes have indicated an ordered mechanism where NAD^+ and saccharopine bind, and lysine, α-ketoglutarate, and NADH are released in this order.

$$H_2N\text{–}(CH_2)_4\text{–}CH(NH_2)\text{–}COOH$$

L-lysine

It has taken the overall sum:

$$\text{Glu} + \text{NADPH} + \alpha\text{-aminoadipate semialdehyde} \rightarrow \alpha\text{-KG} + NAD^+ + \text{Lys}$$

and two enzymatic reactions to transform an aldehyde into an amine!

In *S. cerevisiae,* the enzymes of α-aminoadipate synthesis are located in the mitochondria, whereas a second aminoadipate aminotransferase and the last three enzymes of lysine biosynthesis are cytoplasmic.

The following scheme summarizes the biosynthesis of the amino acids of the glutamic acid family in bacteria:

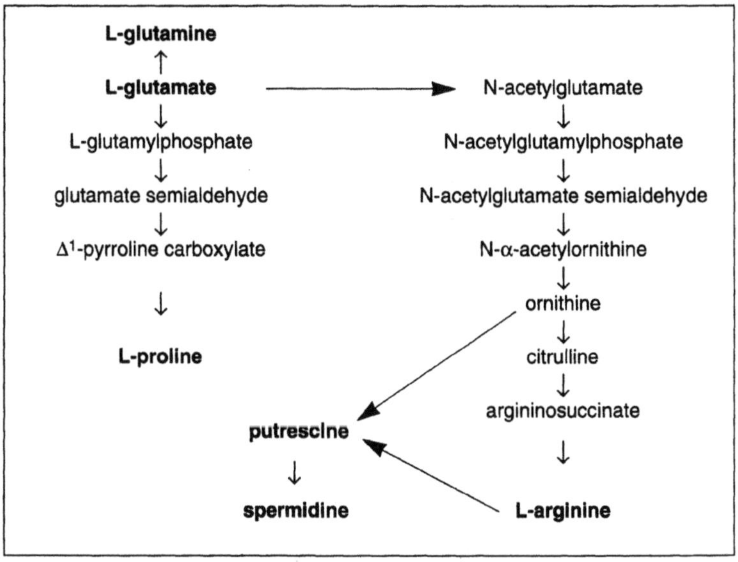

SELECTED REFERENCES

Glutamine synthetase activity and its regulation by covalent modification. Structure

S. G. Rhee, P. B. Chock & E. R. Stadtman, *Advances in Enzymology* **62**, 36-92 (1989).

R. J. Almassy, C. A. Johnson, R. Hanlyn, N. H. Xuong & D. Eisenberg, *Nature* **323**, 304-309 (1986).

R. C. Valentine, B. M. Shapiro & E. R. Stadtman, *Biochemistry* **7**, 2143-2152 (1968).

Glutamine synthetase. Regulation of gene expression

S. Kustu, K. Sei & J. Keener, in *Regulation of Gene Expression – 25 years on,* I. R. Booth & C. F. Higgins eds., pp. 139-154. Cambridge University Press, Cambridge (1986).

A. J. Ninfa & B. Magasanik, *Proc. Natl. Acad. Sci. US* **83**, 5909-5913 (1986).

J. Keener & S. Kustu, *Proc. Natl. Acad. Sci. US* **85**, 4976-4980 (1988).

Glutamate synthase

R. E. Miller & E. R. Stadtman, *J. Biol. Chem.* **247**, 7407-7419 (1972).

A. R. Rendina & W. H. Orme-Johnson, *Biochemistry* **17**, 5388-5393 (1978).

Arginine biosynthesis and regulation

R. Cunin, N. Glansdorff, A. Piérard & V. Stalon, *Microbiol. Revs.* **50**, 314-352 (1986).

D. Lim, J. Oppenheim, T. Eckhardt, & W. K. Maas, in *Gene Expression and Regulation: The Legacy of Luigi Gorini*, pp. 55-63. Excerpta Medica, Amsterdam (1988).

CHAPTER XX

Biosynthesis of amino acids derived from phosphoglyceric acid and pyruvic acid

I. BIOSYNTHESIS OF GLYCINE AND SERINE

The use of appropriate mutants and an adequate enzyme analysis made it possible to decide which of these amino acids is synthesized first, since the final step in the pathway, the interconversion of serine and glycine, is catalyzed by a single enzyme, serine hydroxymethyltransferase, and since mutations affecting the biosynthesis of serine result in a nutritional requirement which can be satisfied by either of them. Mutations deficient in hydroxymethyltransferase, however, require glycine specifically and cannot grow with serine.

The first enzyme of the pathway, phosphoglycerate dehydrogenase (*serA*, 63 mn on the *E. coli* chromosome) catalyzes a transformation of D-3-phosphoglycerate, an intermediate of the glycolytic cycle, into 3-phosphohydroxypyruvate:

$$H_2O_3P\text{–}O\text{–}CH_2\text{–}CHOH\text{–}COOH + NAD^+ \rightarrow H_2O_3P\text{–}O\text{–}CH_2\text{–}CO\text{–}COOH + NADH + H^+$$

3-phosphoglycerate 3-phosphohydroxypyruvate

The enzyme is a homotetramer whose subunit has a molecular weight of *ca.* 41,000. In both *E. coli* and *S. typhimurium*, phosphoglycerate dehydrogenase is the target of an allosteric inhibition by serine, which inhibits it non-competitively. In these organisms, this inhibition constitutes the main mode of regulation of serine biosynthesis, since the levels of the dehydrogenase and of the other enzymes of serine biosynthesis are not increased by serine starvation, or decreased by excess serine. Glycine is also an allosteric inhibitor, but it needs to reach a concentration more than a 100 times higher to reach the same degree of inhibition. It is not known whether glycine and serine bind at the same site or not. The conformational change resulting from serine binding is cooperative.

The 3-phosphohydroxypyruvate is the substrate of a transamination reaction, catalyzed by phosphoserine aminotransferase (*serC*, 20 mn), yielding phosphoserine:

$$H_2O_3P\text{–}O\text{–}CH_2\text{–}CO\text{–}COOH + Glu \rightarrow H_2O_3P\text{–}O\text{–}CH_2\text{–}CH(NH_2)\text{–}COOH + \alpha\text{–}Kg$$

phosphoserine

A specific serine phosphate phosphatase hydrolyzes this compound to serine:

$$\begin{array}{c} CH_2OH\text{–}CH\text{–}COOH \\ | \\ NH_2 \end{array}$$

L-serine

As stated above, the interconversion of serine and glycine is carried out by serine hydroxymethyltransferase. This enzyme has been partially purified from *E. coli* and its apparent molecular weight estimated by molecular sieving to be 170,000. The relevant gene (*glyA*, 55 mn) has been cloned and sequenced, and corresponds to a polypeptide of 417 residues (MW = 45,265). Thus, the native enzyme appears to be a homotetramer. The enzyme is not subject to feedback inhibition, at least in *Enterobacteriaceae,* by any of the numerous compounds that have been tested.

The interconversion can essentially be written on paper as follows:

$$\underset{\text{serine}}{\begin{array}{c} CH_2OH\text{–}CH\text{–}NH_2 \\ | \\ COOH \end{array}} \rightarrow \underset{\text{glycine}}{\begin{array}{c} CH_2\text{–}NH_2 \\ | \\ COOH \end{array}} + \underset{\text{formaldehyde}}{HCHO}$$

Actually, the C_1 fragment obtained from serine is a tetrahydrofolate derivative of formaldehyde, 5-10 methylene tetrahydrofolic acid and the reaction is, in fact:

$$\text{serine} + FH_4 \rightarrow \text{glycine} + CH_2OH\text{–}FH_4$$
$$CH_2OH\text{–}FH_4 \rightarrow HCHO + FH_4$$

Sum:

$$\text{serine} \rightarrow \text{glycine} + HCHO$$

Serine hydroxymetyhyltransferase catalyzes many other reactions involving aldehydes other than formaldehyde, but only the serine-glycine interconversion requires tetrahydrofolate. Compelling evidence shows that formaldehyde introduces constraints not encountered with other aldehydes and that the primary role of H_4-folate is to add and remove formaldehyde from the active site.

In addition to tetrahydrofolate, the reaction requires pyridoxal phosphate. Extensive spectroscopic studies suggest that it participates in the reaction through the formation of a Schiff's base between the amino group of serine and the formyl group of pyridoxal phosphate (external aldimine). The resulting system of conjugated double bonds extends from the α-carbon of the amino acid to the nitrogen atom of the pyridine ring, thus labilizing the hydroxymethyl group of serine and facilitating its cleavage.

All forms of the enzyme for which a primary structure is known have five threonine residues near the active -site lysyl residue that forms the internal aldimine with pyridoxal phosphate. The nine-residue conserved sequence is VVTTTTHK (pyr)-T. In *E. coli*, each of the five threonines has been changed to an alanine by site-directed mutagenesis. The five resulting mutant proteins have been purified and characterized with respect to affinity of coenzymes and substrates. Only the enzyme with the threonine in bold face in the sequence plays a critical role: the protein with an alanine replacing that threonine has normal affinities for the substrates and coenzymes but displays only 3% of the catalytic activity of the wild type enzyme.

Under certain conditions of growth, it has been reported that *E. coli* and *Clostridium pasteurianum* are able to obtain glycine from threonine by the action of an enzyme called threonine aldolase:

$$\underset{\text{threonine}}{CH_3\text{–}CHOH\text{–}CH(NH_2)\text{–}COOH} \rightarrow \underset{\text{glycine}}{H_2N\text{–}CH_2\text{–}COOH} + \underset{\text{acetaldehyde}}{CH_3\text{–}CHO}$$

Finally, let us recall that the methylenetetrahydrofolate produced by serine hydroxymethyltransferase is the substrate of the reductase yielding the methyltetrahydrofolate which is the source of the methyl group of methionine. This is why the β-carbon of serine is the source of that methyl group.

Regulation of serine hydroxymethyltransferase at the transcriptional level

The activity of serine hydroxymethyltransferase produces one-carbon units which are utilized in the synthesis of methionine, purines, and thymine. The control of serine hydroxymethyltransferase synthesis is complex. Several compounds related to one-carbon metabolism (serine, glycine, methionine, purines, thymine) are all apparently involved in a kind of cumulative repression. In spite of the partial control by methionine, the regulation does not involve the methionine repressor, coded by *metJ*: the controlling element is MetR which activates slightly the transcription of *glyA* in the presence of homocysteine. The nucleotide component of the control is PurR. Much remains to be investigated in this very complex regulation.

II. BIOSYNTHESIS OF CYSTEINE

The biosynthesis of cysteine is important not only because cysteine is a constituent of proteins, but also because it is involved in methionine, thiamine, biotin, pantetheine, lipoic acid, glutathione, and other cellular compounds.

There auxotrophic mutants of *E. coli* and *S. typhimurium* that cannot utilize sulfate as a sulfur source, but can grow on sulfate. Other mutants cannot utilize either sulfate or sulfite, but can grow on thiosulfate[27]. These mutants were characterized by nutritional studies long before genetic mapping.

27. Thiosulfate has been eliminated as an intermediate; its utilization by the mutants mentioned is due to a reductive cleavage of the ion yielding sulfite and sulfide, the latter being the true intermediate.

Cysteine biosynthesis is a branched convergent pathway where one branch leads from sulfate to sulfide, the other branch being the conversion of serine to O-acetylserine. The convergence point is the reaction of O-acetylserine with sulfide to yield cysteine.

The sole pathway of sulfate assimilation in bacteria and fungi involves an activation of sulfate with ATP, catalyzed by sulfate adenylyltransferase (or ATP sulfurylase) whose gene, *cysD*, is located at 59 mn on the *E. coli* chromosome:

$$\text{ATP + sulfate} \rightleftharpoons \underset{\text{(APS)}}{\text{adenosine-5'-phosphosulfate}} + \text{PPi}$$

The equilibrium constant of this reaction does not favor the synthesis of APS; it follows that either pyrophosphatase or the next reaction, or both, are necessary to drive the sulfur further toward cysteine synthesis.

In *Enterobacteriaceae*, APS cannot be reduced directly to sulfide[28] and the next reaction catalyzed by adenylylsulfate kinase (or APS kinase, *cysC*, 59 mn; 603 nucleotides corresponding to a protein of 22321 Da) produces a derivative phosphorylated at the 3'-position of ribose:

$$\text{ATP + APS} \rightarrow \underset{\text{(PAPS)}}{\text{3'-phosphoadenosine-5'-phosphosulfate}} + \text{ADP}$$

In the absence of APS, ATP phosphorylates the enzyme with a rate constant which is equivalent to the V_{max} for the overall APS kinase reaction, suggesting that the phosphorylated enzyme is on the reaction path.

PAPS can now be reduced. This is achieved by a two-step process catalyzed by a specific PAPS sulfotransferase (*cysH*, 59 mn). This enzyme transfers the sulfonyl moiety of PAPS to thioredoxin to form an acceptor $-S-SO_3^-$ derivative. The intervention of a thiol such as glutathione or the second sulfhydryl group of the thioredoxin adduct eventually liberates free sulfite (for a discussion of thioredoxin, see p. 287-289).

The sulfite reductase of *E. coli* is a very complex enzyme, able to achieve the 6-electron reduction of sulfite to sulfide without release of intermediates. It is an $\alpha_8\beta_4$ protein. The two component polypeptides, α and β (specified by *cysJ* and *cysI*, both at 59 mn) have molecular weights, respectively, of about 60,000 and 55,000. By treating the holoenzyme with urea, one obtains a flavoprotein with an α_8 structure and a monomeric hemoprotein. The flavoprotein contains 4 FAD and 4 FMN and is endowed with NADPH-cytochrome *c* reductase activity. The hemoprotein contains an Fe_4S_4 cluster and a heme group called siroheme, a product of uroporphyrinogen metabolism. The hemoprotein is able to reduce sulfite with reduced methylviologen as the electron donor.

The physiological electron flow is believed to be as follows:

$$\text{NADPH} \rightarrow \text{FAD} \rightarrow \text{FMN} \rightarrow Fe_4S_4 \rightarrow \text{siroheme} \rightarrow \text{sulfite}$$

28. Yeasts and plants reduce APS directly.

One cannot detect free flavoprotein or free hemoprotein in wild-type extracts, but they can be demonstrated in extracts of *cysJ* or *cysI* mutants unable to synthesize one or the other protein. The two genes have been sequenced.

Another mutation leads to the incapacity for reducing sulfite. It occurs at another locus, called *cysG* (73 mn). These mutations are pleiotropic and also lead to a deficiency in nitrite reductase (a siroheme-containing enzyme) and cobalamin synthesis. It is probable that *cysG* (a misnomer) specifies a methylase necessary for siroheme and cobalamin synthesis[29].

O-acetylation of serine

In all the plants and microorganisms that have been examined, O-acetylserine rather than serine is the precursor of the carbon moiety of cysteine. Serine transacetylase (*cysE,* 80 mn) catalyzes its synthesis from serine and acetyl CoA:

$$\text{serine + acetyl CoA} \rightarrow \text{O-acetylserine + CoA}$$

In crude extracts, the transacetylase is associated as a complex with the enzyme of the pathway that condenses O-acetylserine and sulfide to cysteine. This complex (MW = *ca* 300,000) dissociates in the presence of O-acetylserine to free serine transacetylase and O-acetylserine sulfhydrylase (*cysK,* 52 mn). The dissociation is prevented by sulfide. The multifunctional complex is formed by a homotetrameric serine transacetylase (MW = 160,000) and two molecules of a dimeric O-acetylserine sulfhydrylase composed of two identical subunits of 36,000 daltons each. The complex has been called cysteine synthase.

$$H_2S\text{–}CH_2\text{–}CH(NH_2)\text{–}COOH$$

cysteine

Because mutants devoid of serine transacetylase cannot be derepressed for the other cysteine biosynthetic enzymes, it is assumed that O-acetylserine plays the role of an internal inducer[30].

Allosteric regulation of cysteine synthesis

As shown by isotope competition experiments, cysteine biosynthesis ceases if cells are grown in excess cysteine. Serine transacetylase is allosterically inhibited by cysteine at the exceedingly low concentration of 10^{-6} M.

There is no physiologically significant regulation by cysteine of any other member of the biosynthetic pathway or of the sulfate transport.

29. It has been shown that the products of *cysG* and of the *cob* genes involved in the methylations leading to vitamin B_{12} (see chapter XXVIII) share considerable homology.
30. See *cysB* below.

Regulation of cysteine synthesis at the genetic level

Some words should be said about the organization of the various *cys* genes on the chromosome of *E. coli*[31].

The gene specifying serine transacetylase, *cysE*, is not linked to any other *cys* gene. The genes for sulfate permease, *cysA*, and for O-acetylsulfhydrylase, *cysK*, although both located at 49 mn, are separated by 9 kb, and are not part of a single transcription unit.

In contrast, *cysC* and *cysD*, coding respectively for APS kinase and ATP sulfurylase, are adjacent in *S. typhimurium*; the fact that they catalyze two consecutive reactions suggests that they may constitute a single transcription unit (an operon). There is a well-documented operon, the *cysJIH* transcriptional unit (coding for PAPS sulfotransferase, the apoflavoprotein and the apohemoprotein), defined by polarity effects and a promoter mutation.

Maximal expression of the cysteine regulon, encompassing all these genes, requires three factors: sulfur limitation, O-acetylserine availability, and a protein coded by a positive regulatory gene, *cysB* (28 mn in *E. coli*).

Most *cysB* strains require cysteine for growth and, like *cysE* mutants, cannot be derepressed for the biosynthetic pathway. Merodiploids *cysB*$^+$/*cysB* grow without added cysteine. The protein coded by *cysB* is a *trans*-acting element of positive control. The protein of *S. typhimurium* has been purified to homogeneity: it is a homotetramer of 4 x 36,000 daltons, devoid of any known enzymatic activity. Its hypothetical model of action is that a complex of the cysB protein and O-acetylserine facilitates initiation of transcription at the various *cys* promoters, in the way described for other positive regulators.

III. BIOSYNTHESIS OF ALANINE

No real alanine auxotroph has ever been isolated. This indicates that there must be several ways to synthesize this amino acid.

Rudman and Meister have made a survey of transaminases in *E. coli*. Their work has been continued by others. At least two enzymes can synthesize alanine, but there may be as many as four.

A glutamate-pyruvate transaminase

$$\underset{\text{glutamate}}{HOOC{-}(CH_2)_2{-}\underset{\displaystyle NH_2}{\underset{|}{CH}}{-}COOH} + \underset{\text{pyruvate}}{CH_3{-}\underset{\displaystyle O}{\underset{||}{C}}{-}COOH} \rightarrow \underset{\alpha\text{-ketoglutarate}}{COOH{-}(CH_2)_2{-}\underset{\displaystyle O}{\underset{||}{C}}{-}COOH} + \underset{\text{alanine}}{CH_3{-}\underset{\displaystyle NH_2}{\underset{|}{CH}}{-}COOH}$$

31. The exact gene locations are slightly different in *S. typhimurium*, as is the order of the genes within a given cluster, but this does not affect the conclusions.

is induced by growth with glycerol or with alanine in the growth medium. This suggests a degradative function rather than a biosynthetic one.

Rudman and Meister discovered another transaminase, transaminase C, which catalyzes the following reaction

valine + pyruvate → α-ketoisovalerate + alanine

Since another transaminase, transaminase B catalyzes the reaction

glutamate + α-ketoisovalerate → valine + α-ketoglutarate,

the sum of the two reactions results in the conversion of pyruvate to alanine, with glutamate as the amino donor, and serves a biosynthetic function this time.

In several microorganisms other than *Enterobacteriaceae,* pyruvate can undergo a reductive amination to form alanine:

$$CH_3\text{–}CO\text{–}COOH + NH_3 + NADH + H^+ \rightarrow CH_3\text{–}CH(NH_2)\text{–}COOH + NAD^+ + H_2O$$

In addition, an aspartate-β-decarboxylase has been found in a number of microorganisms (among others, *Alcaligenes faecalis, Desulfovibrio desulfuricans*, some species of *Achromobacter*, and *Pseudomonas*) and its occurrence has been reported in the silkworm, the lobster, and the crayfish.

It catalyzes the following reaction:

$$HOOC\text{–}CH_2\text{–}CH(NH_2)\text{–}COOH \rightarrow CH_3\text{–}CH(NH_2)\text{–}COOH + CO_2$$

The enzyme from *Alcaligenes faecalis* has been obtained in the pure state. It is made of 12 subunits of 57,000 kDa each, organized as six dimers of molecular weight 114,000. Each monomer contains one mole of pyridoxal phosphate. The enzyme is allosterically inhibited by α-keto acids. Its catalytic and inhibition mechanisms have been exhaustively analyzed by Meister.

IV. BIOSYNTHESIS OF VALINE

This biosynthetic route is exactly parallel to that of isoleucine and uses the same enzymes. α-aceto-α-hydroxyacid synthetase which, as we have seen, catalyzes the condensation of one molecule of acetaldehyde (arising from pyruvate) with a molecule of α-ketobutyrate (deriving from threonine), is also able to carry out the condensation of a molecule of acetaldehyde with a second molecule of pyruvate. The enzyme requires the presence of Mg^{2+} and thiamine pyrophosphate for its activity. The reaction takes place in two steps: first, a molecule of hydroxyethylthiamine pyrophosphate or active acetaldehyde is formed, and then it condenses with a second molecule of pyruvate to produce α-acetolactate (or with a molecule of α-ketobutyrate to produce α-aceto-α-hydroxybutyrate, in the isoleucine pathway)

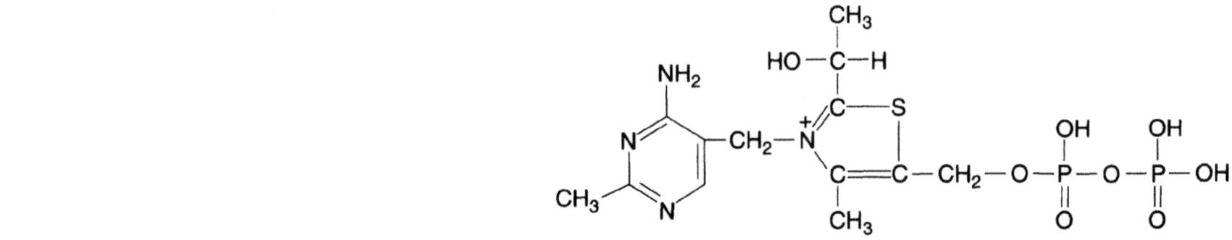

α-hydroxyethyl thiamine pyrophosphate or "active acetaldehyde"

The reaction can be summarized as follows:

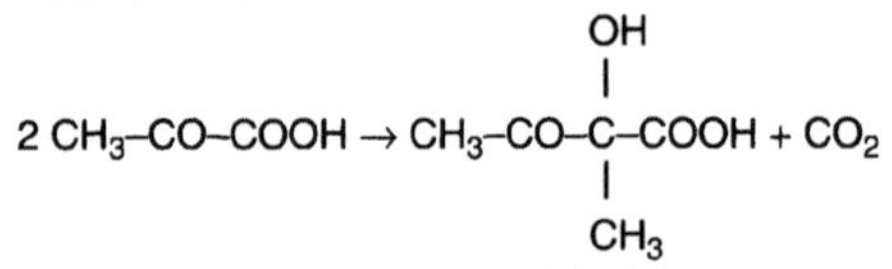

α-acetolactate

A rearrangement similar to the one already seen in isoleucine biosynthesis occurs under the influence of the reductoisomerase. It involves a reduction of the carbonyl group of α-acetolactate and the production of α,β-dihydroxy-β-methylbutyrate:

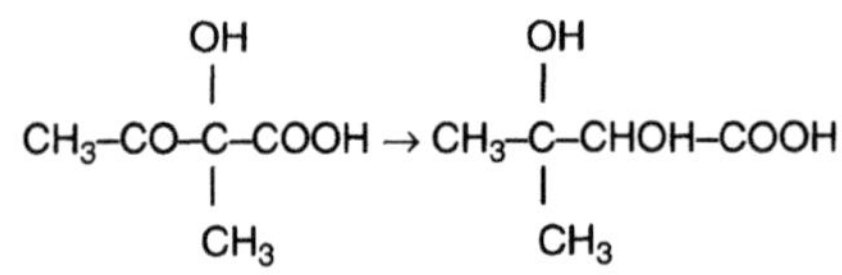

α,β-dihydroxy-β-methylbutyrate

The dehydratase specific for α,β-dihydroxyacids catalyzes the loss of one molecule of water, and α-keto-β-methylbutyrate is obtained:

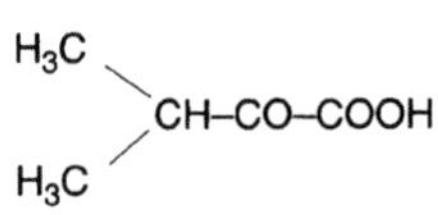

The same transaminase that was involved with isoleucine (*ilvE*) acts on this α-keto acid to give L-valine:

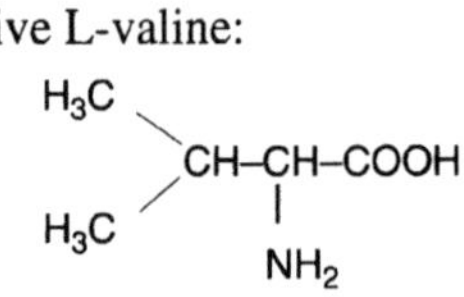

The implications of the existence of a series of enzymes common to the biosynthesis of valine and isoleucine in microorganisms become obvious when we examine the scheme at the top of the next page, which applies to *Neurospora* as well as to *E. coli.*

Every mutant deficient in threonine dehydratase will be isolated as an isoleucine auxotroph; every mutant deficient in α-aceto-α-hydroxyacid synthetase, reductoisomerase, dehydratase, or transaminase (reactions A, B, C, and D) will show a double requirement (valine *plus* isoleucine) for growth. We have seen that isoleucine is an allosteric feedback inhibitor of

L-threonine		
↓	(threonine dehydratase)	
α-ketobutyrate		pyruvate
↓	A	↓
α-aceto-α-hydroxybutyrate		α-acetolactate
↓	B	↓
α,β-dihydroxy-β-methylvalerate		α,β-dihydroxy-β-methylbutyrate
↓	C	↓
α-keto-β-methylvalerate		α-keto-β-methylbutyrate
↓	D	↓
L-isoleucine		L-valine

threonine dehydratase; this creates no problem for growth of the mutants since the end product, isoleucine, has to be added to the medium. But, in contrast, an effective feedback inhibition of α-aceto-α-hydroxyacid synthetase by valine will present difficulties, since it automatically creates a deficiency in isoleucine. This is in fact the case in many organisms, including the K12 wild-type strain of *E. coli,* whose growth is inhibited by valine, the inhibition being removed in the presence of isoleucine.

V. BIOSYNTHESIS OF LEUCINE

α-keto-β-methylbutyrate, the immediate precursor of valine, is also the direct precursor of leucine. It is acetylated by α-isopropylmalate synthetase (*leuA*) which catalyzes the reaction of the keto acid with a molecule of acetyl CoA to yield α-isopropylmalate:

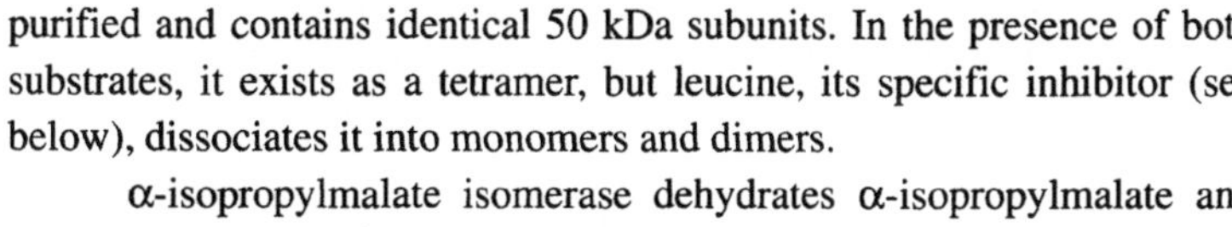

α-isopropylmalate

The α-isopropylmalate synthase from *S. typhimurium* has been purified and contains identical 50 kDa subunits. In the presence of both substrates, it exists as a tetramer, but leucine, its specific inhibitor (see below), dissociates it into monomers and dimers.

α-isopropylmalate isomerase dehydrates α-isopropylmalate and then rehydrates it to β-isopropylmalate:

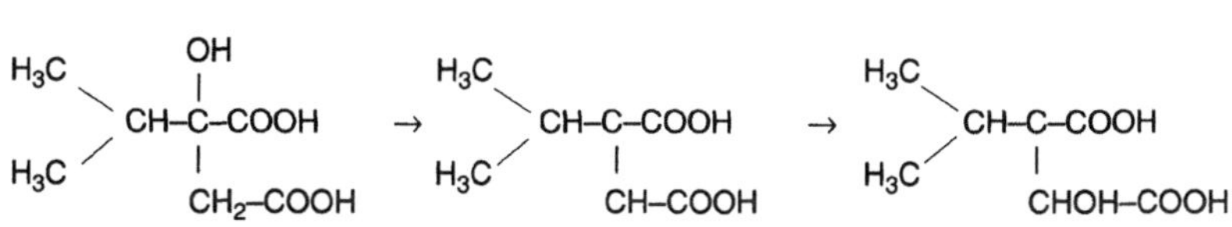

β-isopropylmalate

This reaction is accomplished by isopropylmalate isomerase, an unstable protein; genetic data predict that it is composed of two different subunits, respectively specified by the two genes *leuC* and *leuD*. The reaction catalyzed by the isomerase is reminiscent of the reaction catalyzed by aconitase in the tricarboxylic acid cycle, and may involve the intermediate formation of dimethylcitraconate

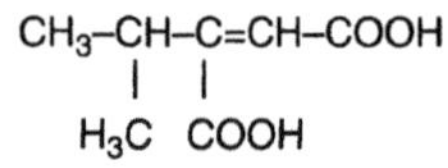

dimethylcitraconate

An oxidative decarboxylation of β-isopropylmalate, catalyzed by the NAD^+-dependent β-isopropylmalate dehydrogenase *(leuB)*, leads to the formation of α-ketoisocaproic acid:

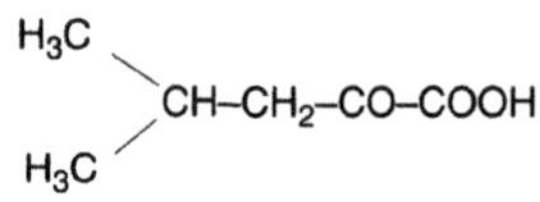

α-ketoisocaproate

The enzyme from *S. typhimurium* has been purified and is a dimer of 35 kDa subunits, requiring a monovalent cation for activity. The comparison of its sequence to that of ten other fungal or bacterial β-isopropylmalate dehydrogenases, and to several isopropylmalate isomerases, turns up a broadly-conserved motif which might delineate the binding site for β-isopropylmalate, the common ligand of these two enzymes.

Finally, a transamination reaction produces leucine:

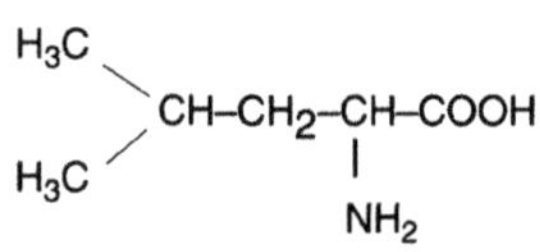

L-leucine

The conversion of α-ketoisocaproate to leucine can be carried out either by the transaminase encoded by *ilvE*, or by the aromatic transaminase (*tyrB*, 92 mn).

The *leuABCD* cluster of genes constitutes (in that order) an operon, situated at 2 mn on the *E. coli* circular chromosome. The *leu* operon has a single promoter, followed by a leader region and an attenuator (see below).

In a later chapter, we shall see that α-keto-β-methylbutyric acid is also the precursor of the vitamin pantothenic acid.

Like many others (more and more often discovered), the leucine-forming enzymes display a limited specificity. The pathway has therefore the potential of increasing the chain length not only of α-keto-β-methylvalerate, but also of pyruvate, α-ketobutyrate, and even α-ketoisocaproate – through one or several cycles of elongation – to yield longer keto acids and their aminated derivatives. Thus, mutants of *Serratia marcescens* have been described which accumulate α-aminobutyrate, norvaline, norleucine, and homoleucine. In some cases, where several copies of the leucine genes are present (for example on multicopy plasmids), these endogenously synthesized unnatural amino acid analogues may be incorporated in *E. coli* proteins (see also chapter VII).

VI. REGULATION OF VALINE, ISOLEUCINE, AND LEUCINE BIOSYNTHESIS

We have already seen that threonine dehydratase is subject to feedback inhibition by isoleucine, and α-aceto-α-hydroxyacid synthetase by valine. In both cases, the target of allosteric inhibition is the first enzyme specific to the synthesis of the particular amino acid. The branch leading to leucine does not deviate from this rule: α-isopropylmalate synthetase is subject to allosteric feedback by L-leucine in coliform bacteria and in yeast.

The controls, by repression, in *Enterobacteriaceae* are more complex. It should be remembered that each of the last four steps in the biosynthesis of valine and isoleucine are catalyzed by a common enzyme. Repression of their synthesis presents a delicate problem, since they bring about the production of two essential metabolites. In addition, the first reaction leading to the biosynthesis of leucine uses an intermediate in the synthesis of valine as a substrate.

The solution to this problem is the following: except for the reductoisomerase (which is induced by α-aceto-α-hydroxyacids), valine, leucine, and isoleucine are *all* required to repress the synthesis of the other enzymes in the biosynthetic pathway leading to valine and isoleucine. In contrast, leucine alone is enough to repress the synthesis of the enzymes of its own pathway. Table I, adapted from Freundlich, Burns, and Umbarger, shows the effect, on several of the enzymes involved, of limiting each of the amino acids during growth in a chemostat.

Table I. Multivalent repression of the synthesis of branched-chain amino acids.
The results are expressed as specific activities, μmoles of product formed per mg protein per hour.

Limiting amino acid	Threonine deaminase	Dihydroxyacid dehydratase	α-isopropylmalate synthase
None	4	5.8	1.2
Isoleucine	74	30	3.6
Valine	77	37	2.4
Leucine	58	24	12

This phenomenon has been called *multivalent repression* by the authors; we have encountered another case of it when we studied the divalent repression of the synthesis of the enzymes of the threonine operon.

Again, the repression occurs through attenuation. Here, the sequence of the leader peptide involved in attenuation is:

Met Thr Ala **Leu Leu** Arg **Val Ile** Ser **Leu Val Val Ile** Ser **Val Val Val Ile Ile Ile** Pro Pro Cys Gly Ala Ala Leu Gly Arg Gly Lys Ala

and is seen to contain several leucine-, valine-, and isoleucine-grouped residues where the uncharged specific tRNAs can cause stalling of the ribosomes translating the RNA, favoring the transcription of the structural genes (see chapter XI, Attenuation).

In contrast, the leader peptide involved in the regulation of the leucine operon – which is also under attenuation control and repressed by leucine alone – contains four tandem leucine residues:

Met Ser His Ile Val Arg Phe Thr Gly **Leu Leu Leu Leu** Asn Ala Phe Ile Val Arg Gly Arg Pro Val Gly Gly Ile Gln His

The following scheme summarizes the reactions involved in the biosynthesis of the amino acids derived from phosphoglyceric acid and pyruvic acid.

phosphoglycerate	→	pyruvate	
↓		↓	
phosphohydroxypyruvate		α-acetolactate	
↓		↓	
phosphoserine		α, β-dihydroxy-β-methylbutyrate	
↓		↓	
L-serine	↘	α-keto-β-methylbutyrate →	α-isopropylmalate
↓		↓	↓
glycine	**L-cysteine**	**L-valine**	β-isopropylmalate
			↓
			α-ketoisocaproate
			↓
			L-leucine

SELECTED REFERENCES

Serine hydroxymethylase

L. Schirch, *Advances in Enzymology* **53**, 83-112 (1982).

G. V. Stauffer, in *Amino Acids: Biosynthesis and Genetic Regulation,* K. M. Herrmann & R. L. Somerville eds., pp. 103-113. Addison-Wesley Publishing Company, Inc. Reading, Mass (1983).

Sulfite reductase

L. M. Siegel, M. J. Murphy & H. Kamin, *J. Biol. Chem.* **248**, 251-261 (1973).

Valine and Leucine

H. E. Umbarger, in *Escherichia coli and Salmonella typhimurium, Cellular and Molecular Biology,* J. L. Ingraham, K. B. Low, B. Magasanik, M. Schachter & H. E. Umbarger eds., pp. 352-367. American Society for Microbiology, Washington, D.C. (1987).

Aspartate-β -decarboxylase

S. S. Tate & A. Meister, *Advances in Enzymology* **35**, 503-543 (1971).

CHAPTER XXI

Biosynthesis of aromatic amino acids and its regulation

Since glucose is an aliphatic substance, the synthesis from it of phenylalanine, tyrosine, and tryptophan presents the biochemist with the problem of the biosynthesis of the aromatic nucleus.

The key to this problem was found, some forty years ago, by the isolation of polyauxotrophic mutants which required the presence of the three aromatic amino acids in the medium for growth and, in addition, *p*-aminobenzoic acid (a precursor of folic acid). The existence of this multiple requirement as a result of a single mutation, immediately suggested that the genetic block had to do with the inability to synthesize a precursor common to all four essential metabolites. After trying 55 cyclic and hydroaromatic derivatives without success, B. D. Davis discovered that shikimic acid could replace the four growth factors. Moreover, growth was proportional to the amount of shikimate added. Other bacterial mutants, in which the lesion lies at a later step in aromatic biosynthesis, accumulate shikimic acid in the medium. It was therefore concluded that shikimic acid was the precursor sought; however, its identification was based only on chromatographic analysis and the response given by mutants. Later, the compound was isolated and characterized by the classical methods of organic chemistry.

Shikimic acid has been known since the nineteenth century as a rare constituent of certain plants. We shall now examine how this compound is formed in *E. coli*, and how it leads to chorismic acid – from which the different aromatic compounds diverge.

I. COMMON PATHWAY (SHIKIMIC PATHWAY)

Formation of shikimic acid

A mutant strain accumulating shikimic acid was grown on glucose labeled at specific positions as sole carbon source. The distribution of carbon atoms in the different carbon atoms of shikimate was then studied. The results showed that the carboxyl group and carbon atoms 1 and 2 of shikimic acid were derived from a three-carbon intermediate of glycolysis (C1 to C3, or C4 to C6 of glucose) and that the four remaining carbons could come from a tetrose (C3 to C6 of glucose, with some scrambling due to cycling in the hexose monophosphate shunt). This suggested that a seven-carbon intermediate might be formed by the condensation of a C3 derivative with a C4 coming from the pentose phosphate pathway.

A few years later, extracts of *E. coli* were shown to catalyze the conversion of erythrose-4-phosphate (E4P) and phosphoenolpyruvate (PEP) to a seven-carbon sugar derivative which was identified as 3-deoxy-D-arabino-heptulosonate-7-phosphate (DAHP), with the elimination of inorganic phosphate:

$$\begin{array}{llllll}
CHO & & & & COOH & \\
| & & & & | & \\
CHOH & & COOH & & C{=}O & \\
| & & | & & | & \\
CHOH & + & C{-}O{-}PO_3H_2 & \rightarrow & CH_2 & + Pi \\
| & & \| & & | & \\
CH_2OPO_3H_2 & & CH_2 & & (CHOH)_3 & \\
 & & & & | & \\
 & & & & CH_2OPO_3H_2 & \\
\text{E4P} & & \text{PEP} & & \text{DAHP} &
\end{array}$$

As in the case of aspartokinases, a study of the inhibition and repression of 3-deoxy-D-arabino-heptulosonate-7-phosphate synthesis showed that there were at least two isofunctional enzymes catalyzing this reaction. One was inhibited and repressed by phenylalanine, the second was inhibited and repressed by tyrosine. They were called, respectively, DAHP synthase (phe) and DAHP synthase (tyr). A few years later, a third minor isofunctional enzyme, DAHP synthase (trp), whose activity was inhibited and whose synthesis was repressed by tryptophan, was identified.

The three enzymes from *E. coli*, and DAHP (tyr) from *S. typhimurium*, have been obtained in the pure state, cloned, and sequenced.

DAHP (phe) (*aroG*, 17 mn) is a tetramer of identical subunits of 350 residues each, whereas DAHP (tyr) (*aroF*, 57 mn) is a homodimer whose subunits comprise 356 amino acids. The third isoenzyme, DAHP (trp) (*aroH*, 37 mn), is also a homodimer whose subunits have a molecular weight of about 40,000. All three enzymes contain one atom of iron per mole of native enzyme. For the two more abundant isofunctional proteins, the reaction mechanism has been analyzed and found to be ordered and sequential, with PEP being the first ligand bound.

A single enzyme, 3-dehydroquinate synthase (*aroB*, 75 mn), catalyzes the transformation of DAHP to 3-dehydroquinic acid. The reaction requires NAD^+ and cobaltous ions. The requirement for NAD^+ long remained a mystery, since the reaction does not involve a net transfer of electrons. The problem was finally solved when isotope experiments showed that all the tritium of labeled DAHP is conserved in dehydroquinate. This is due to an oxidation at C5 by NAD^+, the hydrid transfer from the enzyme-bound NADH, using the same hydrogen atom as was taken from DAHP. The following scheme summarizes these facts.

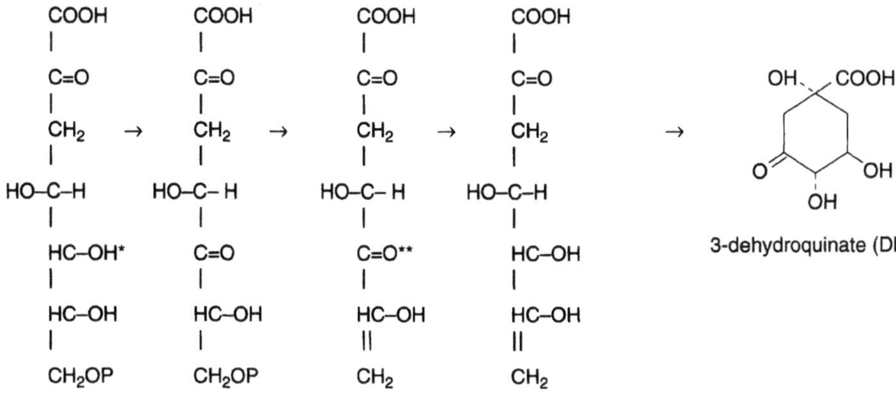

DAHP

3-dehydroquinate (DHQ)

* Site of oxidation by NAD^+ ** Site of reduction by NADH

Cloning techniques have provided large amounts of DHQ synthase and the enzyme has been obtained in the pure state. It is a monomeric protein of molecular weight *ca.* 40,000.

Dehydroquinate dehydratase, also called dehydroquinase (*aroD*, 37 mn), removes one molecule of water from DHQ and introduces the first double bond of the aromatic ring. The gene for the enzyme has been cloned and sequenced. The protein has been overexpressed and purified to homogeneity. The native structure of the pure protein is predominantly dimeric, but it exists in equilibrium with monomeric subunits. The enzyme is inactivated by diethylpyrocarbonate; an equilibrium mixture of product and substrate protect against the inactivation which affects one histidine residue, a good candidate for the general base which is postulated to participate in the mechanism of the enzyme. The product of the reaction, which is reversible, is 3-dehydroshikimate.

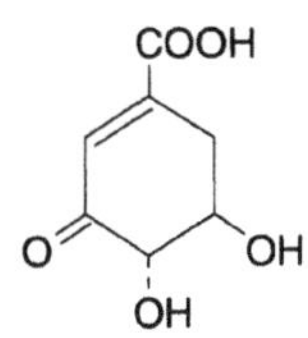

3-dehydroshikimate

This compound is acted upon by shikimate dehydrogenase (formerly known as dehydroshikimate reductase, *aroE*, 72 mn) which reduces it to shikimic acid with help from NADPH. The reaction is stereospecific and transfers the hydrogen from the A side of NADPH only. Homogeneous shikimate dehydrogenase has been obtained: it has a molecular weight of 32,000.

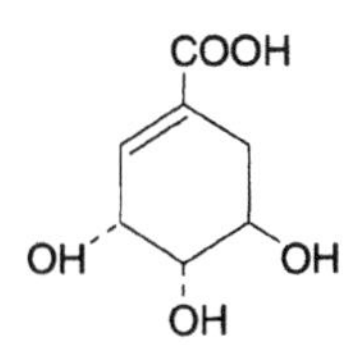

shikimic acid or
3, 4, 5, trihydrocyclohexene-1-carboxylic acid

Formation of chorismic acid

Shikimate kinase (*aroL*, 9 mn) catalyzes the phosphorylation of shikimic acid by ATP. No mutants lacking this activity were found among all the aromatic auxotrophs tested. The reason resides in the existence of a second shikimate kinase of low affinity for shikimic acid, for which no satisfactory rationale has been found. It may have another major function concerned with some other essential pathway.

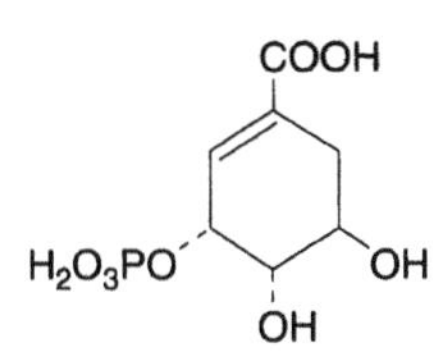

shikimate-3-phosphate

Aromatic auxotrophs blocked in the last step of the common pathway were found to accumulate 5-enolpyruvoylshikimate. This is the dephosphorylated form of the actual intermediate, 5-enolpyruvoyl-shikimate-3-phosphate (EPSP), which is synthesized from shikimate-3-phosphate and phosphoenolpyruvate by 5-enolpyruvoylshikimate-3-phosphate synthase (*aroA*, 20 mn). The gene has been cloned and over-producing strains have been obtained, from which the pure enzyme has been obtained and studied. It is a monomer of *ca.*. 50,000 daltons whose amino acid sequence has been established.

The mechanism is ordered, shikimate-3-phosphate binding first. The reaction catalyzed introduces the three-carbon fragment which is to become the side chain of phenylalanine and tyrosine, and to be removed in the synthesis of tryptophan.

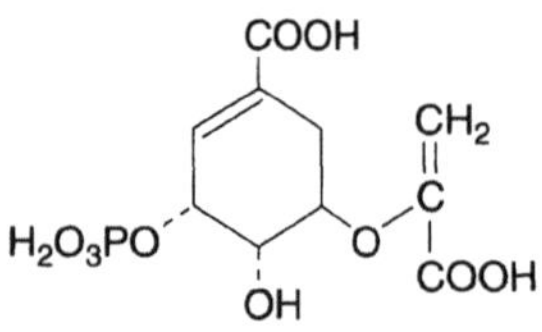

EPSP

The chorismate synthase (*aroC*, 51 mn) introduces a second double bond into the aromatic ring system and removes the phosphate group from EPSP, to yield chorismic acid (from the Greek χορισμοσ, separation) which, as we shall see, is the common precursor of phenyl-alanine, tyrosine, tryptophan, *p*-aminobenzoic acid, *p*-hydroxybenzoic acid, and of ubiquinone, menaquinone, and enterochelin, and of many plant alcaloids, cinnamic acid derivatives, and lignin.

The chorismate synthase from *E. coli* is sensitive to oxygen and has been only partially purified; only a tentative molecular weight has been proposed.

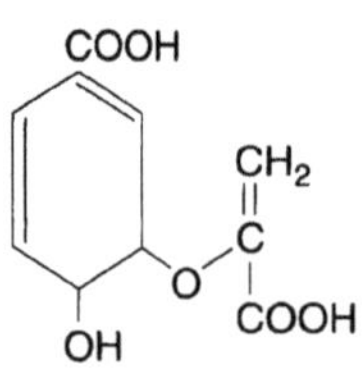

chorismic acid

Physiological aspects of the regulation of the common pathway

Whereas inhibition of DAHP synthase (phe) and DAHP synthase (tyr) by phenylalanine and tyrosine can be almost total, the inhibition of DAHP (trp) by tryptophan never exceeds 60%, even at saturation.

The feedback inhibition of the three DAHP synthases has already been discussed. The situation is curiously similar to what we have already encountered in another branched chain, the biosynthesis of amino acids derived from aspartate. Here again, the presence of the three isofunctional enzymes, whose activities and rates of synthesis are affected by individual amino acids, allows the modulation of the overall rates of synthesis in response to their availability in the medium or in the cytoplasm.

DAHP synthase (tyr) is not repressed solely by tyrosine, but also by very high levels of phenylalanine; DAHP synthase (phe) is repressed by phenylanine, but also by tryptophan;the synthesis of the third isofunctional enzyme is repressed only by tryptophan.

When *E. coli* grows in minimal medium, DAHP synthase (phe) represents more than 80% of the total activity; however, under conditions where growth is limited by the aromatic amino acids (minimal medium supplemented with all the amino acids except phenylalanine, tyrosine, and tryptophan), DAHP synthase (tyr) is derepressed and becomes the major enzyme. The third enzyme, DAHP synthase (trp), reaches only 5%

of the total activity when fully derepressed by specific tryptophan limitation, perhaps due to an inefficient promoter for the *aroH* gene.

The molecular basis of the modulation of the rates of synthesis of the three isofunctional enzymes will be discussed later in this chapter.

DHQ synthase and dehydratase, shikimate dehydrogenase and enolpyruvoylshikimate-3-phosphate synthase, are synthesized constitutively, *i.e.*, their level is insensitive to the concentrations of aromatic amino acids in the growth medium. In contrast, shikimate kinase is derepressed up to 10-fold when cells are starved for tyrosine and tryptophan or carry an inactive *tyrR* regulator gene[32].

Characteristics of the common pathway in other organisms

With some minor variations, the control pattern of the common pathway is similar to that of *E. coli* in the *Enterobacteriaceae* examined. *Neurospora crassa* also has the three DAHP synthases.

Yeasts contain two DAHP synthases, one inhibited by phenylalanine, the other by tyrosine.

Bacillus subtilis has a single DAHP synthase. The enzyme is bifunctional and is endowed with chorismate mutase activity. It is composed of four identical subunits of molecular weight 38,500. In some strains, the bifunctional enzyme forms a trifunctional complex with shikimate kinase which, however, is a separate protein. The DAHP synthase activity is insensitive to the aromatic amino acids: an efficient inhibition by any one of the three metabolites would be detrimental to the organism. The regulation occurs through its inhibition by chorismate and prephenate.

In *Brevibacterium flavum*, the subunit of the tetrameric enzyme has a molecular weight of 55,000. The enzyme is synergistically inhibited by phenylalanine and tyrosine. However, a second protein is required for chorismate mutase activity. This dimeric (2 x 13,000) protein stimulates DAHP synthase activity. The complex exists as an $\alpha_4\beta_8$ quaternary structure, where chorismate mutase activity is inhibited by phenylalanine and tyrosine, whereas tryptophan activates chorismate mutase but not DAHP synthase. .

In *Neurospora crassa*, *Euglena* and some yeasts, including *S. cerevisiae*, the five central activities of the shikimate pathway are carried out by a pentafunctional protein, which catalyzes the conversion of DAHP to 5-enolpyruvoylshikimate-3-phosphate without the intermediates ever leaving the protein. The multifunctional protein from *N. crassa* is a homodimer (2 x 165,000), in which four of the five activities are activated by DAHP.

N. crassa chorismate synthase has been obtained pure as a homodimer (2 x 55,000).

32. See below.

II. BIOSYNTHESIS OF PHENYLALANINE AND TYROSINE FROM CHORISMIC ACID

It is quite plausible that the first precursor identified in the biosynthesis of these two amino acids, prephenic acid, arises by the rearrangement of chorismic acid:

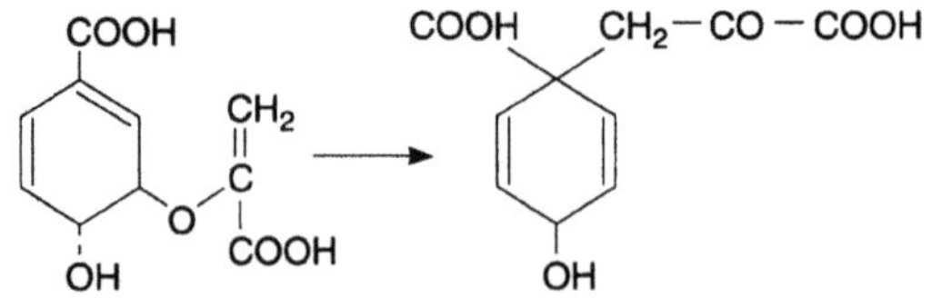

chorismic acid prephenic acid

The enzyme catalyzing this reaction has been called chorismate mutase.

During the purification of the *E. coli* enzyme, chromatography of the extracts on DEAE-cellulose yielded two separate peaks of chorismate mutase activity. Prephenate dehydratase was associated with the first peak and prephenate dehydrogenase with the second. Subsequent studies have shown that in each case, both of the activities are carried out by a single, bifunctional enzyme.

Chorismate mutase-prephenate dehydratase is the product of the *pheA* gene, situated at 57 mn on the *E. coli* chromosome, whereas chorismate mutase-prephenate dehydrogenase is encoded by the *tyrA* gene (57 mn). The product of the first bifunctional enzyme, which involves a decarboxylation and a dehydration, is phenylpyruvic acid. The second bifunctional enzyme also catalyzes a decarboxylation and, in addition, a dehydrogenation requiring the presence of NAD^+ as electron acceptor. Its product is *p*-hydroxyphenylpyruvic acid.

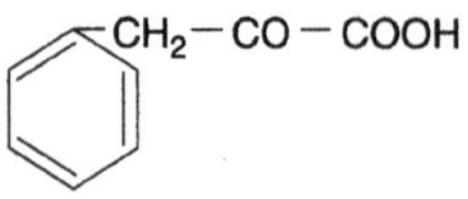

phenylpyruvic acid

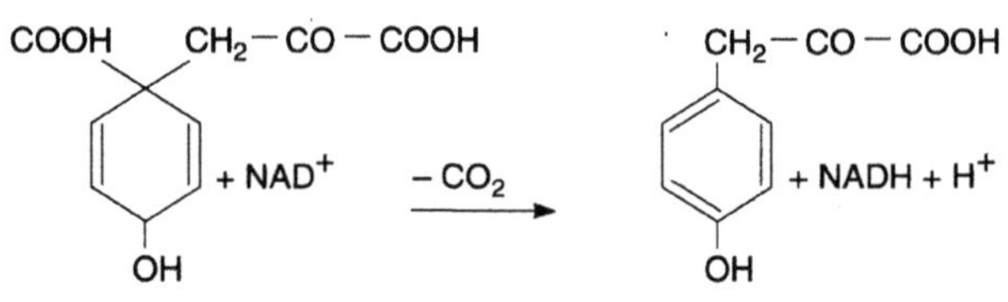

prephenate *p*-hydroxyphenylpyruvate

The two enzymes are homodimers with subunit molecular weights of about 40,000.

In the case of chorismate mutase-prephenate dehydratase, a number of studies involving loss of only one of the activities, differential inhibitions and inactivations point to two independent catalytic sites, whereas for chorismate mutase-prephenate dehydrogenase it seems likely that the two reactions are catalyzed at closely situated and interacting active sites.

Each of the two α-keto acids is the substrate for transamination reactions which yield, respectively, phenylalanine and tyrosine.

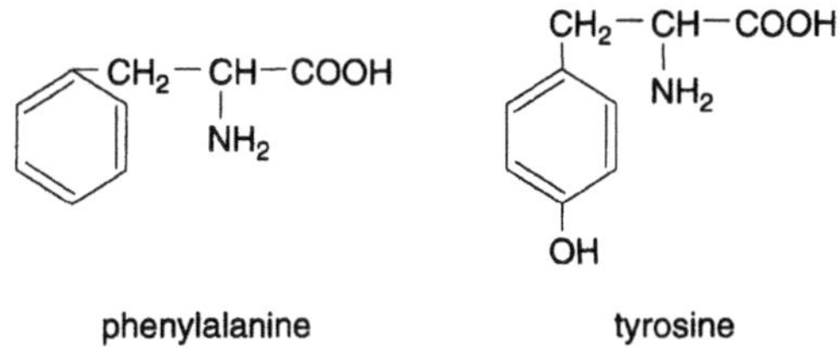

In both cases, glutamate is the amino donor. The early studies of Rudman and Meister, already referred to several times, showed that there were a number of transaminases with a rather broad specificity in *E. coli.* Phenylalanine and tyrosine can be formed by at least three aminotransferases, the aromatic aminotransferase (*tyrB*, 92 mn), the aspartate aminotransferase (*aspC*, 21 mn), and the branched-chain aminotransferase specified by *ilvE.* Mutations in *tyrB* and *aspC* result in a tyrosine requirement, whereas lesions in the three genes lead to a tyrosine *and* phenylalanine auxotrophy. Based upon relative K_m values of the keto acids for the various enzymes, it seems that under normal physiological conditions the synthesis of the two amino acids is primarily carried out by the aromatic aminotransferase, the product of the *tyrB* gene. The coding region of this gene corresponds to a polypeptide of 397 amino acid residues. The native enzyme is a homodimer, whose activity is repressed by tyrosine and derepressed as a result of a mutation in the *tyrR* regulatory gene.

tyrR regulon

The *tyrR* gene was first identified as being responsible for the resistance of a mutant possessing only the DAHP (tyr) enzyme to the tyrosine analogue 4-aminophenylalanine. Since the analogue acts as a false co-repressor of the synthesis of this enzyme, but is not a feedback inhibitor as tyrosine, derepressed mutants were isolated readily. The mutations were shown to be recessive in diploids and caused the derepressed synthesis of DAHP synthase (tyr) *(aroF)*, chorismate mutase-prephenate dehydrogenase *(tyrA)*, and of the aromatic aminotransferase *(tyrB)*. *tyrR* codes for an aporepressor which acts on specific operator targets upstream the *aroF* and *tyrA* structural genes, which are part of a single transcription unit (57 mn).

Later, it was found that *tyrR* controls not only the expression of tyrosine-specific genes, but also that of shikimate kinase, DAHP synthase (phe) and of the TyrR protein itself. In those cases however, the corepressor status is different. Whereas the repressor is tyrosine in the case of *aroF*, *tyrA* and *tyrB*, the synthesis of shikimate kinase is repressed by any of the three aromatic amino acids, that of DAHP synthase (phe) by phenylalanine and tryptophan and that of the TyrR protein does not appear to require the presence of an effector molecule.

The TyrR protein has been cloned and sequenced. Its deduced molecular weight is 53,099. The operator loci with which the TyrR protein interacts have been characterized and a consensus "Tyr box" has been identified, existing in one or more copies upstream one or the other structural genes.

TyrR is endowed with an ATPase activity which is stimulated by tyrosine. However, it would be premature to conclude that transcriptional repression is mechanistically connected to the binding and hydrolysis of ATP.

Regulation of *pheA* gene by attenuation

We have just seen that the expression of chorismate mutase-prephenate dehydrogenase is controlled by the TyrR protein combined to tyrosine. In contrast, it appears that the expression of chorismate mutase-prephenate dehydrogenase is mainly regulated by an attenuation mechanism. An analysis of the leader region preceding *pheA* reveals it to have characteristics similar to those reported for the *trp* operon. The leader peptide is 15 amino acids long and contains seven phenylalanine residues.

Met Lys His Ile Pro **Phe Phe Phe** Ala **Phe Phe Phe** Thr **Phe** Pro

Other organisms. Arogenate pathway of phenylalanine and tyrosine biosynthesis

In contrast to bacteria and plants, knowledge of the regulatory properties of the single chorismate mutase from yeasts is not extensive: generally, the enzyme is activated by tryptophan, and in some cases can be negatively regulated by phenylalanine and/or tyrosine, this inhibition being reversed by tryptophan.

We have seen that in *B. subtilis*, chorismate mutase, and DAHP synthase coexist on a common single protein. This physical arrangement is the molecular basis for the sensitivity of DAHP synthase to inhibition by chorismate and prephenate in a pattern of allosteric control which has been called sequential feedback inhibition.

An alternative pathway of phenylalanine biosynthesis was first discovered in *Pseudomonas aeruginosa*, and of tyrosine biosynthesis in cyanobacteria. Transamination in this pathway precedes the dehydration-decarboxylation and the dehydrogenation-decarboxylation step respectively. This pathway uses a prephenate transaminase:

prephenate + glutamic acid →

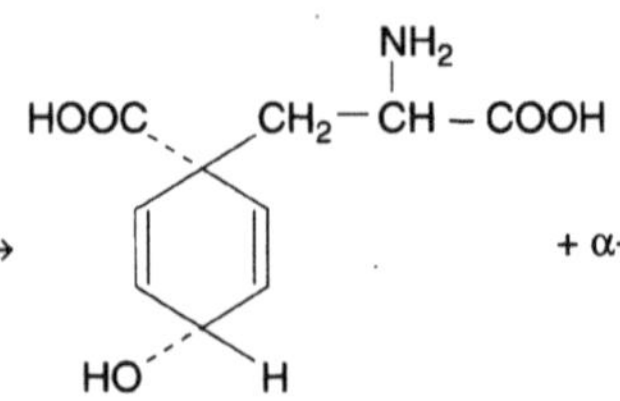

+ α-ketoglutaric acid

arogenate

An arogenate dehydratase leads then to phenylalanine, and an arogenate dehydrogenase to tyrosine.

Euglena gracilis synthesizes phenylalanine exclusively by the arogenate pathway. In contrast, activities for both pathways coexist in *Pseudomonas aeruginosa,* where a single dehydrogenase acts upon prephenate and arogenate. In addition, *Pseudomonas* and some other bacterial species can directly hydroxylate phenylalanine to tyrosine.

In *Erwinia herbicola*, a monofunctional prephenate dehydrogenase has been obtained by genetic engineering: it retains the dehydrogenase activity but lacks the mutase activity, demonstrating the separability of the two catalytic domains.

In higher plants as in bacteria, different patterns prevail in different species. For example, *Phaseolus* can utilize the two pathways whereas, in *Zea mays*, the arogenate pathway is the only route to phenylalanine synthesis.

III. BIOSYNTHESIS OF TRYPTOPHAN FROM CHORISMIC ACID

Early studies showed that some microorganisms requiring tryptophan for growth could also grow if indole or anthranilic acid were added to the culture medium in place of tryptophan. The reactions involved in the biosynthesis of tryptophan as well as the enzymes catalyzing these reactions are now known in detail.

Anthranilate synthase-anthranilate phosphoribosyltransferase

This enzyme, synthesizes anthranilate from chorismic acid and glutamine

chorismic acid + glutamine →

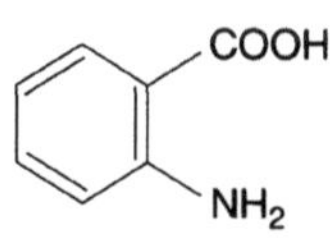

+ glutamate + pyruvate

anthranilic acid

A γ-glutamylenzyme is formed as an intermediate in the reaction and hydroxylamine, which inhibits anthranilate synthase, also inhibits the glutaminase activity.

In *E. coli* and *S. typhimurium*, the enzyme is composed of two molecules each of the polypeptides encoded by the *trpE* and the *trpD* genes, referred to as components I and II respectively. Component I catalyzes the formation of anthranilate from chorismate with ammonia as the amino donor. However, for the reaction to proceed with glutamine as the amino donor, component II is also necessary. Component II, in addition to its glutamine binding site and to its glutamine amidotransferase activity, alone catalyzes the anthranilate phosphoribosyltransferase activity.

anthranilate + phosphoribosylpyrophosphate → phosphoribosylanthranilate + PPi

Anthranilate synthase is the target of feedback inhibition in the tryptophan branch of the aromatic pathway. Component I binds tryptophan, and since the inhibition is competitive with respect to chorismate, this determines the rate of chorismate channeling toward tryptophan biosynthesis. Component II has the binding site for glutamine (it contains the radioactivity when the enzyme complex is treated with radioactive 6-diazo-5-oxo-norleucine, an analogue of glutamine which binds covalently; this covalent affinity labeling depends on the presence of chorismate and is inhibited by tryptophan).

The enzyme has an $\alpha_2\beta_2$ structure. Component I has 520 amino acids and is devoid of tryptophan, and Component II has 531 amino acids.

Proteolytic digestion of the *S. typhimurium* complex leads to the loss of the phosphoribosyltransferase activity and to a notable loss of molecular weight, resulting from the reduction of the molecular weight of Component II to a 19,000 dalton fragment. This small N-terminal piece still carries the glutamine binding site and still interacts with Component I to yield a functional glutamine-dependent anthranilate synthase activity indistinguishable from that of the native enzyme, but devoid of phosphoribosyltransferase activity, which belongs to the C-terminal piece.

If we screen the different associations between anthranilate synthase (AS) and anthranilate phosphoribosyltransferase (PRT), we find that in pseudomonads, *B. subtilis* and *Serratia marcescens*, glutamine amidotransferase (GAT) is carried by a polypeptide chain which is *not* fused to PRT (or to AS, for that matter). In these organisms, the conversion of chorismate to phosphoribosylanthranilate is catalyzed by three different proteins. Figure 1 summarizes these findings.

It is seen clearly that the sum of the molecular weights of GAT plus PRT is approximately the same, whether the activities reside on the same polypeptide or not.

Phosphoribosylanthranilate isomerase-indoleglycerol phosphate synthase

The next two steps are catalyzed in *E. coli* and *S. typhimurium* by a single enzyme, phosphoribosylanthranilate isomerase-indoleglycerol phosphate synthase, encoded by the *trpC* gene. Phosphoribosylanthranilate is a ribonucleotide and the isomerase converts it to 1-(O-carboxyphenylamino)1-deoxyribulose 5-phosphate by an Amadori rearrangement. This compound is then decarboxylated concomitantly with a ring closure to yield indoleglycerol phosphate, the immediate precursor of tryptophan.

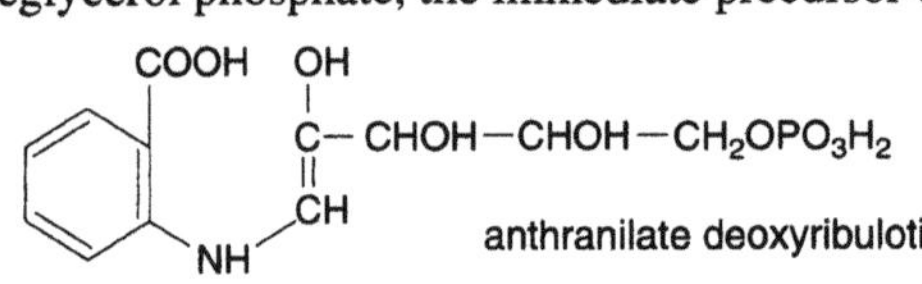

anthranilate deoxyribulotide (CDRP)

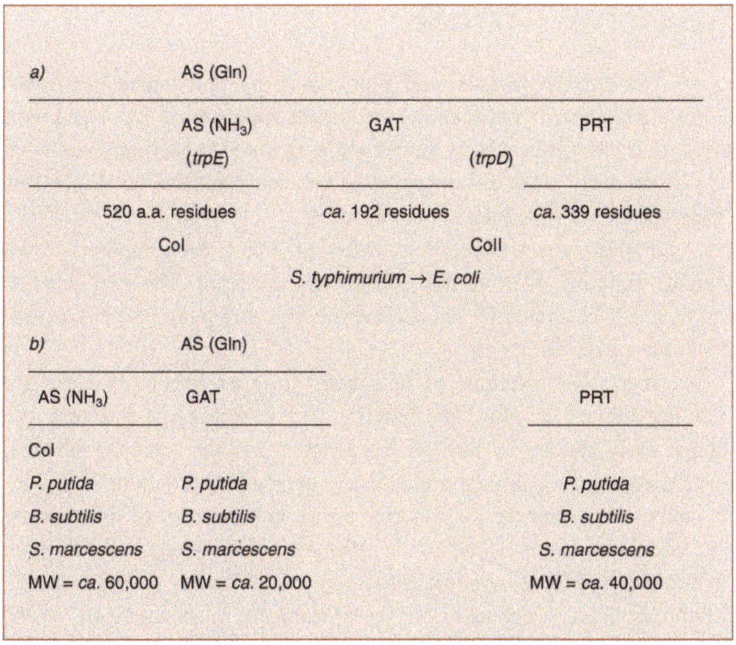

Figure 1. Conversion of chorismate to phosphoribosylanthranilate in different microorganisms.

CDRP was identified in the non-phosphorylated form as a product accumulating in certain tryptophan mutants of *E. coli* and *Aerobacter aerogenes* starved for tryptophan.

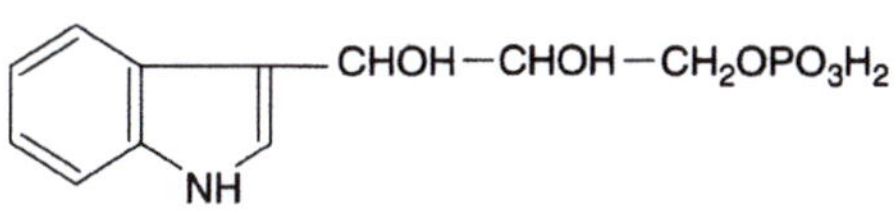

indoleglycerol phosphate (IGP)

The active centers of the two reactions are distinct and do not overlap.

The bifunctional enzyme is a single polypeptide of 45,000 daltons. Studies carried out with mutants of the enzyme and using limited proteolysis on the wild-type enzyme have established that the indoleglycerol phosphate synthase reaction is carried out by the amino terminal half of the polypeptide while the carboxy terminal half forms the isomerase domain. The protein has been crystallized.

In some non-enteric bacteria and in *Saccharomyces cerevisiae*, the two activities are carried by two distinct polypeptides.

In *Euglena*, the anthranilate phosphoribosyltransferase, the phosphoribosylanthranilate isomerase, and the indoleglycerol phosphate synthase activities are associated with the tryptophan synthase activity in a single homodimeric protein (2 x 160,000 daltons).

In yeast, the synthesis of anthranilate synthase and indoleglycerol phosphate synthase is under the control of a single genetic locus, but phosphoribosylanthranilate isomerase is a quite separate enzyme.

Tryptophan synthase

The enzyme from *E. coli* was chosen for the studies that established the colinearity between the nucleotide sequences in DNA and their translation into sequences of amino acids. It has also been an invaluable tool in the early work on mutagenesis and the genetic code. It presents characteristics which deserve our attention.

In procaryotes, tryptophan synthase is a tetramer composed of two separate subunits, with an $\alpha_2\beta_2$ quaternary structure. The two subunits can be easily separated by ion exchange chromatography into two α subunits and a β_2 subunit.

Tryptophan synthase of *E. coli* has been an important model for studying subunit structure and function in a multienzyme complex. Although early studies focused on the genetics and the separate subunits, further work (using limited proteolysis, chemical modification and kinetic studies) has concentrated on the compared properties of the separate and associated subunits and on the changes occurring upon association. The cofactor of the β_2 subunit, pyridoxal phosphate has served as a useful chromophore probe to study the environment of its active site in the isolated β_2 and when reassociated with α. An extensive X-ray diffraction study has provided an incomparable help to the understanding of the mechanism of tryptophan synthase.

The native enzyme ($\alpha_2\beta_2$) catalyzes the pyridoxal phosphate dependent condensation of indoleglycerol phosphate (IGP) and serine to yield L-tryptophan, and glyceraldehyde-3-phosphate, the physiological reaction (reaction *a*).

The reactions catalyzed by the isolated subunits take place at a much lower rate than the ones carried out by the intact enzyme. It ap-

(*a*)

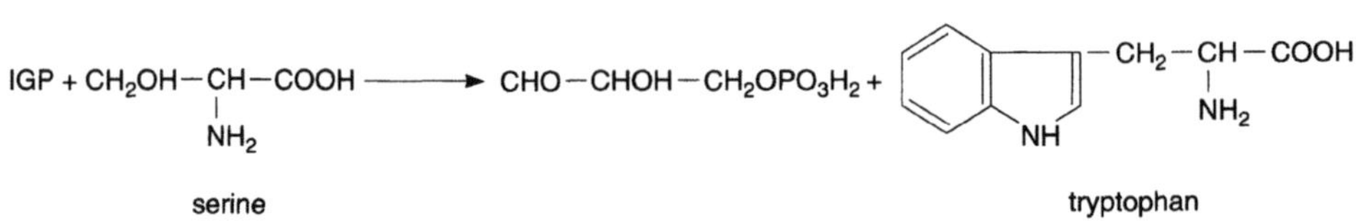

(*b*)

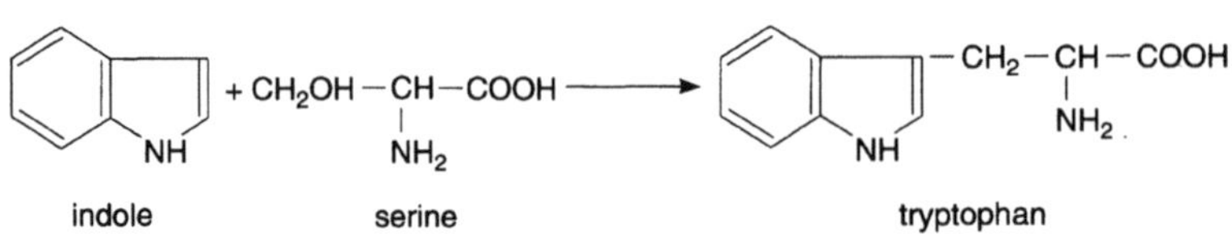

(*c*)

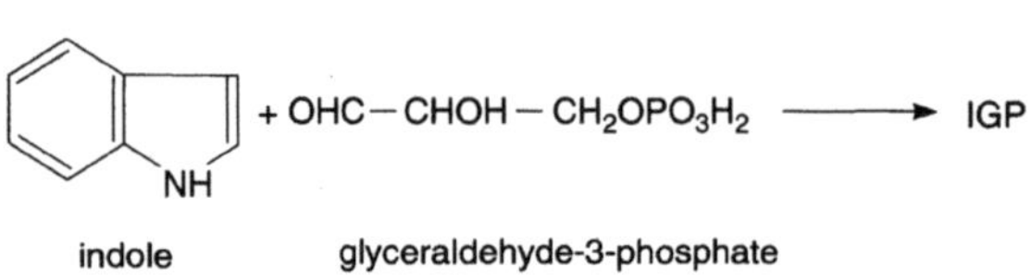

pears that each of the two chains has a specific combining site for indole, specific for the pyridoxal-phosphate dependent reaction *b* in the case of the β chain, and for reaction *c* in the case of the α chain.

The reassociation of the two subunits results in the activation of the two individual reactions and the creation of a catalytic site for reaction *a*. The active sites of both chains are certainly reorganized in a very special manner in the $\alpha_2\beta_2$ complex, since we know that indoleglycerol phosphate is converted to tryptophan without the free indole participating as an intermediate.

Additional reactions are catalyzed by the isolated subunits: among others, the β_2 subunit can deaminate serine to pyruvate and ammonia (serine dehydratase activity), or condense serine to 2-mercaptoethanol to give S-hydroxyethylcysteine. The existence of these non-physiological reactions has led to an understanding of the mechanism of action of the β_2 subunit: all the reactions in which serine is a substrate occur through a series of Schiff base intermediates formed between enzyme-bound pyridoxal phosphate and serine or derivatives consistent with the general mechanism of β-elimination or β-replacement pyridoxal phosphate-catalyzed reactions.

$$R{-}CH_2{-}CH(NH_2){-}COOH + H_2O \rightarrow CH_3{-}CO{-}COOH + RH + NH_3 \quad (\beta\text{-elimination})$$
$$R{-}CH_2{-}CH(NH_2){-}COOH + R'H \rightarrow R'{-}CH_2{-}CH(NH_2){-}COOH + RH \quad (\beta\text{-replacement})$$

The three-dimensional structure of the $\alpha_2\beta_2$ complex of tryptophan synthase from *S. typhimurium* has been determined by X-ray crystallography at 2.5 Å resolution. The four polypeptide chains are arranged nearly linearly in an αββα order forming a complex 150 Å long. The overall peptide fold of the α subunit (which cleaves indoleglycerol phosphate, reaction *c)* is that of an 8-fold α/β barrel. Its active site has been located by difference Fourier analysis of the complex with indolepropanol phosphate, a competitive inhibitor and a close analogue of the natural substrate. The pyridoxal phosphate-dependent β_2 subunit is composed of two domains of approximately equal size, folded into similar helix/sheet/helix structures. The binding site for pyridoxal phosphate lies deep within the interface between the two β subunit domains. The active sites of the two subunits are separated by 25 Å but are directly connected by a tunnel[33] with a diameter matching that of the intermediate substrate indole, facilitating its diffusion from its point of production in the α subunit active site to the site of tryptophan synthesis in the β subunit active site and preventing its escape to the solvent during catalysis. Rapid kinetic studies have confirmed that the tunnel is the preferred route of entry of indole into the β active site in the indole + serine → tryptophan reac-

33. Globular proteins are usually solidly packed and interior holes occur very rarely. Although all the known protein atoms have been accounted for in the electron density map, several weak density features remain in the tunnel which might be occupied by loosely bound water molecules (which by the way should not facilitate the diffusion of the hydrophobic indole!).

tion and that indole appears to be channelled between the α and β sites in the overall physiological reaction. The knowledge of the tridimensional structure and of the location of the active site, as well as the availability of site-directed mutagenesis allow rapid progress in both the detailed catalytic mechanism and the laws governing the reciprocal communications between the α and the β subunit in the $\alpha_2\beta_2$ complex.

In *N. crassa*, tryptophan synthase is a homodimeric protein of a molecular weight of 150,000. In *S. cerevisiae*, it is also a homodimer of 707 residues per monomer with fused α and β domains. The N-terminal domain (239 residues) is highly homologous to the α subunit of *E. coli* and a C-terminal 389 residue segment is homologous to the β chain of *E. coli.* A 28-residue fragment, with no homology to either subunit, connects the two domains. We have seen that in *Euglena,* tryptophan synthase is part of a multifunctional polypeptide.

The following scheme summarizes the steps identified in the biosynthesis of the aromlatic amino acids in *E. coli.*

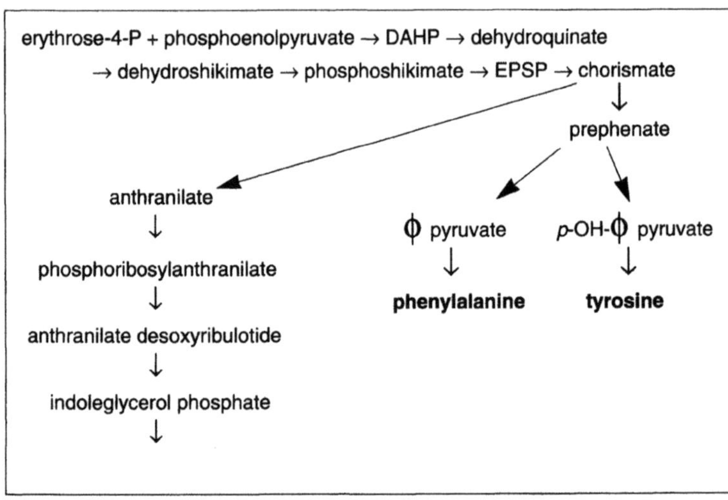

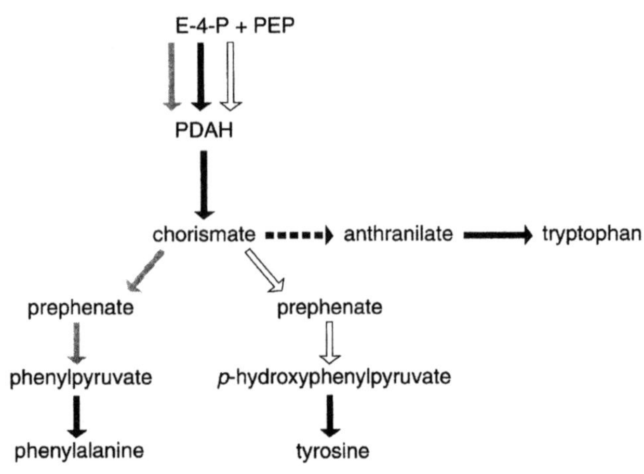

The scheme at the bottom of the preceding page summarizes the existing state of our knowledge about feedback inhibition in the biosynthesis of these amino acids in *E. coli.*

The reactions shown in gray are inhibited by phenylalanine; those in white by tyrosine. The reaction shown by a dotted arrow is inhibited by tryptophan.

In contrast, in *B. subtilis* there is only one DAHP synthase, which is inhibited by chorismate and prephenate:

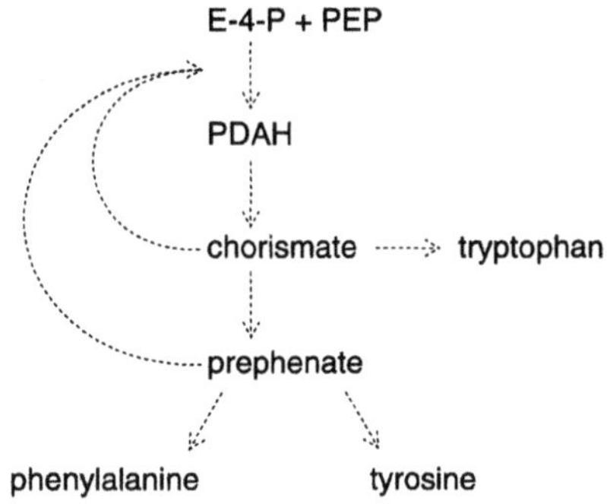

Regulation of tryptophan biosynthesis at the genetic level. Tryptophan repressor

Under physiological conditions, the five enzymes of tryptophan biosynthesis, coded by genes organized into a single operon, are found in roughly equimolar amounts in the cell. When the synthetic rate is altered by physiological means, plots of the activity of one enzyme against the other show coordinate control[34].

Since the Trp repressor was the first repressor to be characterized in a biosynthetic pathway, it appears useful to describe the rationale of its discovery.

What are the properties to be expected from derepressed mutants? If the synthesis of the enzymes specific for tryptophan biosynthesis are no longer under the control of the size of the intracellular tryptophan pool, one expects to find these enzymes in a greater amount in a derepressed mutant; furthermore, the addition of tryptophan should not reduce appreciably the enzyme content. The greater amount of tryptophan enzymes could be correlated with an increased synthesis of tryptophan. This increased synthesis might lead to a resistance to a tryptophan analogue.

The following experiment was carried out by Cohen and Jacob in 1959.

A suspension of *E. coli* is spread onto a solid medium containing 10 mM 5-methyltryptophan, a toxic analogue of tryptophan. Some resis-

34. For details of coordinate *vs.* parallel control, see the chapter on histidine biosynthesis.

tant mutants give rise to colonies, which belong to two categories: a) those which do not excrete tryptophan; b) those which excrete tryptophan, as evidenced by a syntrophy test with an indicator strain auxotrophic for tryptophan. Among the excreting bacteria, one finds some which synthesize greater quantities of tryptophan synthase than the wild type bacteria, this greater amount being unaffected by the presence of tryptophan in the medium (table I).

Table I. Tryptophan synthase activity of *E. coli* grown with or without tryptophan.

Extracts	Tryptophan synthase (units)		Repression (%)
	Culture		
	with Trp	without Trp	
Wild type	250	66	74
trpR mutant	1467	1449	0

It was found later that the *trpR* mutation affects all the tryptophan biosynthetic enzymes in a similar manner.

The *trpR* gene has been localized at 0 mn on the circular *E. coli* bacterial chromosome, far from the structural genes (28 mn) coding for the biosynthetic enzymes, *trpEDCBA* (transcribed in that order, which is also the order of the corresponding biochemical reactions). Its relative location, as well as that of many other repressors with regard to the location of the structural genes they control, is very different from that of the *i* gene *vs.* the *lac* operon, which may be entirely fortuitous.

The experiment which allows one to decide whether $trpR^-$ (derepressed) is dominant or recessive with regard to $trpR^+$ (wild type) follows: a cross is made between a male Hfr $trpR^-$ Sm^S and a female F^- $trpR^+$ Sm^R; at various times after the injection of the male chromosome, the zygotes are spread onto a solid medium containing streptomycin and 5-methyltryptophan. Neither of the two parental strains can grow on this medium, the males because they are sensitive to streptomycin (Sm^S), the females because their growth is inhibited by 5-methyltryptophan. The gene for streptomycin sensitivity is practically never injected, being located at more than 70 mn. Only the zygotes having the constitution

$$\frac{trpR^-}{trpR^+\ Sm^R}$$

can give rise to colonies if $trpR^-$ is dominant. The recombinants $trpR^-$ resistant to streptomycin (Sm^R) can also give rise to colonies, but they appear much later. The kinetics of appearance of colonies in function of the time of spreading shows that the zygotes are sensitive to 5-methyltryptophan and do not give rise to colonies. In other words, *trpR*$^+$ is dominant over *trpR*$^-$. As in the case of the *lac* operon, the hypothesis can be advanced that *trpR*$^+$ codes for a cytoplasmic repressor which is not synthesized or synthesized in an inactive form by the *trpR*$^-$ mutants.

One would have guessed that feedback inhibition of anthranilate synthase by tryptophan would prevent tryptophan biosynthesis even at high enzyme levels, but this is clearly not the case.

Most of the physiological and genetic studies on the tryptophan repressor are due to Yanofsky, Crawford, and their associates.

The promoter of the *trp* operon is about 200 nucleotides upstream of *trpE*, the first structural gene. The *trp* operator, a 22-base pair target, is within this promoter. When this sequence is bound to the repressor, RNA polymerase cannot form a productive complex with the promoter. The binding occurs only in the presence of tryptophan.

In addition to preventing the transcription of the genes of the tryptophan operon, the tryptophan repressor exerts the same effect on the transcription of *aroH*, the gene coding for DAHP (trp), and on its own transcription. Examination of the sequences upstream the *aroH* and *trpR* genes and comparison with the operator of the *trp* operon have permitted the identification of a consensus "Trp box" for the binding of the repressor.

Among the 5-methyltryptophan mutants excreting tryptophan, there was another class, which belonged to the constitutive dominant type. This dominant constitutivity affects in merodiploids all the genes of the tryptophan operon in *cis* position with regard to the mutation, but not those in *trans*. All these mutations were localized in a segment adjacent to the operon, but upstream the first structural gene. They affected the operator.

Unitary model for induction and repression

This experiment establishes the basis for a unitary theory of the mechanism of repression in catabolic and biosynthetic systems: in the case of the *lac* operon, the free repressor blocks transcription and is inactivated upon combination with the inducer. In the biosynthetic systems, the free repressor is inactive and active only in combination with its ligand, the corepressor.

Isolation of *trp* repressor

Zubay and his colleagues used the same *in vitro* protein synthesis system they had first used with β-galactosidase, but this time the source of DNA was a segment isolated from a transducing defective bacteriophage carrying a a fusion of the *lac* and *trp* operons (λ-d*trp*-*lac*) (fig. 2). It is therefore possible to assess the presence of the aporepressor in the *trpR*$^+$ strains, and its absence in the *trpR*$^-$ strains, by measuring the *in vitro* synthesis of β-galactosidase, which is now under the control of the

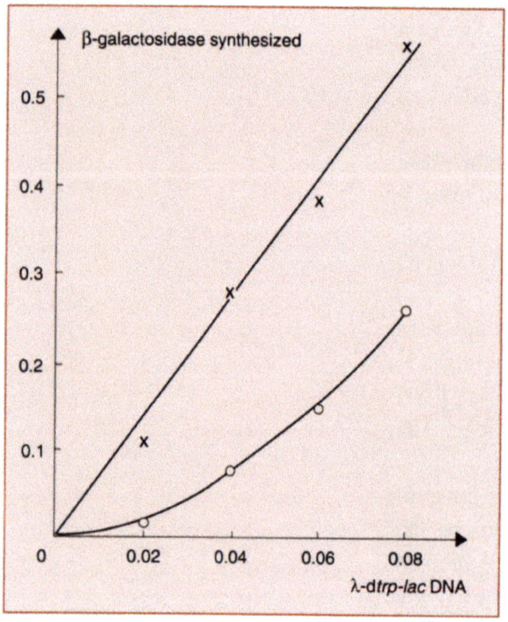

Figure 2. Synthesis of β-galactosidase as a function of the concentration of λ-d*trp-lac* DNA. Experiments performed with *trpR*⁻ supernatants (x) and with the same supernatants (○) where a solution of partially purified Trp repressor (54 μg/ml) was added.
(From G. Zubay, D. E. Morse, W. J. Schrenk & J. H. M. Miller (1972). With permission from Dr Geoffrey Zubay.)

trp promoter and operator. The synthesis of β-galactosidase is progressively reduced by increasing the amount of *trpR*$^+$ cellular extracts. Such a repression is not observed when β-galactosidase synthesis is programmed by normal *lac* DNA (fig. 3).

This specific and highly specific detection has helped the purification of the Trp repressor which, like many proteins interacting with nu-

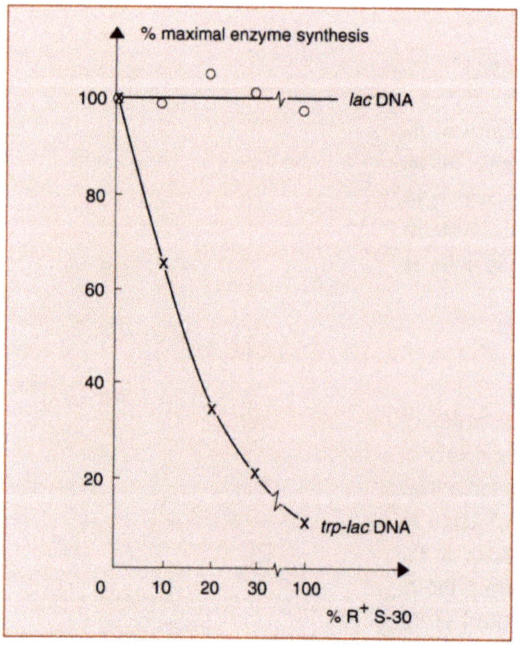

Figure 3. Synthesis of β-galactosidase as a function of the origin of the supernatants. Experiments performed with lac DNA or trp-lac DNA. The supernatants (S-30) originate from either *trpR*⁺ or *trpR*⁻ strains and they are used either alone, or in variable proportions. Abscissa: percentage of *trpR*⁺ extract.
(From G. Zubay, D. E. Morse, W. J. Schrenk & J. H. M. Miller (1972). With permission from Dr Geoffrey Zubay.)

cleic acids, is strongly adsorbed by phosphocellulose. It is important to recall that tryptophan is required for β-galactosidase synthesis and that it is always present in this *in vitro* DNA-programmed protein synthesis

The synthesis of anthranilate synthase and of tryptophan synthase has also been measured directly by an analogous *in vitro* system, but programmed by *trp* DNA instead of *trp-lac* DNA. Anthranilate synthase and the α and β subunits of tryptophan synthase are synthesized in equimolar quantities, which suggests that the regulation of translation is similar *in vitro* and *in vivo*. Furthermore, the synthesis of the enzymes follows a temporal sequence corresponding to the relative sequence of the corresponding genes on the DNA.

In order to show the effect of tryptophan, the aporepressor, the synthesis of the specific messenger RNA has been measured. This type of experiment where the synthesis of protein is not measured has shown that tryptophan is indispensable to the effective repression of the expression of the *trp* operon in the presence of the aporepressor.

The *trpR* gene has been cloned and overproduced by bacteria containing genetically engineered plasmids. It codes for a 107-residue polypeptide normally existing as a dimer, whether complexed or not with tryptophan.

In equilibrium binding experiments, the K_D of high affinity binding of Trp repressor to a 43-bp segment containing the natural *trp* promoter operator region was determined to be 2×10^{-10} M. Loss of one or both tryptophan molecules from the complex destabilizes the complex immediately. A heterodimeric aporepressor with a single tryptophan binding site has been constructed: its affinity for operator, when the single site is saturated, is 20- to 25-fold higher than that of the natural aporepressor, but 20- to 25-fold lower than that of the natural tryptophan-saturated repressor.

The Trp repressor has been purified and crystallized with and without tryptophan. The ternary complex aporepressor-tryptophan-operator sequence has also been crystallized. The results of their X-ray crystallographic analysis will be presented in the chapter on protein-DNA interactions.

IV. SYNTHESIS OF *p*-AMINOBENZOIC ACID

As anthranilate synthase, *p*-aminobenzoate synthase, uses glutamine and chorismate as substrates. It also produces glutamate and pyruvate, but instead of anthranilate (*o*-aminobenzoate), its specific product is *p*-aminobenzoate. The two aromatic products differ only in the disposition of the amino substituent on the benzene ring.

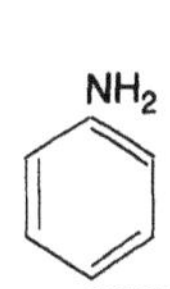

p-aminobenzoic acid

Like anthranilate synthase, *p*-aminobenzoate synthase is composed of two different subunits. The larger component, CoI (MW *ca.* 50,000) catalyzes the synthesis of *p*-aminobenzoate from chorismic acid and ammonia, while the second subunit CoII (MW *ca* 20,000) is a glutamine amidotransferase which confers upon the enzyme complex the ability to utilize the amide group of glutamine as the nitrogen source. In some microorganisms *(B. subtilis, Acinetobacter calcoaceticus, Pseudomonas acidovorans)*, this glutamine amidotransferase is able to form a functional complex with Component I of both anthranilate and *p*-aminobenzoate synthases.

SELECTED REFERENCES

The common pathway. Biosynthesis of phenylalanine and tyrosine

See the reviews by K. M. Herrmann & J. Pittard, pp. 301-350 in *Amino acids: Biosynthesis and Genetic Regulation,* K. M. Herrmann & R. Somerville eds. Addison-Wesley Publishing Company, Reading, Mass. US (1983).

Tryptophan synthase.

E. W. Miles, *Methods in Enzymology* **49**, 127-186 (1979).

E. W. Miles, *Methods in Enzymology* **64**, 93-172 (1991).

C. G. Hyde, S. A. Ahmed, E. A. Padlan, E. W. Miles & D. R. Davies, *J. Biol. Chem.* **263**, 17857-17871 (1988).

Tryptophan repressor. Functional aspects

G. N. Cohen & F. Jacob, *Compt. Rend.* **248**, 3490-3492 (1959).

G. Zubay, D. E. Morse, W. J. Schrenk & J. H. M. Miller, *Proc. Natl. Acad. Sci. US* **69**, 1100-1103 (1972).

P. H. Pouwels & J. Van Rotterdam, *Proc. Natl. Acad. Sci. US* **69**, 1786-1790 (1972).

J. K. Rose & C. Yanofsky, *Proc. Natl. Acad. Sci. US* **71**, 3134-3138 (1974).

CHAPTER XXII

Biosynthesis of histidine and its regulation

All the genes of histidine biosynthesis are adjacent to each other, and form an operon located at 44 mn on the *E. coli* chromosome.

The enzyme called phosphoribosylphosphate-ATP-pyrophosphorylase (PR-ATP-pyrophosphorylase for short, *hisG*) catalyzes a Mg^{2+} dependent reaction between ATP and 5-phosphoribosyl-1-pyrophosphate (PRPP)[35] to give N1-(5'-phosphoribosyl)-ATP:

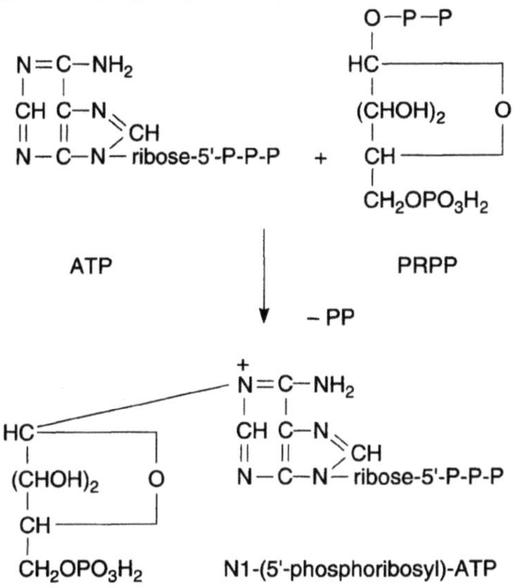

The enzyme has been purified and is a homohexamer under catalytic conditions (subunit MW = 33,000). The enzyme from wild type binds six molecules of histidine and undergoes a conformational change upon binding. It is inhibited non-competitively by histidine with respect to both substrates at physiological histidine concentrations. Other factors may be involved in modulating enzyme activity *in vivo*: the *in vitro* inhibition by histidine is enhanced by AMP, ADP, and guanosine-5'-diphosphate-3'-diphosphate (ppGpp).

The second intermediate in the biosynthetic chain leading to histidine has been identified as N1-phosphoribosyl-AMP, formed from the preceding compound by PR-ATP-pyrophosphohydrolase *(hisE)*, which removes a pyrophosphate residue. Another enzyme, PR-AMP-1,6-cyclohydrolase *(hisI)*, converts this compound to phosphoribosylformiminoaminoimidazolecarboxamide ribotide:

35. PRPP synthesis will be considered when we study the synthesis of nucleotides.

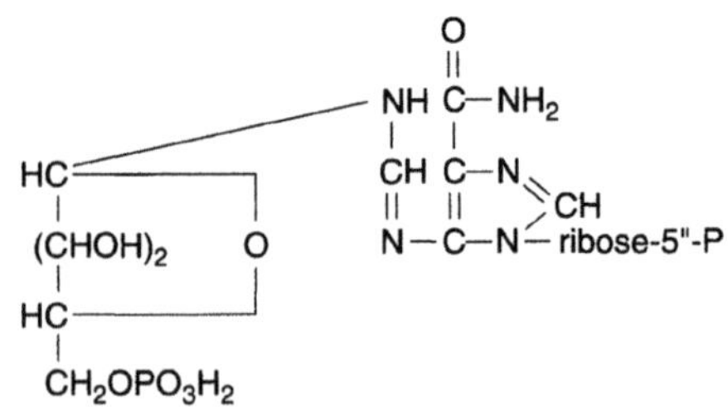

N-(5'-phosphoribosylformimino)-5-amino-4-imidazolecarboxamide-(5"-phosphoriboside)

An examination of the nucleotide sequences of the *E. coli* and *S. typhimurium hisE* and *hisI* genes indicates that the two corresponding activities are contained in a single bifunctional protein (subunit MW = 22,800), an observation confirmed by their cosedimentation as a homodimer.

An isomerase, coded by *hisA*, a monomer of 245 amino acid residues, converts this compound to the corresponding ribuloside.

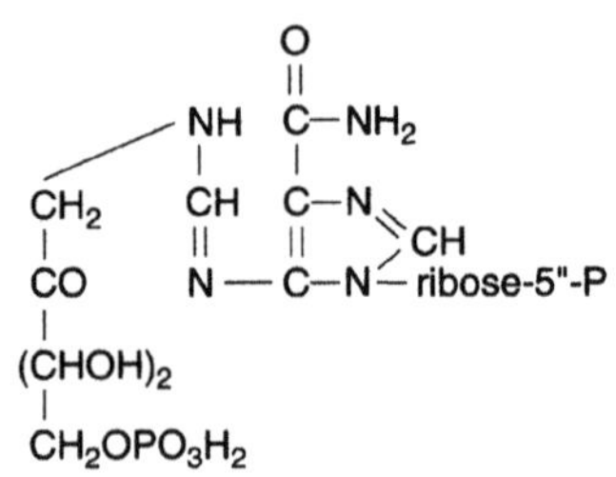

Glutamine provides the amino group for the next reaction, catalyzed by an amido transferase *(hisH)* and imidazoleglycerol phosphate is formed by a cyclase *(his F)*. The amido transferase is built from two identical subunits of 196 amino acid residues each; the cyclase also appears to be a homodimer (2 x 258 residues). The two proteins essential for the biosynthesis of histidine in *E. coli* have been overexpressed and purified to apparent homogeneity. The protein encoded by *hisF* has an ammonia-dependent activity that results in the conversion of the biosynthetic intermediate N^1-[(5'-phosphoribulosyl)formimino]-5-aminoimidazole 4-carboxamide ribonuleotide (PRFAR) to imidazole glycerol phosphate (IGP) and 5-aminoimidazole-4-carboxamido-1-β-D-ribofuranosyl-5'-monophosphate (AICAR). The second protein, encoded by *hisH*, exhibits no detectable catalytic properties with the biosynthetic intermediate PRFAR, glutamine, or ammonia. In combination, the proteins are capable of stoechiometric conversion of glutamine and PRFAR to form AICAR, IGP, and glutamate. Neither protein alone is able to mediate a conversion of the nucleotide substrate to a free metabolic intermediate. The HisF and HisH proteins form a 1:1 dimeric complex which constitutes the IGP synthase holoenzyme. In the holoenzyme, glutamine is 1000-fold more efficient than ammonia.

In *S. cerevisiae*, the two activities are encoded by a single bifunctional protein, whose N-terminal moiety corresponds to HisH, and the C-terminal moiety to HisF. The physical uncoupling of the two activities as observed in *Enterobacteriaceae* is also found in the three eubacteria *Lactococcus lactis*, *Azospirillum brasilense* (a diazotroph), *Streptomyces coelicolor*, and in the archaebacterium *Methanobacterium vannielii*.

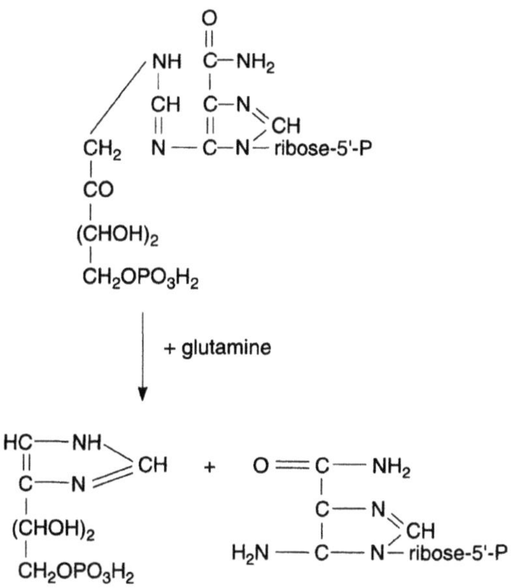

The other product of the reaction, 5-amino-4-imidazolecarboxamide ribonucleotide (AICAR), is itself a precursor of ATP. It is in this way that the purine nucleus of ATP is recycled during the biosynthesis of histidine.

Imidazoleglycerolphosphate dehydratase *(hisB)*, a homotetramer (4 x 40,000), removes a molecule of water and transforms its substrate into imidazoleacetol phosphate.

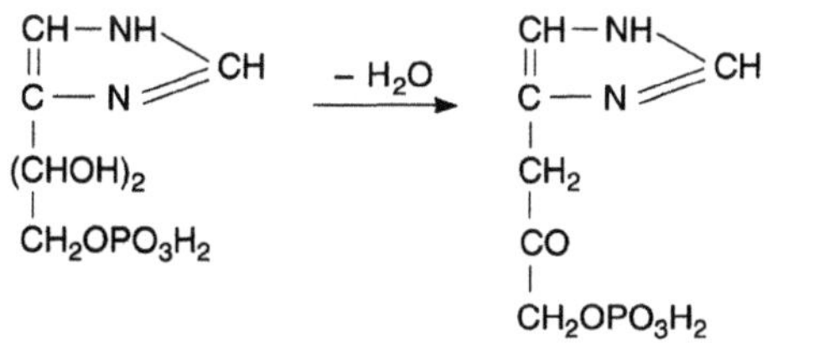

imidazoleglycerol phosphate imidazole acetol phosphate

Histidinol phosphate aminotransferase, a homodimer of 2 x 356 residues *(hisC)*, catalyzes a transamination between imidazoleacetol phosphate and glutamate and leads to the production of histidinol phosphate, which is then subject to a specific histidinol phosphatase (forming a bifunctional enzyme with imidazoleglycerol phosphate dehydratase, both enzymes being coded by *hisB*). The histidinol thus formed is then oxidized to histidine by histidinol dehydrogenase *(hisD)*, a homodimer of 2 x 434 amino acid residues.

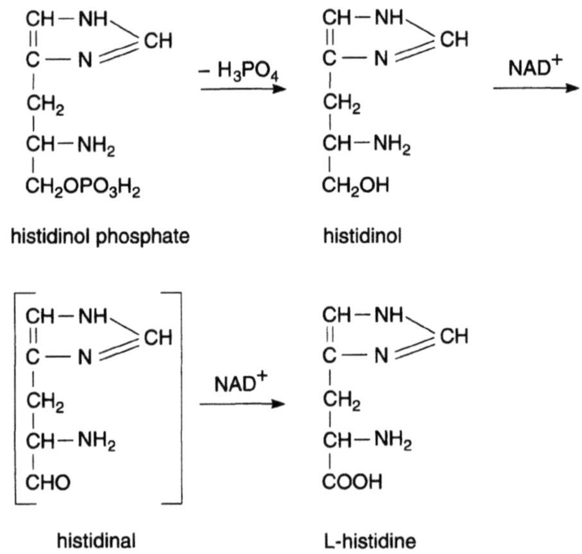

The oxidation of the primary alcohol group of histidinol to the carboxyl group of histidine requires two molecules of NAD^+. In an analogous case of successive reductions of a carboxyl group (aspartic acid to homoserine), two separate enzymes are responsible for the formation of the intermediate aldehyde and for its reduction. Here the aldehyde histidinal is indeed the intermediate, but probably remains enzyme-bound, although there are auxotrophic mutants that can grow on histidinal.

Regulation of histidine biosynthesis at the genetic level

Table I represents the histidine operon of *Salmonella typhimurium*[36]. One can see at first glance that, in contrast to the threonine and the tryptophan operons, the order of the genes on the chromosome (*hisOGDCBHAFIE*, *hisO* being the sequence coding for an operator) is notably different from that of the reactions catalyzed by the corresponding enzymes.

The question raised by Ames and Garry in 1959 was the following: does the repression of the synthesis of each one of the enzymes of the biosynthetic pathway occur in a coordinate fashion? In other terms, is the variation in the cellular content of each of these enzymes identical under conditions where the intracellular histidine pool is varied ?

To answer their own query, Ames and Garry used mutants in which one or the other of the enzymes of the biosynthetic chain were in-

36. The position of the operon is at 44 mn on the *E. coli* map, and at 42 mn on the *S. typhimurium* map.

Table I. The histidine operon of *Salmonella typhimurium*.
Note that the order of genes differs from the order of enzyme reactions. *hisB* and *hisI, E* code for bifunctional enzymes. *hisO* is the operator which controls the expression of the operon.

Gene	G	D	C	B
Reaction	1	10	8	7, 9
Enzyme	PR-ATP pyro-phosphorylase	histidinol dehydrogenase	transaminase	dehydratase, phosphatase
Gene	H	A	F	I, E
Reaction	5	4	8	3,2
Enzyme	amido-transferase	isomerase	cyclase	PR-AMP hydrolase, PR-ATP pyrophosphohydrolase

activated. As a result, their growth required the presence of exogenous histidine in the growth medium. This could be added as histidine itself. The variations of the concentrations of exogenous histidine do not affect the growth rate, but only the final cellular yield. The specific activity of the various enzymes of the pathway is low and identical to those found in the wild-type strain growing with or without histidine. This shows that the synthesis of those enzymes is maximally repressed in the wild type. However, in certain mutants, the specific activity rises when histidine becomes limiting and the enzyme extraction is done several hours after growth has stopped. In that case, the enzymes studied are synthesized at the expense of the limiting histidine originating from the slow bacterial protein turnover.

Similarly, histidine can be provided by the addition of substances liberating histidine by slow hydrolysis, carnosine (β-alanylhistidine), histidylhistidine or N-formyl-α-L-histidine, the latter giving the more interesting results.

Figure 1 shows the effects of histidine and of various concentrations of formylhistidine on the growth of an histidine auxotroph of *S. typhimurium.*

Figures 2 a, b, and c show the specific activities of three enzymes of the pathway, compared to a fourth enzyme of the same pathway, histidinol phosphate phosphatase. It is immediately apparent that the ratio of the specific activities is constant, independently of the auxotroph used and of the growth conditions. Figure 2 d shows that this relationship does not apply if the relative contents of histidinol phosphate phosphatase and glutamate dehydrogenase, not belonging to the histidine operon, are compared.

The above results have been extended to all the enzymes of the operon. They show that histidine affects equally the synthesis of all the

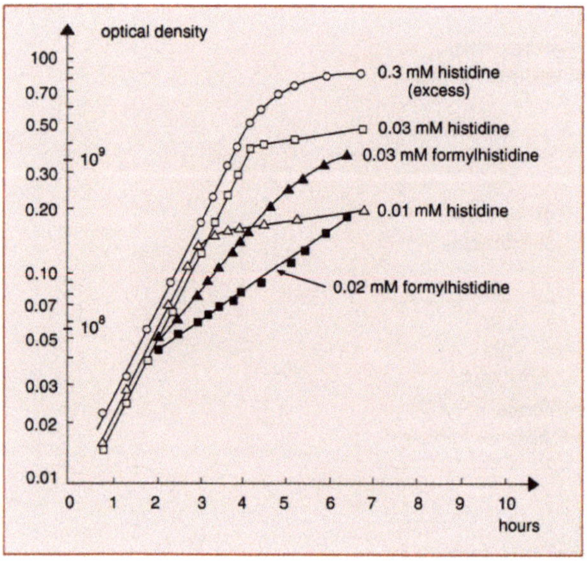

Figure 1. Growth of a mutant requiring histidine. A 30-fold variation in the exogenous concentration of histidine does not affect the growth rate, but only the yield. In contrast, the growth rate on N-α-formyl-L-histidine depends greatly on its concentration in the medium. The situation is equivalent to an internal chemostat. If the concentration of formylhistidine is lowered to 0.01 mM, there is no growth at all.
(From B. N. Ames & B. J. Garry, *Proc. Natl. Acad. Sci. US* **45**, 1453-1461 (1959). With permission from Dr. Bruce Ames.)

enzymes of the operon. Ames and Garry have named this phenomenon *coordinate repression.*

This coordinate repression is in keeping with the fact that transcription of the histidine operon leads to the synthesis of a single polycistronic messenger and implies that, at least in the absence of so-called polar mutations, the translation of this messenger leads to equimolar amounts of the different enzymes of the histidine biosynthetic pathway.

In 1972, a classic study by Lewis and Ames showed that there was an inverse correlation between the intracellular amount of charged histidinyl-tRNA (not of free histidine), and the level of the enzymes of the *his* operon in derepressed mutants and a physiologically derepressed strain.

The mutants genetically derepressed have been obtained, as in the case of the tryptophan pathway, by selecting colonies resistant to the toxic effects of a histidine analogue. The derepressed mutants are distinguished from non-derepressed mutants thanks to a peculiar morphology of the colonies. The mutations are grouped in six independent loci of the *S. typhimurium* linkage map[37].

a) *hisO* (44 mn) was first believed to be an operator locus. Its mutations cause a dominant constitutive phenotype only expressed in *cis* in merodiploids. It is thus formally analogous to the lactose and tryptophan operators we have previously studied. The term operator in its formal acceptation does not apply in this particular case, and *hisO* mutants are actually promoter mutants. The term operator is now reserved for defining a repressor binding site. The next classes of derepressed mutants

37. The original observations were made with *S. typhimurium*. The values given in the text correspond to the *E. coli* linkage map, except for *hisU* and *hisW*, where only *S. typhimurium* data are available.

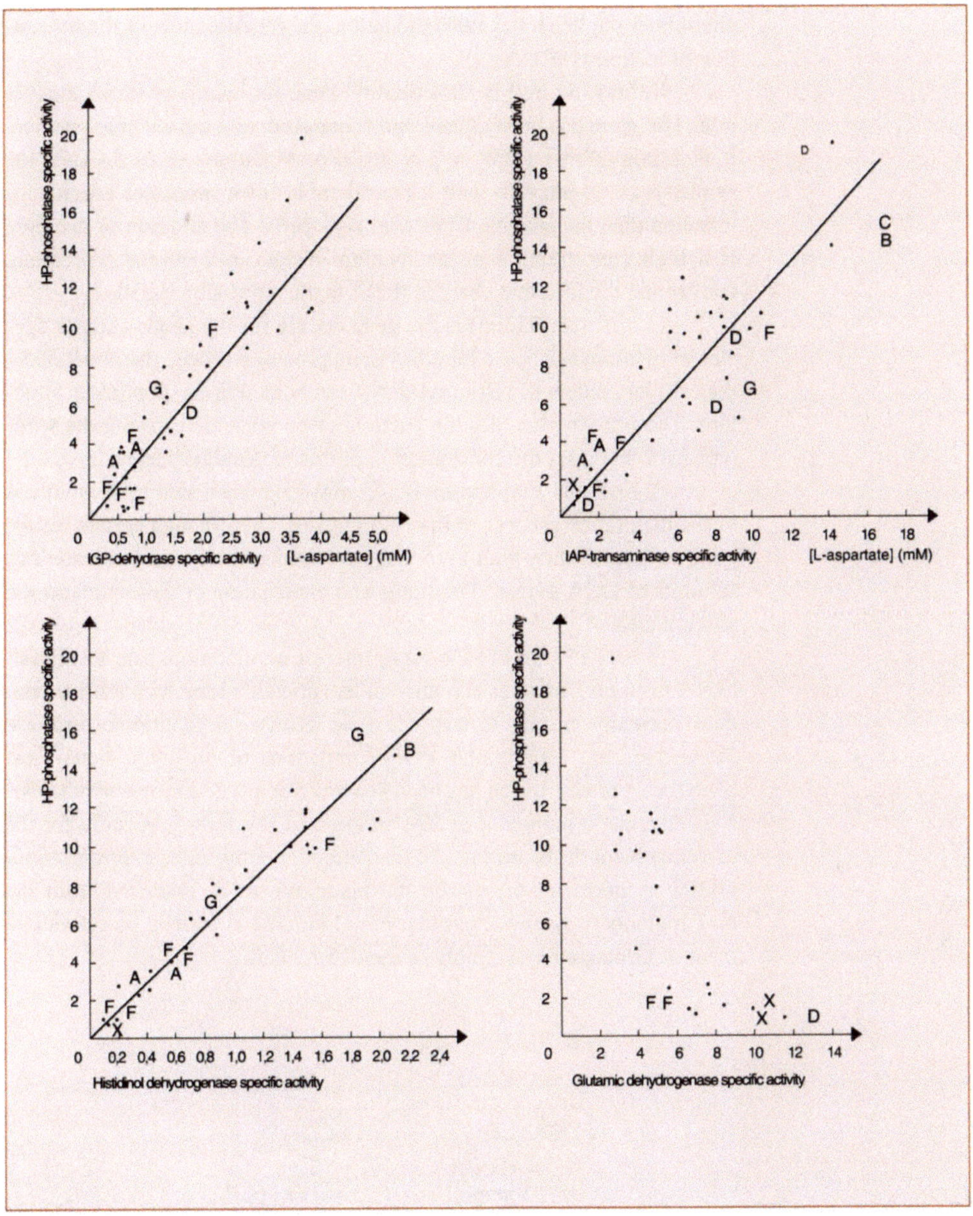

Figures 2. a, b, c, & d, specific activities of some enzymes involved in histidine biosynthesis and of glutamate dehydrogenase in various extracts of *S. typhimurium*.
Each point represents the specific activities of two of the enzymes in the crude extracts of wild type and of various auxotrophic mutants grown under different conditions. x, wild-type extracts; •, extracts of mutants lacking PR-ATP pyrophosphorylase. Extracts of other mutants are symbolized by letters corresponding to their genes (see fig. 1). When a given mutant is lacking one of the four activities examined, the specific activity of the corresponding enzyme is missing from the graph.
(From B. N. Ames & B. J. Garry, *Proc. Natl. Acad. Sci. US* **45**, 1453-1461 (1959). With permission from Dr. Bruce Ames.)

alter either the level, the aminoacylation, the modification, or the maturation of histidinyl-tRNA.

b) *hisS* (54 mn) is the structural gene for histidinyl-tRNA synthetase. The gene has been cloned and sequenced, and the enzyme purified. It is a homodimer (MW = 2 x 40,000). Mutations in *hisS* cause the synthesis of an enzyme with a considerably lower apparent affinity for histidine than the enzyme from wild-type cells. The addition of histidine at a high concentration in the medium results in a normal repression, confirming the idea that charged tRNA is the repression signal.

c) *hisR* (*ca.* 85 mn) is the gene coding for the single cellular species of histidinyl-tRNA. Mutations in its promoter reduce (by about 50%) the cellular content of histidinyl-tRNA, without altering its primary structure. The introduction into the bacteria of an episome carrying the wild-type *hisR* increases the concentration of this molecular species.

d) *hisU* (81 mn) and *hisW* (45 mn) have been said to be involved in the maturation process of the histidinyl-tRNA, although certain results point to their identity with *gyrB* and *gyrA* respectively, coding for the two subunits of DNA gyrase. The status and explanation of these mutants are still not clear.

e) *hisT* (50 mn) is the more interesting mutation. The histidinyl-tRNA from *hisT* mutants contains, in its anticodon loop, two uridine residues normally present in the wild type as two pseudouridine residues. Since there are suppressible *amber* mutations of *hisT*, this gene must code for a protein, which has been shown to be pseudouridine synthetase. This enzyme, a monomer of 270 amino acid residues, is specific for the rearrangement of the uridines of the anticodon loop, since a third pseudouridine is normally present in the histidinyl-tRNA extracted from the *hisT* mutants (it is now clear that pseudouridine is formed by rearrangement of uridine after assembly of the tRNA chain).

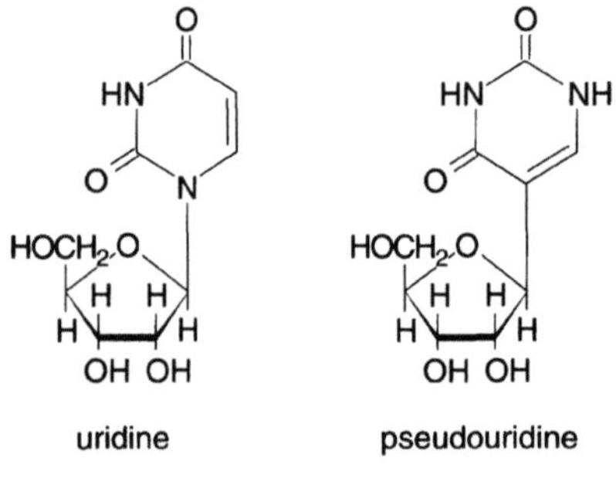

The growth of the *hisT* mutants is practically not affected, indicating that the presence of the two pseudouridines in the anticodon loop is necessary for repression, but dispensable for global protein synthesis. Actually, the histidinyl-tRNA extracted from the mutants has the same characteristics as wild-type histidinyl-tRNA in the aminoacylation reaction, be it for reaction rate and extent or affinity for the cognate aminoacyl-tRNA synthetase.

A word of caution: although the above studies concluded that the pseudouridine modification is dispensable for cellular growth, they were based on a limited number of experiments with unspecified *hisT* mutants that might have retained residual activity. These conclusions were recently reexamined with strains that were absolutely devoid of pseudouridine synthase. These strains have a uracil requirement which interferes with cell division, and lead to the concept that the loss of certain tRNA modifications can actually have catastrophic effects.

Despite the impact of the repressor-operator model established mainly in the *lac* and *trp* systems, the group led by Bruce Ames considered gene regulatory mechanisms involving charged histidinyl $tRNA^{His}$ in its normal role in protein synthesis and imagined that there was a His-codon-containing RNA "operator" sequence overlapping with the starting point of translation of the first structural gene. In the presence of an excess charged histidinyl-$tRNA^{His}$, a ribosome would initiate translation of this sequence,proceed to its termination codon, and somehow interfere with the initiation by ribosomes at the start of the structural genes. Repression would result, and the His-codon-containing peptide would be released into the cytoplasm. When histidinyl-$tRNA^{His}$ is limiting, the ribosome would stall at the histidine codons of the operator sequence, allowing access to the initiation codons of the structural gene. This prophetic theory prefigures the attenuation model, beautifully developed by Yanofsky, for the tryptophan operon. Actually, in 1974, it was shown that a mutation upstream the structural genes elicits a constitutive *his* operon expression independent of histidinyl-tRNA control.

The attenuation model is described elsewhere in detail in this volume. Here is the leader peptide sequence of the histidine attenuator, which is made of 16 amino-acid residues with seven histidine residues in a row:

Met Thr Arg Val Gln Phe Lys **His His His His His His His** Pro Asp

SELECTED REFERENCES

Histidine biosynthesis and its regulation

B. N. Ames & B. J. Garry, *Proc. Natl. Acad. Sci. US* **45**, 1453-1461 (1959).

S. W. Artz & D. Holzschu, in *Amino Acids: Biosynthesis and Genetic Regulation*, K. M. Herrmann & R. L. Somerville eds., pp. 379-404. Addison-Wesley Publishing Co, Reading, Mass. (1983).

PR-ATP pyrophosphorylase

B. N. Ames, R. G. Martin & B. J. Garry, *J. Biol. Chem.* **236**, 2019-2026 (1961).

R. G. Martin, *J. Biol. Chem.* **238**, 257-268 (1963).

Attenuation

W. M. Barnes, *Proc. Natl. Acad. Sci. US* **75**, 4281-4285 (1978).

CHAPTER XXIII

Biosynthesis of nucleotides

I. BIOSYNTHESIS OF PYRIMIDINE NUCLEOTIDES

Synthesis of 5-phosphoribosyl-1-pyrophosphate (PRPP)

The synthesis of PRPP from ATP and ribose-5-phosphate is catalyzed by phosphoribosylpyrophosphate synthetase as follows:

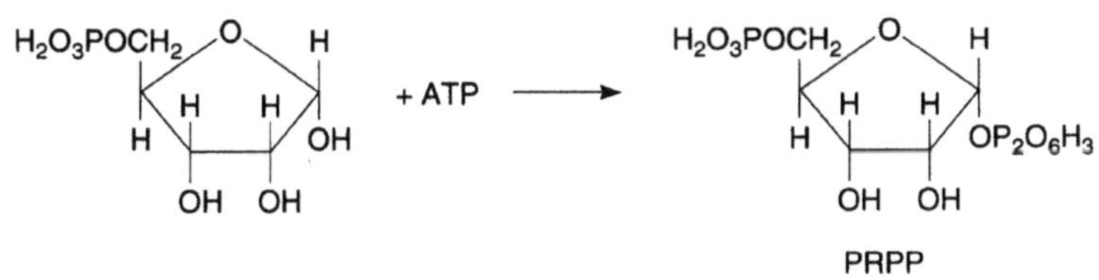

The equilibrium constant favors PRPP synthesis.

PRPP is involved in pyrimidine and purine nucleotide, histidine, tryptophan, and pyridine nucleotide syntheses. Ten known enzymes compete for PRPP as substrate, the higher quantitative requirement being for pyrimidine and purine nucleotides.

The PRPP synthetase subunit (encoded by *prs,* 26 mn) has a molecular mass of 31 kDa. Inorganic phosphate is essential for the stability of the enzyme and is required for its activity, although its function in the catalysis remains obscure. Mg^{2+} ions are required both as free cations and as an ATP chelators. The synthesis of the enzyme is derepressed upon uracil starvation, but the nature of the repressing metabolite remains unknown. The other metabolites deriving from PRPP do not seem to exert an influence on the level of PRPP synthetase.

The intervention of PRPP in the synthesis of pyrimidines occurs at the level of orotidine-5'-phosphate synthesis (see below).

Synthesis of carbamylphosphate

In bacteria, the first step in the synthesis of pyrimidines is the synthesis of carbamyl phosphate from ammonia or an ammonia donor and CO_2, at the expense of two molecules of ATP:

$$NH_3 + HCO_3^- + 2\ MgATP \rightarrow \underset{\text{carbamyl phosphate}}{H_2N\text{–}COO\text{–}PO_3H_2} + 2\ ADP + Pi$$

This reaction is catalyzed by carbamyl phosphate synthetase, encoded by the genes of the *carAB* operon (1 mn, called *pyrA* in *S. typhimurium*). The operon is oriented from A to B. The enzyme is a monomer made of two subunits which have been cloned and sequenced. The heavy subunit (MW = 117,710), coded for by *carB,* is able to catalyze the reaction with ammonia as the amino donor, whereas the light subunit, coded for by *carA (*MW = 41,270) is actually an amidotransferase which allows the enzyme to function with glutamine (cf. anthranilate synthase for a similar situation). The light subunit binds glutamine and has glutaminase activity.

Carbamyl phosphate synthetase is an allosteric enzyme: both the heavy subunit alone and the complex display a sigmoidal MgATP saturation curve, whereas the kinetics are hyperbolic with respect to glutamine or bicarbonate. The activity of the enzyme is inhibited by UMP and activated by ornithine, IMP and, in the case of the *Salmonella* enzyme, by PRPP. All these effectors act by altering the affinity of the enzyme for MgATP.

The mechanism of carbamyl phosphate synthesis by the large subunit is known to proceed through carboxy phosphate. The intermediate is formed by an initial phosphorylation of bicarbonate resulting in an enzyme-bound bicarbonate

$$E + ATP + HCO_3^- \rightarrow E.\ CO_3PO_3^{3-} + ADP + H^+$$

In the second step, the bound carboxy phosphate undergoes a nucleophilic attack by NH_3 yielding enzyme-bound carbamate:

$$E.\ CO_3PO_3^{3-} + NH_3 \rightarrow E.\ CO_2\text{-}NH_2 + Pi$$

Carbamate is phosphorylated by the second molecule of ATP to form carbamyl phosphate in the terminal step:

$$E.\ CO_2\text{-}NH_2 + ATP \rightarrow E + \text{carbamyl phosphate} + ADP$$

In addition to the synthesis of carbamyl phosphate, the enzyme catalyzes three partial reactions: a bicarbonate-dependent hydrolysis of ATP:

$$MgATP + H_2O \xrightarrow{HCO_3^-} MgADP + Pi$$

a synthesis of ATP from carbamyl phosphate and ADP:

$$Mg\ ADP + \text{carbamyl phosphate} \rightarrow MgATP + CO_2 + NH_3$$

and the glutaminase activity referred to above, catalyzed by the light subunit

$$\text{glutamine} + H_2O \rightarrow \text{glutamate} + NH_3$$

There is considerable evidence that each of the MgATP molecules involved in the overall reaction not only act in mechanically discrete steps, but also at separate sites.

Three conserved histidine residues, His243, 781, and 788, are located within the large subunit of carbamyl phosphate synthetase from *E. coli* and were identified by sequence identity comparisons. They were individually mutated to asparagine residues. The mutation at position 243 blocks the formation of the carbamate intermediate from carboxy phosphate and ammonia. The mutant modified at position 781 has an order of magnitude reduction for both carbamyl phosphate and ATP syntheses, consistent with the proposal that the C-terminal portion of the large sub-unit is primarily involved in the phosphorylation of the carbamate intermediate. The mutation at position 788 synthesizes carbamyl phosphate at a rate comparable to that of the wild type, but ATP synthetase and bicarbonate-dependent ATPase activities of the enzyme are reduced 10-fold.

The most common phenotype among the mutants affecting the *carAB* operon is a double requirement for uracil and arginine. Since carbamyl phosphate is also an intermediate in arginine biosynthesis, it follows that mutants unable to synthesize it require both uracil and arginine. The activation of the enzyme by ornithine is also understandable on a teleonomic basis, since an excess of ornithine would require more carbamyl phosphate to direct the pathway toward arginine.

This enzyme is different from a synthetase, found in vertebrate liver, which uses ammonia exclusively and is activated by N-acetylglutamate (required for the active enzyme conformation; it does not play a role in CO_2 activation). This enzyme is localized in mitochondria and is used for the synthesis of arginine and urea.

In both *Neurospora crassa* and yeast, there are two carbamyl phosphate synthetases, for pyrimidine and arginine synthesis respectively, designated P and A. The A enzyme from *Neurospora* consists of two different polypeptides, products of unlinked genes. A large subunit is capable of using NH_3, and a small subunit imparts the ability to use glutamine but has no intrinsic glutaminase activity – as opposed to the small subunit of the *S. typhimurium* enzyme. Mutants lacking the small subunit are auxotrophic for arginine. Thus, the glutamine-dependent reaction is the physiological one. Arginine represses the synthesis of both subunits, which is insensitive to pyrimidines. The reverse is true of carbamyl phosphate synthetase P: the mutants devoid of it are pyrimidine auxotrophs.

Synthesis of cytidine and uridine triphosphates

All the *E. coli* genes mentioned in this paragraph have been cloned and sequenced.

Aspartate transcarbamylase (ATCase, *pyrB* and *pyrI*, 97 mn) is an enzyme that carries out the condensation of one molecule of aspartate with a molecule of carbamyl phosphate. The reaction is formally similar to the ornithine transcarbamylase (OTCase) reaction (see chapter XIX):

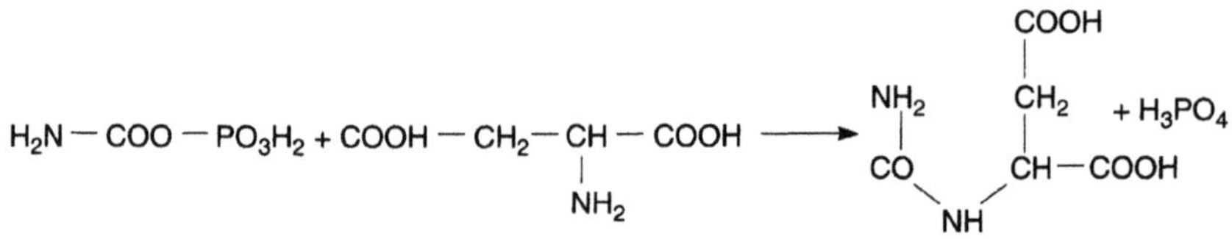

carbamylaspartate
(or ureidosuccinate)

The structure and regulation of ATCase activity deserve special attention for historical reasons, and because ATCase is one of the most thoroughly-studied allosteric enzymes. It will be examined in detail in the next paragraph.

Dihydroorotase (*pyrC*, 23 mn), a homodimeric dehydratase (2 x 38,500) containing one atom of tightly bound zinc per subunit, catalyzes the next step, a reversible cyclization forming the pyrimidine ring:

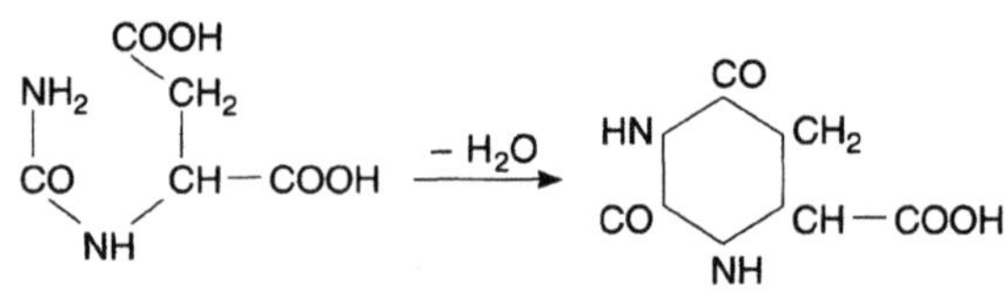

carbamylaspartate dihydroorotate

Dihydroorotate oxidase (*pyrD*, 21 mln) is membrane-bound and linked to the electron transport system of the cell. It consists of two identical flavin-containing subunits of molecular weight 37,000 each, and catalyzes the formation of orotic acid, the physiological electron acceptor appearing to be ubiquinone.

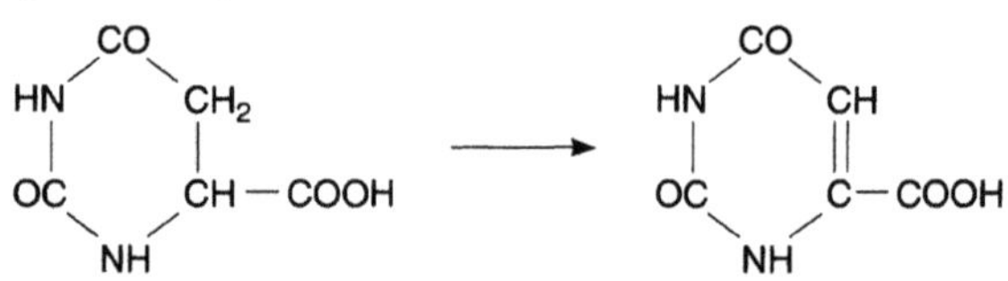

dihydroorotate
orotate
(or uracil-4-carboxylate)

The normal pathway of biosynthesis does not involve a direct decarboxylation of orotate to uracil, but a reaction of orotate with 5-phosphoribosyl-1-pyrophosphate (PRPP) catalyzed by orotate phosphoribosyltransferase (MW = 2 x 23,000; *pyrE*, 82 mn) to produce orotidine-5'-phosphate, a nucleoside (OMP). This compound is then converted into uridine-5'-phosphate (UMP) by a specific decarboxylase (MW= 2 x 27,000; *pyrF*, 28 mn) (see schema opposite).

A specific kinase, UMP kinase (*pyrH*, 5 mn), phosphorylates the nucleoside monophosphate UMP to the diphosphate UDP and nucleoside diphosphokinase (*ndk*, 53 mn in *S. typhimurium*), which catalyzes the

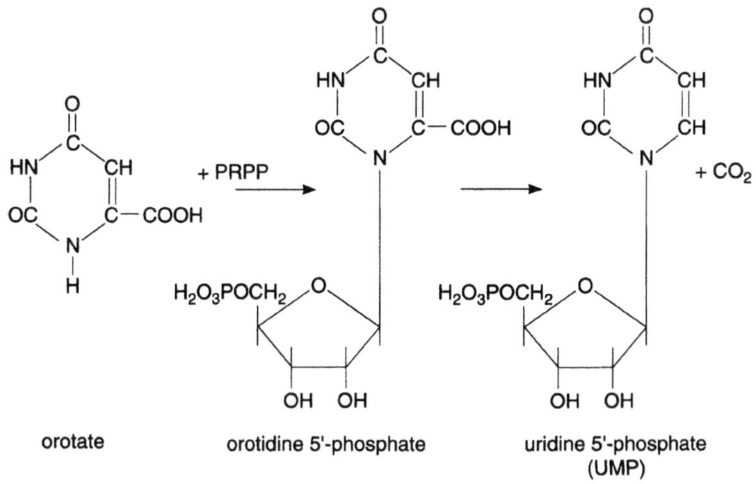

transfer of the γ-phosphate of any nucleoside-5'-triphosphate to any nucleoside-5'-diphosphate

$$N_1TP + N_2DP \rightarrow N_1DP + N_2TP$$

is responsible for the synthesis of UTP, at the expense of ATP for example. Nucleoside diphosphokinase activity involves a high-energy phosphohistidine intermediate.

Finally, the second pyrimidine nucleotide, cytidine triphosphate, is formed from UTP by the action of CTP synthetase (*pyrG*, 60 mn).

$$UTP + NH_3 + ATP \rightarrow CTP + ADP + Pi$$

CTP synthetase exists as a tetramer of identical subunits (50 kDa each) under assay conditions, and dissociates to a dimer in the absence of UTP and ATP. Glutamine is the preferred amino donor although, like in many other amidotransferases we have encountered, ammonia reacts too, albeit less efficiently. The reaction mechanism involves a glutamylation of a reactive cysteine on the enzyme, with the liberation of ammonia which then reacts with UTP. ATP and UTP bound to the enzyme and each ligand induces conformational changes in the protein, resulting in tetramerization and also in allosteric activation of the glutamylation reaction.

Direct utilization of pyrimidines and of their derivative

In microorganisms, exogenous bases and nucleosides can be utilized by means of numerous enzymes, called the "salvage" enzymes, which enable the cells to utilize preformed bases and nucleosides for nucleotide synthesis.

For example, UMP can be synthesized by uracil phosphoribosyltransferase (*upp,* 54 mn), an enzyme activated by GTP and inhibited by ppGpp and uracil nucleotides. The expression of *upp* is controlled by pyrimidine compounds and the corresponding homotrimeric enzyme (3 x 23,500 Da) increases considerably upon pyrimidine starvation.

Second, uridine can be synthesized through the action of uridine phosphorylase

$$\text{uracil} + \text{ribose-1-phosphate} \rightleftharpoons \text{uridine} + \text{Pi}$$

However, due an unfavorable equilibrium, this reaction can occur only if either uracil or ribose-1-phosphate is present at a high concentration. Uridine can then be processed by uridine kinase.

As we shall see, there are also "salvage" enzymes for the utilization of preformed purines.

Aspartate transcarbamylase of *E. coli*

This enzyme has been crystallized from a pyrimidine-requiring mutant of *E. coli* grown under conditions of physiological derepression which enable crude extracts to be obtained with this protein representing more than 10% of the total soluble proteins. It is to date one of the most studied and best-known allosteric enzymes and some of the early results obtained by Gerhardt, Pardee, Schachman, and their colleagues have been the foundation for many ideas about metabolic regulation and its mechanism.

The enzyme saturation curve as a function of aspartate concentration is sigmoid, indicating – as we have seen several times – that more than one substrate molecule can complex with each molecule of enzyme, and that the binding of the first molecule facilitates the binding of a second.

Cytidine triphosphate, the end product of the biosynthetic pathway starting with the carbamoylation of aspartate, is an inhibitor of the enzyme. The inhibition by CTP is competitive with respect to aspartate

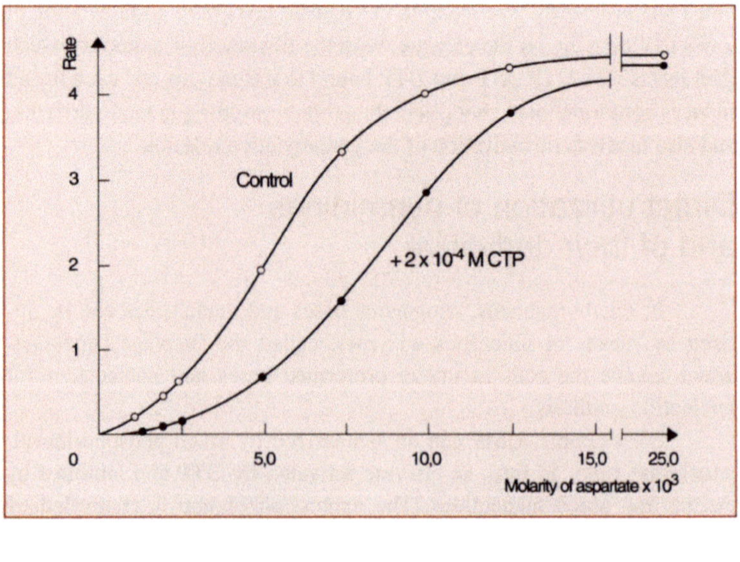

Figure 1. Displacement of the sigmoid saturation curve of aspartate transcarbamylase by CTP. The experiment was carried out with a fixed concentration of carbamyl phosphate and variable aspartate concentrations.

and appears as a displacement of the aspartate concentration giving 50% of the inhibition towards higher values of concentration (fig. 1). In other words, the apparent affinity of the enzyme for aspartate decreases in the presence of CTP, but the cooperativity between molecules of substrate is still visible. There is no detectable effect on the value of V_{max}.

The maximum inhibition by CTP can reach up to 75% at saturation of this effector and is increased (by up to 100%) by synergism between CTP and UTP, the second pyrimidine nucleotide, which has no effect by itself. By an original method using continuous flow dialysis, England and Hervé have shown that UTP enhances the affinity of CTP 5-fold and, conversely, CTP enhances the affinity for UTP 80-fold. The isolated regulatory subunits (see below) bind the two pyrimidine nucleotides following the same pattern as the entire enzyme. Observations strongly suggest that the synergistic inhibition mechanism relies entirely on interactions between two independent allosteric sites belonging to the same regulatory subunit.

The inhibitory effect of CTP is countered by ATP, an antagonist of ATP. ATP also has an intrinsic activating effect on the uninhibited enzyme; the activation can reach 100% and is associated with an increase of the apparent affinity of the enzyme for aspartate and a normalization of the saturation curve so as to follow the Henri-Michaelis law. This suggests that ATP induces a conformational change resulting in the loss of cooperative interactions between substrate molecules. It is assumed that the antagonism between ATP and CTP/UTP plays a role in maintaining a balance between the intracellular pools of purine and pyrimidine nucleotides.

Although the inhibition by CTP is competitive with respect to aspartate, early experiments established unequivocally that aspartate and CTP combine at different sites on the enzyme surface. This was shown not only by desensitization of the enzyme towards its allosteric inhibitor, but also by physical separation between different subunits, one of which contains the catalytic site and the other the site with affinity for the allosteric inhibitor.

Physical and chemical measurements have shown that the enzyme has a molecular weight of *ca.* 310,000. Low concentrations of urea, or treatment by heat or with mercurials, abolishes the sensitivity to CTP, without the modifying the catalytic activity. This desensitization is accompanied by major changes in the molecular weight of the enzyme. It can be shown that the action of *p*-mercuribenzoate dissociates the enzyme into two components which can be separated by centrifugation in a sucrose gradient (fig. 2).

Whereas the native enzyme has a sedimentation coefficient of 11.8S, the two components have coefficients, respectively, of 5.8S and 2.8S. The 5.8S fraction has the catalytic activity; it displays a greater specific activity than the native enzyme but differs in that it is completely insensitive to CTP or ATP. It is therefore sufficient for catalytic activity but

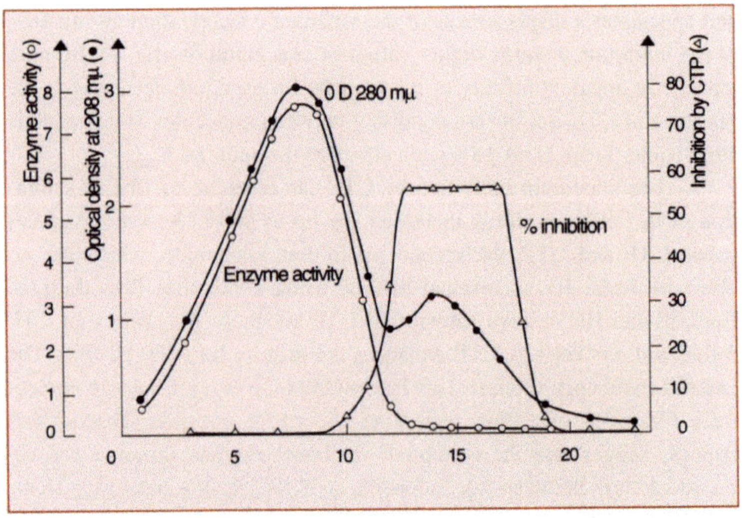

Figure 2. Separation by sucrose gradient centrifugation of the two components of aspartate transcarbamylase. Seven milligrams of p-MB treated enzyme were centrifuged at 38,000 rpm for 20 hours at 10 °C. The centrifuge tubes were then pierced, and fractions of twelve drops collected. The black circles show the protein profile; the empty circles indicate catalytic activity, measured after addition of 2-mercaptoethanol. The regulatory protein is identified by adding the test fraction to fraction 6, and measuring the resulting activity in the presence and in the absence of 4 x 10^{-4} M CTP.

not for its regulation. Its kinetics, as a function of substrate concentration, strictly follow the Michaelis law. The addition of 2-mercaptoethanol to the lighter fraction does not cause any catalytic activity to appear, but allows the fraction to regain the ability to bind CTP lost by the enzyme when treated with the mercurial. The 2.8S component is therefore the carrier of the enzyme's capacity to be inhibited by CTP, and plays no catalytic role.

If 2-mercaptoethanol is added to the unseparated mixture of the two fractions produced by treatment with *p*-mercuribenzoate, nearly all the enzyme is recovered with its native properties of sedimentation and inhibition. The 2.8S component is indispensable for the reaggregation; neither component alone can aggregate upon removal of the mercurial.

Metal ions bound to the regulatory subunit are essential for reconstitution from isolated subunits; the naturally-ocurring metal is zinc with a stoechiometry of six atoms per enzyme molecule. Zinc plays no role in the catalysis, it merely stabilizes the dimeric form R_2 of the regulatory subunit and promotes its recombination with the catalytic subunit.

The catalytic subunit, coded by *pyrB*, and the regulatory subunit, coded by *pyrI*, have been cloned and sequenced. The catalytic monomer (C) is composed of 311 amino acid residues and the regulatory monomer of 153 residues. Knowing the weight contributions from the 5.8S and the

2.8S components, calculations clearly show a structure of the type R_6C_6 for aspartate transcarbamylase. The mercurial agent cleaves the enzyme into two C_3 units and three R_2 units. X-ray crystallography confirms that the enzyme has a 2-fold and a 3-fold symmetry axis.

A 3% significant ligand-promoted decrease in the sedimentation coefficient was demonstrated, which is in accord with diffusion and X-ray scattering measurements, indicating a swelling of the enzyme. Subsequent crystallographic data obtained by Lipscomb showed that the distance along the 3-fold symmetry axis increases by approximately 12 Å upon the binding of N-phosphonacetyl L-aspartate (PALA), a bisubstrate analogue which has some of the structural features expected from a transition analogue. In addition to this moving apart of the two catalytic trimers, there is a reorientation of each of them and of the three R dimers.

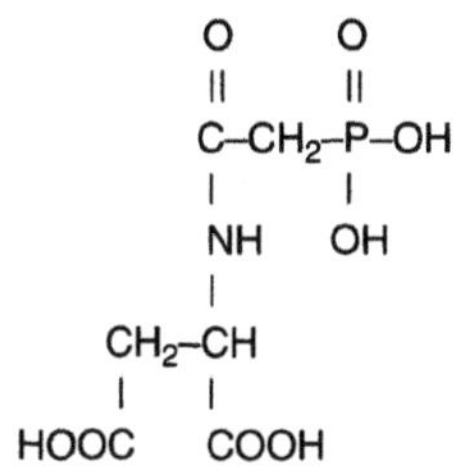

PALA

(In this formula, the structures of aspartate and carbamylphosphate are recognizable.)

A quantitative interpretation of the cooperativity of ATCase is dependent upon the knowledge of the energetics of the allosteric transition. Since unliganded ATCase is predominantly in the T conformation and since the R state is detectable only when the enzyme is partially or fully saturated with active site ligands, there is no direct method of determining the T $\rightleftharpoons$ R equilibrium constant for unliganded enzyme.

According to the Monod-Wyman-Changeux model, unliganded ATCase can exist in either of two quaternary conformations, R or T, and the ratio T/R is the allosteric equilibrium constant L. Ligands are assumed to bind reversibly to either form with dissociation constants K_R and K_T Because the equilibria are linked, the addition of a ligand which will bind preferentially to the R form must promote the T $\rightarrow$ R transition, the shift of the equilibrium caused by the preferential binding of the substrates to the R form being responsible for the homotropic, cooperative, interactions. Effectors binding to the regulatory chains may also show preferential interactions. If, for example, ATP binds more tightly to the R conformation, activation would occur and, conversely, preferential binding of CTP to the less active T state would cause inhibition (heterotropic effect). Accordingly, CTP and ATP must affect the T $\rightleftharpoons$ R equilibrium. The model is simple in that ligand affinity is determined by the quaternary state of the molecule, T or R, and not by the number, location, or nature of bound ligands.

Thus, the first models proposed, to account for the regulatory properties of ATCase, postulated that the nucleotides act directly on the transition involved in the homotropic cooperative interactions between the catalytic sites for aspartate binding. Numerous indications have accumulated over the years showing that the situation is more complex. The effects of the nucleotides are best explained by local conformational changes, which either alter the affinity of the catalytic sites for aspartate or modulate the stability of the interfaces between regulatory and catalytic chains, thus facilitating (ATP) or hindering (CTP) the transition from T to R state upon substrate binding. X-ray solution-scattering experiments and equilibrium isotope exchange kinetics in fact show that ATP and CTP do not act *directly* on the $T \rightleftharpoons R$ equilibrium and alter the rate constant for the binding of aspartate to the catalytic sites.

The crystallographic structures of the T and R states, unliganded or liganded with either ATP or CTP, have been solved at high resolution by Lipscomb. The protein exhibits several levels of organization. It is composed of two catalytic trimers held together by three regulatory dimers. The catalytic sites lie at the interface between two catalytic chains belonging to the same trimer and involve amino acid side chains belonging to these two catalytic chains. The regulatory sites binding competitively ATP and CTP. are located at the extremities of the regulatory dimers.

Each of the two types of chains comprises two domains. One can distinguish a carbamylphosphate and an aspartate binding site in the catalytic chains. Similarly, in the regulatory chains, there is an allosteric domain where the regulatory sites binding ATP and CTP lie, and a zinc domain. This region contains a zinc atom bound to four cysteine residues localized in the C-terminal part of the regulatory chain. These zinc atoms are indispensable for the association of the catalytic and regulatory chains. The effect of mercurials on these cysteines is to provoke the dissociation between the two types of chains as described above.

The ATP-ligated ATCase is in the T state and the interactions characteristic of this state are not disrupted upon ATP binding to the unliganded enzyme, indicating that, contrary to the simple model prediction, ATP does not promote the T to R transition. The distance between the two catalytic trimers increases by only 0.4 Å upon ATP binding, compared to the 11 Å increase in that distance occurring during the T to R transition.

Although precise structural information for the T and R conformations is of inestimable value, an understanding of the allosteric transition is not yet at hand and requires physico-chemical studies on solutions of wild-type and altered (by chemical modification or by site-directed mutagenesis) enzyme, and an explanation of the results obtained in terms of the structural information available. For example, hybrid ATCase-like molecules have been obtained, containing chromophores at well-defined locations, and have provided information about the effects of various ligands on the $T \rightleftharpoons R$ conversion.

Site-directed mutagenesis may cause slight perturbations in the structure of individual chains at regions remote from the site of amino-acid substitutions affecting the T $\rightleftharpoons$ R equilibrium. Therefore, the attribution of a change in regulatory properties to the direct effect of a single amino-acid replacement may often be an overinterpretation. Let us give an example.

The binding site for CTP and ATP is located at the contact region between regulatory chains, and it seemed likely from the crystallographic data that a change in Lys60 would affect nucleotide binding. Indeed, if this Lys is replaced by a histidine, the capacity to bind nucleotides is abolished and the enzyme appears to be kinetically "frozen" in the R state. However if, instead of histidine, Lys60 is replaced by glutamine or arginine, the enzyme retains both homotropic and heterotropic effects. If an alanine residue is substituted for Lys60, these effects are altered: ATP activates normally, but the enzyme is now insensitive to CTP. It is thus hazardous – and often unwarranted – to attribute to a change in a single amino acid residue the properties of this residue in the wild-type enzyme.

Other residues are involved in the regulatory process. The regulatory chains are folded in two domains, the effector-binding domain and the zinc-binding domain, making contacts with the catalytic chains. It follows that the CTP and ATP regulatory signals must necessarily be transmitted through the zinc domain. At the interface between the two domains, Leu32 and Leu76 of the allosteric domain are in contact with Val106 and Leu151 of the zinc domain, forming a hydrophobic pocket into which the side-chain of Tyr77 is inserted. When this tyrosine residue is replaced by a phenylalanine, the effect of ATP on the activity of the enzyme is completely inverted; it becomes an inhibitor instead of an activator, while the inhibition by CTP is maintained. This shows that the interface between the two domains of the regulatory chain plays an important role in the discrimination between the two regulatory signals.

Site-directed mutagenesis has also been applied to the catalytic chains, taking advantage of knowledge of the tridimensional structure for the choice of the substitutions. There is an intersubunit interface between a catalytic chain in one C3 trimer and a catalytic chain in the other trimer. This limited contact is observed in the T conformation, but is disrupted in the R conformation. It involves an interaction between Lys164 and Glu239 residues. If these two residues are replaced respectively by a Glu at position 164 and a Lys at position 239, the mutant ATCase exhibits hyperbolic kinetics, no inhibition by CTP or activation by ATP, and there is no cooperativity in binding PALA: the unliganded enzyme is in the "frozen" R conformation.

The catalytic trimers of *E. coli* can be transformed into the allosteric form not only by association with the regulatory dimers, but also by site-directed mutagenesis by substituting Arg105 by an alanine residue.

Aspartate transcarbamylase of other organisms

Native *E. coli* ATCase provides the allosteric model for feedback inhibition of the pyrimidine pathway. It shows sigmoidal substrate saturation curves with 100% activation by ATP, 75% inhibition by CTP and 100% inhibition by CTP/UTP synergism. In *Erwinia herbicola,* another member of *Enterobaceriaceae* family, the bicistronic operon also encodes a dodecamer holoenzyme $3R_2$-$2C_3$ with 153 residues per R monomer and 311 residues per C monomer, like in *E. coli*. However, the sequence diverges by 20% in the R monomers and 18% in the C monomers. In spite of conservation of the residues directly implicated in the active site and in the effector-binding sites, the *Erwinia* enzyme has diverged with respect to homotropic and heterotropic responses as exhibited by hyperbolic saturation curves, lack of activation by ATP, and only slight inhibition by CTP. Nonetheless, the UTP/CTP synergistic inhibition approaches 90% (similar to the *E. coli* value) and suggests that the *Erwinia* enzyme is a naturally-occurring R-state enzyme in contrast to the *E. coli* enzyme, whose T/R equilibrium approximates an *L* value of 250.

Other ATCases have been purified; those from *Streptococcus faecalis* and *Bacillus subtilis* do not appear to possess a feedback control mechanism, since their activity is affected neither by ATP nor by CTP. The *B. subtilis* enzyme is trimeric, like the catalytic chain of the *E. coli* enzyme with which its 33,500 Da polypeptide chain shows 35% homology. Its crystallographic structure is more like that of the T form than the R form of the *E. coli* enzyme. By site-directed mutagenesis, if Arg99 (corresponding to Arg105 in *E. coli*) is replaced by an alanine residue, the *B. subtilis* enzyme retains its activity, but the regulatory activity of homotropic substrate cooperativity for aspartate molecules is now observed. Thus, a small number of substitutions could transform a simple metabolic step into a regulatory juncture. It has been proposed that that this possible evolutionary potential could explain, in part, the natural abundance of oligomeric enzymes. Although the wild-type *B. subtilis* enzyme does not show cooperativity for either substrate, and is insensitive to both purine and pyrimidine nucleotides, it is developmentally regulated: regulation occurs in the stationary bacterial growth phase, prior to sporulation, where the enzyme is degraded by an ATP-dependent protease.

The activity of the enzymes from several pseudomonads are influenced by several nucleotides. In no case has a study been made as extensive as the one carried with the *E. coli* enzyme.

Regulation of pyrimidine nucleotide synthesis at the genetic level

The genes *carAB* and *pyrBI,* which specify the two enzymes made from two different subunits, *i. e*., carbamylphosphate synthetase and aspartate transcarbamylase each form an operon. All the other genes are scattered on the chromosome of *E. coli* or *S. typhimurium.* However, *pyrE* and *pyrF* constitute small bicistronic operons, each encompassing proteins of unknown function.

The expression of all six genes is non-coordinately regulated by the availability of pyrimidines, increased by starvation and reduced by addition of pyrimidines.

The synthesis of carbamylphosphate synthetase, which provides the common precursor to uracil and arginine, is subject to cumulative repression by arginine and a pyrimidine. The effect of arginine is mediated by the arginine repressor-arginine complex.

The nucleotide sequence of the entire *pyrBI* operon has been determined. The genes are transcribed in the order promoter-leader sequence-*pyrB-pyrI.* The structural genes are preceded by a *rho*-independent terminator, itself preceded by a thymidylate-rich cluster. The existence of these structures has led to the proposal that there is a tight coupling between the transcribing RNA polymerase and the ribosomes translating a "leader peptide" required for transcription through the attenuator, in a way analogous to the attenuation of transcription observed in the case of the regulation of the biosynthesis of some amino acids. The same mechanism seems to apply to the regulation of the synthesis of orotate phosphoribosyltransferase *(pyr E).*

The genes coding for dihydroorotase and dihydroorotate oxidase, *pyrC* and *pyr D,* do not contain attenuator-like structures, but the leader sequences contain regions of high symmetry which might have some regulatory function.

II. BIOSYNTHESIS OF PURINE NUCLEOTIDES

Biosynthesis of 5-amino-4-imidazole carboxamide ribonucleotide

Eight successive reactions are necessary to synthesize this compound, the precursor of the purines. Since most of the intermediates in this pathway are not easily available, few thorough enzymological studies have been made.

First, the amide group of glutamine displaces the pyrophosphate group of PRPP to form 5-phosphoribosylamine. This reaction is catalyzed by PRPP amidotransferase (*purF*, 50 mn), an enzyme whose subunit has a molecular weight of 53,000. The quaternary structure, trimeric or tetrameric is still doubtful. The gene has been cloned and sequenced.

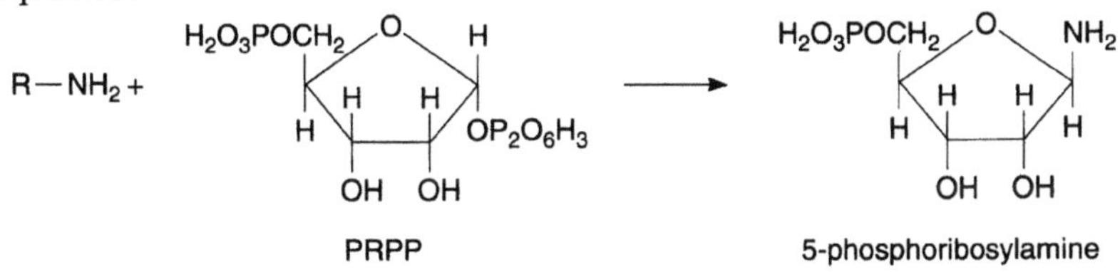

During the reaction, the configuration of the C1 of ribose is inverted from α to β.

As is the case for many amidotransferases, PRPP amidotransferase can also use ammonia as the amino group donor. Reaction with 6-diazo-5-oxonorleucine, an analogue of glutamine, modifies covalently the N-terminal cysteine of the enzyme which is thus the active site residue required for the glutamine transfer reaction. This enzyme, the first committed step in the biosynthesis of the purine nucleus, is subject to feedback inhibition by AMP and GMP, which act synergistically and are competitive inhibitors of PRPP. The hyperbolic PRPP saturation curve becomes sigmoidal in the presence of these allosteric effectors.

The amino acid glycine is added to the phosphoribosylamine molecule in a reaction which uses ATP as the energy source, and which produces 5'-phosphoribosyl-1-glycinamide (glycinamide ribonucleotide, GAR), ADP, and Pi. The enzyme catalyzing this reaction, phosphoribosyl glycinamide synthetase, is encoded by *purD* (90 mn).

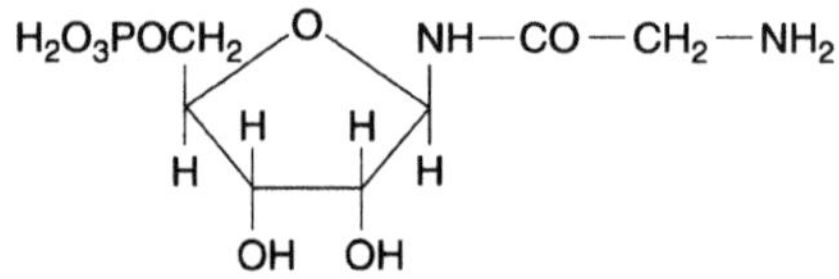

GAR

The free amino group of glycine is formylated by phosphoribosyl formylglycinamide synthetase (also called 10-formyltetrahydrofolate: 5'-phosphoribosylglycinamide formyltransferase, *purL*, 55 mn) to yield a compound (abbreviated to FGAR), the ribonucleotide of formylglycinamide. The formyl donor is N^{10}-formyltetrahydrofolic acid.

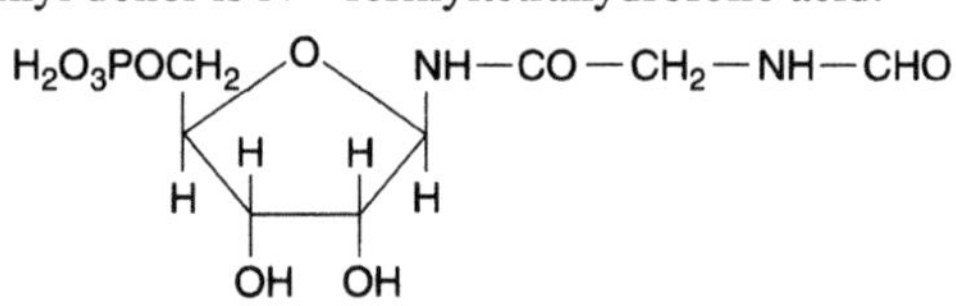

FGAR

There is an additional enzyme, GAR transformylase T encoded by *purT* which can utilize formate directly: formate is generated by pyruvate formate lyase during anaerobic growth, whereas an enzyme, encoded by *purU*, provides the major source of formate used in aerobiosis.

The three-dimensional structure of the enzyme has been solved both as an apoenzyme and as a ternary complex with the substrate glycinamide ribonucleotide and a folate inhibitor. The structure is a modified doubly-wound α-β sheet with flexibility in the active site, included a disordered structure in the apo structure, which is ordered in the ternary complex. This enzyme is a target for anti-cancer therapy and for structure-based drug design.

Another amidotransferase, FGAR amidotransferase (*purG*, 54 mn;

MW of the native enzyme = 130,000) transfers, in the presence of ATP, the amide group of glutamine that becomes the future nitrogen atom 3 of the purine ring, the other products of the reaction being glutamate, ADP, and Pi. The product of the reaction that interests us here is FGAM, the ribonucleotide of formylglycinamidine:

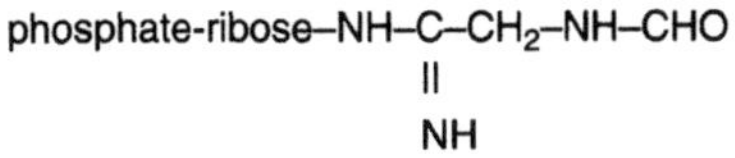

FGAM

An ATP-dependent cyclization completes the formation of the imidazole ring and produces aminoimidazole ribonucleotide (AIR):

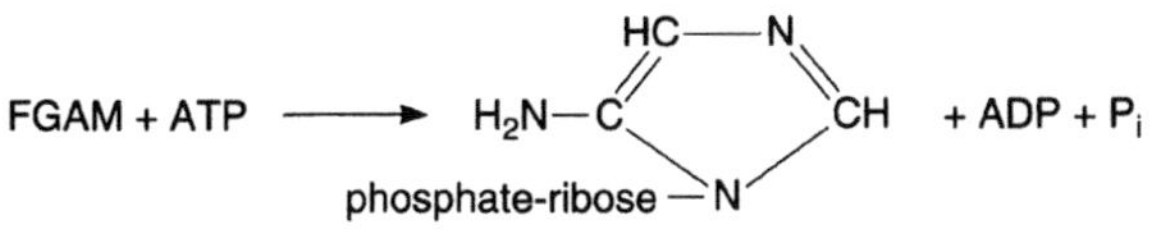

AIR[38]

The gene coding for AIR synthetase (*purI*, 54 mn) has been cloned and sequenced. It is a homodimer (2 x 38,000 kDa). The gene *purE* (12 mn) codes for phosphoribosylaminoimidazole carboxylase, responsible for the formation of the ribonucleotide of 5-amino-4-imidazolecarboxylic acid (CAIR). This reaction results in the introduction of carbon atom 6 of the future purine ring:

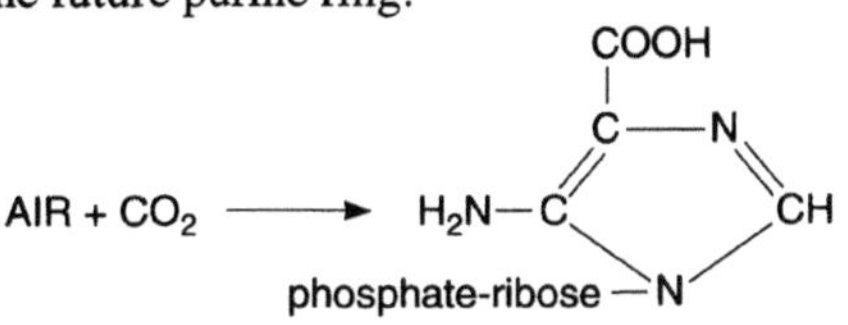

CAIR

This is followed by the addition of a molecule of aspartic acid in the presence of ATP to yield 5-amino-4-imidazole-N-succinylcarboxamide ribonucleotide (SAICAR), a reaction catalyzed by SAICAR synthetase (*purC*, 53 mn); its quaternary structure is not known, but it has a subunit mass of 27 kDa.

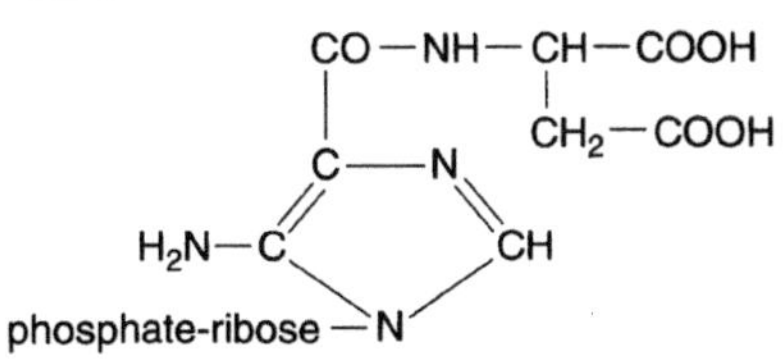

SAICAR

The ribonucleotide of 5-amino-4-imidazolecarboxamide (AICAR) is reached after an elimination of fumaric acid, catalyzed by an adenylosuccinate lyase, which also functions in the synthesis of AMP from IMP (*purB*, 25 mn).

38. Note that AIR is also a precursor of thiamine.

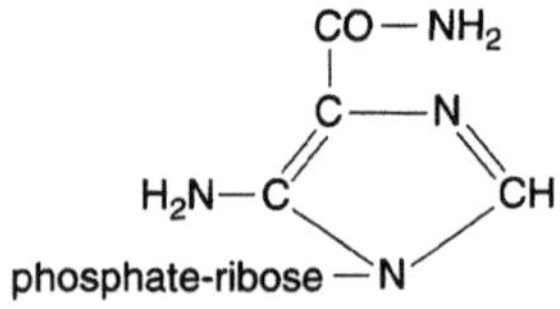

AICAR[39]

This is the second time we have encountered this curious amination procedure, which consists of adding an aspartic acid molecule and then removing a fumaric acid molecule (cf. the synthesis of argininosuccinate, the intermediate between citrulline and arginine). We will soon see a third example.

This part of the purine biosynthetic pathway is regulated mainly by feedback inhibition of PRPP amidotransferase and by regulation of the expression of the *pur* genes. This expression is repressed when exogenous purines are added to the medium, and derepressed upon purine starvation. The nature of the purine effector is not known with certitude.

Synthesis of inosinic acid

Carbon atom 2, the final carbon of the purine ring, is introduced by transferring the formyl group of N^{10}-formyltetrahydrofolate to the 5-amino group of AICAR to yield 5'-phosphoribosyl-4-carboxamide-5-formamidoimidazole (FAICAR), a reaction catalyzed by the product of *purH* (90 mn).

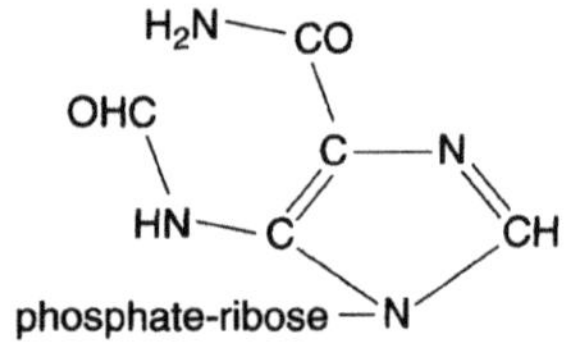

FAICAR

The last step, which gives rise to the completed purine ring, is catalyzed by the product of *purJ* (89 mn in *S. typhimurium*), IMP cyclohydrolase; a molecule of water is eliminated and the second ring is closed in a reaction which does not require ATP:

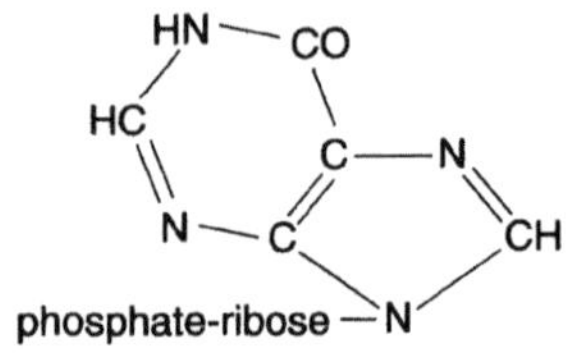

inosinic acid (IMP)

39. AICAR is also formed during the synthesis of imidazoleglycerol phosphate (see the synthesis of histidine).

Synthesis of guanylic and adenylic acids

Guanylic acid is derived from IMP in two steps. First, IMP is oxidized in position 2 by IMP dehydrogenase, an NAD-dependent enzyme (*guaB*, 54 mn). The *E. coli* enzyme is a homotetramer (4 x 54,000 Da), containing a cysteine at the IMP binding site; it is inhibited by GMP and ppGpp. The corresponding gene has been cloned and sequenced. The product of the reaction is xanthylic acid (XMP). Xanthylic acid is then aminated at carbon 2 by an amidotransferase called guanylate synthetase (*guaA*, 54 mn), a homodimer of subunit molecular weight 58,000. As usual, the amidotransferase can use either glutamine or ammonia as the amino donor and requires ATP, which is cleaved to AMP and pyrophosphate; the activity is irreversibly lost when the enzyme is treated with a covalently acting glutamine analogue. Here again, the gene has been cloned and sequenced.

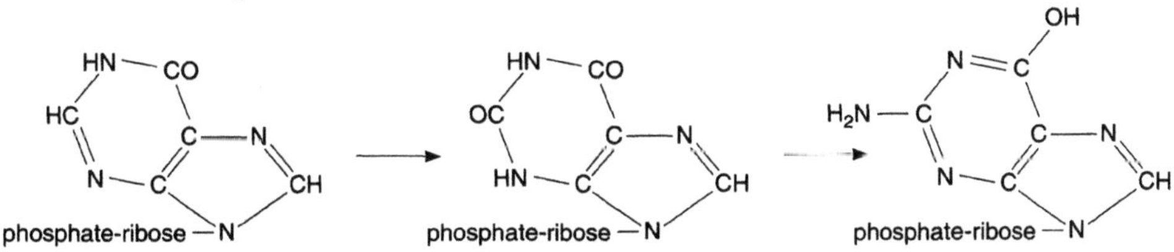

Adenylic acid (AMP) derives also from inosinic acid in two steps. First, aspartate condenses to position 6 of IMP, in the presence of GTP (which is hydrolyzed to GDP and Pi during the reaction). The enzyme responsible for this reaction is adenylosuccinate synthetase (*purA*, 95 mn). This enzyme is homodimeric; the molecular weight of its subunit is 48,000. It contains an N-terminal glycine-rich sequence which is homologous to a conserved sequence found in many other guanine nucleotide-binding proteins. which appears to play a role in enzyme catalysis. Its crystal structure has been determined. The enzyme is inhibited by AMP, GDP, and ppGpp.

$$\text{IMP + aspartate + GTP} \rightarrow \text{adenylosuccinate + GDP + Pi}$$

The same adenylosuccinate lyase *(purB)* responsible for the transformation of SAICAR to AICAR, catalyzes the removal of a molecule of fumaric acid and adenylic acid (AMP) is produced.

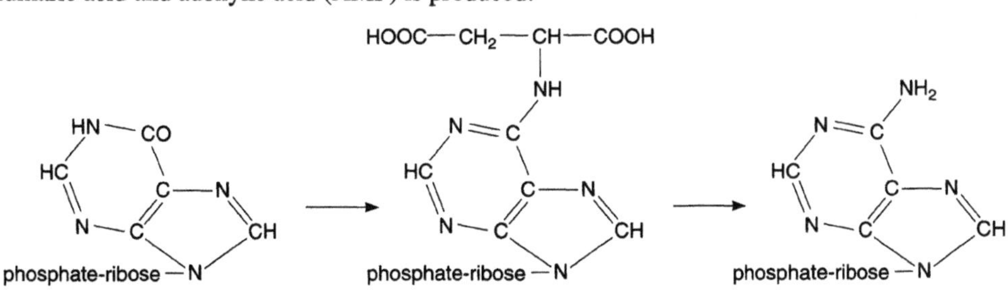

Remarks on the control of purine nucleotide biosynthesis

Study of control mechanisms is complicated by the fact that during the biosynthesis of the purine nucleus, a by-product is a precursor of histidine. Also, in addition to "salvage" enzymes responsible for the direct formation of nucleotides from preformed bases or nucleosides, hypoxanthine and xanthine (purine bases not found in RNA), can be respectively converted to inosinic and xanthylic acids, precursors of the two "natural" nucleotides.

In addition to PRPP amidotransferase, a certain number of reactions are subject to allosteric inhibition effects. We shall examine only some of them here.

The reactions converting xanthylic acid to guanylic acid, adenylic acid to inosinic acid by deamination, and guanylic acid to inosinic acid, are practically irreversible. In the absence of a rigid control, the three enzymes (xanthosine-5'-phosphate aminase, IMP dehydrogenase and GMP reductase) would theoretically be able to catalyze an irreversible futile cycle of reactions leading to IMP through GMP to IMP again. This cycle would have no result other than the wastage of ATP by the hydrolysis accompanying the amination of XMP. The same difficulty would arise if the conversion of IMP to AMP via adenylosuccinate was coupled with the hydrolytic deamination which transforms AMP to IMP.

These difficulties are resolved by the specific inhibition of IMP dehydrogenase by GMP:

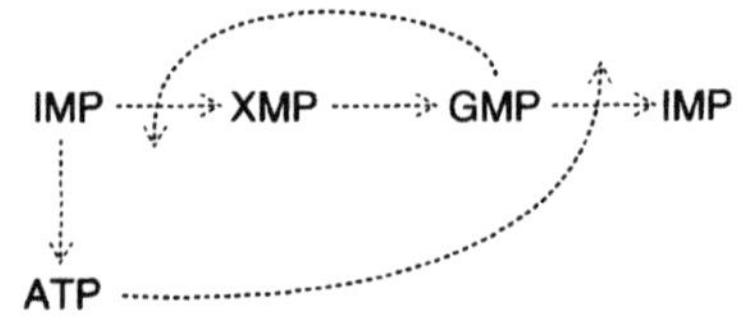

In the same way, adenylosuccinate synthetase in *E. coli* is inhibited by ADP.

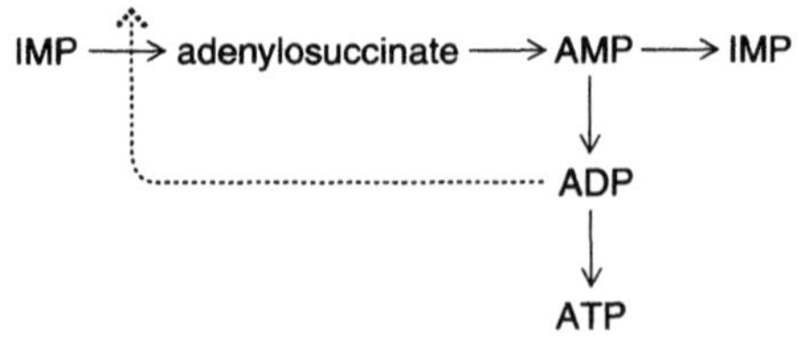

The purine phosphoribosyltransferases catalyze the entry into the cycle of exogenous purine bases, which are not normal intermediates in the synthesis of nucleotides. They catalyze the following type of reaction:

base + PRPP → nucleoside monophosphate + pyrophosphate

E. coli has two of these enzymes, a guanine phosphoribosyltransferase (*gpt*, 6 mn; sequence determined) and a hypoxanthine phosphoribosyltransferase (*hpt*, 3 mn). Both enzymes use guanine and hypoxanthine as substrates, but at different efficiencies, and are inhibited by ppGpp. The product of *gpt* can also salvage xanthine.

In *Bacillus subtilis*, all these reactions are inhibited by a large number of purine nucleotides. From the standpoint of cellular control, it is significant that the nature of the inhibitors and the compared sensitivity of the different reactions can be understood in terms of intracellular equilibria. For example, xanthine phosphoribosyltransferase is especially sensitive to the guanine nucleotides GTP and dGTP; so an excess of either of these prevents the accumulation of XMP, the direct precursor of the guanine nucleotides. Incidentally, we have just seen that the alternative pathway for the synthesis of XMP from IMP is also inhibited by GMP, making the guanine nucleotides the regulators of every step in their synthesis.

The gene *purR* from *E. coli* encodes a 38 kDa repressor protein which regulates the expression of ten *pur* regulon genes, required for the formation of IMP and of *guaBA*, which in turn is required for conversion of IMP to GMP. The purine repressor also contributes to the coregulation of at least five additional genes: *pyrC* and *pyrD*, involved in the synthesis of pyrimidine nucleotides; *glyA*, required for generation of one-carbon units; *codA* (8 mn) which encodes cytosine deaminase, a salvage enzyme; and *purR* itself, which it autoregulates. Recently, *glnB* (coding for protein P_{II}, involved in nitrogen regulation), *prsA* (coding for PRPP synthetase) and *speA* (coding for arginine decarboxylase), have been added to this list. All these genes have functions related to nucleotide metabolism. Hypoxanthine and guanine are the corepressors for binding to a 16-bp conserved operator sequence present in the promoter region or, for the coregulated genes, in the coding regions. The purine repressor is dimeric, it has been overproduced and appears as a complex allosteric protein, having an N-terminal domain of 52 residues for DNA binding and a C-terminal domain of 289 residues for corepressor binding. Binding of the corepressor is accompanied by a conformational change in the corepressor binding domain. The repressor-operator control is independent of *purB* promoter strength as well as of other 5' flanking sequences. It appears that the binding of repressor to the *purB* operator inhibits transcription elongation by a steric inhibition "roadblock" mechanism.

From nucleoside monophosphates to nucleoside diphosphates and triphosphates

All naturally occurring nucleoside triphosphates can be derived from mono- and diphosphates by successive phosphorylations catalyzed by nucleotide kinases.

In one organism or the other, enzymes specific for AMP (adenylate kinase, *adk*, 11 mn), GMP, CMP, UMP, and dTMP have been resolved, and some of them purified to homogeneity. They all use MgATP as the preferred phosphate donor. The first three are active on the deoxynucleoside monophosphates, whereas UMP kinase is specific for UMP. dUMP kinase can use dUMP but not TMP as substrate.

Once the nucleoside diphosphates obtained, they are phosphorylated to the triphosphates by the non specific nucleoside diphosphate kinase.

SELECTED REFERENCES

Carbamylphosphate synthetase

P. M. Anderson & A. Meister, *Biochemistry* **5**, 3157-3163 & 3164-3167 (1966).

P. Trotta, M. E. Burt, R. H. Haschemeyer & A. Meister, *Proc. Natl. Acad. Sci. US* **68**, 2599-2603 (1971).

PRPP synthetase

R. L. Switzer, in *The Enzymes*, P. D. Boyer ed., 3rd edition, **9**, pp. 607-628. Academic Press, New York (1973).

Aspartate transcarbamylase

G. B. Jacobson & G. R. Stark, in *The Enzymes*, P. D. Boyer ed., 3rd edition, **9**, pp. 225-308. Academic Press, New York (1973).

H. K. Schachman, *J. Biol. Chem.* **263**, 18583-18586 (1988).

G. Hervé, in *Allosteric Enzymes*, G. Hervé ed., pp. 61-79. CRC, Boca Raton, Florida (1989).

E. R. Kantrowitz & W. N. Lipscomb, *Trends Biochem. Sci.* **15**, 53-59 (1990).

Nucleoside diphosphokinase

R. P. Agarwal, B. Robinson & R. E. Parks Jr., *Methods in Enzymology* **51**, 376-386 (1978).

CHAPTER XXIV

Biosynthesis of deoxyribonucleotides

Ribonucleotide reductases play a central role in DNA biosynthesis. They catalyze the conversion of NDPs (NTPs) to dNDPs (dNTPs) concomitantly with oxidation of thiols within their active sites. A single active site can accommodate reduction of both purine and pyrimidine nucleotides. In addition to the transcriptional and translational controls, and to the allosteric regulation of the enzyme activity, the overall regulation may imply the regulation of the production and destruction of auxiliary cofactors and proteins.

Formation of deoxyribonucleoside diphosphates from ribose nucleoside diphosphates

Ribose is not directly reduced to deoxyribose. In *E. coli,* the reduction occurs at the level of nucleoside diphosphates which are reduced to the deoxynucleoside diphosphates. The enzyme catalyzing this reaction is called ribonucleoside diphosphate reductase. The same enzyme acts on all four nucleoside diphosphates. The substrates ADP, GDP and UDP are intermediates in the synthesis of the nucleoside triphosphates, whereas CDP is not, since CTP originates from UTP under the action of CTP synthetase (chapter XXIII). CDP must derive from nucleoside diphosphokinase (which catalyzes a reversible reaction) acting on CTP or from CMP produced during phospholipid synthesis and mRNA turnover.

Ribonucleoside diphosphate (NDP) reductase system of *E. coli*

THIOREDOXIN AND THIOREDOXIN REDUCTASE

Early observations established that crude extracts required NADPH for the reduction of ribonucleoside diphosphates. During purification of the system, a preparation was obtained which would no longer catalyze the reduction of CDP in the presence of NAPDH but instead required reduced lipoic acid. Certain other observations then suggested that dihydrolipoate was not the natural hydrogen donor, but a "model substance" and that the true hydrogen donor had been eliminated during the purification. This substance was shown to be a heat-stable protein of low molecular weight, given the name thioredoxin. Thioredoxin can be isolated from *E. coli* in an oxidized disulfide formed referred to as thioredoxin (S_2)

The molecular weight of thioredoxin is 12,000. It consists of a single polypeptide chain of 108 amino acid residues with two cysteine residues at positions 32 and 35 which form the disulfide bond in thioredoxin (S2). The sequence -Trp-Cys-Gly-Pro-Cys of the active site exposed at the surface of the molecule is significantly conserved in the thioredoxin of most species studied (fig 1).

S*DKIIHL* TDDSFDTDVLKADG*AILVDFWA* EW**C**GP**C**KMIAP
ILDEIADEYQGK*LTVAKL* NIDQNPGTAPKYGIRGIP*TLLL*
*F*KNGEVAA*TKV* GALSKGQLKEFLDANLA

Figure 1. Primary structure of *E. coli* thioredoxin.
The residues participating to the β-sheets are in italics; those pertaining to helical segments are underlined. The two cysteines are in bold characters. The conserved sequence is WCGPC (residues 31-35).

The tridimensional structure of S_2 has been determined by X-ray crystallography at 2.8 Å resolution and then refined to 1.68 Å. Figure 2 shows that four α-helices surround the central core formed by a five-stranded β-sheet limited by non-polar residues. Three strands of the β-sheet are parallel and two are antiparallel. The active site forms a bulge at the surface of the protein.

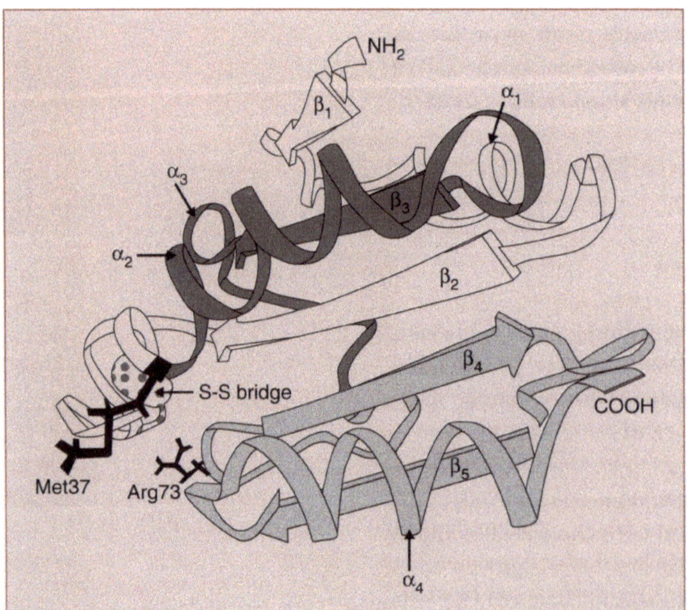

Figure 2. Three-dimensional structure of *E. coli* thioredoxin-S2 from X-ray crystallography at 2.8 Å resolution. The code (color, 1-37; dark grey, 38-73; and light grey, 74-108) refers to fragments generated by selective cleavage at Met37 or Arg73.
(From A. Holmgren, with permission from the *Journal of Biological Chemistry*.)

Conversion by NADPH to reduced thioredoxin, thioredoxin $(SH)_2$, is catalyzed by thioredoxin reductase. The reductase is a flavoprotein which reduces the single cystine of thioredoxin to two cysteine residues. In the presence of catalytic amount of thioredoxin and its reductase, other substances with disulfide bridges are reduced (for example, oxidized lipoate, glutathione, insulin and other proteins). The thioredoxin-thioredoxin reductase system has therefore a wider role than the reduction of ribonucleoside diphosphates. It is thus involved in the oxidoreduction of the disulfide bonds of several proteins, and as such participates in the regulation of the enzymes of the Calvin cycle during photosynthesis according to the light/dark alternance.

Characterization of viable *E. coli* mutants devoid of thioredoxin resulted in the discovery of the glutaredoxin- glutathione system as a hydrogen donor for ribonucleotide diphosphate reductase and has called into question the unique role of thioredoxin. Thioredoxin and glutaredoxin (85 residues) are small proteins containing an active site with redox-active disulfide; they both function in electron transfer via the reversible oxidation of two vicinal -SH groups to a disulfide bridge.

In *E. coli* glutaredoxin, the active site contains a redox-active disulfide/dithiol of the sequence Cys11-Pro12-Tyr13-Cys14. The Cys14 → Ser14 glutaredoxin, obtained by site-directed mutagenesis, is completely inactive with ribonucleotide reductase, demonstrating that dithiol glutaredoxin is the hydrogen donor for the reductase.

The following figure schematizes the mechanism of reduction of ribonucleoside diphosphates in this complex system:

NADPH + H^+	→	$NADP^+$
	glutathione reductase	
GSSG		2 GSH
glutaredoxin $(SH)_2$		glutaredoxin (S_2)
	ATP, Mg^{2+}	
NDP	→	dNDP
	NDP-reductase system	
thioredoxin $(SH)_2$		thioredoxin (S_2)
	thioredoxin reductase	
$NADP^+$	→	NADPH + H^+

This scheme emphasizes the fact that thioredoxin (or glutaredoxin) fulfills a catalytic role in the presence of thioredoxin reductase (or glutathione reductase) and excess NADPH. These two systems also provide electrons for the reduction of phosphoadenosine phosphosulfate (see cysteine biosynthesis, chapter XX) and of methionine sulfoxide.

RIBONUCLEOSIDE REDUCTASE

This enzyme requires ATP and Mg^{2+}.

The ribonucleoside diphosphate reductase from *E. coli* has an $\alpha_2\beta_2$ structure. It contains two different dimeric subunits (B_1, 172 kDa and B_2, 87 kDa), each being composed of two identical protomers, that can be separated during purification. The structural genes of these subunits (*nrdA* and *nrdB* respectively, 49 mn) have been cloned, sequenced, and overexpressed.

Each polypeptide chain of the small subunit B_2 contains a stable radical located on tyrosine 122. The radical lacks one ring electron and thus is an oxidized tyrosine residue. An inactive form of B_2 contains a normal tyrosine 122. In addition to the tyrosyl radical, both the active and inactive form of B_2 contain a binuclear iron center, in which Fe(III) ions are antiferromagnetically coupled to each other by a μ-oxo bridge. Both irons are coordinated to N and O ligands and nuclear magnetic resonance studies indicate that at least one of the ligands is histidine. This center stabilizes the radical and is also involved in its generation.

It has been possible to substitute manganese for iron in protein B_2. Its study could provide a model for the native diferrous form of the protein and allows us to propose a model for the reduction of the iron center.

Protein B_2 has been crystallized, and a refined three-dimensional structure is now available. The large subunit B_1 contains the effector binding sites directing the activity and substrate specificity of the enzyme. The catalytic center is located at the interface between the two subunits, with B_1 contributing the redox active dithiols, in equilibrium with the reduced or oxidized forms of thioredoxin (or glutaredoxin).

The radical is absolutely required for enzyme activity and is thus believed to participate in the activation of the sugar moiety during ribonucleotide reduction. *E. coli* contains an enzyme system that introduces the radical into the protein. This system consists of four proteins and requires reduced pyridine nucleotides, flavin, dithiothreitol and oxygen. Three of these proteins have been obtained in the pure form and characterized as superoxide dismutase, catalase and NAD (P)H: flavin oxidoreductase. The catalase and superoxide dismutase are required to scavenge harmful oxygen radicals formed by the NAD (P)H: flavin oxidoreductase. The fourth protein can be replaced by ferrous ions. This system provides the cell with the ability to regulate the activity of ribonucleotide reductase through the amount of radical present in B_2.

The radical is found buried within the protein. However, it is known to react with a large number of radical scavengers, antioxidants and reductants. This has been the basis for numerous studies on inhibition of ribonucleotide reductase. In particular, hydroxyurea and hydroxylamine are excellent scavengers of the tyrosyl radical and lead to the inactive form of the enzyme[40,41]. In contrast, the iron center was thought to

40. The susceptibility of B_2 to inactivation by hydroxyurea is a sensitive and easy method for detecting changes in the B_1B_2 complex. Actually, the difference in the susceptibility of the radical to hydroxyurea is, to date, the only direct evidence that a change in the B_1B_2 complex accompanies the binding of substrate during normal reduction.

41. Nitric oxide synthase is responsible for the synthesis of NO and citrulline from arginine

be chemically inert. Iron is tightly bound to the polypeptide chain, and in order to prepare apoB_2, the apoprotein, extensive dialysis against a strong iron chelator is required. Under anaerobic conditions, substituted hydrazines and hydroxylamines are good reductants of the iron center. The bacterial radical introducing enzyme system has also been shown to function as a protein B_2-diferric reductase and might be the physiological reductant. The formation of ferrous B_2 is a key step during the activation of protein B_2 since the tyrosyl radical can only be generated during reoxidation of the ferrous center by oxygen.

In *E. coli*, the genes for both B_1 and B_2 form a single operon and coordinately transcribed.

Regulation of the activity of ribonucleoside diphosphate reductase

The large subunit is endowed with a special class of allosteric sites which determine the substrate specificity of the enzyme. Of the four dNTPs, only dCTP does not bind to these sites and its accumulation has therefore no effect on the reductase. As we shall see, dCTP is an allosteric activator of dCMP deaminase, which provides the dUMP required for dTTP synthesis.

ATP stimulates the reduction of the pyrimidine nucleoside diphosphates, CDP and UDP, and has practically no effect on the reduction of purine nucleoside diphosphates; the reduction of the latter is stimulated on the other hand by dTTP and dGTP. dATP strongly inhibits both reductions, its effect being reversed by ATP but not by dTTP.

The current explanation of these interactions, which seem so complicated at first, is as follows: different nucleotides behave as allosteric effectors stabilizing certain states of ribonucleoside diphosphate reductase and so determine the substrate specificity of the enzyme. Thus, ATP stabilizes a state favorable to the reduction of pyrimidine ribonucleotides; dGTP stabilizes a state favorable to the reduction of purine ribonucleotides; dTTP stabilizes a state favorable to both types of reduction, and dATP stabilizes an inactive state. These effects are thought to be one of the foundations of a physiological mechanism for obtaining a balanced supply of the different substrates required for the enzymatic synthesis of DNA (fig. 3).

Figure 4 summarizes the different interrelations listed above.

All these studies have been carried out with purified systems. One can surmise however that they are operative in whole cells. This has been shown by experiments where the concentrations of deoxyribonucleotides has been perturbed by the addition of nucleosides or inhibitors, or by mu-

by activated macrophages. The tyrosyl radical of the *E. coli* enzyme is efficiently scavenged by donors generating nitric oxide. This may be the molecular basis for the inhibition of DNA synthesis induced in tumor cells by activation of the NO synthase pathway.

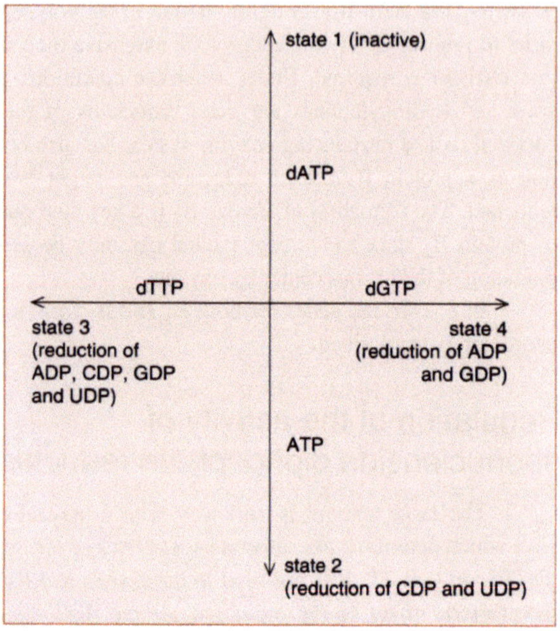

Figure 3. Schematic interpretation of the allosteric effects in the reduction of ribonucleoside diphosphates.

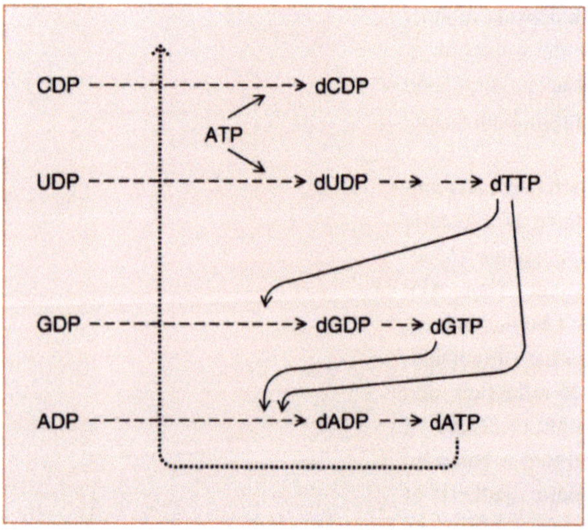

Figure 4. Scheme for the physiological regulation of the synthesis of deoxyribonucleoside diophosphates. The black arrows indicate stimulations; the vertical arrow shows the inhibition by dATP. Synthesis of dCDP and dUDP requires ATP as allosteric effector. Other enzymes transform dUDP to dTTP. dTTP occupies a central position in this scheme: in the presence of ATP, low concentrations of dTTP can cause an additional stimulation of the reduction of pyrimidine ribonucleoside diphosphates, but higher concentrations are inhibitory; this is not shown in the scheme, as the principal effect of dTTP is the stimulation of the synthesis of dGDP. The dGTP formed from this in turn stimulates the formation of dADP. The dADP is transformed to dATP, which is the allosteric inhibitor of the four reduction reactions catalyzed by ribonucleoside diphosphate reductase. The extent of inhibition depends on the concentration of ATP, which reverses the effect of dATP on the reduction of the two pyrimidine ribonucleoside diphosphates.

tations in the sites affecting the regulation of the enzyme. The enzyme of certain organisms, like some bacteriophages or herpes viruses lack the allosteric regulation: the loss of this control could be at the origin of the toxic and mutagenic disequilibria between the different concentrations of deoxyribonucleotides observed during the viral infections.

dCMP deaminase and thymidylate synthase

The above figure shows that ribonucleoside diphosphate reductase itself, even with its allosteric properties, does not allow the relative reduction of CDP and UDP. It appears that the relative concentrations of dCDP and dTTP is maintained by dCMP deaminase.

dCMP deaminase, also called desoxycytidylate aminohydrolase, catalyzes the deamination of deoxycytidylate (dCMP) to desoxyuridylate (dUMP), thus providing the nucleotide substrate for thymidylate synthase. It has been found in bacteria, sea urchin eggs and in the tissues of warm-blooded animals:

$$dCMP + H_2O \rightarrow dUMP + NH_3$$

It is stimulated by dCTP and inhibited by dTTP. The latter inhibits thus its own production and leads to an accumulation of dCTP, thus reestablishing the physiological equilibrium between the two deoxyribonucleotides.

Thymidylate synthase catalyzes the transformation of dUMP into deoxythymidine monophosphate (dTMP). The cofactor of this reaction is 5-10 methylene tetrahydrofolic acid (CH_2THF), member of a family of compounds encountered whenever a methyl or hydroxymethyl group has to be transferred. It is a dimer of identical subunits of about 35 kDa each, and structural studies have shown that the dimer interface of each monomer is composed of five stranded β-sheets, of which about 29 residues interact with the other monomer. Arg178 and Arg179 are donated from one monomer to the active site of the other, and these residues interact with the phosphate moiety of the substrate dUMP, The enzyme catalyzes the transfer of the methylene group to C5 of dUMP and reduction of the transferred one-carbon unit to a methyl group, utilizing a hydride equivalent provided by tetrahydrofolate. Its tridimensional structure is known: the O4 and N3 of dUMP form hydrogen bonds with asparagine177 of the enzyme. This Asn residue is invariant in all thymidylate synthetases. If by site-directed mutagenesis, it is replaced by an aspartic acid residue, dCMP is now the preferred substrate (10^6 times the turnover number of that observed with dUMP). Thus Asn177 may be a major determinant of the nucleotide specificity of the enzyme. It is interesting to note that bacteriophage T4 dCMP hydroxymethylase, which is structurally homologous to thymidylate synthetase and catalyzes a similar reaction with the opposite substrate specificity possesses an aspartic acid residue at the position corresponding to Asn177!

dUTPase

This enzyme catalyzes the hydrolysis of dUTP to dUMP and PPi and maintains a low concentration of dUTP so that uracil cannot be incorporated into DNA. The dUTPase from *E. coli* is strictly specific for its dUTP substrate, the active site discriminating between nucleotides with respect to the sugar moiety as well as the pyrimidine base. Its tridimensional structure has been determined at a resolution of 1.9 Å. It is a symmetrical homotrimer, not showing the "classical" nucleotide binding domain and whose subunit is composed of 152 amino acid residues.

It has been shown in *S. cerevisiae* that the gene encoding dUTPase is essential for viability. In its absence, cell death occurs from the incorporation of uracil into DNA and the attempted repair of this damage by an excision-repair mechanism. Death can be avoided by providing exogenous thymidylate under conditions where the cell is rendered permeable to this metabolite.

Ribonucleoside phosphate reductase of other organisms

The enzyme from *E. coli* is the prototype of the iron-tyrosyl radical enzyme, which is found in mammalian cells, in yeast and in some procaryotes.

Some other procaryotes *e.g. Brevibacterium ammoniogenes* contain a manganese dependent enzyme, which has an $\alpha\beta_2$ structure (α = 80 kDa and β = 100 kDa) instead of $\alpha_2\beta_2$. It appears to possess a binuclear manganese center and to have also a tyrosyl radical.

A third class of reductases depend on the presence of cobalamin, and *Lactobacillus leichmanii* is characteristic of this group. The enzyme from *L. leichmanii* has been purified to homogeneity. It is a monomer with a molecular weight of 76,000, the activity of which is completely dependent on the presence of 5'-deoxy-5'-adenosylcobalamin, which functions as a radical generator during catalysis and plays the role of the *E. coli* B_2 subunit. The enzyme uses ribonucleoside triphosphates and not diphosphates as substrates. The reaction mechanism of this reductase involves also paramagnetic species intermediates as part of a radical-dithiol system.

The recent isolation of the mammalian and yeast genes coding for the ribonucleoside phosphate reductase has shown that, in addition to the well-documented allosteric regulation, the synthesis of the enzyme is also tightly regulated at the level of transcription. The mRNAs for both subunits are cell-cycle regulated and in yeast, inducible by DNA damage. Furthermore, yeast encodes a large subunit gene that is expressed only after DNA damage. This regulation probably facilitates DNA replication repair processes.

Novel ribonucleoside triphosphate reductase reaction in *E. coli* grown under anaerobic conditions

An additional class exists which has been detected in the strict anaerobe *Methanobacterium thermoautotrophicum* and studied in *E. coli* grown under anaerobic conditions. DNA must be obviously synthesized in the absence of oxygen and the enzyme activity that has been characterized in anaerobic *E. coli* is due to a different protein, not coded for by either the *ndrA* or the *ndrB* gene. This enzyme reduces triphosphates and not diphosphates. Its allosteric regulation is not very different from that of the aerobic enzyme. A major difference between the two reductases is the rapid inactivation of the anaerobic enzyme by air and its moderate sensitivity to hydroxyurea, two properties it shares with the *Methanobacterium* enzyme. Since mutants unable to synthesize cobalamin are not inhibited in their anaerobic growth, it is unlikely that a derivative of vitamin B_{12} is involved as a cofactor, as in *L. leichmanii*.

There is strong evidence that in anaerobic *E. coli*, S-adenosylmethionine (SAM) together with a metal participates in the generation of the radical required for the reduction of the carbon 2' of the ribosyl moiety of nucleoside triphosphates. Systems have been described where SAM plus a metal, and adenosylcobalamin are interchangeable. The overall reaction occurs in two steps. The first is an activation of the reductase by an auxiliary protein and a boiled extract of *E. coli*: in addition to SAM, this step requires the presence of NADPH, dithiothreitol, and K^+ ions. The second step, the activated reductase reduces CTP to dCTP with ATP acting as an allosteric effector. During activation, SAM is reductively cleaved to methionine and 5'-deoxyadenosine.

The apparent molecular weight of the purified anaerobic reductase is *ca.* 150,000, as estimated by gel exclusion chromatography. It is a homodimer.

The anaerobic reductase is of special interest from the point of view of evolution and may have been a prerequisite for the transition of an RNA world to a DNA world, at a time when the geochemists tell us that oxygen was scarce. An anaerobic ribonucleotide reductase would then have been an absolute requirement.

Synthesis of deoxyribonucleoside triphosphates from the diphosphates

The phosphorylation of deoxynucleoside diphosphates to the triphosphates is accomplished by the non specificic nucleoside diphosphate kinase. The origin of dTTP deserves special attention however. dCDP is transformed to dCMP by the reversible dCMP kinase. dCMP is then deaminated by dCMP deaminase and dUMP is formed; dUMP is methylated to dTMP by dTMP synthase, an enzyme utilizing 5,10-methylene tetrahydrofolate as the one-carbon donor. Thymidylate kinase phosphorylates dTMP to dTDP and nucleoside diphosphokinase catalyzes the synthesis of dTTP.

The sigmoid form of the substrate saturation curves for dCMP deaminases obtained from a great variety of sources, ranging from chicken to bacteria infected with bacteriophage T4, as well as their sensitivity to allosteric inhibition, indicates that dCMP deaminase is an important point of regulation in deoxynucleotide syntheses. The enzymes of every source are inhibited by dTTP and activated by dCTP (hydroxymethyl dCTP for the phage enzyme). When dCMP deaminase is lost by mutation, the cells undergo a decreased DNA synthesis and growth, which is normalized by the addition of pyrimidine deoxynucleosides to the medium (another example of salvage).

SELECTED REFERENCES

Thioredoxin and glutaredoxin

A. Holmgren, *J. Biol. Chem.* **264**, 13963-13966 (1989).

S. Pille, *Ann. Institut Pasteur/Actualités* **3**, 34-50 (1992).

Ribonucleoside diphosphate and triphosphate reductases

L. Thelander & P. Reichard, *Ann. Rev. Biochem.* **48**, 133-58 (1979).

P. Reichard, *Biochemistry* **26**, 3245-3248 (1987).

P. Reichard, *Ann. Rev. Biochem.* **57**, 349-374 (1988).

J. Stubbe, *Advances in Enzymology* **63**, 349-419 (1990).

C. Gerez and M. Fontecave, *Biochemistry* **31**, 780-786 (1992).

M. Fontecave, V. Nivière & E. Mulliez, *Ann. Institut Pasteur/Actualités* **3**, 159-165 (1992).

P. Reichard, *J. Biol. Chem.* **268**, 8383-8386 (1993).

CHAPTER XXV

Biosynthesis of some water-soluble vitamins and of their coenzyme forms

Biosynthesis of thiamine and cocarboxylase

Thiamine, also known as vitamin B_1, can be considered to be made up of two parts, one being 2-methyl-6-amino-5 hydroxymethyl pyrimidine ("pyrimidine" for the simplicity of presentation) and the other 4-methyl-5-hydroxyethylthiazole ("thiazole") (fig. 1).

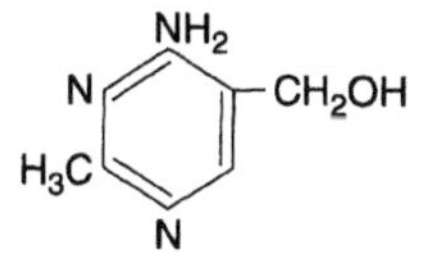

"pyrimidine"

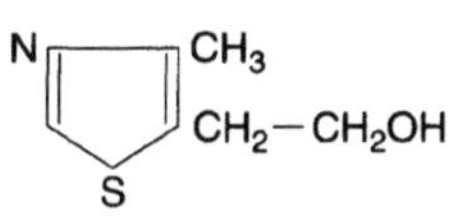

"thiazole"

In *S. typhimurium*, the first five steps in purine biosynthesis also serve as the first steps of the pyrimidine moiety of thiamine.

The pyrimidine originates from AIR, a molecule we have already encountered in the biosynthesis of the purine ring. Although the details of the conversion of AIR to pyrimidine are still unknown, studies with radioactive precursors have shed some light on this difficult problem. Formate is incorporated exclusively in C2, whereas atoms C1 and C2 of glycine become respectively atoms C4 and C6 of the pyrimidine. The nitrogen atom of glycine becomes N1 of the pyrimidine. It thus becomes clear that the imidazole ring of AIR must open between C4 and C5 and that a two-carbon unit must be inserted to yield C5 of pyrimidine and the hydroxymethyl group. It appears now that these two carbon atoms also originate from AIR. Figure 1 summarizes all these observations.

However, it has been demonstrated in *S. typhimurium* that some strains unable to synthesize phosphoribosylamine can nevertheless synthesize the pyrimidine moiety of thiamine. This alternative route of B_1 synthesis appears to require the other enzymes of the purine pathway and to require exogenous pantothenate.

The elucidation of the origin of the thiazole proceeds at a much slower pace, be it for the carbon or for the sulfur atoms. A report has been made according to which the S-CH_3 unit of methionine is incorporated as a whole into the thiazole.

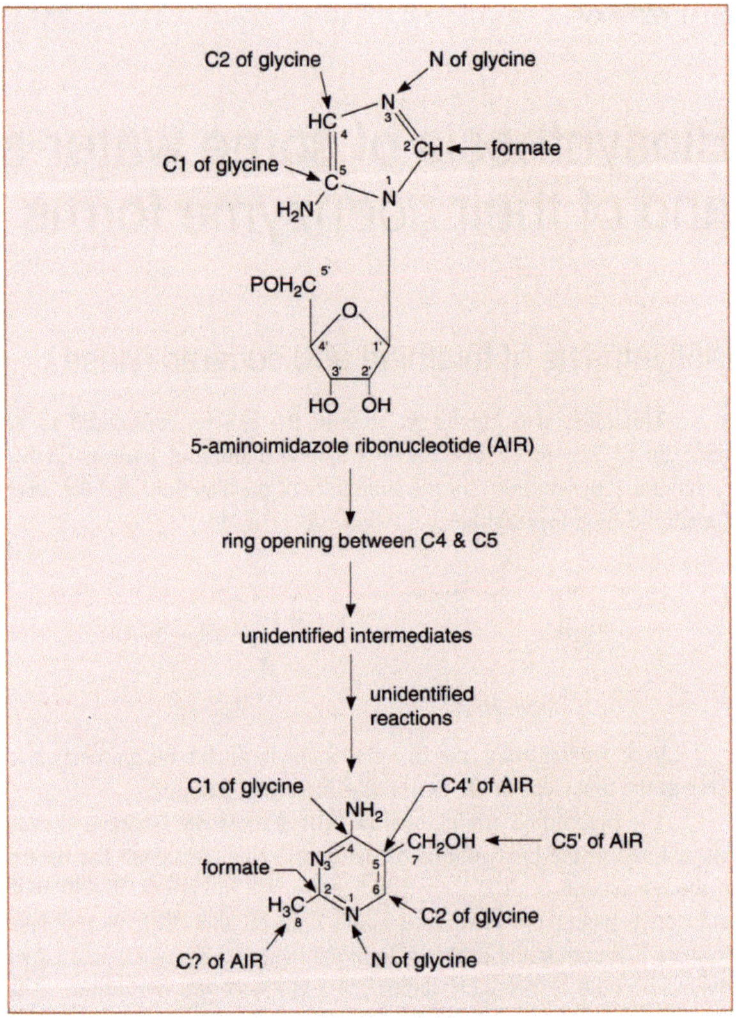

Figure 1. Biosynthesis of the pyrimidine component of thiamine.
(From G. M. Brown & J. M. Williamson, in *Escherichia coli and Salmonella typhimurium, Cellular and Molecular Biology* (J. L. Ingraham, K. B. Low, B. Magasanik, M. Schachter & H. E. Umbarger eds.), vol. I, pp. 521-530, American Society for Microbiology, Washington, DC, 1987. With permission from the American Society for Microbiology.)

In contrast, the condensation of the two parts of the thiamine molecule have been well characterized in yeast and bacteria. In naturally thiamine-requiring microorganisms or in auxotrophic mutants, three groups can be distinguished:

1) organisms requiring pyrimidine only and therefore able to synthesize the thiazole moiety;

2) mutants requiring thiazole and therefore able to synthesize the pyrimidine moiety;

3) finally, those requiring preformed thiamine for growth and unable to condense the two parts when they are provided.

The study of these organisms justifies the conclusion that both parts are synthesized separately and then connected. Experiments carried out with cell-free extracts support the following scheme to account for the synthesis of thiamine from its component parts:

pyrimidine + ATP → pyrimidine pyrophosphate

thiazole + ATP → thiazole monophosphate

pyrimidine pyrophosphate + thiazole monophosphate → thiamine phosphate + pyrophosphate

thiamine phosphate + H_2O → thiamine + Pi

The synthesis of cocarboxylase has been reported to occur in yeast by the action of thiaminokinase which acts on thiamine and ATP, whereas in *E. coli,* thiamine phosphate is the substrate from which the cofactor is synthesized.

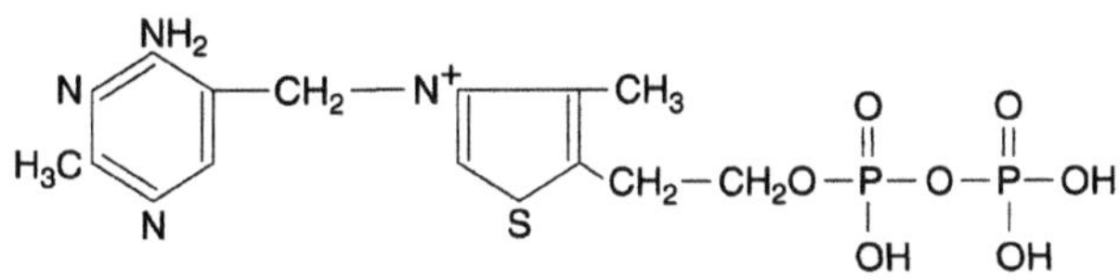

thiamine diphosphate or cocarboxylase

Control of thiamine biosynthesis

The pathway of thiamine synthesis differs from most pathways we have studied so far, in that whereas the latter can be represented in the form of linear or divergent sequences, the thiamine pathway consists of two converging biosynthetic pathways each of which synthesizes products which do not normally accumulate. In addition, the amounts of the compounds synthesized are several orders of magnitude below those of the amino acids and nucleotides.

The control mechanism found in *Salmonella typhimurium* is summarized in the following figure.

The gray arrows interrupting the biosynthetic reactions represent repressive controls, while the white arrow represents control by feedback

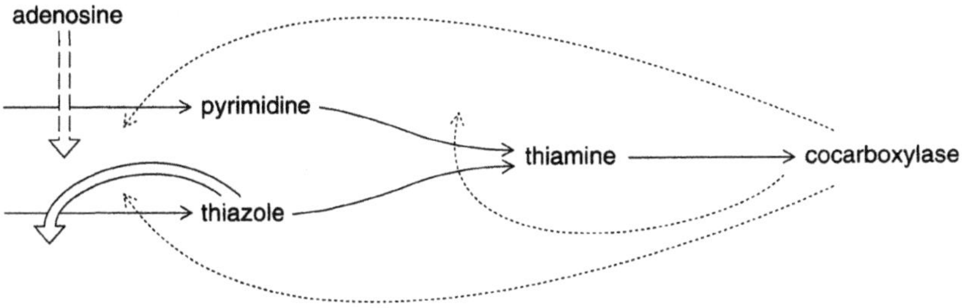

inhibition of enzyme activity. The final product, cocarboxylase, can be seen to repress the synthesis of the enzymes involved in both the initial biosyntheses as well as that of the enzyme which condenses the pyrimidine and thiazole. In addition, thiazole exerts an inhibitory effect on an enzyme (not identified as yet) involved in its biosynthesis.

An interesting effect of adenosine is found in this system: preincubation of the organism with this compound depletes its reserve of intracellular thiamine, so causing an enormous derepression of the synthesis of thiamine.

Biosynthesis of riboflavin

Some lower fungi, like *Eremothecium ashbyii* or *Ashbya gossipyii* synthesize so much riboflavin that the compound crystallizes in the growth medium. It is obvious that such organisms are the tools of choice for studying the synthesis of riboflavin (which is also known as vitamin B_2).

Many experiments have shown that addition of purines to the growth medium stimulates the synthesis of riboflavin. Experiments carried out with compounds labelled in different carbon atoms suggest that the purine carbon skeleton, with the exception of carbon 8, is incorporated as such into the isoalloxazine ring of riboflavin. This points to 4,5-diaminouracil or a ribityl derivative of the same, as a possible intermediate; furthermore, this compound has been found to accumulate in a mutant of *Aspergillus nidulans* which requires riboflavin.

It has been since proposed and demonstrated that GTP (1) is actually transformed to N-phosphoribosyl-diaminouracil (2) by a specific cyclohydrolase and that this compound is then deaminated and reduced to 5-amino-2,6-dioxy-4-(5'-phosphoribitylamino) pyrimidine (3).

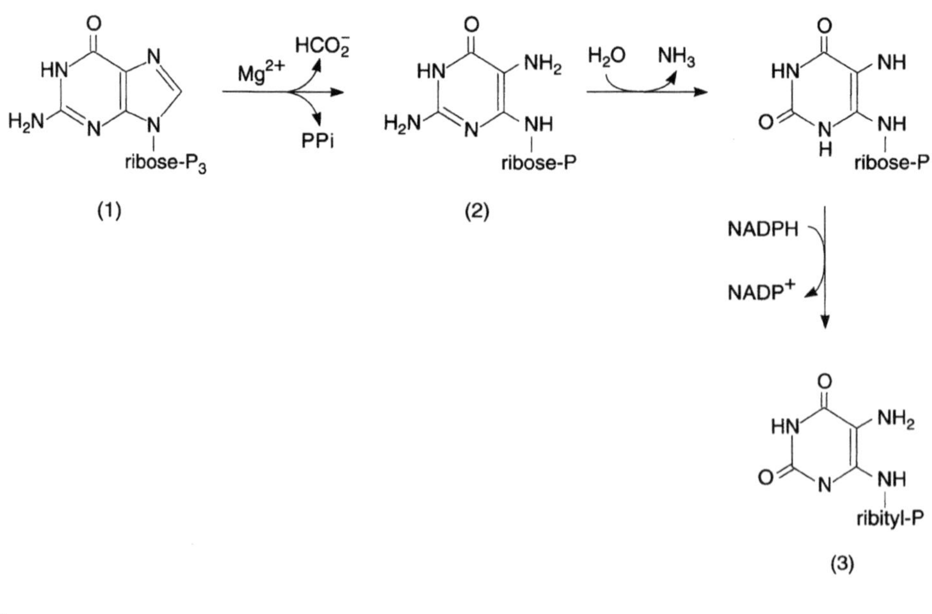

The next precursor of riboflavin to have been identified with certainty is 6,7-dimethyl-8-ribityllumazine which is synthesized from the previous compound and from carbon atoms 1, 2, 3 and 5 of a pentose. The conversion of the lumazine to riboflavin by riboflavin synthase had been previously demonstrated.

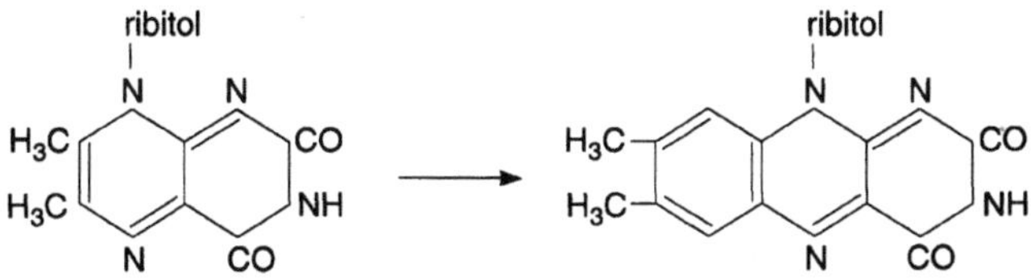

6,7-dimethyl-8-ribityllumazine riboflavin

Riboflavin synthase catalyzes in a single enzymatic step the conversion of two molecules of 6,7-dimethyl-8-ribityllumazine to one molecule of riboflavin and one molecule of a ribitylpyrimidine compound. One molecule of the lumazine derivative acts as a donor of a 4-carbon unit and the other is the acceptor.

In different animal and plant tissues and in bacteria, the existence of a flavokinase which catalyzes the phosphorylation of riboflavin at the terminal position of the ribitol chain has been shown:

$$\text{riboflavin} + \text{ATP} \rightarrow \text{riboflavin-5'-phosphate} + \text{ATP}$$

Riboflavin-5'-phosphate is also called flavin mononucleotide, abbreviated to FMN.

An enzyme has been purified from yeast which catalyzes the formation of flavin adenine dinucleotide (FAD) from FMN and ATP:

$$\text{FMN} + \text{ATP} \rightarrow \text{FAD} + \text{pyrophosphate}$$

The synthesis of flavins is not adjusted as precisely to the physiological needs of bacteria as the synthesis of major metabolites, as is shown by the following results, obtained with *E. coli.*

1) Flavins are produced in large excess during exponential growth; the amount excreted can reach as much as eight times the intracellular concentration.

2) The synthesis is not linked closely to growth and can continue long after cessation of growth in *E. coli*; conversely, the growth of lactic acid bacteria which require riboflavin can continue for quite a long time after they have been deprived of the growth factor. It can be shown that flavins have no feedback inhibitory effect on their synthesis; through control by repression, the amounts of the enzymes necessary for the synthesis of flavins do not vary by more than a factor of two.

Bacteria would probably not benefit from very strict control of flavins and certain other coenzymes. The wastage of matter and energy involved in the uncontrolled synthesis of coenzymes is very slight indeed. To avoid the small excretion observed (small in absolute terms), it would be necessary to have more rigid control: *i. e.*, lower concentrations of the end product would be required to inhibit its own synthesis (by repression or allosteric inhibition). Lower intracellular concentrations would then be able to affect the synthesis of macromolecules in which enzymes with flavin coenzymes are involved at different points. To give an order of magnitude, the internal concentration of free flavins in *E. coli* is already lower than 4×10^{-6} M.

One case has been analyzed in some detail in *B. subtilis*, where the genes coding for the enzymes catalyzing the reactions from guanosine-5'-triphosphate to riboflavin are organized in a single operon. Expression of the operon is under the control of the *ribC* gene, coding for an aporepressor whose effectors are riboflavin, FMN and FAD. Several *cis*-dominant mutations have been obtained in a region, called *ribO*, which could be the repressor binding site. These mutations, like the *ribC* mutations are expressed as riboflavin overproducers, resulting in accumulation in the growth medium (the highest concentration observed is of the order of 20 mg/l).

Biosynthesis of nicotinamide, NAD^+ and $NADP^+$

The details of the biosynthesis of the pyridine nucleotides in procaryotes have been mainly worked out with *Escherichia coli.* Thirty years ago, it was shown that glycerol and a dicarboxylic acid could be the precursors of the nicotinamide moiety of the pyridine nucleotides.

In fact, the actual precursors are iminoaspartate and dihydroxyacetone phosphate (DHAP).

In the first committed step of nicotinamide biosynthesis, the iminoaspartate is formed by aspartate oxidase, encoded by *nadB* (56 mn).

$$\text{L-aspartate} + \text{FAD} \rightarrow \text{iminoaspartate} + \text{FADH}_2$$

The reduced form of FAD donates its electrons to oxygen. Iminoaspartate is an unstable intermediate which would spontaneously give rise to oxaloacetate and ammonia in the absence of conversion to quinolinic acid, the next intermediate in the pathway. Quinolinate synthase, coded for by *nadA* (17 mn) condenses iminoaspartate and DHAP, according to

$$\text{iminoaspartate} + \text{DHAP} \rightarrow \text{quinolinate} + \text{Pi} + 2\,H_2O$$

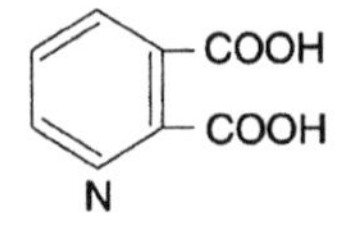

quinolinic acid

nadC (3 mn) codes for a quinolinate phosphoribosyltransferase which catalyzes the reaction of quinolinic acid with phosphoribosylpyrophosphate to yield nicotinic acid mononucleotide (deamido-NAD^+).

$$\text{quinolinic acid} + \text{PRPP} \xrightarrow{Mg^{2+}} \text{deamido-NAD}^+ + \text{PPi} + CO_2$$

NAD^+ synthetase (*nadE*) adds an amide group to the nicotinic acid moiety of deamido-NAD^+

$$\text{deamido-NAD}^+ + \text{ATP} + NH_3 \xrightarrow{Mg^{2+}} NAD^+ + \text{PPi} + \text{AMP}$$

which completes the synthesis of NAD^+.

The synthesis of $NADP^+$ is catalyzed by NAD^+ kinase which transfers a phosphate group from ATP to carbon 2 of the ribose moiety adjacent to adenine in NAD^+.

The structural formulas of NAD^+ and $NADP^+$ are set out below.

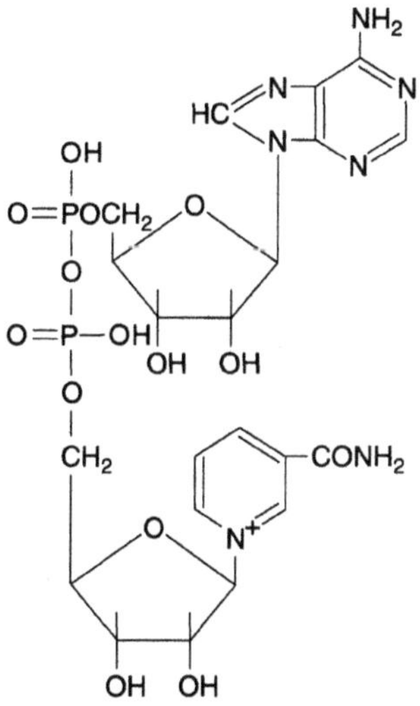

nicotinamide adenine dinucleotide (NAD^+)

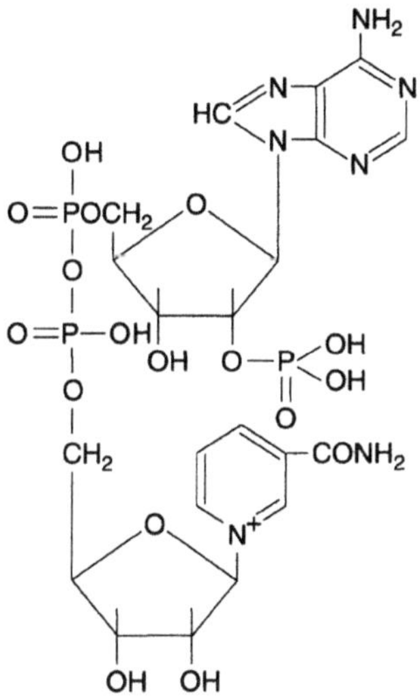

nicotinamide adenine dinucleotide phosphate ($NADP^+$)

Bacteria which result from mutation, or naturally occurring ones, often require nicotinic acid or nicotinamide for their growth. These compounds are transformed through specific phosphoribosyl transferases which in the case of nicotinic acid yield deamido-NAD^+ and in the case of nicotinamide, nicotinamide mononucleotide, which is the substrate of an adenylyltransferase converting it to NAD^+.

It is interesting to note that deamido-NAD^+ reacts with 5,6-dimethylbenzimidazole to yield nicotinic acid and α-ribazole, one of the components of cobalamin (see under biosynthesis of vitamin B_{12}).

In animals and fungi, the precursor of nicotinic acid is tryptophan. Tryptophan pyrrolase, a heme enzyme, catalyzes the opening of the pyrrole ring by an oxygenation reaction which adds molecular oxygen with the production of N-formylkynurenine:

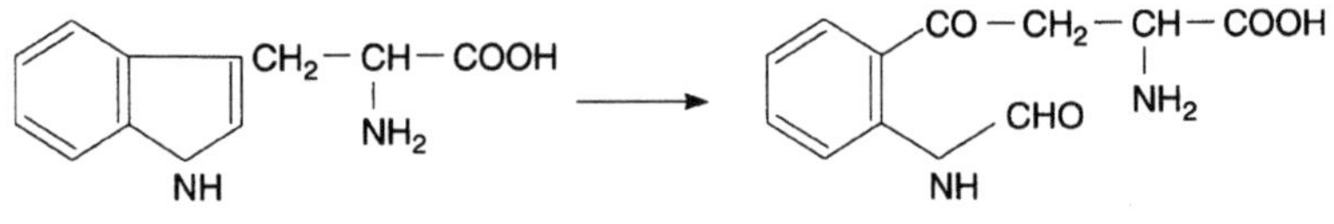

L-tryptophan N-formylkynurenine

The compound N-formylkynurenine loses its formyl group by the action of a specific formamidase and the kynurenin obtained is hydroxylated by kynurenine 3-hydroxylase, the oxygen atom of the hydroxyl group deriving from molecular oxygen and not from water. For an unknown reason, this hydroxylase requires NADPH to function.

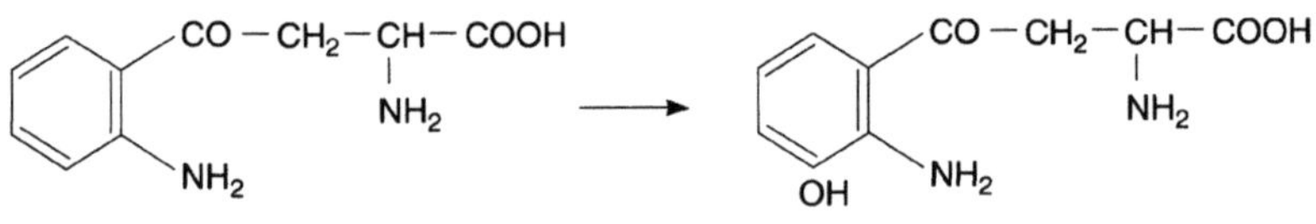

kynurenine 3-hydroxykynurenine

Kynureninase is an enzyme with pyridoxal phosphate as cofactor; it hydrolyzes 3-hydroxykynurenine to 3-hydroxyanthranilic acid and alanine. The benzene ring of 3-hydroxyanthranilic acid is then opened by the action of an oxygenase which has been highly purified from beef liver and is activated by ferrous ions and sulfhydryl compounds. The oxidation product is 1-amino-4-formyl-1,3-butadiene-1,2-dicarboxylic acid:

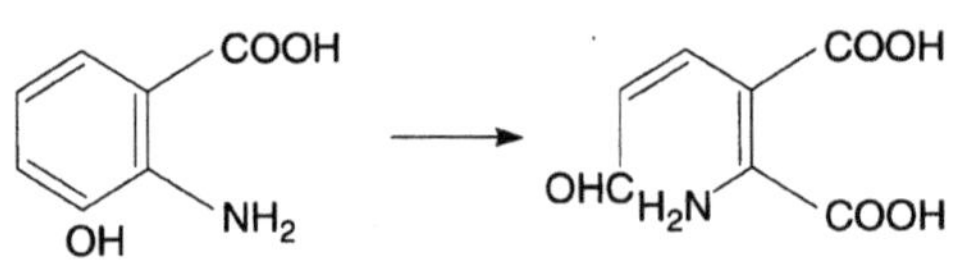

3-hydroxyanthranilic acid 1-amino-4-formyl-1,3-butadiene-1,2-dicarboxylic acid

Specific enzymes carry out the cyclization of this compound to form quinolinic acid, which can either be decarboxylated to nicotinic acid

quinolinic acid nicotinic acid

or react with PRPP to yield deamido-NAD^+ which is then subjected to the same reactions as those which have been described above for *Escherichia coli.*

Regulation of the biosynthesis of nicotinamide and its derivatives

The *nad* genes are scattered on the *E. coli* and *S. typhimurium* chromosomes and thus do not correspond to a single unit of transcription (operon).

Aspartate oxidase (*nadB*) is under the repressive control of the product of the *nadR* gene with NAD as the probable corepressor molecule. Moreover, L-aspartate oxidase is inhibited by the end-products NAD and NADH, but not by NADP and NADPH. Thus, aspartate oxidase is, both at the transcriptional and activity level, the primary site of regulation of the pathway. In addition to *nadB*, the *nadA* gene, specifying quinolinate synthetase is also under the repressive control of *nadR*.

NAD^+ and the ADP-ribosylation of proteins

The secreted form of diphtheria toxin is a single polypeptide chain containing two disulfide bridges and no free sulfhydryls. Trypsin cleaves the intact toxin at an arginine residue located within one of the disulfide loops and generates two large fragments A and B (Mr = 21,000 and 39,000 kD, respectively) still linked by a disulfide bridge. This nicked form is still toxic but devoid of enzymatic activity. After reduction, fragment A is now endowed with ADP-ribosyltransferase activity; the role of fragment B is to facilitate attachment and entry of fragment A into the eukaryotic target of the toxin.

The donor of the ADP-ribosyl moiety is NAD^+, according to

$$\text{N-ribose-P-P-ribose-A} + \text{protein} \rightarrow \text{protein-ribose-P-P-ribose-A} + \text{N} + \text{H}^+$$

(N, nicotinamide; A, adenine)

In the case of diphtheria toxin, the protein target is the eukaryotic elongation factor EF2 and the effect of the toxin is inhibition of protein synthesis.

After diphtheria toxin, other cases of protein ADP-ribosylation have been found in other toxins which play a role in disease. Several ADP-ribosylating toxins have been cloned and some of the crystal structures have been solved. The toxins have significant similarities in structure with some identity in the amino acids which participate in the catalytic site of ADP-ribosylation. They differ however in the amino acids which they modify (table I).

ADP-ribosylation is a reversible modification of proteins. We have seen that an ADP-ribosylation cycle regulates dinitrogen reductase in *Rhodospirillum rubrum*. (As in the case of the reversible covalent adenylylation of glutamine synthetase, the inactivating and the reactivating reaction are different.)

Table I. Targets of ADP-ribosylation.

Toxin	Modified amino acid
Cholera	arginine
Pertussis	cysteine
Botulinum	asparagine
Diphtheria	diphthamide*

*An unusual amino acid residue of EF2, of unknown function, formed by posttranslational modification of histidine.

Biosynthesis of para-aminobenzoic acid, of folic acid and its derivatives

Examination of the structural formula of folic acid

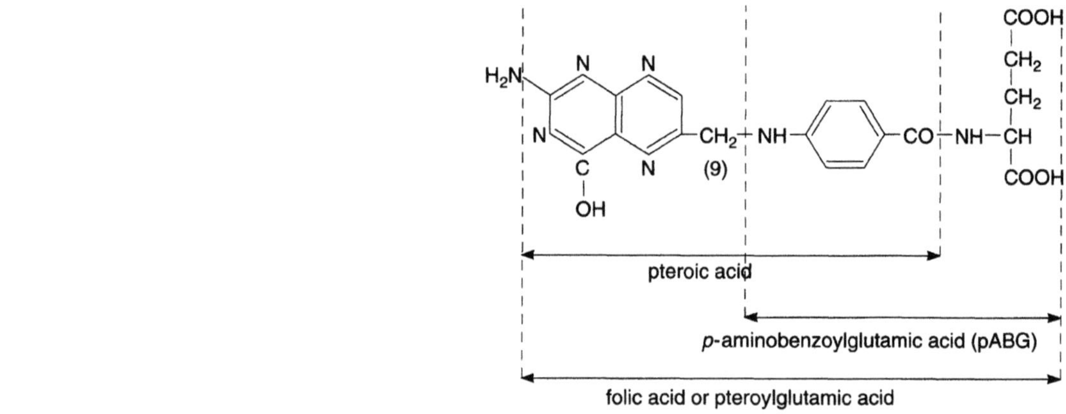

shows that this compound is constituted of three elements, *p*-amino-benzoic acid (*p*-AB), 2-amino-4-hydroxypteridine and glutamic acid. The combination of pteridine and *p*-AB through the methylene bridge, numbered 9 on the formula, is called pteroic acid. Chorismic acid, the common precursor of the three aromatic amino acids, is also the precursor of *p*-AB[42]. Prephenic acid and para-hydroxybenzoic acid are inactive as substrates for the cell-free system which synthesizes *p*-AB from chorismate. The nitrogen atom of *p*-AB is derived from the amide nitrogen of glutamine.

Folic acid is a member of a great family of natural pteridines. Before any enzymological work had been done on its biosynthesis, indications had accumulated according to which one purine was a precursor of pteridines. These indications came from experiments done with whole organisms where the incorporation in pteridines of various ^{14}C possible precursors was followed. It was thus observed that guanosine or a nucleotide of guanine was the precursor of the pteridine part of folic acid and that during its incorporation, carbon 8 of guanine was lost, as in the case of the biosynthesis of riboflavin. These observations led to propose a biosynthetic pathway that had already been put forward for the synthesis of the insect pteridine pigments. These hypotheses have now been confirmed by an extensive enzymological work.

GTP is opened by a cyclohydrolase which catalyzes four successive reactions, including the opening of the 5-membered heterocycle of GTP, the loss of formate (corresponding to carbon 8 of guanine), an

42. For the sake of simplicity, the synthesis of *p*-AB has been examined with the biosynthesis of aromatic amino acids.

Amadori rearrangement and a ring closure, the final product being dihydroneopterin triphosphate

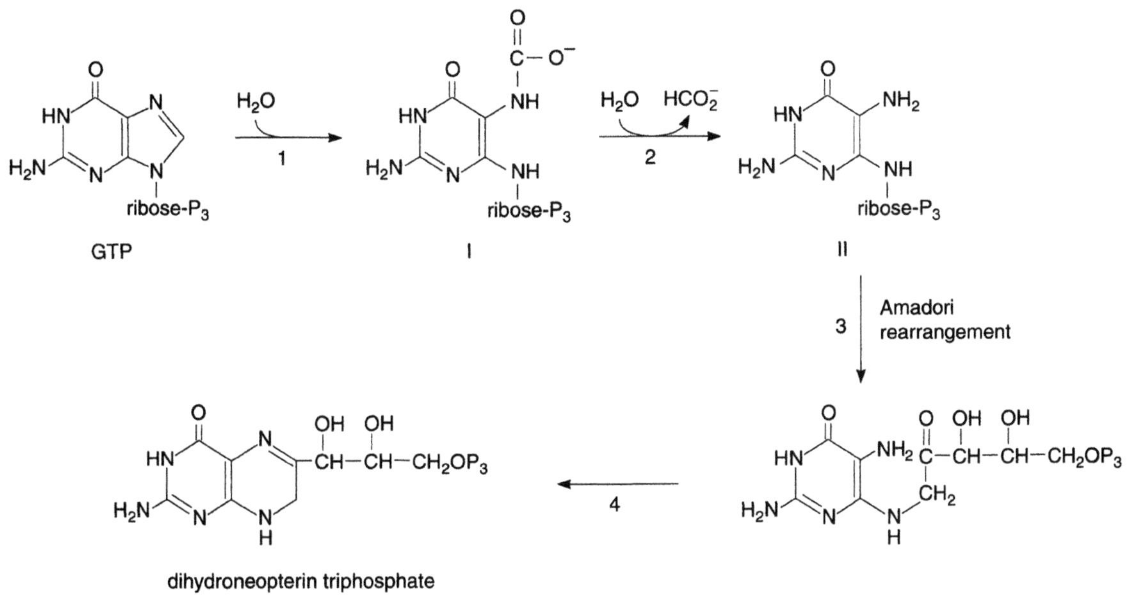

The GTP cyclohydrolase has been purified to homogeneity. It has a molecular weight of 210,000 and is composed of eight identical subunits.

A specific enzyme removes the two distal phosphoryl groups of dihydroneopterin triphosphate, the last phosphate being removed by a non-specific phosphomonoesterase. Thus, dihydroneopterin is obtained.

The next step is catalyzed by dihydroneopterin aldolase and leads to 2-amino-4-oxy-6-hydroxymethyl-7,8-dihydropteridine, abbreviated as dihydropterin-CH_2OH

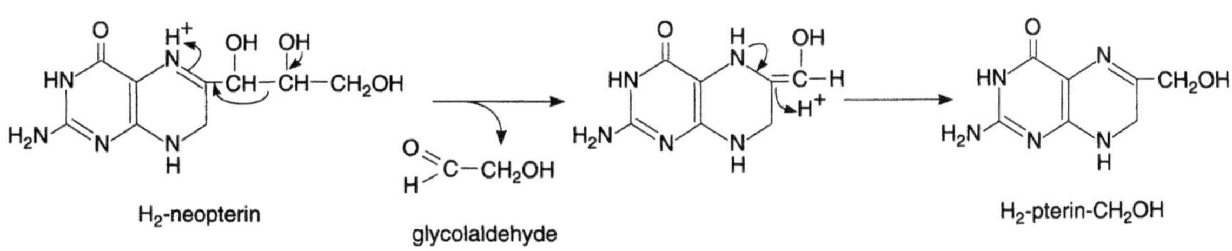

The next enzyme utilizes ATP in the presence of Mg^{2+} to synthesize the pyrophosphoryl ester dihydropterin-CH_2OPP. The enzyme has been purified to homogeneity from *E. coli* and is a monomer of molecular weight 25,000, whose sequence of the 28 first amino acids has been determined.

Dihydropterin-CH_2OPP reacts then with *p*-AB with the elimination of pyrophosphate and the formation of dihydropteroate, a reaction catalyzed by dihydropteroate synthase in the presence of Mg^{2+}. The enzyme from *E. coli* has been purified to homogeneity, has an apparent molecular mass of 52 kDa and is a dimer. Although this enzyme is capable of using *p*-aminobenzoylglutamate (*p*-ABG) instead of *p*-AB, there are many facts to indicate that *p*-ABG is not a biosynthetic intermediate.

Dihydrofolate synthase couples dihydropteroic acid and glutamic acid in the presence of ATP to produce dihydrofolate.

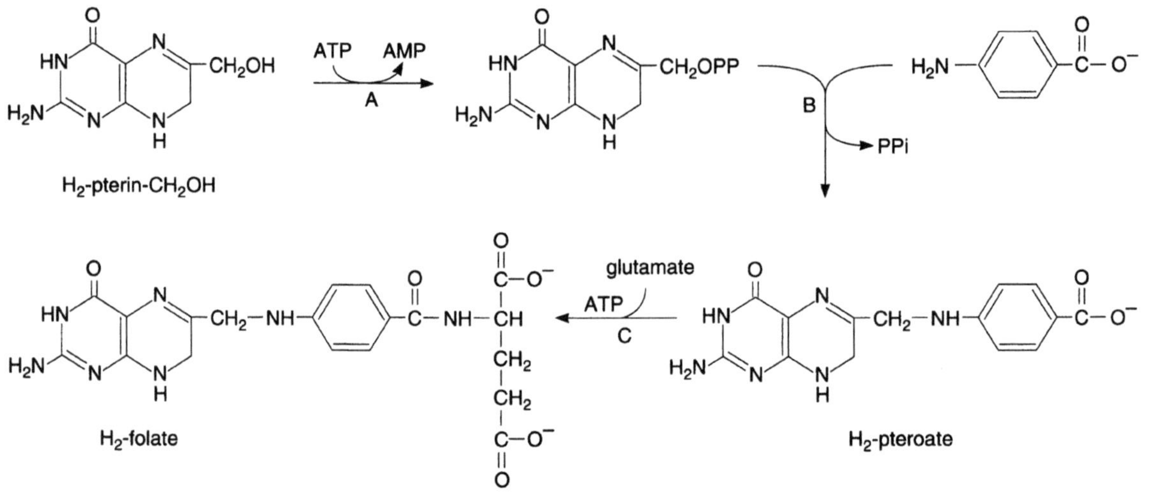

Folic acid is not an intermediate in the synthesis of tetrahydrofolic acid. It is isolated in the oxidized form from natural sources, because of the ease of oxidation of the reduced species. Fortunately for the economy of the cell, there are enzymes to reduce folic acid, the dihydrofolate and tetrahydrofolate dehydrogenases, which use NADH or NADPH.

The tetrahydrofolate dehydrogenase is also called dihydrofolate reductase: it is a monomeric protein of about 180 residues. High-resolution crystal structures have been reported for the enzyme from *E. coli* and *Lactobacillus casei*, as well as for the enzymes from chicken, mouse and man. In each of these structures, the amino and carboxyl termini are located at the protein surface and in adjacent positions not expected on a purely random basis. The structure of several binary complexes with synthetic antifolate drugs or with $NADP^+$, and of a ternary complex with folate and $NADP^+$ has also been determined, so that the catalytic mechanism is known to considerable stereochemical detail.

Other derivatives of folic acid are known: we have already encountered N^5-methyltetrahydrofolic acid and N^5,N^{10}-methylenetetrahydrofolic acid, in connection with the biosynthesis of glycine and methionine where these compounds act as carriers of one-carbon units. Also known are other kinds of tetrahydrofolate derivatives, such as the one used in the transfer of formimino groups (–CH=NH), or such as the 4-formyl, 10-formyl, 10-hydroxymethyl, 5,10-methenyl.

Other naturally-occurring forms are the triglutamyl and hepta-glutamyl forms of folic acid. In these, the glutamyl groups are linked through their γ-carboxyl groups. Sulfonamides, which were known since the pioneer work of D.D. Woods in 1940 to be antagonists for the growth of *p*-AB requiring organisms, are competitive inhibitors of *p*-AB in the reaction catalyzed by dihydropteroate synthase.

Biosynthesis of vitamin B_6 pyridoxine, and of its derivatives, pyridoxal, pyridoxamine and pyridoxal phosphate

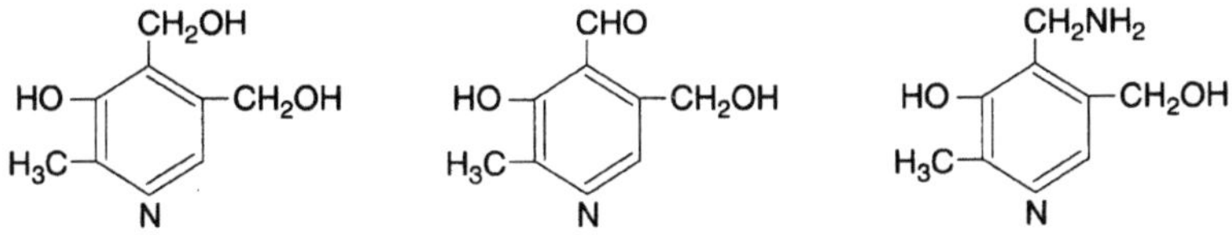

As all these three forms of vitamin B_6 are equally active for the growth of vitamin B_6-requiring organisms, it is to be expected that they are readily interconvertible. Pyridoxal phosphokinase, which catalyzes the reaction

pyridoxal + ATP → pyridoxal phosphate + ADP

also catalyzes the phosphorylation of pyridoxine and pyridoxamine. Pyridoxine phosphate and pyridoxamine phosphate are transformed to pyridoxal phosphate respectively by a specific transaminase and oxidase.

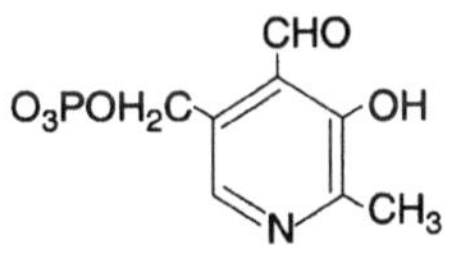

pyridoxal phosphate

The study of the biosynthesis of this group of compounds has not yielded unambiguous data as to its obligatory precursors. This is partly due to the small amounts of pyridoxine derivatives in the cell, but mainly because pyridoxineless mutants are not always mutants that lack a gene involved in pyridoxine biosynthesis. As an example, it suffices to mention a mutant which grows on either pyridoxine or threonine, due to the fact that its threonine synthase, a pyridoxal phosphate dependent enzyme, has a greatly diminished affinity for its cofactor, and not to the fact that threonine is a precursor of pyridoxine. The literature contains many such examples. Approximately 250 genuine mutants *(pdx)* have been genetically analyzed and fall into five unlinked complementation groups on the *E. coli* chromosome. Their study has not allowed a definite conclusion. On the other hand, nutritional experiments support the idea of a tenuous relationship between threonine and branched chain amino acid biosynthesis and that of pyridoxine. The best evidence comes from a careful series of experiments where mutants were fed possible precursors and the labeled synthesized pyridoxine was degraded carbon by carbon or by obtaining derivatives with two carbon atoms. This painstaking work has unfortunately not led to the unequivocal identification of any precursor but has showed distribution patterns that limit the possibilities to relatively few.

When glycerol is fed, *E. coli* makes specifically labeled pyridoxines, with no trace of randomization in the label. Glycine contributes both its methylene carbon and its nitrogen in two adjacent atoms of the pyridoxine, a result that contradicts however data obtained from genetic studies. Other similar experiments have been made with serine, glycolaldehyde, pyruvate and pyruvaldehyde, and did not lead to results it is possible to evaluate.

The synthesis of compounds of this family seems to be under control by feedback inhibition: addition of pyridoxine at a concentration of 4×10^{-7} M to exponentially growing cultures of *E. coli* immediately stops all synthesis of pyridoxine derivatives. The conclusion must be verified when the specific enzymes of the biosynthetic pathway will have been identified.

Biosynthesis of biotin, biotin~CO_2 and biocytin

Biotin can be thought of formally as a thiophen ring, with an *n*-pentanoyl side-chain and a urea moiety:

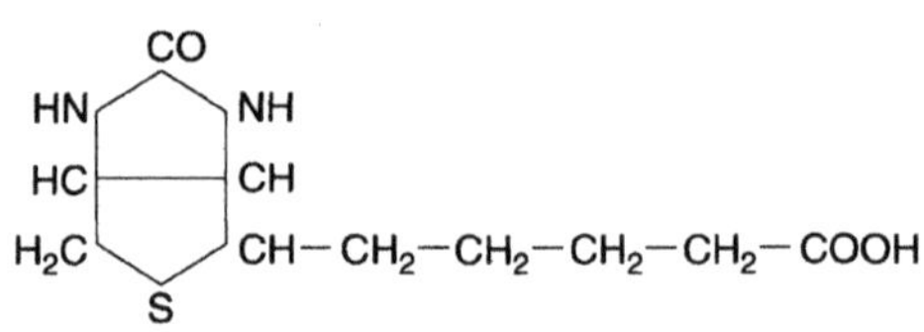

biotin

Studies carried out more than fifty years ago had shown that certain biotin requiring mycobacteria could utilize pimelic acid in its place:

$$HOOC{-}CH_2{-}CH_2{-}CH_2{-}CH_2{-}CH_2{-}COOH$$

pimelic acid

Although the mode of formation of this acid is not known as yet, one knows that pimeloyl CoA is an intermediate and that some of the enzymes involved in in the biosynthesis or in the degradation of fatty acids are concerned.

The next step implies a condensation of pimeloyl CoA with L-alanine, catalyzed by a pyridoxal phosphate enzyme called for short KAP-synthetase:

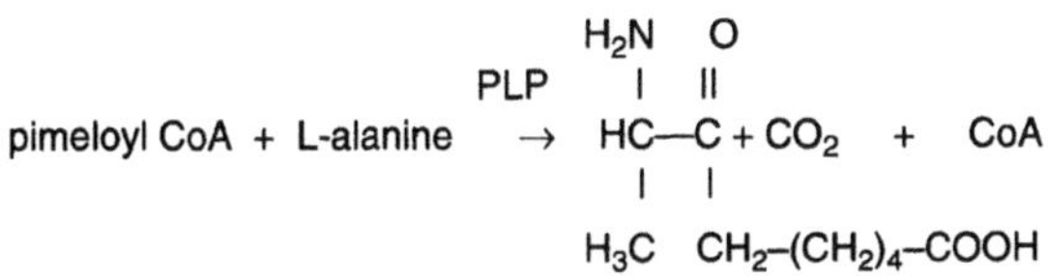

7-keto-8-aminopelargonate (7-KAP)

7-KAP had been observed in the culture media of various microorganisms and biochemically characterized as an open chain structure devoid of sulfur, before having been chemically identified. Its synthesis was demonstrated in cell-free extracts of an *E. coli* mutant auxotrophic for biotin and it was established that the sole reactants were pimeloyl Coa, L-alanine and pyridoxal phosphate (PLP). The molecular weight of the purified enzyme has been obtained by filtration studies and is of 45,000. Genetic complementation studies indicate that it is composed of two or more identical subunits.

The next reaction is a transaminase having as substrates 7-KAP and S-adenosylmethionine (SAM) and leading to 7,8-diaminopelargonic acid (DAPA).

$$\text{7-KAP + SAM} \xrightarrow{\text{PLP}} \text{DAPA + S-adenosyl-2-oxo-4-methylbutyrate}$$

It has been suggested by du Vigneaud as early as 1942 that DAPA could be a precursor of biotin; he had observed that this compound had an activity equivalent to 20% of that of biotin for an auxotrophic yeast requiring the vitamin for growth. Pontecorvo had made similar observations with *Aspergillus nidulans.* But these findings did not retain the attention and DAPA was accepted as an obligatory intermediate only when it was isolated from the culture medium of a biotin-requiring auxotroph. In the first *in vitro* experiments, it was established that, in addition to 7-KAP and PLP, methionine was required as the donor of the amino group. However, the methionine requirement was due to the test and it was later shown that SAM was the true amino donor. DAPA synthetase has been purified and the kinetics of the reaction has been thoroughly analyzed. DAPA-synthetase has been purified to near homogeneity: its average molecular weight is 94,000 and it is composed of two apparently identical subunits as shown by genetic studies which show intragenic complementation among the *bioA* mutants.

Another precursor of biotin had been discovered in the early 1940s, called dethiobiotin,

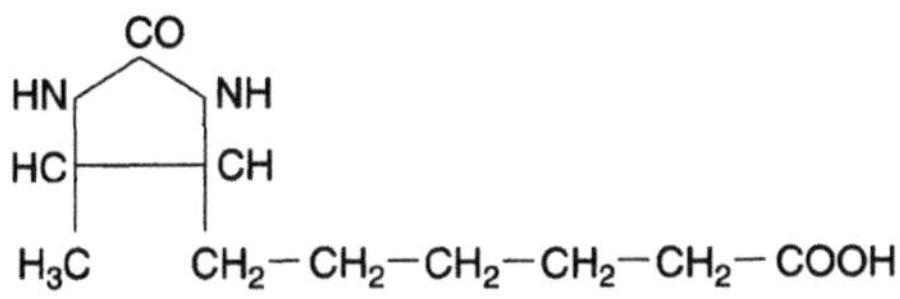

dethiobiotin

where the pimelic acid can be recognized. The precursor-product relationship of DAPA and dethiobiotin has been established through the genetic analysis of a great number of *E. coli* mutants auxotrophic for biotin. However, it was only later that this relationship was established on a biochemical basis when dethiobiotin synthetase, responsible for the transformation, was shown to require ATP and bicarbonate in addition to DAPA. It was found that if ^{14}C bicarbonate was used, the radioactivity was found in the carbonyl of the ureido group of dethiobiotin. A plausible mechanism for the reaction has been proposed:

$$DAPA + ATP + CO_2 \rightarrow \text{dethiobiotin} + ADP + Pi$$

Since du Vigneaud and his colleagues demonstrated in 1942 that dethiobiotin was as efficient as biotin to allow the growth of *Saccharomyces cerevisiae*, a great number of studies were made in vain on the postulated reaction

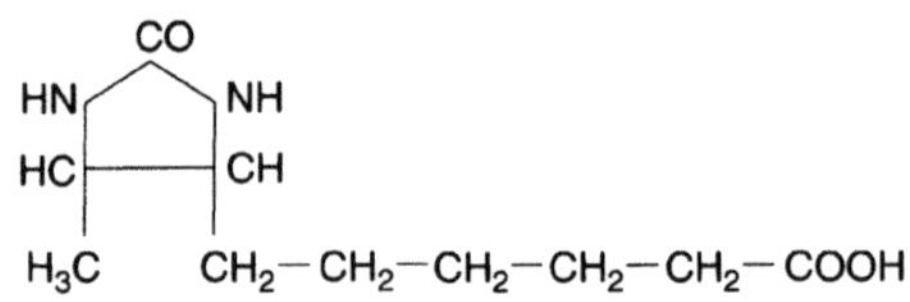

dethiobiotin → biotin

in whole cells of various organisms, due to the absence of a cell-free system able to carry the reaction. The mechanism of incorporation of sulfur is still an open question.

Acetyl CoA carboxylase and other carboxylases contain biotin. The activation of CO_2 necessary for the reaction results from the carboxylation of biotin to form an energy-rich compound, biotin~CO_2

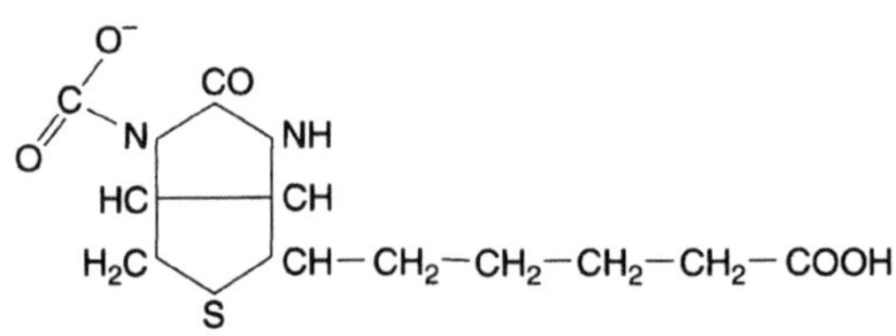

biotin~CO_2

A combined form of biotin has been known for a long time, called biocytin (ε-N-biotinyllysine). This suggested that biotin was bound to the carboxylase apoenzyme through the ε-amino group of a lysine residue. Using radioactive biotin, propionylapocarboxylase and ATP, the radioactive holoenzyme can be synthesized by a specific enzyme. Radioactive biocytin has been isolated after proteolytic digestion of the holoenzyme. We can then depict a carboxylated holocarboxylase as follows:

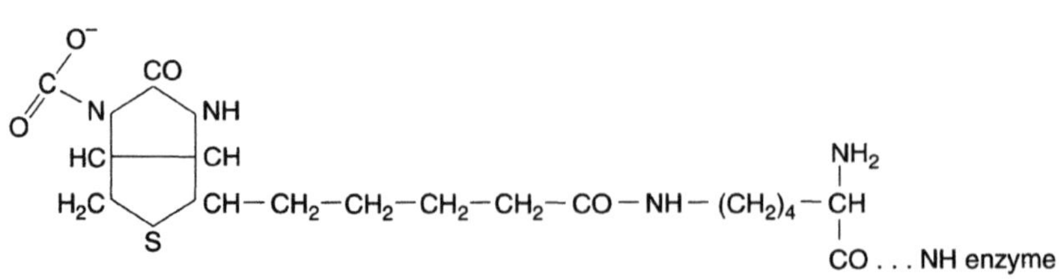

a carboxylated holocarboxylase

The reaction synthesizing holocarboxylase involves the intermediate formation of biotinyladenylate, as we shall see in the following paragraph.

Biotin operon and its repressor

All the biotin genes are grouped in a single operon which is located at 17 mn on the chromosome of *E. coli*. The genes of the operon *bioA, BFCD* are transcribed divergently from a single regulatory region located between the *bioA* and the *bioB* genes. Transcription in both directions is co-repressed by biotinyl-5'-adenylate and the biotin repressor. The following table shows the respective function of these genes:

bioC and *bioH*	unidentified steps in pimeloyl CoA biosynthesis
bioF	7-keto-8-aminopelargonic acid synthetase
bioA	7,8-diaminopelargonic acid synthetase
bioD	dethiobiotin synthetase
bioB	biotin synthetase (introduction of the S atom)

We have no precise information on the function of the *bioC* and *bioH* genes, other that a mutation in either gene results in no excretion of a known intermediate in the pathway. The products of these genes have therefore been assigned to some early steps before 7-KAP synthesis. All the biotin genes are coordinately repressed when biotin in excess of 1 ng/ml is added to the growth medium.

The repressor is a very interesting bifunctional protein of 321 amino acid residues, which acts at two different levels: in addition to its repressor function, it is endowed with acetyl CoA carboxylase biotin holoenzyme synthetase activity. This protein activates biotin to biotinyl-5'-adenylate and transfers the biotin to acceptor proteins. As soon as these proteins are totally biotinylated, biotinyl-5'-adenylate accumulates and serves as the corepressor of the biotin operon. Thus, we have here the case of an enzyme synthesizing its own repressor, a unique property so far among DNA-binding proteins. Mutations in the corresponding gene inactivate the repressor function partially or totally and alter also the enzymatic function. This repressor has recently been crystallized. It binds to a 40-bp symmetric biotin operator site to prevent transcription of the biotin biosynthetic genes. The structure of the repressor is highly asymmetric and consists of three domains. The N-terminal domain is mostly α-helical, contains a helix-turn-helix binding motif and is loosely connected to the remainder of the molecule. The central domain consists of a seven-stranded mixed β-sheet with α-helices covering one face. The other side of the sheet is largely exposed to the solvent and contains the enzyme active site. The C-terminal domain comprises a six-stranded antiparallel β-sheet sandwich. The location of biotin is consistent with mutations affecting the enzyme activity

Two molecules of the monomeric repressor bind cooperatively to one molecule of operator in the nanomolar range of concentration. The data support a model where two molecules of repressor bind with the two operator half sites, and form a dimer only when they bind. Since the complex between repressor and DNA has not been crystallized yet, details of its structure remain an open question.

Biosynthesis of lipoic acid

Lipoic acid is a growth factor for *Streptococcus faecalis* and *Butyribacterium rettgeri.*

The chemistry and structure of lipoic acid have been elucidated quite early. It is an 8-carbon unbranched fatty acid (octanoic acid) that contains a terminal 6-8 disulfide (dithiolane) functional group:

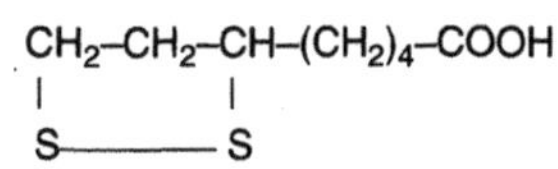

lipoic or 6,8-thioctic acid

In spite of this, information about the detailed enzymic mechanism of its biosynthesis has been delayed due to the minute amounts present in the cells and to an inability to use mutant techniques, although auxotrophs have been obtained in *E. coli* and *S. typhimurium.* Most of our knowledge comes from studies made with putative precursors labeled with radioactive or stable isotopes. When octanoic acid is fed to *E. coli,* it appears to be incorporated as a unit. The present scanty data support the following biosynthetic pathway:

acetyl CoA → octanoic acid → 8-mercaptooctanoic acid → thioctic acid

Sulfur is directly inserted at the C6 and C8 positions with inversion of configuration at the prochiral center. Cysteine appears to serve as the source of sulfur, although it has not been determined whether the two sulfur atoms are inserted simultaneously or sequentially.

Lipoic acid and biotin share many common features.

Both biosynthetic pathways involve enzymes of fatty acid biosynthesis. Both function only when covalently linked through formation of an amide linkage to an ε-amino group of lysine of their respective apoenzymes (see the section on biotin; the disulfide group of lipoic acid participates in the acyl transfer reactions of the pyruvate oxidase and α-ketoglutarate dehydrogenase multienzyme complexes).

Biosynthesis of pantothenate and coenzyme A

SYNTHESIS OF PANTOTHENIC ACID

Mutants of *E. coli* unable to synthesize pantothenic acid fall into three classes: those which can synthesize pantoic acid, but require β-alanine; those which can synthesize β-alanine, but need pantoic acid for their growth; and finally those which are able to synthesize the two moieties but are unable to join them. The mutants from the latter group require preformed pantothenate for their growth.

β-alanine is formed by decarboxylation of L-aspartic acid, under the action of a specific decarboxylase coded by *panD* (3 mn):

$$HOOC\text{–}CH_2\text{–}CH(NH_2)\text{–}COOH \rightarrow HOOC\text{–}CH_2\text{–}CH_2\text{–}NH_2 + CO_2$$

β-alanine

Pantoic acid derives from α-keto-β-methylbutyric acid, which is also a precursor of valine (see p. 234). An hydroxymethyl group is added to this compound by an enzyme, coded by *pan B* (3 mn), ketopantoate hydroxymethyltransferase or ketopantoaldolase, which depends on the presence of tetrahydrofolate for its activity. The enzyme has been purified to homogeneity. Its product, called ketopantoate for simplicity, is α-keto-β, β-dimethyl-γ-hydroxybutyric acid, which can be utilized for their growth by certain mutants auxotrophic for pantoate.

Ketopantoate is then reduced to pantoate by a reductase, coded by the yet unmapped *panE*, utilizing NADPH as the electron donor.

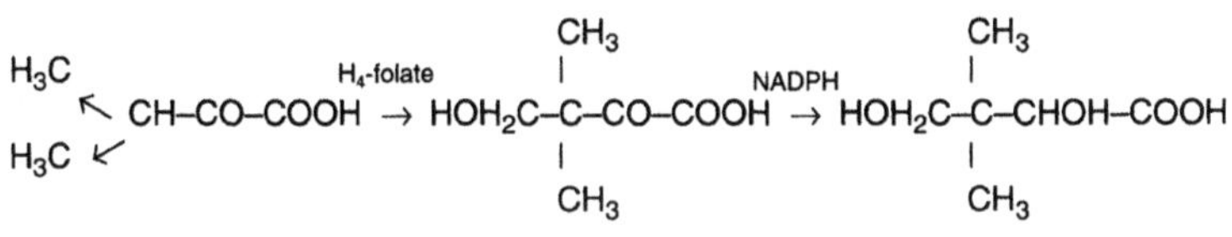

α-ketoisovalerate ketopantoate pantoate

Pantothenate synthetase, coded by *panC* (3 mn), then couples β-alanine and pantoate to yield pantothenate, in the presence of both a divalent (Mg^{2+} or Mn^{2+}) and a monovalent (K^+ or NH_4^+) cation. The enzyme has been purified to homogeneity.

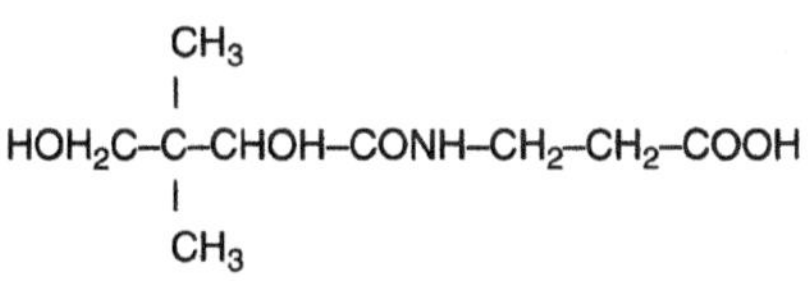

pantothenic acid

The activity of the enzyme is ATP-dependent. In addition to pantothenate, the other products of the reaction are AMP and pyrophosphate. This reaction has a certain historical importance, since it is the first instance where a net synthesis of a peptide bond has been obtained *in vitro* and where an enzyme substrate complex has been clearly demonstrated. This reaction can be decomposed as follows:

enzyme + ATP + pantoate → enzyme-pantoyladenylate + PPi (1)

enzyme-pantoyladenylate + β-alanine → pantothenate+AMP + enzyme (2)

SYNTHESIS OF COENZYME A FROM PANTOTHENIC ACID

The first reaction of the synthesis of coenzyme A (CoA) is a phosphorylation of the hydroxymethyl group of pantothenate by pantothenate kinase. The phosphopantothenate thus formed is then coupled to cysteine by phosphopantothenylcysteine synthetase; the reaction requires ATP. A specific decarboxylase leads to phosphopantetheine, where the cysteine residue of the substrate is replaced by by a mercaptoethylamine group:

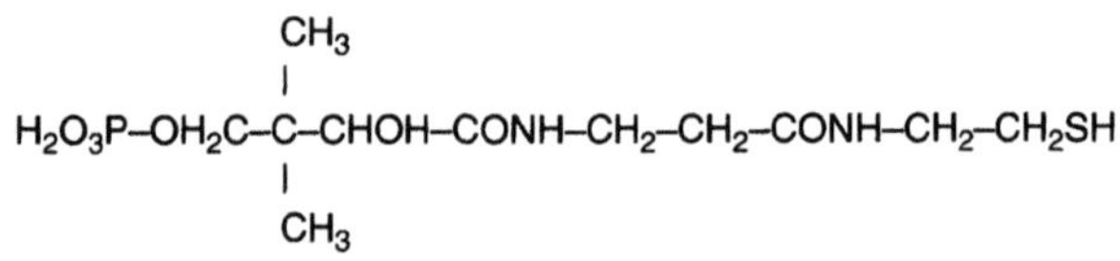

4'-phosphopantetheine

The structural gene for *E. coli* pantothenate kinase (*coa* A, 90 mn) has been cloned and sequenced. Its polypeptide product has a molecular weight of 35,400. Strains containing the *coaA* gene on a multicopy plasmid express 76 times more pantothenate kinase activity, but their steady-state level of coenzyme A and acyl carrier protein remain unchanged or only slightly increased, suggesting that the critical factor controlling the intracellular CoA concentration is regulation of pantothenate kinase activity by feedback inhibition. It seems that other bacteria, such as *Acetobacter suboxydans* and *Lactobacillus helveticus* can carry pantetheine synthesis without prior phosphorylation of pantothenate; in these organisms, the first phosphorylation occurs at the pantetheine level.

An adenylyltransferase catalyzes the transfer of an adenylyl group from ATP to phosphopantetheine, the products of the reaction being pyrophosphate and dephosphocoenzyme A, to which a phosphate group is added to one of the ribose hydroxyls by dephosphocoenzyme A kinase, thus completing the synthesis of coenzyme A.

None of the enzymes catalyzing the series of reactions leading

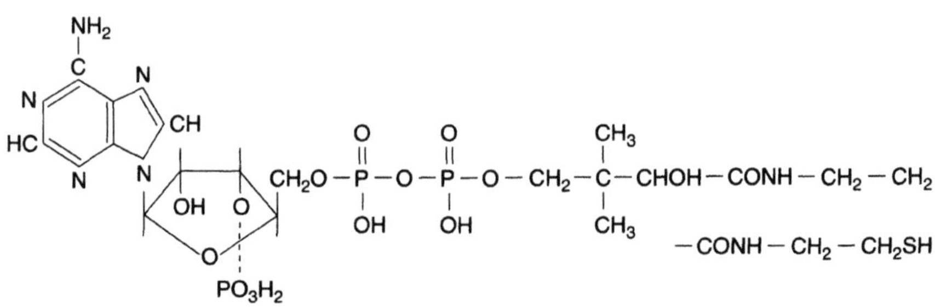

coenzyme A

from pantothenate to coenzyme A has been purified to a significant extent. Although many mutants unable to synthesize pantothenate have been isolated, there is only one report of a mutant blocked in the conversion of pantothenate to coenzyme A. It is a temperature-sensitive mutant of *Salmonella typhimurium* that cannot phosphorylate pantothenate at the non-permissive temperature. Its genetic locus is at 89 mn.

It is the free sulfhydryl group of CoA which reacts with the acyl groups (acetyl, succinyl, malonyl, etc.) in numerous syntheses, in particular in fatty acid biosynthesis and in the tricarboxylic acid cycle.

ACYL CARRIER PROTEIN

This protein has been considered under the chapter dealing with fatty acid biosynthesis.

Biosynthesis of inositol

If one compares the structures of glucopyranose and inositol, the hypothesis is readily made that inositol is derived from glucose by cyclization, accompanied by a change of configuration at C-5. Synthesis by animals seems to be limited and this is the reason why inositol has often been ranged among the vitamins:

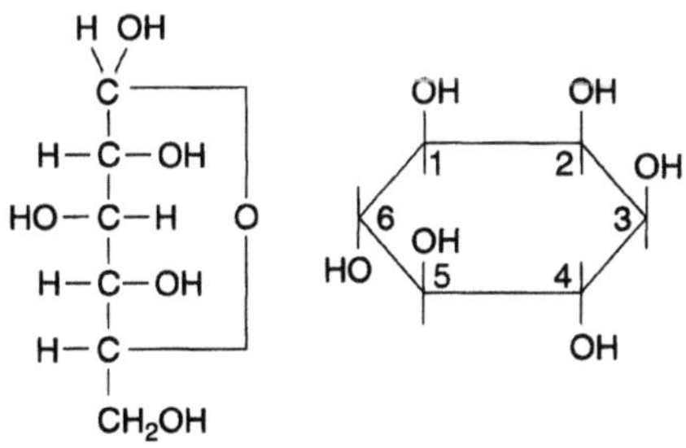

glucose inositol

An enzyme system, *myo*-inositol-1-phosphate synthase, which converts glucose-6-phosphate to *myo*-inositol-1-phosphate has been clearly demonstrated in yeast. This system has an absolute requirement for NAD^+. In a second step, a phosphatase which requires Mg^{2+} ions hydrolyzes inositol phosphate. The same results have been obtained with testicular extracts from mammals.

Biosynthesis of pyrroloquinoline quinone

Pyrroloquinoline quinone (PQQ) is the cofactor of numerous dehydrogenases located in the cytoplasmic membrane or in the periplasm of Gram-negative aerobic bacteria (the two known exceptions are the methanol dehydrogenases of *Clostridium thermoautotrophicum* and *Nocardia*, two Gram-positive organisms). The role of these quinoproteins is to catalyze the primary oxidation step of non phosphorylated substrates such as alcohols, aldehydes and aldoses. *E. coli* possesses all the genetic information required for PQQ biosynthesis and most of the responsible genes have been identified.

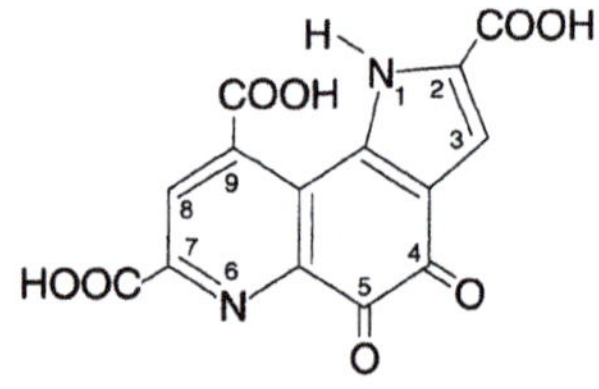

pyrroloquinoline quinone

The two putative precursors of PQQ are glutamate and tyrosine. This result has been obtained by specific ^{13}C labeling of the various possible precursors followed by the analysis of the mass and nuclear magnetic resonance spectra of the products obtained. The part of the PQQ molecule containing N6, C7,8,9 and the two carboxyls C7' and C9' derive directly from glutamate; the rest of the molecule originates from tyrosine. An hypothetical assembly mechanism has been proposed (fig. 2), but the enzymatic steps between the two precursors and PQQ are actually unknown. Cross-feeding experiments between different categories of *Methylobacterium organophilum* and the efforts by various laboratories to isolate intermediates led to no results.

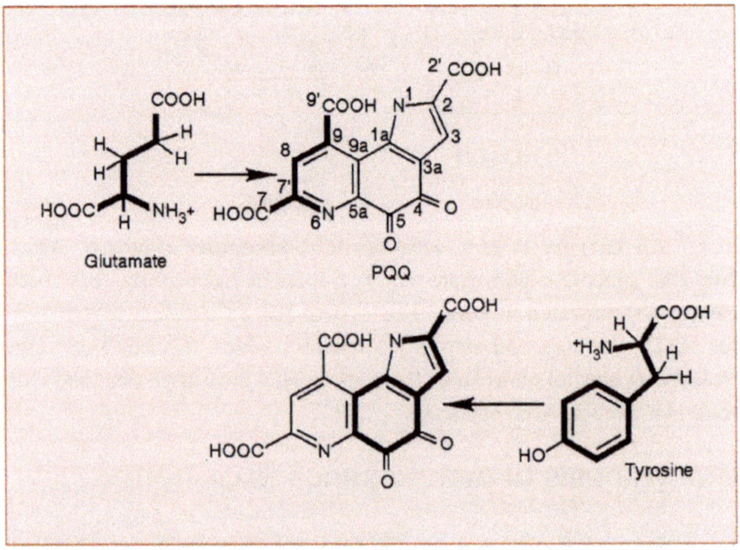

Figure 2. Proposed assembly of the two precursors of PQQ.

The genetic analysis of PQQ biosynthesis rests on the observations of the restored function of biochemical activities (glucose dehydrogenase from *Acinetobacter calcoaceticus* and methanol dehydrogenase of *Methylobacterium organophilum*) rather than on the direct estimation of PQQ. A 5085 bp DNA fragment from *Acinetobacter* complements all the known *pqq* mutants. This fragment contains contiguous open reading frames (ORF) coding respectively for 29.7, 10.8 and 43.6 kDa polypeptides. Following immediately the last ORF, there is an ORF not involved in PQQ biosynthesis and then a small open reading frame coding for a 24 amino acid polypeptide whose integrity is essential to the synthesis of PQQ.

In the methylamine dehydrogenase from *Thiobacillus versutus* and several other bacteria where the cofactor was believed to be PQQ, it was found that it was actually tryptophyl tryptophanquinone (TTQ, fig. 3). Nothing is known about the biochemistry of its formation, except that the two tryptophan residues originate from modifications of residues present in the amino sequence of the light chains (L) of the heterotetrameric L_2H_2 methylamine dehydrogenase.

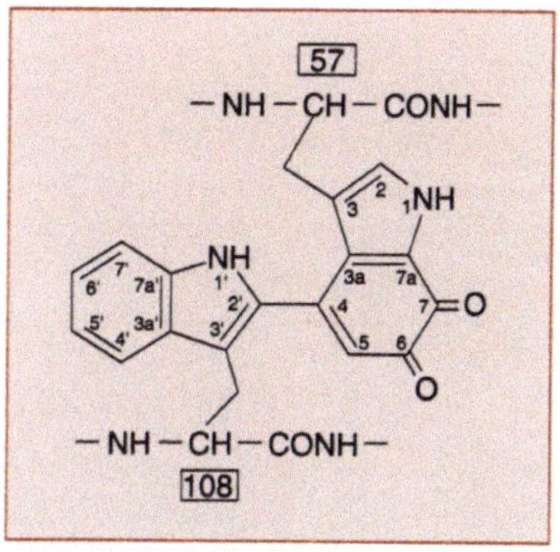

Figure 3. Tryptophyl tryptophanquinone cofactor of methylamine dehydrogenase from *T. versutus*. The precursor tryptophans are at positions 57 and 108 in the amino acid sequence derived from the gene sequence of *T. versutus* methylamine dehydrogenase.
(From E. G. Huizinga, B. A. M. vanZanten, J. A. Duine, J. A. Jongejean, F. Huitema, K. S. Wilson & W. G. J. Hol, *Biochemistry* **31**, 9789-9795 (1992). With permission from the American Chemical Society.)

SELECTED REFERENCES

Consult the articles by Bentley and Meganathan, Brown and Williamson, Dempsey, Eisenberg: in *Escherichia coli and Salmonella typhimurium, Cellular and Molecular Biology,* (J. L. Ingraham, K. B. Low, B. Magasanik, M. Schachter & H. E. Umbarger eds.), pp. 512-550. American Society for Microbiology, Washington, D.C., (1987).

Biosynthesis of water-soluble vitamins
G. M. Brown, *Ann. Rev. Biochem.* **32**, 419-462 (1963).

CHAPTER XXVI

Biosynthesis of carotene, vitamin A, sterols, ubiquinones and menaquinones

Synthesis of the common precursor

The enzyme β-ketothiolase condenses two molecules of acetyl CoA to form acetoacetyl CoA (see p. 131).

In a manner analogous to the condensation of acetyl CoA with oxaloacetate to give citrate, β-hydroxy-β-methylglutaryl CoA synthase catalyzes the condensation of acetoacetyl CoA with a new molecule of acetyl CoA :

$$CH_3{-}CO{-}CH_2{-}COSCoA + CH_3COSCoA + H_2O \rightarrow$$

$$CH_3{-}\underset{\underset{CH_2\text{-}COOH}{|}}{\overset{\overset{OH}{|}}{C}}{-}CH_2{-}COSCoA + CoASH$$

β-hydroxy-β-methylglutarate

The true intermediate in the reaction appears to be acetoacetyl-ACP. A specific reductase reduces β-hydroxy-β-methylglutarate to mevalonic acid, with the aid of two molecules of NADH (or NADPH).

$$CH_3{-}\underset{\underset{CH_2{-}COOH}{|}}{\overset{\overset{OH}{|}}{C}}{-}CH_2{-}CH_2OH$$

mevalonic acid

Mevalonate kinase catalyzes the phosphorylation of mevalonic acid at position 5:

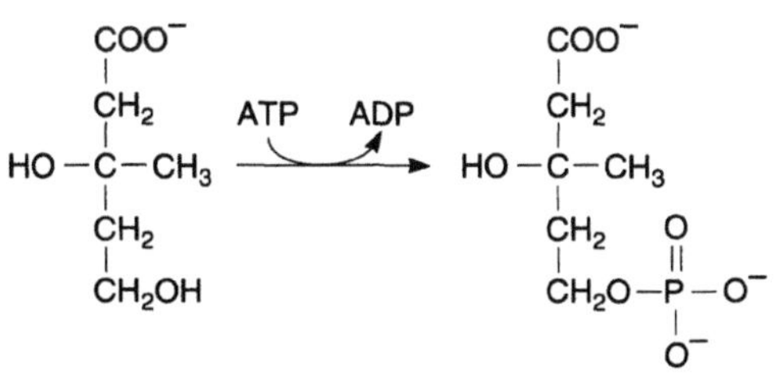

5-phosphomevalonate

A different kinase adds a second phosphoryl group to this molecule:

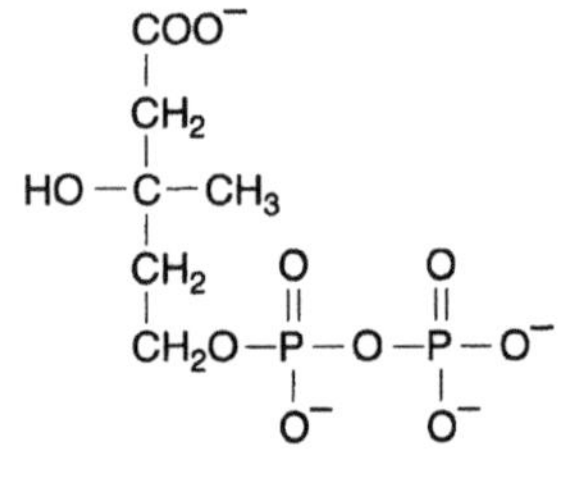

5-pyrophosphomevalonate

A specific anhydrodecarboxylase catalyzes the simultaneous removal of a molecule of CO_2 and the formation of a double bond; the product is an intermediate common to the biosynthesis of carotenoids and sterols, isopentenylpyrophosphate

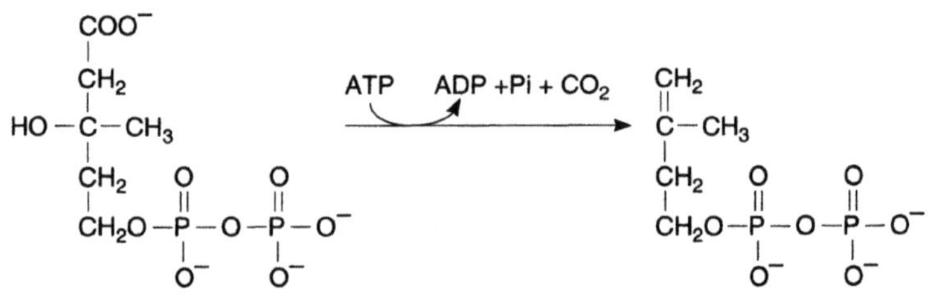

5-pyrophosphomevalonate isopentenyl pyrophosphate

An isomerase is an essential step early in the isoprenoid biosynthetic pathway. It provides the potent electrophile dimethylallyl pyrophosphate (or prenyl pyrophosphate)

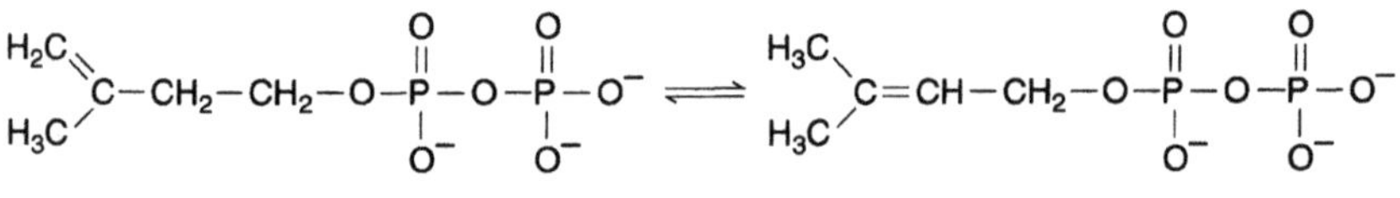

isopentenyl pyrophosphate dimethylallyl pyrophosphate

required for the subsequent prenyl transfer reactions. The pathway is ubiquitous: more than 20,000 naturally occurring isoprenoid metabolites have been identified. End products of the pathway include carotenoids, sterols and ubiquinones. In addition, prenylation is an essential posttranslational modification for many proteins that play key roles in membrane structures, signal transduction across membranes and transport between organelles.

Synthesis of β-carotene, carotenoids and vitamin A

Carotenoids constitute one of the most widespread classes of pigments found in bacteria, fungi, and photosynthetic organisms. The majority of them have 40 carbon atoms and are thus members of the tetraterpene family. Since they arise biogenetically from isoprene units, they have the methyl branches typical of isoprenoid compounds. They share the common biosynthetic pathway with related compounds until the condensation of two molecules of geranylgeranylpyrophosphate forming the first carotenoid, phytoene.

SYNTHESIS OF THE CAROTENOIDS

In the structural formula of carotene reproduced below, the isoprene skeleton of isopentenylpyrophosphate is recognizable:

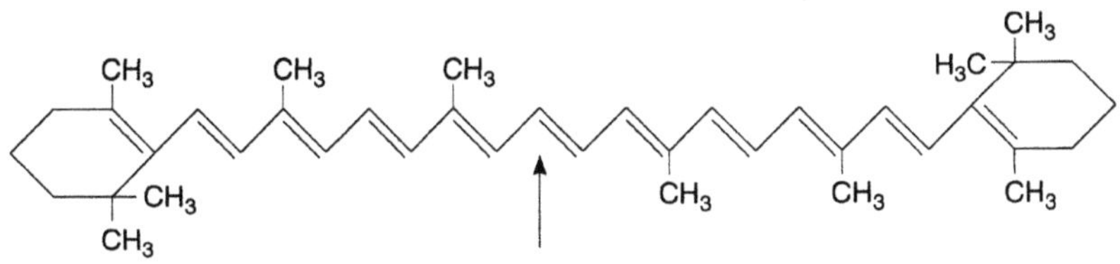

β-carotene $C_{40}H_{56}$

Eleven conjugated trans *double bonds, four methyl branches, and two terminal β-ionone rings. The α- and γ-isomers differ at one end.*

The isoprenoid nature and the symmetry of the carotene skeleton indicate that it is formed from eight isopentenyl units condensed in the following way:

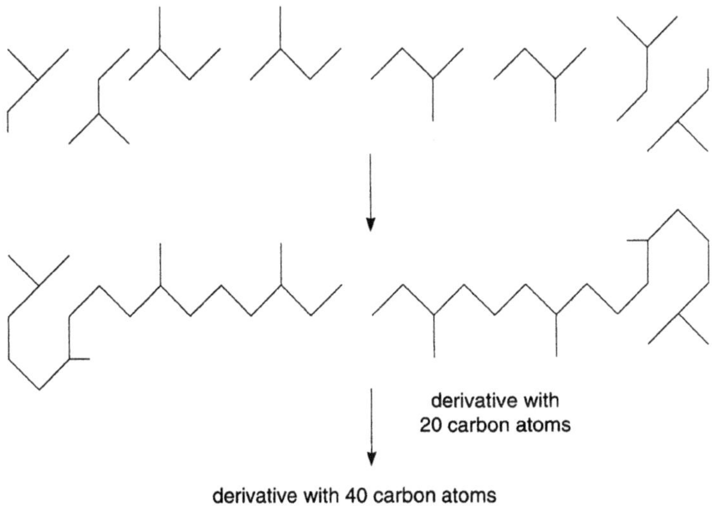

There are actually many experiments showing that bacterial extracts can produce phytoene from mevalonate, pyrophosphomevalonate and isopentenylpyrophosphate. From the last substrate, the synthesis does not require ATP and occurs equally well under aerobic and anaerobic conditions.

It can be concluded from various experiments carried out with different organisms that the synthesis of C_{40} derivatives occurs in steps involving the following intermediates: dimethylallylpyrophosphate, geranylpyrophosphate (C_{10}), farnesyl pyrophosphate (C_{15}), geranylgeranyl-pyrophosphate (C_{20}), followed by a condensation of two C_{20} derivatives to give prephytoene pyrophosphate, which is dephosphorylated to phytoene by phytoene synthase:

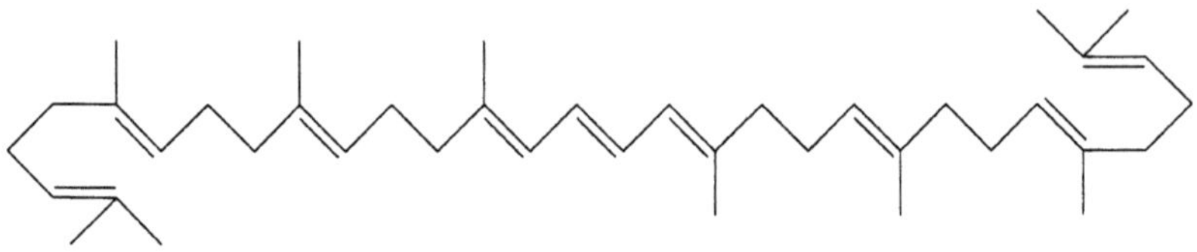

phytoene

Phytoene contains nine double bonds, three of them conjugated, six branch methyl groups and two terminal isopropylidene groups.

In plants, the prephytoene pyrophosphate synthase and phytoene synthase activities are carried by a bifunctional protein.

Phytoene is the first carotene in carotenoid biosynthesis. Depending on the diversity of the end products of the carotenogenic pathways, a variable number of double bonds is introduced into this colorless hydrocarbon.

Desaturation of phytoene proceeds by a dehydrogenase-electron transferase mechanism. The enzyme involved is feedback-inhibited by subsequent carotenoids in various organisms: the phytoene desaturase is the limiting step in the pathway and is a prime target site for herbicides.

Genes coding for the phytoene desaturases have been sequenced from *Rhodobacter capsulatus, Neurospora crassa, Erwinia uredovora* and the cyanobacterium *Synechococcus.* The three first enzymes only share conserved amino acid sequences. The end products of phytoene desaturation found in the various species are different: in the three bacteria, only *Erwinia* and *Synechococcus* produce lycopene, whereas *Rhodobacter* does not. In order to analyze the events, *E. coli* has been transformed by a plasmid of *Erwinia* carrying the genes coding for all the enzymes necessary for the synthesis of 15-*cis* phytoene, the carotenoid precursor, together with a plasmid carrying the desaturase genes of either *Erwinia, Rhodobacter* or *Synechococcus.* The products were then analyzed. The *Rh. capsulatus* desaturase formed *trans*-neurosporene, with three additional double bonds, and two *cis*-isomers. The intermediates with only one additional double bond (phytofluene) and two additional double bonds (three isomers of ζ-carotene) have also been detected. The *Synechococcus* desaturase forms two *cis*-isomers of ζ-carotene and small amounts of *trans*-ζ-carotene. Since this cyanobacterium contains lycopene, it needs an additional ζ-carotene desaturase to form this carotenoid, which possesses four double bonds in addition to those of phytoene. As for the *Erwinia* desaturase, it forms *trans*-lycopene from 15-*cis* phytoene with small amounts of the *cis*-isomer without needing an additional enzyme.

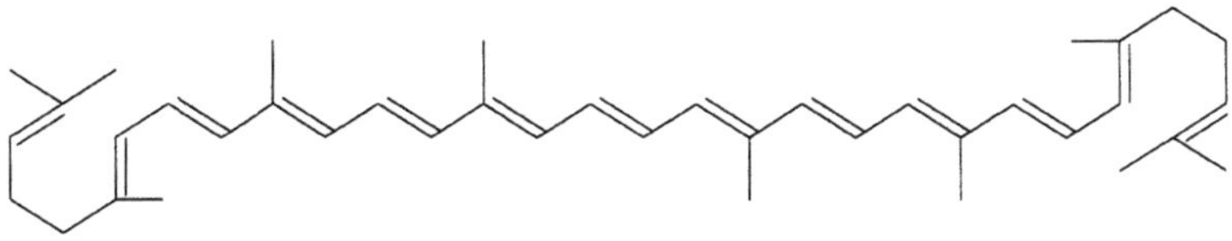

lycopene

Phytofluene contains ten double bonds, five of them conjugated, six branch methyl groups and two terminal isopropylidene groups. As in phytoene, the central double bond between C15 and C16 has cis *configuration.*

Phytoene desaturase is universally a membrane-integrated system. Its purification has been hindered by the practical difficulties associated with assaying the enzyme and the rapid loss of activity upon separation from the membrane. Using molecular biological techniques, the gene coding for the desaturase in *Erwinia uredovora* has been overexpressed in *E. coli*: the enzyme, however, was sequestrated into inclusion bodies requiring urea treatment for solubilization. Purification was finally achieved by standard biochemical techniques. FAD is involved in desaturation, wheras NAD^+ and $NADP^+$ are inhibitory. This contrasts with the desaturase from *Anacystis* where phytoene desaturation is pyridine nucleotide-dependent.

All phototrophic organisms contain carotenoids in their photosynthetic apparatus. These pigments have a dual role: first, they are auxiliary absorbers of light energy which they transfer to chlorophyll. However, photosynthesis cannot take place without chlorophyll, and therefore the carotenoids cannot replace it. A role of the carotenoids might be the protection of chlorophyll against photo-oxidation: a blue-green mutant of *Rhodopseudomonas spheroides* can synthesize colorless phytoene, but is unable to convert it to colored carotenoids. While this mutant can grow normally in conditions of anaerobic photosynthesis, it is rapidly killed by simultaneous exposure to light and oxygen. Death is accompanied by the destruction of bacteriochlorophyll. Further experiments have shown that this destruction is a secondary phenomenon and that the mortality curves are the same if the experiment is carried out at low temperature without destruction of chlorophyll. The mechanism of the protection conferred by carotenoids against photo-oxidation requires further work.

REGULATION OF CAROTENOID SYNTHESIS

Carotenoid synthesis is regulated in different organisms by different endogenous and exogenous factors. Blue light is among the most effective activators of carotogenesis in non-photosynthetic microorganisms. Accumulating evidence points to the possibility of blue light affecting the expression of the carotenoid genes at the transcriptional level. This is certainly the case for the non-photosynthetic gram-negative bacterium *Myxococcus xanthus* and in the fungus *Neurospora crassa*. In the latter case, blue light has been shown to activate specifically the expression of two carotenoid genes. In *M. xanthus*, blue light has been shown to stimulate two unlinked genes of carotenoid synthesis. Chemical analyses of mutant strains suggest the existence of a cluster of structural genes being transcribed from a single, light-inducible promoter under the control of three regulatory genes.

SYNTHESIS OF VITAMIN A

All the carotenes, which are abundant in green plants and fruits, have provitamin A activity. Conversion of carotene to the vitamin occurs mainly in the small intestine, but other organs and serum have a limited conversion activity. The activity is absent in invertebrates. β-carotene is oxidatively cleaved (at the point indicated by the arrow in the top formula p. 323) by the intestinal mucosa into two molecules of retinal (an aldehyde), whereas α- and γ-carotene yield only one molecule of retinal. Alcohol dehydrogenase then reduces retinal to all-*trans* retinol (the vitamin proper), which is then esterified (mainly as palmitate) and stored in the liver.

Retinal is the substrate of an oxidase which transforms it to retinoic acid, which can stimulate growth in vitamin A deficient animals, promote cell differentiation, and suppress carcinogenesis.

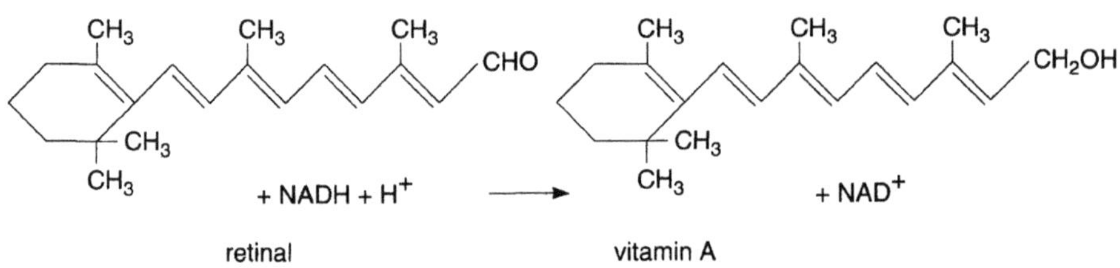

Synthesis of sterols

The regulation of cholesterol synthesis is a complex biological process, and an understanding of it is important for determining the biochemical function of cholesterol in normal and pathological conditions.

The pathway of synthesis of sterols is identical with the synthesis of carotenoids up to the stage of farnesylpyrophosphate, the C15 derivative.

Rat farnesylpyrophosphate synthase catalyzes the sequential condensation of isopentenylpyrophosphate with dimethylallylpyrophosphate and then with the resultant geranylpyrophosphate. Farnesyl pyrophosphate is utilized in the synthesis of squalene, farnesylated proteins, dolichols, coenzyme Q, geranylgeranylpyrophosphate and some other important molecules. Its sequence contains an aspartate-rich motif which has homology with at least nine prenyl transfer enzymes that utilize an allylic prenylpyrophosphate as one of the substrates.

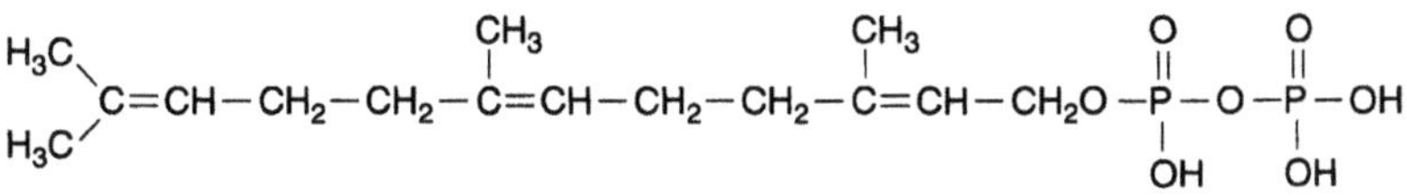

farnesyl pyrophosphate

In a particulate system isolated from yeast and in mammalian liver microsomes, cholesterol is produced from *trans*-farnesyldiphosphate (FPP) with squalene being the first intermediate in this pathway. The formation of squalene is catalyzed by squalene synthase (or farnesyl pyrophosphate: farnesyl pyrophosphate farnesyltransferase) in two separate steps. Two molecules of FPP condense to form the intermediate presqualene pyrophosphate which is then reductively rearranged to form squalene, the reductant being a reduced pyridine nucleotide. The rat liver gene coding for a truncated, active form of squalene synthase has been cloned and expressed in *E. coli*, and sequenced. The yeast enzyme has been purified to homogeneity and is a single polypeptide of M_r 47,000.

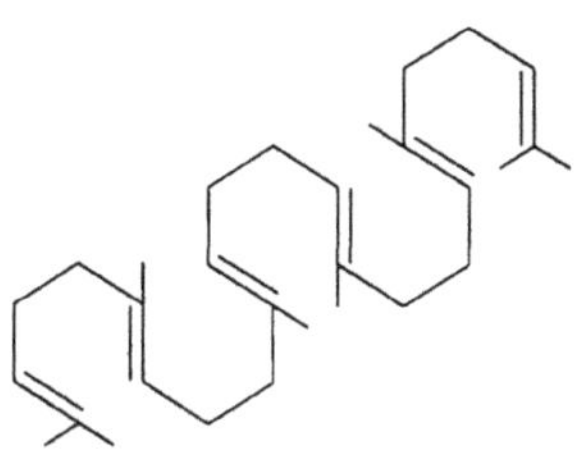

squalene

The conversion of squalene to cholesterol involves first a cyclization requiring molecular oxygen and NADPH: squalene epoxide (2,3-oxidosqualene) is thus formed, which is then cyclized to lanosterol by a cyclase which has been purified from rat liver. Lanosterol is then converted to cholesterol in mammals and to ergosterol in plants and microorganisms. The different steps involved are beyond the scope of this book. Suffice it to say that at least 15 successive reactions are involved in the conversion of squalene to cholesterol.

It has been known for 40 years that the addition of cholesterol to the diet of laboratory animals reduces the rate of synthesis of cholesterol by liver. This reduction occurs at the level of β-hydroxy-β-methylglutaryl reductase synthesis and activity: dietary cholesterol suppresses the synthesis of the reductase at the gene transcription level in the liver and is a feedback inhibitor of the enzyme.

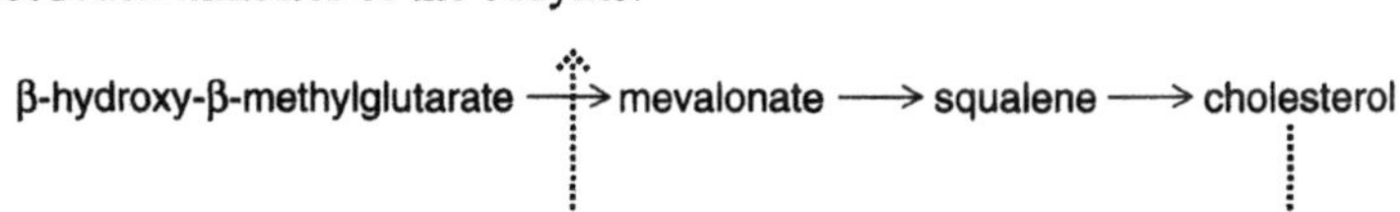

It should be emphasized that the inhibited reaction is the first practically irreversible reaction (committed) in the synthesis of sterols. This feedback inhibition provides a striking example of the absence of structural similarity between an allosteric inhibitor and the substrates of the reaction which it inhibits.

Biosynthesis of ubiquinones and menaquinones

E. coli, together with other Gram-negative facultative aerobic bacteria, contains isoprenoid quinones of both the benzene and the naphthalene series. The major benzoquinone is ubiquinone-8 (with the figure referring to the number of prenyl units present in the side chain), whilst the naphthoquinones are either menaquinone (*me*thylated *na*phthoquinones) or demethylmenaquinone.

Ubiquinone and menaquinone both comprise a nucleus derived from chorismate, a prenyl side chain derived from prenylpyrophosphate and a nuclear methyl group derived from S-adenosylmethionine. In each case, a benzenoid aromatic acid is used as the framework on which the rest of the molecule is constructed.

chorismate ↗ *p*-hydroxybenzoate → ubiquinones
chorismate ↘ *o*-succinylbenzoate → menaquinones

The precursor of benzoquinones is *p*-hydroxybenzoic acid, which itself derives from chorismic acid[43]. The enzyme catalyzing the transformation is called chorismate (pyruvate) lyase and is coded by the *ubiC* gene (92 mn). The prenylation reaction, under control of the *ubiA* gene (92 mn), requires a polyprenylpyrophosphate and results in the release of pyrophosphate. The prenylpyrophosphate, with *n* isoprene units is synthesized through the reaction of dimethylallylpyrophosphate and (n–1) molecules of isopentenylpyrophosphate. The next step (*ubiD*, 86 mn) is a decarboxylation of the prenylated compound to the corresponding 2-polyprenylphenol. Then other hydroxylation reactions, utilizing molecular oxygen in aerobiosis or the oxygen atom of water under anaerobic conditions, and methylation reactions (where all the methyl groups are provided by S-adenosylmethionine) occur.

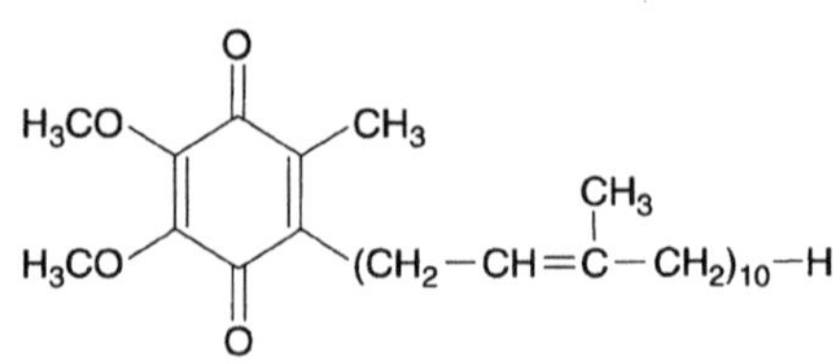

ubiquinone-10

43. See the chapter on the biosynthesis of the aromatic amino acids.

A mutant of *E. coli* unable to grow aerobically on non-fermentable substrates, but which grows anaerobically on fumarate or nitrate, has been isolated. The oxygen consumption by its cell-free extracts is low relative to that of an isogenic control strain, but was restored by adding ubiquinone-1 to the extracts. Transformation with a cloned fragment of chromosomal DNA restored both the cellular quinone content and the ability to grow on succinate. The gene (*ubiG*, 48 mn) responsible was identified to code for the enzyme 2-octaprenyl-3-methyl-5-hydroxy-6-methoxy-1,4-benzoquinone (OMHMB) methyltransferase and catalyzes the terminal step in the biosynthesis of ubiquinone-8. The sequence of the gene and that of the corresponding protein contain a motif which is found in other S-adenosylmethionine-dependent methyltransferases.

The menaquinone called vitamin K_2 plays an essential role in several electron transport systems by serving as the major electron carrier during anaerobic growth in *E. coli* and also mediates electron flow between dehydrogenases and cytochromes in *Bacillus subtilis,* a strict aerobe. Menaquinone is synthesized from a *men*-specific branch of the shikimate pathway, beginning with the isomerization of chorismate to isochorismate. The product of the *menCD* genes are necessary for the conversion of isochorismate to *o*-succinylbenzoate (OSB) a stable intermediate in the menaquinone biosynthetic pathway, which involves at least seven identified enzymes. The OSB synthase system catalyzes the synthesis of OSB from isochorismic acid and α-ketoglutaric acid in the presence of thiamine pyrophosphate (TPP) and Mg^{2+}. The succinyl moiety of OSB is generated by carbon atoms 2 to 5 of α-ketoglutarate, which is decarboxylated by α-ketoglutarate dehydrogenase, the adduct between the resulting succinic semialdehyde and TPP being the source of the succinyl CoA which condenses with isochorismic acid to yield OSB.

$$\alpha\text{-ketoglutarate} + \text{TPP} \rightarrow CO_2 + \text{COOH–}(CH_2)_2\text{–C–(OH)-TPP}$$

"active" succinic semialdehyde

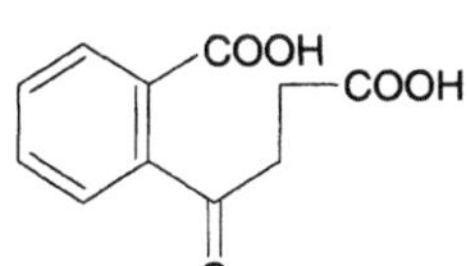

OSB

Another identified step is catalyzed by naphthoate synthase (coded by *menB*), and forms the bicyclic ring system by catalyzing the conversion of *o*-succinylbenzoyl coenzyme A to 1,4-dihydroxy-2-naphthoic acid (DHNA), the precursor of demethylmenaquinone (DMK) and menaquinone (MK).

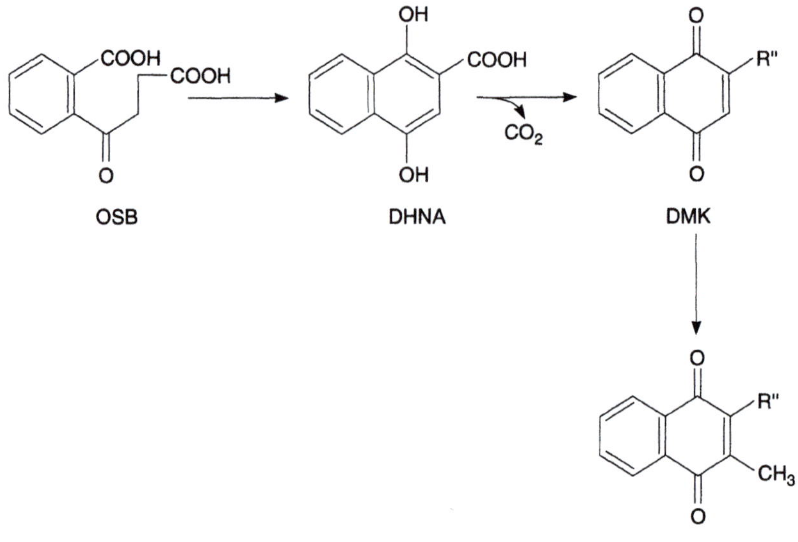

The synthesis of the coenzyme A derivative is due to *o*-succinylbenzoyl coenzyme A synthetase (coded by *menE*) which catalyzes the following reaction:

OSB + CoA + ATP → OSB-CoA + AMP + PPi

The overall conversion DHNA → DMK reaction requires at least two stages: removal of the DHNA carboxyl as CO_2, the attachment of the isoprenoid residue (R). A single enzyme, coded by *menA* (89 mn) apparently catalyzes the three reactions.

The side chain lengths of the menaquinones are probably determined by the availability of the isoprenyl pyrophosphate substrate(s) within the membranes. The prenylation occurs at the level of DHNA (fig. 1).

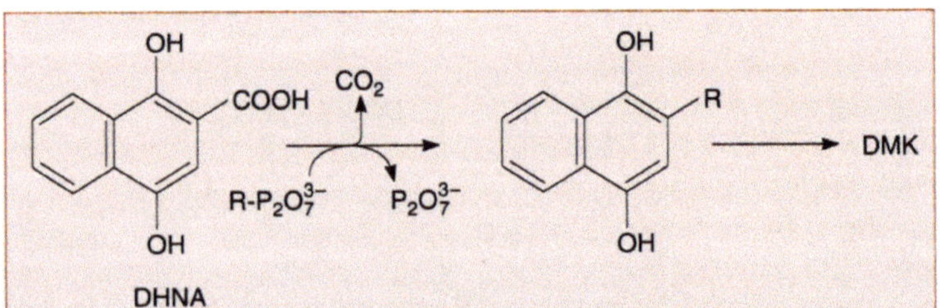

Figure 1. The prenylation of 1,4 dihydroxy-2-naphthoate occurs at the DHNA level. R = prenyl.

With the exception of *menA*, all the other identified genes encoding the menaquinone biosynthetic enzymes are clustered at 48.5 mn on the *E. coli* chromosome. These genes have been cloned and the sequence of two of them, including *menB*, has been reported.

SELECTED REFERENCE

Menaquinones

R. Bentley & R. Maganathan, *Microbiol. Revs.* **46,** 241-280 (1982).

CHAPTER XXVII

Biosynthesis of the tetrapyrrole ring system

The pathways for the biosynthesis of chlorophyll and respiratory pigments of the heme type make use of the same series of reactions up to the protoporphyrin stage. Judicious use of isotopes, of cell preparations catalyzing some of the reactions involved and degradation techniques specially devised for the task, have allowed the brilliant elucidation of the pathway by several groups of workers, among which David Shemin occupies the main place.

Synthesis of protoporphyrin

δ-aminolevulinic acid (ALA) is the central precursor of porphyrin biosynthesis. Eight molecules of ALA provide all the carbon and nitrogen atoms for the porphyrin skeleton of heme and chlorophyll. In some bacteria and in the mitochondria of yeast, birds, and mammals, ALA is formed along the so-called "Shemin route" by the condensation of glycine and succinyl CoA, catalyzed by ALA synthase (*hemA*, 27 mn):

$$COOH{-}(CH_2)_2{-}CO{-}SCoA + NH_2{-}CH_2{-}COOH \rightarrow$$
$$COOH{-}(CH_2)_2{-}CO{-}CH_2{-}NH_2 + CO_2 + CoA$$
ALA

In contrast, in a variety of photosynthetic and non-photosynthetic bacteria (including *E. coli*, *B. subtilis*, and archaebacteria), and in the chloroplasts of higher plants and algae, δ-aminolevulinic acid is formed in a two-step reaction utilizing the five-carbon skeleton of glutamic acid. Glutamate is first activated by an aminoacyl bond to t-RNA^{Glu} by a glutamyl-tRNA synthetase step, indistinguishable from that occurring in protein synthesis: glutamic acid is activated by ligation to glutamyl-$tRNA^{Glu}$ (UUC). An NADPH-dependent glutamyl-tRNA reductase then reduces the tRNA activated glutamic acid to glutamate-1-semialdehyde. Glutamate-1-semialdehyde-2,1-aminomutase (*hemL*) catalyzes a shift in the five-carbon skeleton to produce ALA:

$$HOOC{-}(CH_2)_2CH(NH_2){-}CHO \rightarrow HOOC{-}(CH_2)_2{-}CO{-}CH_2{-}NH_2$$
ALA

This is an interesting transformation, since it is one of the few reactions where a tRNA is utilized in a biosynthetic process other than protein biosynthesis.

ALA dehydratase (*hemB*, 8 mn) catalyzes a reaction between two molecules of δ-aminolevulinic acid to form the pyrrole derivative, porphobilinogen.

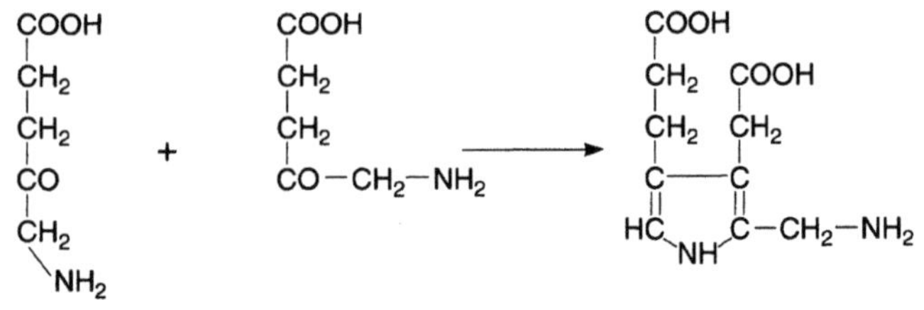

porphobilinogen

ALA dehydratase is composed of eight identical subunits and requires at least four zinc atoms. The first ALA molecule binds to the enzyme through a lysine group and ends up at one side of the final pyrrole, therefore forming the propionate side chain moiety, whereas the acetate side chain occupies the position near the reactive aminomethyl group.

The assembly of the next intermediate, uroporphyrinogen III, from porphobilinogen requires two enzymes, porphobilinogen deaminase – whose sequence of 313 residues has been deduced in *E. coli* from the nucleotide sequence of the corresponding gene (*hemC*, 85 mn) – , and uroporphyrinogen III synthase (also called cosynthase, *hemD*, 85 mn). Porphobilinogen deaminase (which should actually be called hydroxymethylbilane synthase, its product being the substrate for the cosynthase) first catalyzes the tetramerization of porphobilinogen to yield the linear 1-hydroxymethylbilane or pre-uroporphyrinogen III, which, in the presence of the cosynthase, is transformed by ring D inversion and cyclization to uroporphyrinogen III (in ring D, the relative positions of the propionate and acetate side chains are inversed, compared to preuroporphyrinogen; the mechanism of this inversion is a huge enigma for organic chemists, and remains to be solved).

The overall stoechiometry of the reaction is as follows:

$$4 \text{ porphobilinogen} \rightarrow \text{uroporphyrinogen III} + 4\ NH_3$$

The mechanism of hydroxymethylbilane synthase has been extensively analyzed. This enzyme contains a unique nucleophile prosthetic group, a dipyrromethane assembled from two molecules of the substrate porphobilinogen linked together. The enzyme itself is responsible for the assembly and attachment of its own cofactor to a conserved cysteine residue (Cys242). Substitution of this cysteine with serine results in a protein which has no catalytic activity, while a similar mutation of Cys99 to serine does not affect enzymatic activity. The sequences of the yeast *Euglena gracilis* and human enzymes are also known and there is a conserved cysteine residue in all four proteins corresponding to the *E. coli* Cys242 residue. This suggests that the mechanism of action of the enzyme from the other sources is similar to that of the *E. coli* enzyme.

The dipyrromethane cofactor acts as a primer onto which the four substrate molecules are attached sequentially. The enzyme catalyzes the deamination of each substrate, followed by carbon-carbon bond formation and deprotonation. Pre-uroporphyrinogen synthesis proceeds by attaching initially the first substrate molecule (ring A) to the free α position of the cofactor to form a stable enzyme-intermediate complex ES. Subsequent ordered addition of further substrate molecules B, C and D then yield ES2, ES3 and ES4. After the fourth substrate has been added, the enzyme catalyzes the release of pre-uroporphyrinogen by a hydrolysis reaction (fig. 1).

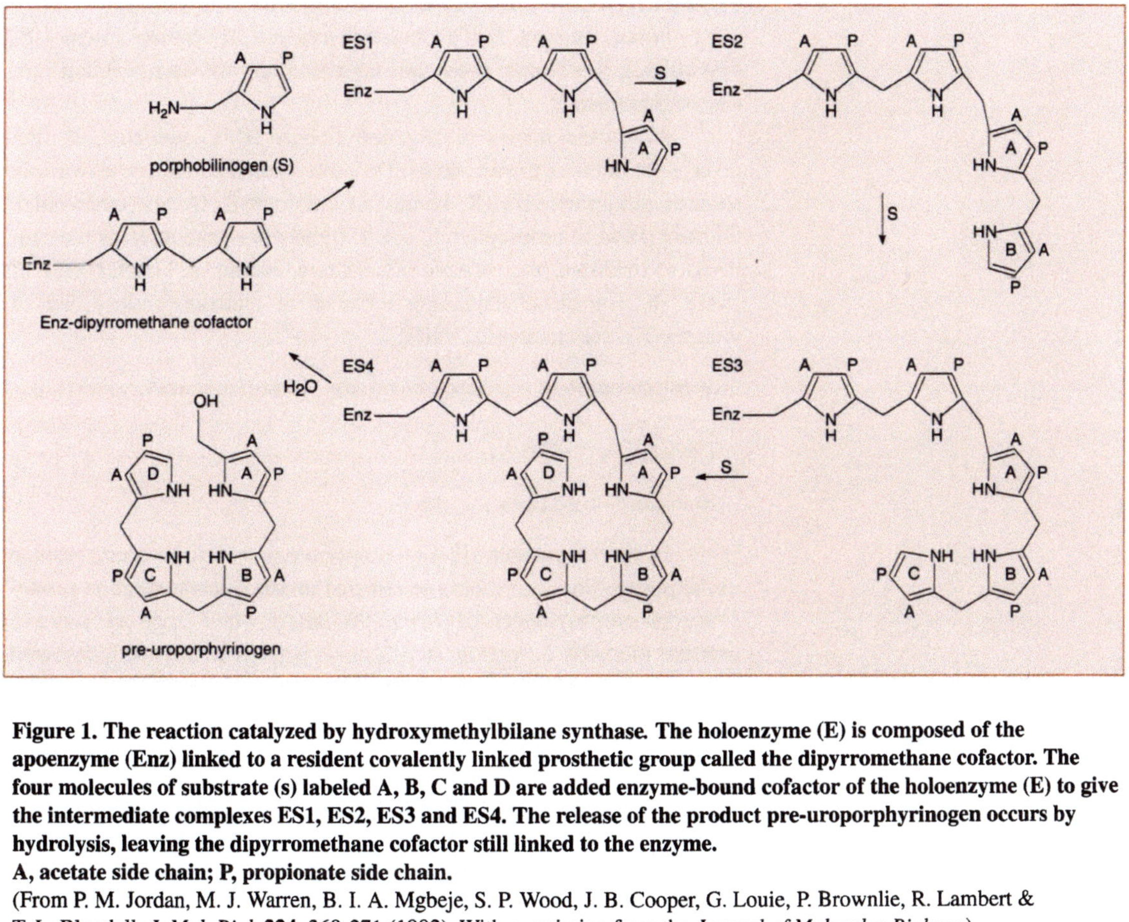

Figure 1. The reaction catalyzed by hydroxymethylbilane synthase. The holoenzyme (E) is composed of the apoenzyme (Enz) linked to a resident covalently linked prosthetic group called the dipyrromethane cofactor. The four molecules of substrate (s) labeled A, B, C and D are added enzyme-bound cofactor of the holoenzyme (E) to give the intermediate complexes ES1, ES2, ES3 and ES4. The release of the product pre-uroporphyrinogen occurs by hydrolysis, leaving the dipyrromethane cofactor still linked to the enzyme.
A, acetate side chain; P, propionate side chain.
(From P. M. Jordan, M. J. Warren, B. I. A. Mgbeje, S. P. Wood, J. B. Cooper, G. Louie, P. Brownlie, R. Lambert & T. L. Blundell, *J. Mol. Biol.* **224**, 269-271 (1992). With permission from the *Journal of Molecular Biology*.)

Evidence suggests that hydroxymethylbilane synthase catalyzes the repetitive deamination and condensation reactions at a single catalytic site. Hence, it must also possess the structural capability of manipulating the cofactor and the growing polymer through the catalytic machinery. This is reminiscent of the mobile 4'-phosphopantetheine arm encountered in the study of fatty acid biosynthesis. The recent crystallization of hydroxymethylbilane synthase from *E. coli* throws further light on this extraordinary enzyme. Its three-domain structure has been defined by X-ray analysis at 1.9 Å resolution. The two first domains structurally resemble the transferrins and the periplasmic binding proteins, while the dipyrromethane cofactor is covalently linked to the third domain and bound by extensive salt bridges and hydrogen bonds within the cleft between domains 1 and 2, at a position corresponding to the binding sites for small ligand molecules in the analogous proteins. The tridimensional structure, as well as results from site-directed mutagenesis, are compatible for a single catalytic site. Interdomain flexibility may aid elongation of the polypyrrole product in the active site cleft of the enzyme.

hemD, the gene for the uroporphyrinogen III synthase proper (the cosynthase), has been cloned and sequenced and the correponding protein overproduced.

After the synthesis of uroporphyrinogen III is completed, the first divergence in the pathway occurs. Decarboxylation commits the pathway towards protoporphyrin IX, heme and chlorophyll. On the other hand, C-methylation of uroporphyrinogen III directs the route towards the synthesis of siroheme (the cofactor of sulfite reductase, see chapter XX), of factor F_{430} (a nickel-containing cofactor of methanogenesis) and of vitamin B_{12} (see chapter XXVIII).

uroporphyrinogen III → protoporphyrinogen → heme (Fe) and chlorophyll (Mg)
↓
precorrin-2 → vitamin B_{12}
↓
siroheme (Fe) and factor F_{430} (Ni)

Uroporphyrinogen III is a tetrapyrrole in which the acetyl residue of the porphobilinogen molecule remains intact. Decarboxylation producing coproporphyrinogen III (*hemE*, 90 mn), in which these acetyls have become methyls, occurs in steps, since porphyrin intermediates with seven, six or five carboxyls have been detected.

An enzyme system (*hemF*, 17 mn), which in ox liver is firmly bound to mitochondria, carries out two oxidative decarboxylation reac-

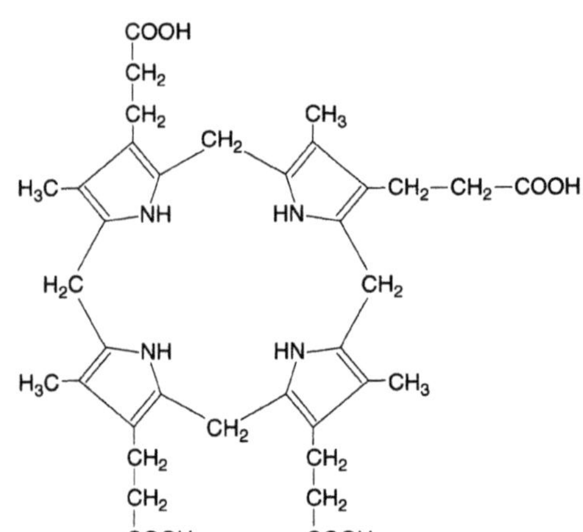

coproporphyrinogen III

tions, the effect of which is to transform two of the propionyl residues of coproporphyrinogen III to vinyl groups, forming protoporphyrinogen which is oxidized to protoporphyrin by protoporphyrinogen oxidase (*hemG*, 86 mn):

$$\text{coproporphyrinogen III} \xrightarrow[-4\,H]{-2\,CO_2} \text{protoporphyrinogen} \xrightarrow{-6\,H} \text{protoporphyrin IX}$$

The porphyrinogens are colorless substances. The color of the porphyrins is due to dehydrogenations which set up a system of conjugated double bonds, a chromophore with strong absorption bands in the near ultraviolet and visible regions of the spectrum.

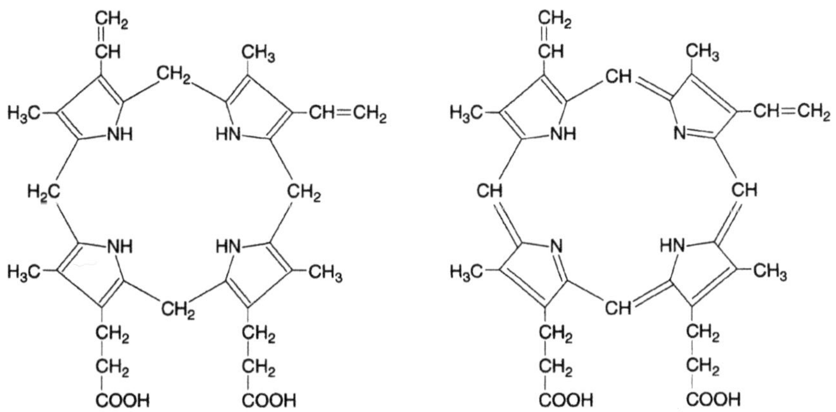

protoporphyrinogen — protoporphyrin IX

The Roman numerals following the names of different porphyrinogens and porphyrins distinguish the different possible isomers. For example, the methyl, vinyl and propionyl chains in protoporphyrin can be arranged to form fifteen different isomers. In fact, a single series of isomers exists in nature, the protoporphyrin IX series, which we shall refer to for now on as protoporphyrin. Other isomers have been isolated from natural sources, but they have no physiological role, arising from disorders in tetrapyrrole metabolism, or from spontaneous chemical reactions among the precursors of tetrapyrroles.

Synthesis of heme from protoporphyrin

The fact that natural complexes of iron and coproporphyrin can be isolated strongly suggests that the metal atom is incorporated into heme at the protoporphyrin stage. Further evidence comes from microorganisms like *Hemophilus influenzae,* where the growth requirement for heme can be satisfied by protoporphyrin. Although the ferrous ion spontaneously forms coordination complexes with protoporphyrin, mutants of *Staphylococcus aureus* requiring hemin are found which cannot grow on protoporphyrin. This suggested that the insertion of the iron atom was due to an enzyme, ferrochelatase. Mutants lacking this enzyme have also been described in *E. coli. (hemH*, 11 mn):

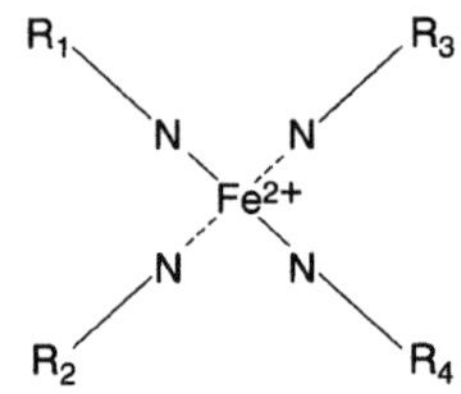

heme or ferrous protoporphyrin

The control of the biosynthesis of the tetrapyrrole ring system occurs at different levels.

Firstly, an iron deficiency in the bacterial growth medium stimulates a considerable build-up of porphyrins exceeding, in *Rhodopseudomonas spheroides* for example, one hundred times the amount of all tetrapyrrole derivatives (bacteriochlorophyll + heme derivatives) normally synthesized in non iron-deficient media.

This phenomenon has been analyzed in detail with cell-free extracts, and it has been shown that the ALA-synthase from this microorganism is very sensitive to hemin, the effect being appreciable even at a concentration of 10^{-7} M. This inhibition is non-competitive with respect to both substrates, succinyl CoA and glycine. Interpretation of the results obtained *in vivo* is as follows: the control of tetrapyrrole biosynthesis in *R. spheroides* occurs by feedback inhibition of ALA synthase by hemin, the formation of hemin itself depending on the concentration of iron in the medium.

The addition of ALA to cultures of *R. spheroides* also causes a severe repression of the synthesis of ALA synthase and ALA dehydratase. It has been suggested that in liver, heme regulates ALA synthase not by altering the rate of transcription of the coding gene, but by decreasing the half-life of the corresponding mRNA.

δ-aminolevulinate dehydratase and ferrochelatase are also feedback inhibited by heme.

R. spheroides, in which the regulation of heme synthesis has been the most carefully studied, is a photosynthetic organism which contains another tetrapyrrole, bacteriochlorophyll. This obviously adds a new parameter in the regulation of the synthesis of tetrapyrroles by the photosynthetic organisms.

Synthesis of chlorophyll from protoporphyrin

The biosynthesis of chlorophyll and heme pigments is a fundamental aspect of photosynthesis, and has been the subject of biochemical and molecular genetics studies for many years. The photosynthetic bacteria are an ideal model, since they are capable of chemoheterotrophic growth in the dark. Then, the synthesis of bacteriochlorophyll can be chemically or genetically disrupted and the properties of non-photosynthetic mutants examined in aerobic cultures grown in the dark. As a result of these studies, it has been possible to identify the series of intermediates in the biosynthesis if bacteriochlorophyll.

Tetrapyrrole biosynthesis proceeds from δ-aminolevulinic acid to protoporphyrin IX. At this point, insertion of Fe^{2+} or Mg^{2+} yields either heme or Mg-protoporphyrin, respectively; in the latter case, subsequent reactions encoded by special genes result in the eventual production of bacteriochlorophyll. Thus, the chelation of Fe^{2+} or Mg^{2+} is assumed to be the decisive step in this sequence (and an important control point) since it is here that the relative levels of the two tetrapyrroles may be determined. These levels vary widely depending on whether the cells are growing chemoheterotrophically with plentiful aeration, when most of the flux is directed toward heme biosynthesis, or whether they are growing under limited oxygen where the flux is largely directed towards bacteriochlorophyll. In the latter case, total tetrapyrrole production increases up to 100-fold, mainly because of the synthesis of high levels of ALA. Another regulatory problem posed by the versatility of *Rhodobacter sphaeroides* is the consequence of anaerobic photosynthetic growth. The decarboxylation of coproporphyrinogen III to form protoporphyrinogen requires molecular oxygen in animals, and in aerobically grown *R. spheroides*; it appears that anaerobically grown cells need to employ a different enzyme which exhibits requirements for Mg^{2+}, ATP, methionine and NAD^+ or $NADP^+$ for activity.

All the intermediates between protoporphyrin IX and bacteriochlorophyll contain magnesium. A mutant of *Chlorella* accumulates the simplest of these compounds, magnesium protoporphyrin, whose structure – except for the metal – is identical with that of heme. A gene has been characterized in the plant *Antirrhinum majus* which shows a high degree of homology to the products of two bacterial genes, *bchH* from *Rhodobacter capsulatus* which codes for an enzyme necessary for one of the first steps unique to the synthesis of bacteriochlorophyll, and *cobN* from *Pseudomonas denitrificans* (see chapter XXVIII) which is necessary for the chelation of cobalt during vitamin B_{12} biosynthesis. It has been proposed that the gene from *Antirrhinum majus* codes for the chelation of magnesium by protoporphyrin IX. We know little about the mechanism of insertion of the magnesium, except that certain mutants accumulate free protoporphyrin and apparently lack the enzyme required for the incorporation of the metal. Another compound which accumulates in another mutant is the monomethyl ester of the preceding one. The chromatophores of *R. spheroides* catalyze the reaction:

Mg-protoporphyrin + S-adenosylmethionine →
Mg-protoporphyrin monomethyl ester + S-adenosylhomocysteine

Examination of the structure of chlorophyll indicates which of the two residues of propionic acid is esterified. This step is the best studied at the enzymatic level between protoporphyrin and chlorophyll. Another mutant accumulates a third compound which can be derived from the preceding one only by a series of reactions. This is Mg-vinylpheoporphyrin a_5, where one of the vinyl groups has been reduced to an ethyl group, the monomethylester of the propionic acid residue has been oxidized and the modified chain cyclized to form a substituted cyclopentanone.

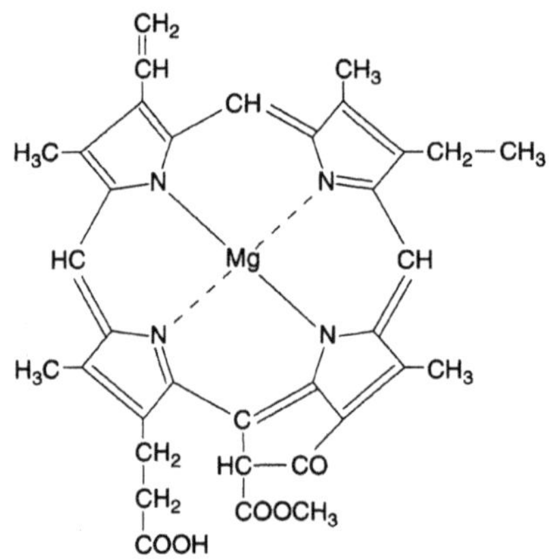

Mg vinylpheoporphyrin a_5
(protochlorophyll)

The reduction of this compound to chlorophyll involves reduction of the tetrapyrrole ring carrying the unchanged propionyl residue and an esterification of this residue by phytol, a C20 diterpene alcohol. The intermediates in these reactions probably never occur in the free state and take place in higher plants and in some algae in the presence of light only.

The structural formulas of chlorophyll *a*, bacteriochlorophyll and phytol are shown at the top of the next page.

Biosynthesis of the phycobilin chromophores. Chromatic adaptation

Cyanobacteria and red algae often grow at about one meter under the surface of water where the light absorbed by chlorophyll has already been absorbed by chlorophyll or by organisms living above.

The phycobiliproteins are their major light-harvesting pigments and we shall now examine some of their properties.

They absorb light in a broad region near the middle of the visible

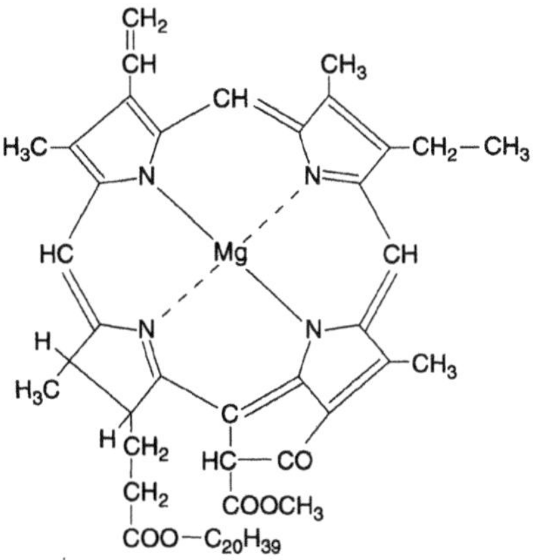

chlorophyll *a*

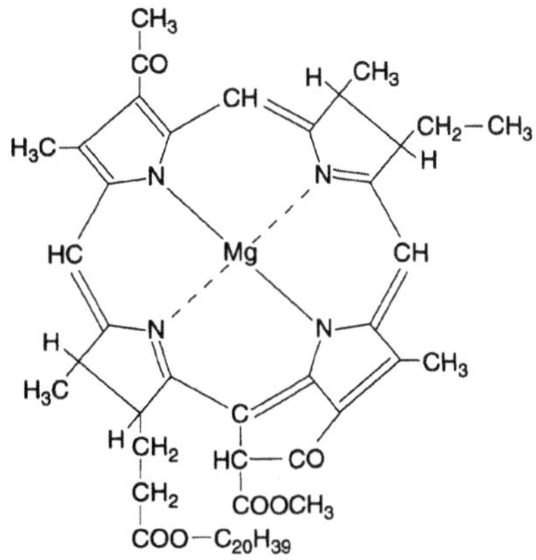

bacteriochlorophyll

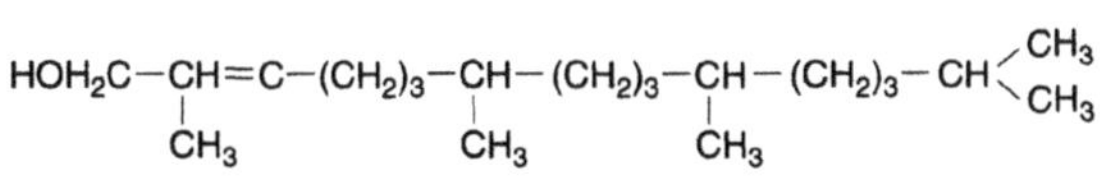

phytol $C_{20}H_{40}$

spectrum, which reaches the depth at which the organisms thrive, and belong to three different spectral classes. Two blue pigments, allophycocyanin and phycocyanin, which have maxima at relatively long wavelengths, occur universally in cyanobacteria and red algae; the red pigment, phycoerythrin, which absorbs at shorter wavelengths, is formed by some, but not all, members of each group. The phycobiliproteins are contained in granules called phycobilisomes which occur in regular arrays on the outer faces of the thylacoids, a system of flattened membranous sacs, near the surface region of the membrane. The overall design of the phycobilisome consists of a hemidisk of six rods radiating fan-like from a central core. Phycobilisomes are made of three principal classes of chromophoric water-soluble proteins (phycobiliproteins): allophycocyanin in the core, and phycocyanin and phycoerythrin in the rods, all composed of α and β subunits. All these chains derive from a unique common ancestor, structurally similar and evolutionarily related to the globin family. Open tetrapyrrole chromophores, the phycobilins, are post-translationally grafted to phylogenetically conserved cysteines on both subunits. Phycocyanobilin is the chromophore of phycocyanin and allophycocyanin; phycoerythrobilin is the chromophore of phycoerythrin (fig. 2).

In addition, phycobilisomes contain auxiliary proteins devoid of chromophore, called the linker polypeptides.

The light energy absorbed by these pigments is transferred with very high efficiency to the chlorophyll-containing photosynthetic reaction centers in the thylacoids. The geometrical arrangement of phycobiliproteins in phycobilisomes as well as their spectral properties (overlap-

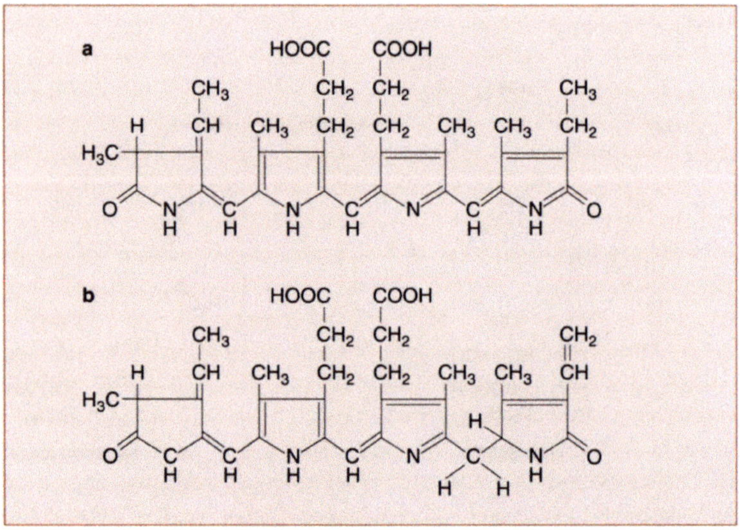

Figure 2. Structures of the chromophores of phycobiliproteins. a, phycocyanobilin, the chromophore of phycocyanin and allophycocyanin; b, phycoerythrobilin, the chromophore of phycoerythrin. Both are covalently linked to the proteins with which they are associated.

ping of the spectra of the energy acceptor and of the energy donor) are such that the overall efficiency of energy transfer from one phycobiliprotein to the other and then to the chlorophyll is very high.

Phycocyanobilin arises from the degradation of protoheme *via* biliverdin IXa. This has been studied in the unicellular red alga, *Cyanidium caldarum*:

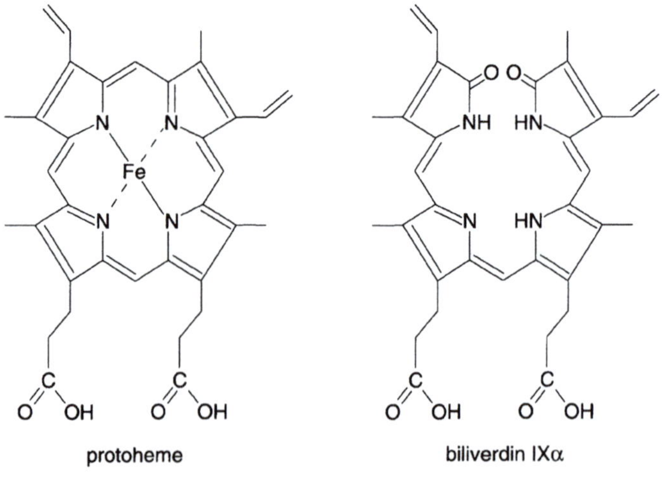

The *in vitro* transformation of protoheme to biliverdin IXα and of the latter to phycobilins requires NADPH, ferredoxin and ferredoxin-$NADP^+$ reductase, as well as heme oxygenase (required for the opening of the tetrapyrrole ring) and phycobilin formation enzymes. The only role of NADPH in both reactions of phycobilin synthesis is to reduce ferredoxin *via* ferredoxin-$NADP^+$ reductase: reduced ferredoxin can directly supply the electrons needed to drive both steps in the transformation of protoheme to phycocyanobilin.

Studies using interposon mutagenesis with *Synechococcus*, a cyanobacterium, suggest that a phycocyanobilin lyase, composed of two different subunits (coded by the *cpcE* and *cpcF* genes of this organism), is specifically required for chromophorylation of a subunit of apophycocyanin. The phycocyanobilin is covalently attached by a thioether linkage to a given cysteine (α-Cys84) of the apoprotein.

In a natural environment and regardless of the geographical location, a great many physico-chemical and biological factors result in quantitative and qualitative variations of the incident solar radiations. This is due, in particular, to the diffusion and absorption properties of the molecules present in the atmosphere, to the circadian rhythms, to the alternance of seasons, to meteorological conditions, and to the vegetation. In an aquatic medium, where cyanobacteria thrive, the selective absorption by water molecules of radiation of a wave length higher than 550 nm and the turbidity due to the presence of organic molecules in suspension, modify considerably the transmission spectrum of the incident light. In most cases, beyond a depth of four meters, the light intensity is greatly diminished and the wave lengths higher than 600 nm are practically absent.

Although it is artificial to dissociate the effect of light intensity from the spectral qualities of the incident light, numerous studies have identified two different responses of the photosynthetic apparatus to the spectral quality of light, namely *inverse chromatic adaptation* and *complementary chromatic adaptation*. The first type concerns all photoautotrophic organisms, whereas the second has been shown to occur only in cyanobacteria.

Inverse chromatic adaptation is characterized by a preferential diminution of chlorophyll *a* in cells grown in red light, *i.e.*, in a region of the visible spectrum mostly absorbed by this pigment. Inversely, in yellow or orange light, mainly absorbed by phycobiliproteins, the synthesis of chlorophyll *a* increases.

As early as 1883, Engelmann had reported a correlation between the distribution of cyanobacteria in a water column and their capacity to synthesize pigments complementary of the spectral quality of the incident light. These observations were extended at the beginning of the century to several cyanobacterial species and it was established that the synthesis of the red pigment, phycoerythrin, was predominant when the organisms were grown in green light whereas the synthesis of the blue pigment, phycocyanin, was predominant when the organisms were grown in red light. Interest in this complementary chromatic adaptation was revived in the 1960s and the following conclusions may be drawn from the various studies performed.

a) Phycoerythrin and phycocyanin are relatively stable proteins which disappear by dilution during growth, when cells are transferred from red light to green light or *vice versa.*

b) When grown in the dark, cells have a pigment composition identical to that of cells grown in red light, *i.e.,* they synthesize only phycocyanin and stop synthesizing phycoerythrin.

c) Only those species of cyanobacteria that have the genetic information to synthesize the two types of phycocyanin (of similar spectral properties) can modulate their content in this pigment in response to the shift from one type of light to the other; in this case, phycocyanin-1 is synthesized whatever the quality of light, while phycocyanin-2 is synthesized only in red light.

d) The synthesis of the linker proteins is coordinately regulated with the synthesis of the phycobiliproteins with which they are associated in the phycobilisome.

e) The light signal is perceived by a photoreversible pigment (PRP) analogous to the plant phytochrome, and whose action spectrum shows two maxima at 540 and 640 nm. The two forms of PRP are interconvertible by red or green illuminations of a few minutes. In this purely photochemical process, only the last illumination received determines the type of phycobiliprotein which will be synthesized.

It has been established that the control of pigment synthesis occurs at the transcriptional level and that the induction of the operons leading to the synthesis of the corresponding pigments in green or red light requires protein synthesis.

These studies are presently in full development.

A new type of chromatic adaptation under conditions of sulfur starvation

According to the biotope, the availability of sulfur can vary considerably (from about 100 μM in freshwater lakes and 0.2 mM in eutrophic lakes to 27 mM in sea water). In hot sulfur springs, H_2S concentration can reach 2 mM. Within the same biotope (except in sea water), sulfur concentrations can vary importantly. Many cyanobacteria have adapted to such ecological niches. In addition to the capacity of certain strains to utilize indifferently H_2O or H_2S, and to be thus able to grow anaerobically or aerobically, cyanobacteria respond in a variety of manners to sulfur starvation.

Three phycocyanin operons have been characterized in a species of *Calothrix.* The operon *cpc1*, which encodes the two phycocyanin 1 subunits, is expressed regardless of the light spectrum, whereas transcription of *cpc2*, which encodes the two phycocyanin 2 subunits and their three associated linker polypeptides, is triggered by the red wavelength component. In addition, a third operon, *cpc3* is transcribed at a very low level, if any.

The *cpc3* operon, which remains silent in cells grown under standard conditions, is induced by sulfur deprivation. It includes five open reading frames correponding to the five genes of the *cpc2* operon. They code for the α and β subunits of a novel phycocyanin and three linker polypeptides respectively. Cells grown in white light, in a medium containing ample sulfate and therefore brown, turned green when they were inoculated in sulfur-poor medium. Spectral analysis revealed the disappearance of phycoerythrin and a concomitant increase of the phycocyanin absorption peak. The green cells could be indefinitely subcultured in this medium, but this could be reversed if ample sulfur was provided. Messenger RNA hybridization with specific DNA probes showed that the *cpc3* operon was highly expressed upon sulfur limitation while the expression of *cpc1* and *cpc2* was switched off.

The peculiarity of the newly transcribed operon is the absence of codons specifying sulfur-containing amino acids, except for the cysteine codons corresponding to the attachment of the phycobilin chromophores on the α and β subunits of phycocyanins and for the five methionine codons necessary for the initiation of translation of the five genes. In addition the N-terminal methionine residues are eliminated in the mature polypeptides. As a result, the mature proteins coded by the *cpc3* operon are completely devoid of methionine.

SELECTED REFERENCES

The tetrapyrroles

S. Granick & S. I. Beale, *Adv. Enzymology* **46**, 33-203 (654 references) (1978).

P. M. Jordan, M. J. Warren, B. I. A. Mgbeje, S. P. Wood, J. B. Cooper, G. Louie, P. Brownlie, R. Lambert & T. L. Blundell, *J. Mol. Biol.* **224**, 269-271 (1992).

Biosynthesis of tetrapyrroles, New Comprehensive Biochemistry **19**, P. M. Jordan ed. 310 pp. Elsevier Science Publishers, New York and Amsterdam (1991).

ALA synthesis

D. Jahn, E. Verkamp & D. Söll, *Trends in Biochem. Sci.* **17**, 215-218 (1992).

Phycobilins

A. N. Glazer, *Ann. Rev. Biochem.* **52**, 125-157 (1983).

G. Rhie & S. I. Beale, *J. Biol. Chem.* **267**, 16088-16093 (1992).

J. Zhou, G. E. Gasparich, V. L. Stirewalt, R. de Lorimier & D. A. Bryant, *J. Biol. Chem.* **267**, 16138-16145 (1992).

Complementary chromatic adaptation

N. Tandeau de Marsac, Chromatic adaptation by cyanobacteria, in *The Molecular Biology of Plastids and Mitochondria,* L. Bogorad & I. K. Vasil eds. Academic Press, New York (1991).

D. Mazel & P. Marlière, *Nature* **341**, 245-248 (1989).

CHAPTER XXVIII

Biosynthesis of cobalamins including vitamin B_{12}

Synthesis of hydrogenobyrinic acid

Vitamin B_{12} can be formally classified with the tetrapyrrole derivatives. It is based on a skeleton called corrin (fig. 1), which differs from the other tetrapyrroles in the absence of a methene bridge between two of the pyrrole rings. In numbering the atoms in this formula, number 20 has been deliberately omitted, in order that all the corrin atoms should correspond to those of the other tetrapyrrole compounds. Some of the biosynthetic intermediates occurring before the formation of hydrogenobyrinic acid are called precorrins.

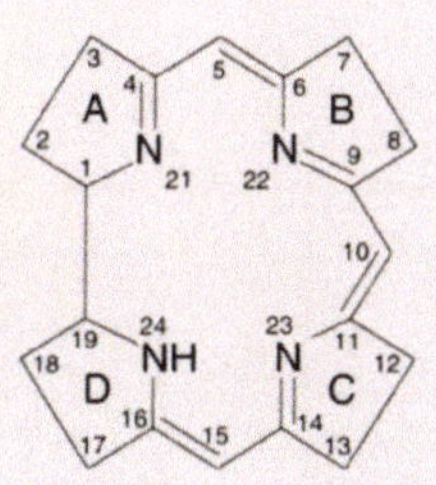

Figure 1. The corrin skeleton.

The structural formula of vitamin B_{12} (or rather of its coenzyme form) is given in figure 2.

The formula shows a centrally bound cobalt atom, coordinated on the one hand by four nitrogen atoms from the tetrapyrrole ring, and on the other, by one of the nitrogens of a benzimidazole derivative. The sixth coordination position is the only known case in biology of an organometallic metal-carbon bond, where the substituent can be a methyl group, or a deoxyadenosyl group (Ado) in the case of the coenzyme forms of vitamin B_{12}. However, the extraction from natural sources usually yields the vitamin in the form of cyanocobalamin, where the substituent is –CN, which can be replaced by –OH (aqua derivative), –Cl, $-NO_2$ or –CNS.

Detailed work has shown that the first steps in the biosynthesis of the vitamin are the same as for hemin and chlorophyll. Glycine, δ-aminolevulinic acid, porphobilinogen and uroporphyrinogen III are readily identified intermediates (see the section on the biosynthesis of the tetrapyrrole ring).

Many steps for the conversion of uroporphyrinogen III into cobyrinic acid remain to be studied in detail, but they include at least eight C-methylations (the methyl groups coming from S-adenosylmethionine), elimination of C20, NADPH-dependent reduction of the macrocycle, decarboxylation of the C12 acetate, methyl migration from C11 to C12, and cobalt insertion. In *Pseudomonas denitrificans*, where a thorough genetic analysis has been made, about 20 genes (the *cob* genes) have been cloned, grouped in four genomic loci corresponding to four distinct complementation groups (from A to D, table 1).

Known precorrins are precorrin-3, which carries three methyl groups at carbons 2 and 7 and 20; precorrin-6x, which carries five methyl groups at C1, C2, C7, C11 and C17, has a contracted macrocycle, still

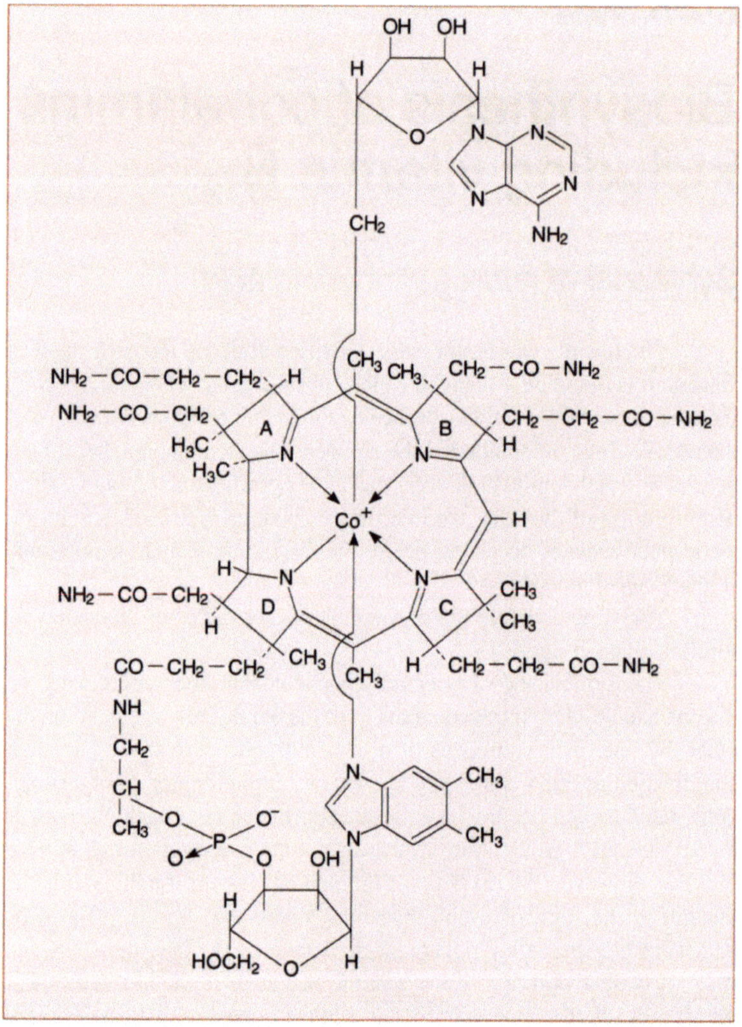

Figure 2.
5'-deoxyadenosylcobalamin.

carries the C12 acetic acid side chain, and stands at the oxidation level of a dehydrocorrin. The structure of precorrin-6x implies that it has undergone ring contraction and extrusion of C20, and that it has to be reduced to the oxidation level of a corrin (precorrin-6y), followed by further methylations at C5 and C15, decarboxylation of the acetic residues at C12 to yield precorrin-8x and methyl migration to C12, leading to hydrogenobyrinic acid. Cobalamin biosynthesis is considered to require between 20 and 30 steps to transform uroporphyrinogen III into cobalamin.

A nice parallel between the arrangement of the *cob* genes and the part of the pathway on which the gene products act can be drawn; it is not known whether this correlation results in common regulation of the expression of genes belonging to the same cluster. It is striking that in *Salmonella typhimurium*, *cob* genes are also organized into three classes

Table 1: Complementation groups of the genes involved in coenzyme B_{12} biosynthesis in *Pseudomonas denitrificans*.

Complementation group A
Methylation at C20, *cobI* (precorrin-2 → precorrin-3)
Methylations at C17, C11 and C1, *cobF* (unidentified enzymes leading from precorrin-3 to precorrin-6x)
Elimination of C20, *cobG*
NADPH-dependent reduction of precorrin-6x to precorrin-6y (transfer of a hydride equivalent to C-19), *cobK*
Methylations at C15 and C5, decarboxylation at C12, precorrin-6y → precorrin-8x, *cobL*
Mutase catalyzing the methyl migration from C11 to C12, precorrin-8x → hydrogenocobyrinic acid, *cob H*

Complementation group B
Insertion of cobalt, cobaltochelatase, *cobST*

Complementation group C
Methylations at C2 and C7, uroporphyrinogen III → precorrin-2, *cobA*
a,c-amidations of hydrogenobyrinic acid, *cobB*
Insertion of cobalt, cobaltochelatase, *cobN*
Cob (I)alamin adenosyltransferase, adenosylation of the *a,c*-diamide of cobyrinic acid, *cobO*
b, d, e, and *g* amidations, *cobQ*
Enzyme system catalyzing the addition of 1-amino-2-propanol to yield cobinamide, *cobC* and *cobD*
Bifunctional cobinamide kinase-cobinamide phosphate guanylyltransferase, yielding GDP-cobinamide, *cobP*
System possibly involved in the reduction of cobalt leading to Co (I) corrinoids, *cobW*[44]

Complementation group D
Nicotinate mononucleotide: 5,6-dimethylbenzimidazolephosphoribosyltransferase, *cobU*
Incorporation of 5, 6-dimethylbenzimidazole-ribose-5'-phosphate, *cobV*

related to three parts of the pathway, such a genetic clustering having been also reported in *Bacillus megaterium*.

Only six methyltransferases are required to perform the eight methylation reactions in *Ps. denitrificans*.

The first methylating enzyme, S-adenosylmethionine: uroporphyrinogen III methyltransferase has been purified to homogeneity. It is a homodimer (2 x 30,000 kDa) and performs the methylations at C2 and C7, leading to precorrin-2. The next reaction, catalyzed by S-adenosylmethionine: precorrin-2 methyltransferase, also obtained as a homogeneous homodimeric protein (2 x 26,000 kDa), is a methylation at C20 on precorrin-2 that generates precorrin-3.

Recently, the nature of the reactions catalyzed by enzymes coded by *cobK*, *cobL* and *cobH* have been elucidated. *cobK* codes for a NADPH-dependent precorrin-6x reductase transforming precorrin-6x to the dihydroderivative precorrin-6y. This reductase has been purified to homogeneity and is a monomer of 263 amino acid residues. The product of *cobL* (subunit of 42,900 kDa, oligomeric structure unknown) methylates at C5 and C15 and decarboxylates the acetic acid side chain at C12, yielding precorrin-8x.

44. The cobalt atom can exist under a + 1, + 2, or a + 3 oxidation state. The Co(III) state is reduced to Co(II) by a flavoprotein reductase, and then to Co(I) by another flavoprotein reductase, NADH being the reductant in both cases.

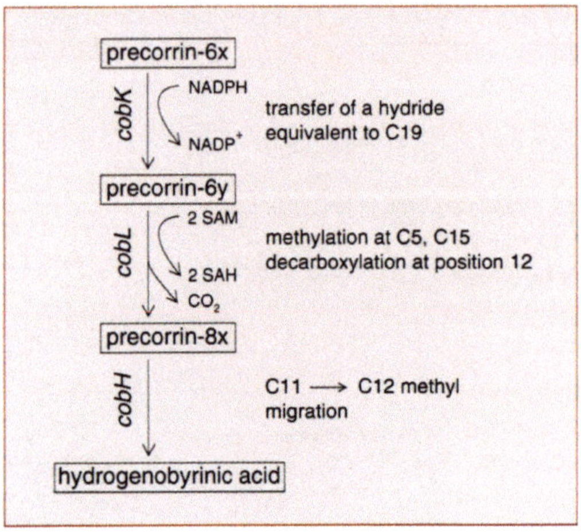

Figure 3. Biosynthetic sequence from precorrin-6x to hydrogenobyrinic acid in *Pseudomonas denitrificans*. SAH, S-adenosylhomocysteine.
(From D. Thibaut, M. Couder, A. Famechon, L. Debussche, B. Cameron, J. Crouzet & F. Blanche, *J. Bacteriol.* **174**, 1043-1049 (1992). With permission from the American Society for Microbiology.)

A mutase, purified 80-fold, coded by *cobH* transfers a methyl group from C11 to C12 and transforms precorrin-8x to hydrogenocobyrinic acid, the cobalt-free counterpart of cobyrinic acid. Figures 3 and 4 summarize the sequence of events and give the structure of cobyrinic acid, the cobalt-containing counterpart of hydrogenobyrinic acid.

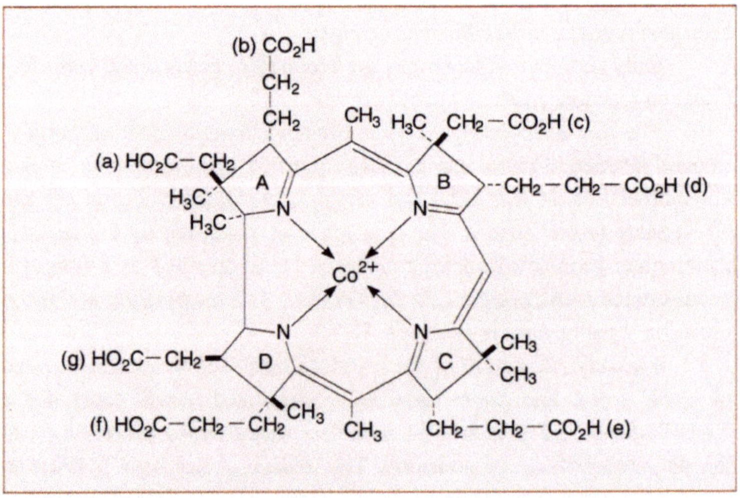

Figure 4. Cobyrinic acid. This acid was first isolated from cultures of *Propionibacterium shermanii*. If we consider that corrin is in the plane of this page, the boldface lines connect substituants above the plane, the dotted lines link substituants below the plane to the corrin nucleus. Hydrogenobyrinic acid is the cobalt-free counterpart of cobyrinic acid. The hexaamide corresponding to the six acid groups is called cobyric acid. When 1-amino-2-propanol is added to cobyric acid, cobinamide is obtained.

Cobinamide biosynthesis

The biosynthesis of cobinamide from hydrogenobyrinic acid involves six stepwise amidations of the peripheral carboxyl groups at positions *a, b, c, d, e* and *g* to yield the corresponding amides and the attachment of (*R*) 1-amino-2-propanol at position *f* (fig. 4).

A number of partially amidated corrinoids have been isolated from cultures of microorganisms or transformed into complete corrinoids during feeding experiments, to discover the sequence for attachment of unsubstituted amide groups and of aminopropanol.

The major corrinoids existing in nature apparently exist in the coenzyme form. The 5'-deoxy 5'-adenosyl (Ado) axial upper ligand of cobalt is presumably introduced after the completion of the corrin ring, but certainly soon after the cobyrinic acid stage, since this compound has been isolated only in the diaqua form, where the two axial ligands of cobalt are water molecules. Co (I) attacks the 5' carbon of ATP and displaces the triphosphate group. S-adenosylmethionine synthetase provides the only other biochemical reaction where a nucleophile displaces as a unit the triphosphate group of ATP.

The first amidation activity along the cobalamin pathway from cobyrinic acid has been attributed to an enzyme purified to homogeneity which catalyzes the stepwise amidations to the *c*-monoamide and to the *a,c*-diamide, called cobyrinic acid *a,c*-diamide synthase. This enzyme is encoded by *cobB* (complementation group C).

The single enzyme (encoded by *cobQ,* complementation group C) that catalyzes amidation of 5'-deoxy-5'-adenosyl cobyrinic acid *a,c*-diamide was purified to homogeneity from extracts of a recombinant strain of *Pseudomonas denitrificans.* Its molecular weight has been estimated by gel filtration to be 97,300 and by gel electrophoresis under denaturing conditions of 57,000. Since stepwise Edman degradation provides a single residue at each step of degradation during sixteen degradation steps, this suggests the enzyme is a homodimer. The enzyme acts on the coenzyme form of substrate only and uses glutamine as the amide donor. This cobyric acid synthase is thus a glutamine amidotransferase, and shares the properties of the enzymes of this group: sensitivity to oxidation, requirement for a reducing agent and replaceability of glutamine by ammonia. It catalyzes the four stepwise amidations of carboxyl groups at positions *b, d, e,* and *g* to yield cobyric acid. The specific order in which the four amidations were carried out were not determined although the tri-, tetra- and penta-amides were clearly characterized.

A series of recent studies suggests that cobalt insertion occurs at the level of hydrogenobyrinic acid *a,c*-diamide in *Ps. denitrificans.* The supporting evidence is the following: a) substantial quantities of hydrogenobyrinic acid, hydrogenobyrinic acid *a,c*-diamide and cobyrinic acid *a,c*-diamide accumulate in cobalamin producing cultures whereas cobyrinic acid has never been detected in these cultures. b) hydrogenobyrinic acid is a much better substrate than cobyrinic acid for the cobyrinic acid *a,c*-diamide synthase (the enzyme specified by *cobB*). c) some *cobB* mutants, which are devoid of the *a,c*-diamide synthase, do not accumulate cobyrinic acid, still synthesize hydrogenobyrinic acid but not the diamide. The enzyme system catalyzing the insertion of cobalt, cobaltochelatase, is complex and consists of two components, respectively of 140 kDa (coded by *cobN*) and 450 kDa. The latter is composed of two polypeptides coded by *cobS* and *cobT*. Each component is inactive by itself, but the cobaltochelatase activity is reconstituted upon mixing the products of *cobN* and *cobST*. The reaction is ATP-dependent and starting from the diamide of hydrogenobyrinic acid, the product has been identified as being the diamide of cobyrinic acid. Insertion of the Ado group takes place at the stage of cobyrinic acid *a,c*-diamide in *Ps. denitrificans*, based on two arguments: 1) cobalamin adenosyltransferase purified to homogeneity from this organism, a homodimer of 2 x 28,000 M_r, does not adenosylate cobyrinic acid, but adenosylates all of the corrinoids where at least positions *a* and *c* are amidated; 2) the *cobO* (complementation group C) mutants, which lack the enzyme, accumulate the diamide in the growth medium.

It has been claimed that in *Propionibacterium shermanii* the incorporation of aminopropanol (deriving from the decarboxylation of threonine) can occur at any of the amidated derivatives from the diamide to the pentaamide. However, in *Ps. denitrificans*, since no partially amidated corrinoids with the aminopropanol have been detected, it is highly probable that an enzyme system (*cobC* and *D*, complementation group C) inserts this residue at the level of cobyric acid (the hexaamide of cobyrinic acid), to finally yield the cobinamide.

Cobinamide kinase and cobinamide phosphate guanylyltransferase activities are carried by a bifunctional enzyme.

The biosynthesis of cobalamin from cobinamide is a four-step process in *P. shermanii*. The first step is the phosphorylation of the hydroxyl group of the aminopropanol residue of cobinamide, to yield cobinamide phosphate which is then transformed into GDP-cobinamide through the addition of a GMP moiety originating from a molecule of GTP.

$$\text{cobinamide + ATP} \rightarrow \text{cobinamide phosphate + ADP}$$
$$\text{cobinamide phosphate + GTP} \rightarrow \text{GDP-cobinamide + PPi}$$

In vitro enzymatic synthesis of cobinamide phosphate by cobinamide kinase and of GDP-cobinamide has been demonstrated with crude protein extracts of cobalamin-producing microorganisms (*P. shermanii*, *Propionibacterium arabinosum* and *Clostridium tetanomorphum*) but it is only recently that a protein endowed with the two activities has been obtained in the pure state from a recombinant strain of *Ps. denitrificans*. When subjected to Edman degradation, the enzyme showed only one sequence for the ten first residues at the N-terminus, sequence which was identical to the one inferred from the nucleotide sequence of the corresponding gene, *cobP* (complementation group C). Moreover, the estimated molecular weight of the denatured protein is identical to the one inferred from the DNA sequence.

Labeling experiments showed that the phosphate group of cobinamide phosphate is quantitatively retained as the inner phosphate of GDP-cobinamide during the guanylyltransferase reaction. However, the data are not sufficient to determine whether the bifunctional enzyme possesses one or two catalytic sites.

From GDP-cobinamide to cobalamin

The introduction of the nucleotide characteristic of vitamin B_{12} occurs at the GDP-cobinamide level. This nucleotide contains 5,6-dimethylbenzimidazole as the base. The details of the biosynthesis of this base are not known in detail, although in certain *Streptomyces* and *Propionibacteria*, it has been reported that all carbon and nitrogen atoms derive from riboflavin, presumably *via* an enzyme-catalyzed conversion of FMN involving the action of an oxygenase. In anaerobic microorganisms, riboflavin is not the precursor of dimethylbenzimidazole and it has been reported that glycine, as well as methionine and the four carbons of erythrose contribute to its synthesis.

An enzyme called nicotinate nucleotide: dimethylbenzimidazole phosphoribosyltransferase – first isolated from *P. shermanii* – catalyzes the following reaction:

5,6 dimethylbenzimidazole + nicotinate mononucleotide

↓

1-α-D-5'-phosphoribofuranosyl-5,6-dimethylbenzimidazole + nicotinate

The characteristic product of the reaction has received the short convenient name of α-ribazole-5'-phosphate. It is coupled to Ado-GDP-cobinamide by an enzyme, cobalamin-5'-phosphate synthase, which requires Mg^{2+} for its activity and has been partially purified from *Ps. denitrificans.* It is coded in this organism by the gene *cobV* (complementation group D). The products of the reaction are cobalamin-5'-phosphate and GMP. Finally a phosphatase removes the phosphate group to produce cobalamin, the coenzyme form of vitamin B_{12}. The enzyme is not specific for the coenzyme form or GDP-cobinamide and could use monocyano- or diaquacobinamide equally well. It can use also α-ribazole as a substrate although less well than α-ribazole phosphate. With the nucleoside as substrate, the reaction product is indeed coenzyme B_{12}.

The reaction sequence in the intermediate stages of the coenzyme B_{12} pathway in *Ps. denitrificans* can now be written:

precorrin-2 → precorrin-3 → → ... precorrin-6x → precorrin-6y → precorrin-8x → hydrogenobyrinic acid → hydrogenobyrinic acid *a,c*-diamide → cob (II)yrinic acid *a,c*-diamide → cob (I)yrinic acid *a,c*-diamide → adenosyl cobyrinic acid *a,c*-diamide → → ...adenosylcobalamin

The very recent characterization of the intermediates between precorrin-3 and precorrin-6x and of the corresponding enzymes coded by *cobG*, *cobJ*, *cobM*, and *cobF* complete the elucidation of the cobalamin pathway in *Ps. denitrificans.*

The time of cobalt insertion in cobyrinic acid-producing bacteria such as *P. shermanii* apparently differs from the one just described. It appears that precorrin-3 is synthesized and converted to the cobalt complex, indicating an early chelation.

Methylcobalamin is involved in methionine biosynthesis and in the methylations of tRNAs, whereas 5'-deoxycobalamin is the coenzyme of various isomerization reactions (glutamate mutase, L-methylmalonyl-CoA mutase, diol dehydratase) and of ribonucleotide triphosphate reductase in *Lactobacillus leichmanii*. In methane production by methanogenic bacteria, the methylcobalamin contains 5-hydroxybenzimidazole instead of dimethylbenzimidazole.

Vitamin B_{12} coenzymes are also involved in the degradation of L-lysine by some species of *Clostridia.*

SELECTED REFERENCES

R. M. Jeter & J. R. Roth, *J. Bacteriol.* **169**, 3189-3198 (1989)

J. Crouzet, F. Blanche, B. Cameron, D. Thibaut & L. Debussche, in *Chemical Aspects of Enzyme Biotechnology*, pp.299-315, T. O. Baldwin *et al.* eds. Plenum Press, New York (1990)

D. Thibaut, M. Couder, A. Famechon, L. Debussche, B. Cameron, J. Crouzet & F. Blanche, *J. Bacteriol.* **174**, 1043-1049 (1992)

Many references to previous work will be found in these last two papers.

CHAPTER XXIX

Interactions between proteins and DNA

DNA-binding proteins

Transcriptional control is the result of a complex cross-talk be-tween regulatory factors and the basic transcription machinery. In addition to the simple regulation by the mere availability of regulatory factors, transcription has been shown, both in procaryotes and in eucaryotes, to be modulated in several instances by posttranslational events, such as covalent modification (see the regulation of transcription of glutamine synthetase, chapter XIX), or redox control.

Regulation of transcription occurs, both in prokaryotes and in eukaryotes, largely through the binding of specific proteins, repressors or activators, that bind to specific DNA sequences. The source of the specificity is recognition by some of the amino acid side chains of the pattern of functional groups on the edges of the base pairs in the DNA major groove, mediated by hydrogen bonding. There is also a contribution from the sequence-dependent conformational preferences of the DNA backbone.

Proteins specifically recognizing certain DNA sequences control gene expression by interfering with the transcription process (for example Lac, Trp, and Met repressors, bacteriophage lambda *cro* and λ repressors, the CAP activating protein, etc.). Some other DNA binding proteins also play structural and catalytic roles in other cellular processes (*e.g.* polymerases and nucleases). How do these proteins recognize a particular base sequence and how do they bind to double-stranded DNA? These questions represent one of the central problems of biochemistry since the beginning of the 1980s and have been studied with physical (crystallography and nuclear magnetic resonance), biochemical, and genetic methods.

Ptashne and his colleagues have shown that the Cro repressor binds to six operator sites on the double-stranded DNA from phage λ; they are grouped in two regions, each containing three 17 bp sites whose sequence is similar but not identical. The affinity of the repressor for these sites varies from 1 to 10. The sequence of each site possesses a binary quasi-symmetry and one can deduce, by comparing all sites, a perfectly symmetrical consensus sequence. The repressor monomer is a protein of 66 residues, but in solution, the protein is dimeric and it is this dimer which binds DNA.

A crystallographic study by Matthews showed, in 1981, that the monomer is composed of three antiparallel β-strands and of three α-helices; the dimer is stabilized by electrostatic interactions between monomers.

The same two groups have studied the λ repressor: the same operator sites as in the case of Cro are recognized, but the affinity of the repressor for these sites varies from 1 to 50, the sites of high affinity for one repressor being sites of low affinity for the other; this difference accounts for the *in vivo* physiological role of the two proteins. The λ repressor is a dimer of two identical subunits of 236 residues each. The N-terminal domain of the polypeptide (1-92) binds specifically the operator sites and reproduces the effects of the intact repressor on transcription.

On the basis of these observations, a model was established by Ptashne in 1986 in a very interesting monograph.

Using the *gal* operon and replacing the two spatially separated *gal* operators by *lac* operators, Choy and Adhya have shown that repression of the operon by the tetrameric Lac repressor involves DNA looping and that repression occurs at a step prior to the formation of the first phosphodiester bond of mRNA.

We have seen that the CAP protein intervenes in the positive regulation of several catabolic operons in *E. coli.* When cAMP is present at a high enough concentration, it forms a complex with CAP and this complex binds to DNA at a specific site. A consensus sequence for the different sites was suggested in 1982, but the individual sites differ considerably from the consensus. It has been shown that the protein is a homodimer whose sequence was established in 1982. The protein is 209 residues long and is composed of two domains: the C-terminal domain binds to DNA while the N-terminal one binds cAMP and is responsible for the main contacts leading to the dimer. McKay, Weber, and Steitz, in 1982, established the crystallographic structure of the intact dimer bound to cAMP. The C-terminal domain comprises three α-helices and two pairs of short antiparallel β-strands, whereas the contacts necessary for the dimer formation are due to the pairing of two identical helices, each belonging to a different subunit, and to some additional contacts. cAMP occupies part of the interface between the two monomers; although it is entirely buried within the N-terminal domain, it contributes – by hydrogen bonds – to the stabilization of the dimer.

Before crystals of the complexes between the regulatory proteins and oligodeoxynucleotides corresponding to the putative binding sequences were available, models were constructed to account for the supposed interactions.

At the level of atomic groups, specific recognition refers to the interaction of amino acid side chains and functional groups of base pairs projecting into the solvent. Crystallographic and genetic studies on DNA-binding proteins and their DNA targets revealed numerous hydrogen bonds or hydrophobic contacts between individual bases and mostly amino acids of the helix-turn-helix motif.

In the case of the three proteins, Cro and λ repressors and CAP, a large number of contacts with DNA involve two helices joined by an abrupt turn (helix-turn-helix motif). The first helix lies above the large groove of DNA near the sugar-phosphate backbone and the second helix adapts partially or totally in the large groove. The structures of these α-helical units are almost identical in the three proteins: 20 to 24 α-carbon atoms are practically superimposable. It appears that the helix-turn-helix motif is specific for DNA binding proteins, since it has not yet been found in other proteins. The three proteins differ in all the other regions: the tertiary foldings are all different; the relative positions of the conserved helices versus the symmetry axis of the dimeric proteins are not identical and the helices in dimers are not as perfectly superimposable as in the monomers; the different orientations of these regions suggest that the different repressors do not dock to DNA in the same manner.

In addition to the homologies in the tertiary structure, sequence homologies are detectable in the totality (repressors from bacteriophages λ, 434, and P22) or only in part of the molecule. Whereas the first case suggests an evolutionary relationship, the homology is restricted in the second case to the helix-turn-helix region in the three proteins whose crystallographic analysis had already been done.

The Trp repressor is a homodimeric protein (2 x 107 residues) consisting of two extensively interlocked subunits (fig. 1). It contains three domains: a central core formed by the N-terminal half of both subunits, and two flexible DNA-reading heads, each formed by the C-terminal half of a subunit. Helices D and E form a helix-turn-helix unit similar to that of the Cro and λ repressors and CAP. The central core serves as a spacer between the reading heads which are 26 Å apart in the aporepressor. X-ray analysis (performed by Paul Sigler) of the repressor crystallized with and without tryptophan have revealed that tryptophan induces the formation of a DNA binding surface by increasing the distance between subunits by 8 Å so that they can fit into adjacent major grooves of DNA.

Previous studies had suggested that protein α-helices could fill the large groove of DNA in its B form. From the above results, and from other examples analyzed since, it appears that this is a common method of DNA recognition: an α-helix with its side chains has a diameter of about 12 Å, whereas the large groove of DNA is about 12 Å wide and 6 to 8 Å deep.

Study of the protein-DNA complexes

The general rules summarized by Pabo and Sauer in 1984 were the following: the structure binding the regulatory proteins is B type DNA; the proteins use the symmetry of their subunits to interact with operator sequences which also exhibit a binary quasi-symmetry. Many proteins that interact with DNA are dimers or tetramers and the corresponding targets are quasi-symmetrical. It was therefore probable that these proteins obeyed the interaction model deduced from the crystallographic study of the above three proteins and that they would form symmetrical complexes. The three proteins use an α-helix to establish the

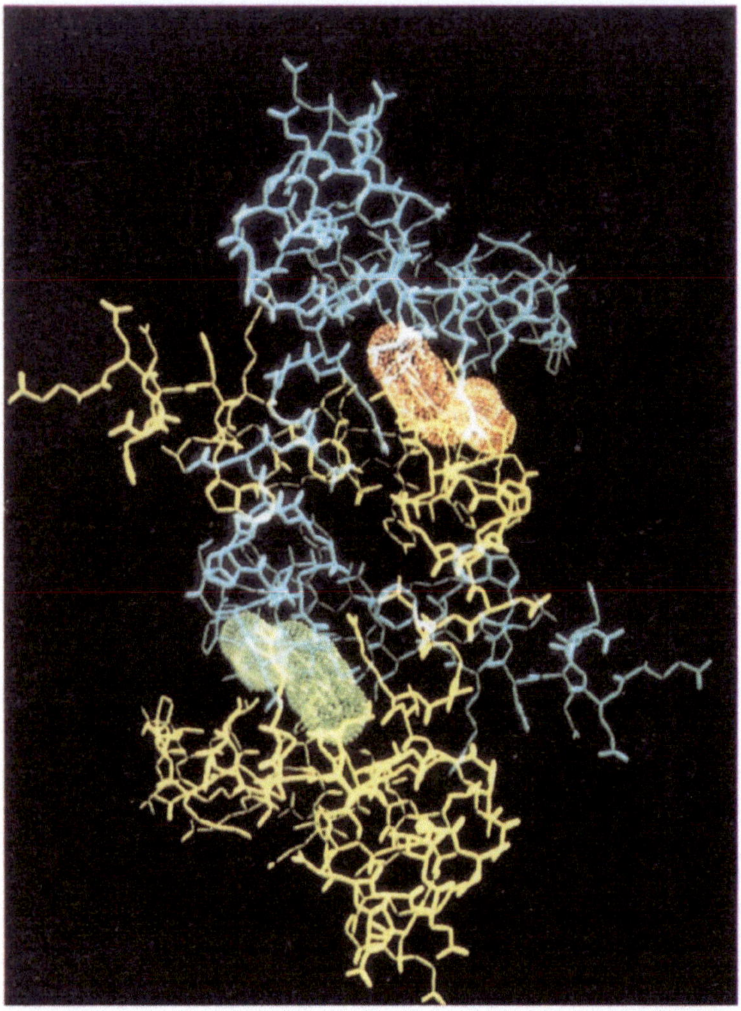

Figure 1. The Trp repressor. One subunit is shown in blue, the other in yellow. The bound tryptophans are shown in green and red. The two subunits are tightly interlocked. (Courtesy of Dr. Paul Sigler.)

majority of the contacts between the protein side chains and the bases of the large groove. The adjacent α-helix contacts the sugar-phosphate backbone of DNA and appears to help the orientation of the first helix. Most contacts seem to occur with a single face of the double helix. However, λ repressor uses a flexible part of the protein to wind around the DNA and to be in contact with both DNA sides. Finally, the recognition of specific base sequences involves hydrogen bonds and van der Waals forces between the protein side chains and the edges of the bases. These specific interactions, added to those with the DNA backbone and to electrostatic interactions stabilize the DNA-protein complexes[45].

Thus, in 1984, the current models derived from the crystal struc-

45. It should be remembered that certain proteins, *e.g. E. coli* RNA polymerase are not composed of subunits presenting a symmetry relationship and that the sequences to which they bind do not present a symmetry relationship.

ture of three proteins were considered as probably correct, but the specialists were aware that model construction, even associated to biochemical and genetic studies, could not give as sure a view as that obtained with real complexes analyzed at high resolution. For example, it is assumed that the operator DNA has a B structure, which is only an approximation. Minor changes in the protein structure, which could occur when the protein binds to DNA, could affect the structure of the complex. It was therefore necessary to study the structure of the binary complexes between repressors and operators, and the ternary complexes formed by the association between operator DNA and activators or repressors (CAP or biosynthetic repressors) in the presence of their specific ligands (cAMP, tryptophan or S-adenosylmethionine, for example).

In 1987 and 1988, the complexes between λ, 434, and Cro repressors and their operators were obtained by Harrison, Pabo, Ptashne and their collaborators. The majority of the above predictions turned out to be correct: the complexes have an order 2 symmetry, the binary axis of the protein coinciding with the binary DNA axis. One admits that the differences in sequence between the two halves of the operator are associated to subtle differences in the protein monomers. DNA appears to be a right-handed helix with the B structure. As predicted, a certain deformation of DNA can be observed when the complex is formed: the same 14 bp-operator adopts a different conformation when it binds the Cro or the 434 repressor. In all the three proteins studied, the second helix of the helix-turn-helix motif is situated in the large DNA groove. Experiments with hybrid proteins where each of the helices comes from a different repressor suggest that the geometry of binding to DNA is similar for many repressors.

The observed complexes favor a DNA-protein recognition by contacts between the protein side chains and the base pairs exposed in the large DNA groove, due to hydrogen bonds and van der Waals forces. It is equally clear that DNA adapts itself in order to increase the complementarity between the two types of molecules: multiple interactions with the DNA phosphate backbone contribute to the global affinity of the binding, but it is essentially the contacts between the exposed base pairs and the residues of the reading helix that allow the repressor to discriminate the operator from the other DNA sequences: everything happens as if the protein were directly "reading" the sequential information coded in the DNA groove. A high-resolution analysis of a cocrystal of the N-terminal domain of λ repressor and λ operator has revealed multiple hydrogen bonds between two lysines of the recognition helix and guanines of the major groove of the operator DNA.

The Trp repressor seems to have an entirely different mode of interaction; there are no direct contacts established by hydrogen or non-polar bonds between the protein and the bases which are known by *in vivo* experiments to be important for the recognition: these bases are either linked to the protein through intermediate water molecules, or directly exposed to the solvent. Paul Sigler and his colleagues, who described the DNA-Trp operator complex in 1988, think that the Trp repressor provides an example of "indirect reading". In other terms, the operator sequence is recognized indirectly through changes induced in the geometry of the phosphate backbone of DNA which allow then the formation of a stable complex.

These results have raised new questions.

a) Is the distortion of the DNA backbone sufficient to allow the repressor to discriminate between the specific and the non-specific sites in the DNA? This question is difficult to answer because the deviations between the operator structure and that of a uniform B DNA appear to result from minor local distortions. On the other hand, the difference in affinities of the Trp repressor for its operator and for a random DNA sequence is only 10^4 fold, a figure much smaller than that observed with many other repressors. The formation of any complex of a protein with DNA involves a non negligible electrostatic component, therefore very dependent on the salt concentration. One must therefore be certain that the complex examined by crystallography is identical to the complex studied in solution.

b) What is the role of water in the recognition process? Sigler and his colleagues think that the hydrogen bonds established between the two macromolecules through the intermediary of water molecules could contribute to the direct reading of three operator base pairs, but admittedly not to that of two other base pairs known to contribute to the recognition.

The presence of water molecules on the surface of DNA and of the protein adds an additional degree of complexity. Furthermore, tryptophan, the corepressor, in addition to the "allosteric" effect it exerts on the protein conformation, appears to interact directly in conjunction with neighboring residues with certain DNA phosphate residues. Recently, a nuclear magnetic resonance study of the Trp repressor-operator complex in solution questions the possibility that water molecules play a role in the recognition process (Jardetzky 1993).

The unique dimeric architecture of the Trp aporepressor built from two highly intertwined chains implies that subunit dimerization is a step on the folding pathway. Despite the complexity of both the architecture and folding kinetics of the Trp repressor, its reversible two-state folding precludes the structural characterization of its folding intermediates. Proteolytic cleavage of each subunit at positions 7 and 71 provides the N and C fragments (8-71 and 72-108, respectively), separating the hydrophobic core into subdomains. Biochemical and biophysical characterization of the purified fragments indicate a native-like dimeric assembly of N fragments and a rather unstructured C fragment. Interestingly, the equimolar mixture of N and C fragments produces an NMR spectrum practically indistinguishable from that of the native protein, indicating a native-like dimeric assembly of TrpR fragments. The results of this proteolytic approach support folding via partially folded dimeric intermediates and illuminates the hierarchy of subdomain folding.

Most of the major families of DNA-binding proteins that have been structurally characterized in procaryotes bind with an α-helix in the major groove. The MetJ repressor, obtained in the homogeneous state by the groups of Greene in the United States, and Cohen at the Pasteur Institute in Paris, has been crystallized in 1988 by the Paris group and by the Phillips group in Leeds. The crystallographic analyses of the aporepressor and of the holorepressor have been carried out by the Leeds group in 1989. As in the case of the Trp repressor, the two subunits are strongly intertwined to form the dimer. S-adenosylmethionine is bound to the dimer at two independent sites, one on each monomer, in a symmetrical manner with regard to the binary symmetry axis of the dimer (fig. 2 & 3).

The purine nucleus of SAM inserts in a hydrophobic pocket normally occupied by a phenylalanine residue in the aporepressor, while the methionine moiety lies at the protein surface (fig. 4).

This explains why methionine does not bind to the aporepressor, while S-adenosylhomocysteine binds with about half the affinity of that for SAM. The trivalent sulfur atom, positively charged lies at the C-terminal end of helix B. In contrast to the Trp repressor, there is no major

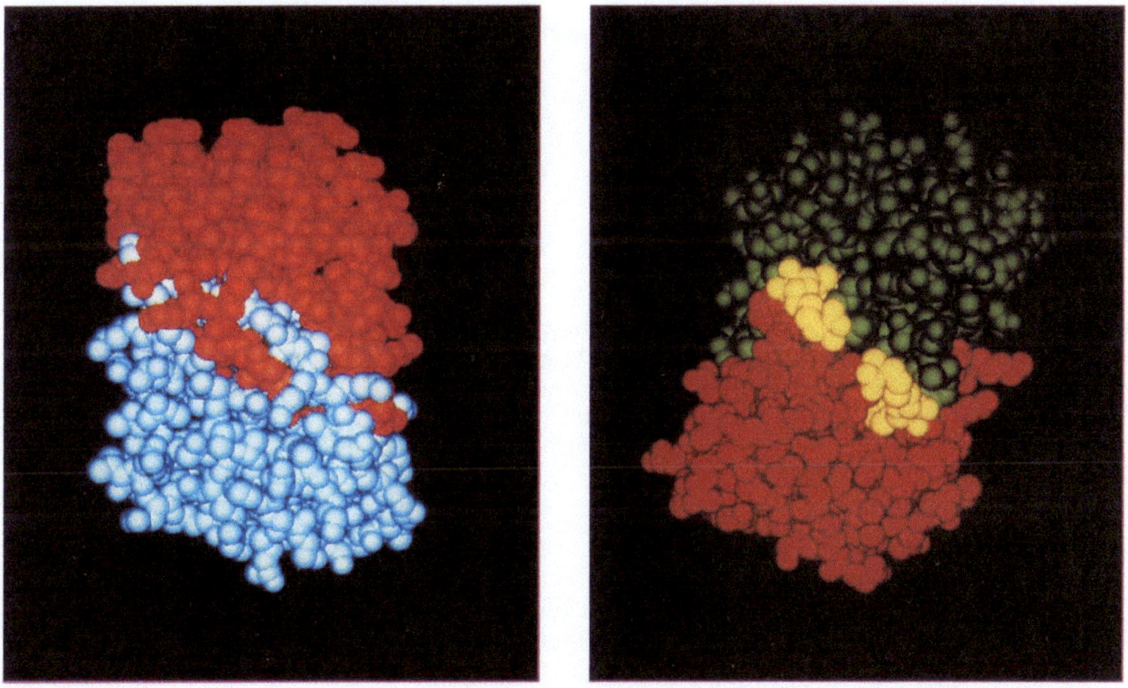

Figure 2. Met repressor space-filling model, with (left) and without (right) S-adenosylmethionine. One subunit is shown in blue or green, the other in red. The SAM molecule is shown in yellow.
(Courtesy of Dr. Simon Phillips.)

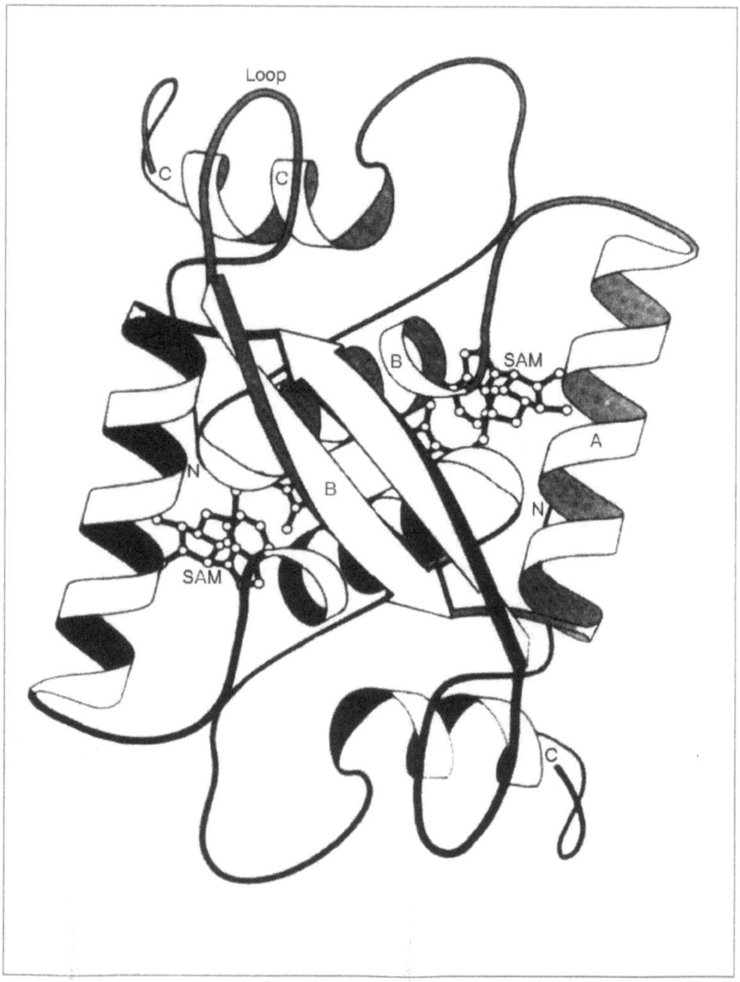

Figure 3. Ribbon representation of the structure of the Met repressor. View along the molecular 2-fold axis, with the SAM molecules as balls and sticks. One subunit is shaded lightly, with the major elements of the secondary structure labeled and the chain termini marked N and C. The other subunit is darkly shaded, with termini indicated by N' and C'. A flexible loop formed by residues 15-20 is also indicated.
(From I. G. Old, S. E. V. Phillips, P. G. Stockley & I. Saint-Girons, *Progress in Biophysics and Molecular Biology* **56**, 145-185 (1991). With permission from Pergamon Press.)

conformational change associated with corepressor binding, and the structures are undistinguishable, except for the absence or presence of SAM. A small conformational change occurs at the N-terminus and at the side chain of Phe65, which occupies the hydrophobic pocket in the absence of corepressor. SAM does not contact DNA in the repressor-operator complex (see below) and its function remains unclear. One possibility is a

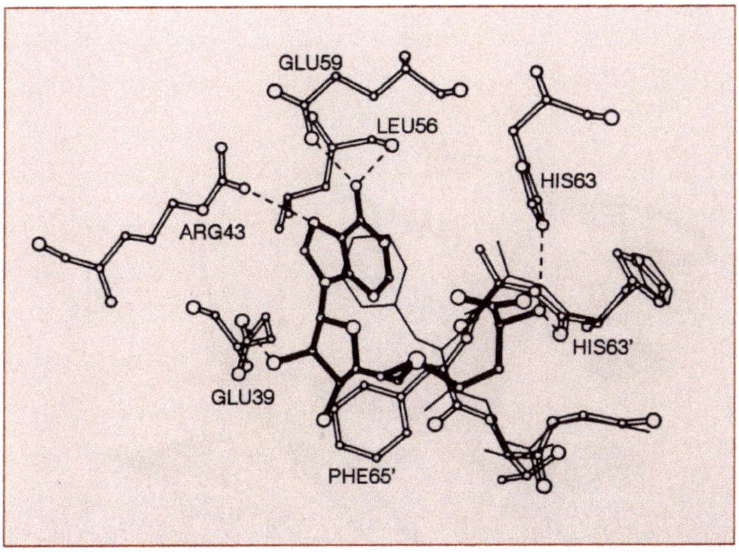

Figure 4. The S-adenosylmethionine binding site. Protein is an open ball and spoke representation for SAM-bound form (=) and as a simple shaded stick representation for the aporepressor form (-), showing the movement of the side chain of Phe 65. The SAM molecule is shown as a shaded ball and spoke representation. The hydrogen bonds are denoted by dashed lines. The two monomers are differentiated by the presence or absence of a prime symbol after residue numbers.
(Courtesy of Dr. S. E. V. Phillips.)

long-range electrostatic effect based on the presence of the positive charge on the sulfur atom.

Each monomer is composed of three α-helices and of a β-strand, accounting for about 50% of the sequence, the rest being accounted for by turns or extended chain. In the protein dimer, the β-strands pair to form an antiparallel β-sheet, while the α-helical regions pack against the sheet and against each other to stabilize the dimer. The molecule lacks a helix-turn-helix motif, which makes it different from all the repressors we have seen so far and from the activator CAP.

The Met repressor has been crystallized with a 19 bp oligonucleotide containing two adjacent 8 bp Met boxes. The analysis of the complex leads to the definition of a new family of prokaryotic regulatory proteins that uses an antiparallel β-sheet for DNA binding (fig. 5).

The methionine repressor binds as two dimeric molecules to two adjacent Met boxes; a dimer binds to each half-site, and each half-site contains a 2-fold symmetry axis that is coincident with the 2-fold axis of the β-sheet (fig. 6).

Sequence specificity is achieved by insertion of double-stranded antiparallel protein β-ribbons into the major groove of B-form DNA, with direct hydrogen-bonding between amino acid side chains on the exposed face of the sheet and the base pairs. Lysine 23 from each β-strand

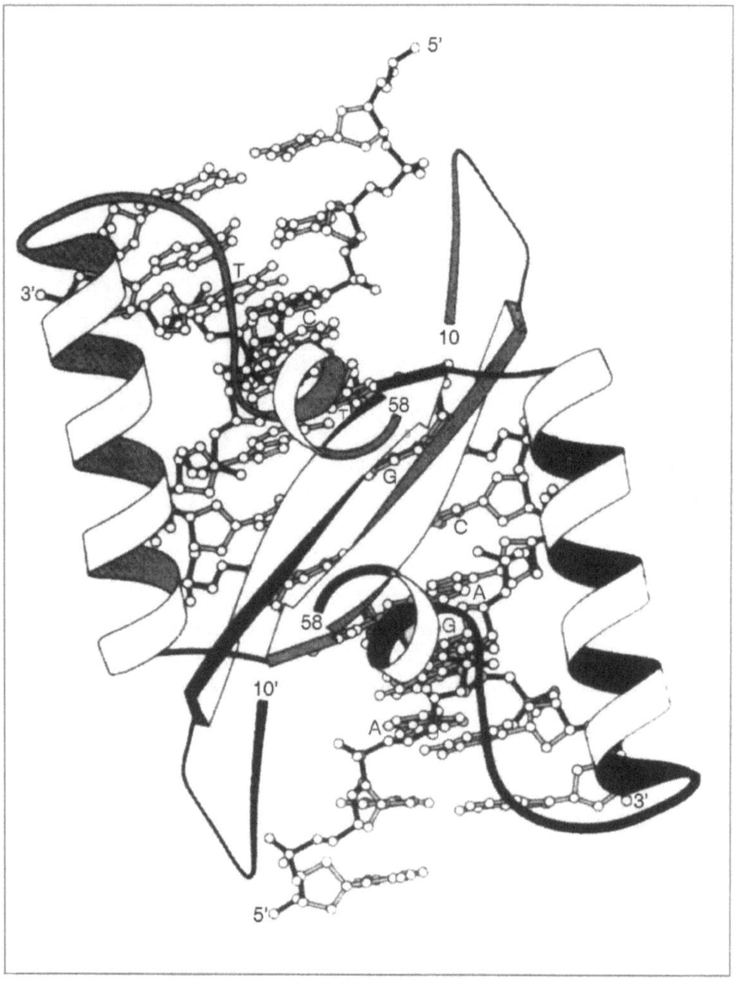

Figure 5. The DNA binding motif, bound to the operator. Viewed as in figure 3. Residues 1-9 and 59-104, and the SAM molecules have been omitted for the sake of clarity. The DNA is shown as ball and stick, with filled bonds for the sugar-phosphate backbone, and open bonds for the remainder of the ribose rings and the bases. The central region of the DNA fragment shown is a Met-box and the bases have been labeled along one strand. The central 2-fold axis of the Met-box coincides with the repressor molecular 2-fold axis. The β-strands lie in the central major groove, which faces the reader, while the minor grooves lie to the upper left and lower right.
(From I. G. Old, S. E. V. Phillips, P. G. Stockley & I. Saint-Girons, *Progress in Biophysics and Molecular Biology* **56**, 145-185 (1991). With permission from Pergamon Press.)

contacts a guanine, and the neighboring threonine 25 on each strand contacts an adenine. Residues from the N-terminal end of α-helix 2 make backbone contacts. The tetramer is stabilized by interactions between α-helix 1 on one dimer and α-helix 1 on the other dimer (see p. 185, the

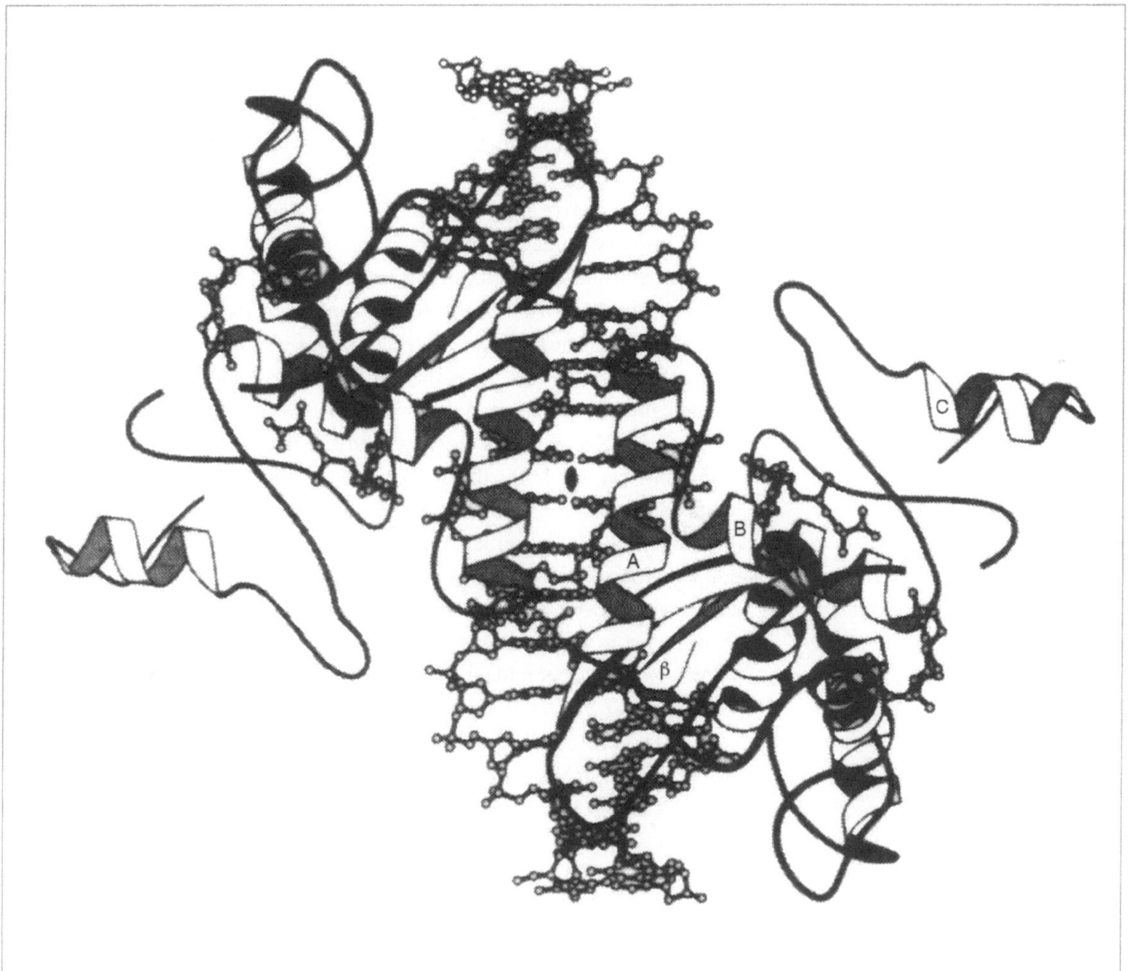

Figure 6. The structure of repressor-operator complex with two repressor dimers, labeled and shaded as in figures 3 & 4. The repressor dimers bind to the same oligonucleotide fragment, one at the lower right, the other at the upper left. They are related by a 2-fold central crystallographic axis shown as a black lozenge, which passes through the center of the oligonucleotide between the two Met boxes. The antiparallel β-ribbons occupy the major groove of the DNA at upper left and lower right, while the α-helices of adjacent dimers form a long, antiparallel protein-protein contact above the minor groove in the center of the diagram. The corepressor molecules lie on the outer surface of the complex, distant from the DNA. The bends in the DNA helix are not very clear in this view, but slight distortion can be discerned.

(From I. G. Old, S. E. V. Phillips, P. G. Stockley & I. Saint-Girons, *Progress in Biophysics and Molecular Biology* **56**, 145-185 (1991). With permission from Pergamon Press.)

sequence of the methionine repressor, which emphasizes the segments involved). The repressor recognizes sequence-dependent distortion or flexibility of the operator phosphate backbone, conferring specificity even for inaccessible base pairs.

Some other types of DNA-binding proteins

A class of DNA-binding proteins has been identified in eukaryotes which makes use of a binding motif called the "leucine zipper". They comprise the transcriptional yeast activator GCN4, the oncoproteins *cjun*, *cfos* and *cmyc* and the C/EBP enhancer. All these protein contain four or five leucine residues situated exactly seven residues apart in the primary sequence and which can be represented as occurring every two spires from an α-helix; it has been proposed that the leucine residues play an important role in the interdigitation of two α-helices, one from each monomer, this being the structural basis of dimerization. In support of this hypothesis, one notes that GCN4, C/EBP and *cjun* bind as dimers to their protein targets, and that *cjun* and *cfos* can form heterodimers able to bind to DNA. It has been shown further that the leucine zipper repetitive motif can serve for the construction of the heterodimers, but that an adjacent domain, different from the leucine zipper, is required for the binding of the heterodimer to DNA. In the case of GCN4, a 60 residues long region containing the leucine zipper is sufficient for dimer formation and for DNA binding. A synthetic peptide corresponding to this region forms in solution essentially helical stable dimers. A leucine mini-zipper motif has been identified in the C-terminal part of the lac repressor (see chap. VI, fig. 7) and in the N-terminal portion of the MetR protein.

The prototype of the zinc finger structure is the transcription factor TFIIIA for RNA polymerase III in the frog *Xenopus laevis*.

This protein is characterized by a particular arrangement of cysteine and histidine residues coordinating a zinc atom. The tridimensional structure of a zinc finger has been determined by X-ray diffraction. The zinc atom is coordinated to two cysteine and two histidine residues and the function is to bring closer an α-helix and a β-sheet. The α-helix of each zinc finger interacts with two bases of the DNA large groove. Pabo and Sauer, who solved the structure of the zinc finger, showed, for the first time, that a protein can contain a periodic structure able to recognize a target (the DNA double helix) whose structure is also periodic. The modular structure of the zinc finger proteins permits the recognition of large targets by adapting the number of fingers to the size of the asymmetric target. All the interactions take place on just one strand of DNA, and there are only a few contacts between the deoxyribose phosphate skeleton of DNA and the transcription factor.

There are two types of zinc fingers: a) the C2H2 type where two cysteine residues face two histidine residues at the base of the finger : the sequence is of the type $hxCx_{2\text{-}4}Cx3hx_5hx_2Hx_{3\text{-}4}Hx_{2\text{-}6}$ (h representing a hydrophobic residue, H a histidine, C a cysteine, and x any amino acid); the proteins of this type generally present several repeated fingers, three for Sp1, nine for TFIIIA, and up to 38 for other factors; b) the C2C2 type where two zinc atoms are coordinated by four cysteine residues, determining two fingers and found in many steroid and thyroid hormone receptors for example. The study by magnetic nuclear resonance of the zinc fingers of the C2C2 type shows that two fingers are necessary for the correct folding of the protein, whereas the folding of a single finger of the C2H2 type can be observed in solution.

SELECTED REFERENCES

M. Ptashne, *A Genetic Switch. Gene Control and Phage l*, 128 p. Blackwell Scientific Publications Ltd, Oxford (1986).

C. O. Pabo & R. T. Sauer, *Ann. Rev. Biochem.* **61**, 1053-1095 (1992).

R. G. Roeder, *Trends Biochem. Sci.* **16**, 402-408 (1991).

Trp repressor. Structural aspects

R. W. Schevitz, Z. Otwinowski, A. Joachimiak, C. L. Lawson & P. B. Sigler, *Nature* **317**, 782-786 (1985).

R. G. Zhang, A. Joachimiak, C. L. Lawson, R. W. Schevitz, Z. Otwinowski & P. B. Sigler, *Nature* **327**, 591-597 (1987).

Met repressor

J. B. Rafferty, W. S. Somers, I. Saint-Girons & S. E. V. Phillips, *Nature* **341**, 705-710 (1989).

S. E. V. Phillips, I. Mainfield, I. Parsons, B. E. Davidson, B. Rafferty, W. S. Somers, D. Margarita, G. N. Cohen, I. Saint Girons & G. Stockley, *Nature* **341**, 711-715 (1989).

W. S. Somers & S. E. V. Phillips, *Nature* **359**, 387-393 (1992).

CHAPTER XXX

Evolution of biosynthetic pathways

Principles of protein evolution

The main goal of the student of biochemical evolution is to reconstruct past events leading to the structure of contemporary proteins, from which the relationships they share with a common ancestor can be defined. A single sequence can give rise to different sequences by speciation and divergence or by gene duplication within a single organism followed by divergence. We shall see examples of these two cases in our survey of the evolution of biosynthetic pathways.

Two theories for the evolution of biosynthetic pathways

Some of the difficulties in understanding the mechanisms whereby multistep metabolic pathways have been constructed is the apparent lack of any selective advantage conferred by individual reactions prior to the establishment of the whole pathway.

In an attempt to solve this paradox, Horowitz proposed, in 1945, that evolution proceeded in a retrograde manner, with individual steps being recruited in the reverse direction relative to the final pathway. The last step in the present-day pathway was acquired first; the penultimate step next, by mutation in one of the two identical genes following a gene duplication allowing the conservation of the first function in the other gene. And so on. This hypothesis supposed that pristine life took place in a primeval soup, a kind of nutritional paradise in which all the essential metabolites and intermediates were available and that the proposed duplications and mutations were selected one after the other as these metabolites became scarce or exhausted. Twenty years later, Horowitz presented several arguments in favor of his theory: 1) the overlap between product and substrate of the enzymes catalyzing consecutive reactions; 2) the existence of allosteric inhibition of the first enzyme of a pathway by its end-product, which could be regarded as a memory of this enzyme for its origin – if one makes the reasonable assumption that the binding sites of the last enzyme and of the first enzyme for this ligand are identical or homologous; 3) the clustering of some biosynthetic genes in bacterial operons, expected for a duplication not followed by a translocation.

This theory, albeit attractive and although it may apply in some cases, postulates the original presence of metabolic intermediates in general, many of which have very short half-lives.

Martinas Ycas has put forward an entirely different hypothesis, according to which the minimal set of enzymes necessary for the viability of an organism are related by descent from a smaller number of ancestral genes. If true, this poses a problem, since it indicates that earlier organisms contained less than the minimal number of enzymic components necessary for present-day organisms; in other words, that these ancient enzymes were endowed with a broader specificity and were able to carry several indispensable reactions. This hypothesis has been reiterated and extended by Jensen.

Progress in molecular biology and the relative facility of gene sequencing have allowed an experimental investigation of this question. Evolutionary relationships between various enzymes can now be traced back by amino acid sequence comparisons, due to the formidable and ever-increasing number of sequences available in the data banks. Amino acid biosynthetic pathways seem especially interesting since, with few exceptions, their different steps are conserved in all organisms where these pathways occur. Therefore, the pathways are probably of very ancient origin and their study potentially applies to the metabolic capacities of the so-called progenote.

We shall now examine some examples.

Methionine and cysteine biosynthetic pathways

In bacteria, methionine biosynthesis involves a total of nine enzymatic reactions in two convergent pathways from aspartate and serine respectively. In particular, the sulfur atom of methionine comes from cysteine by a transsulfuration process involving cystathionine and homocysteine. These compounds are the products of two reactions catalyzed by cystathionine γ-synthase and cystathionine β-lyase, which are encoded by the *metB* and *metC* genes, respectively. These pyridoxal phosphate-dependent enzymes both consist of four subunits, each of molecular weight 42,000 and copurify in many chromatographic systems. It has been shown *in vitro* that cystathionine γ-synthase catalyzes the synthesis of homocysteine from O-succinylhomoserine and H_2S, thus by-passing the cystathionine intermediate.

An understanding of the biochemical similarities shared by these two enzymes came from the comparison of the sequences of the *metB* and *metC* gene products: about one-third of the amino acid residues is conserved between the two sequences (fig. 1). This strong similarity, which encompasses the entire length of the two polypeptides, suggests that these enzymes have a common evolutionary origin. This hypothesis is reinforced by the identification of the lysine residues that bind pyridoxal phosphate in the two enzymes and which are present at the same place in the alignment of the two sequences.

```
C : MADKKLDTQLVNAGRSKKYTLGAVNSVIQRASSLVFDSV-EAKKHATRNRANGEL
       +*  *  *  *           *  *   *   *+  *    * *
B :   MTRKQATIAVRSGLNDDEQYGCVVPPIHLSSTYNFTGFNEPRAHD---------


      FYGRRGTLTHFSLQQAMCELEGGAGCVLFPCGAAAVANSILAFIEQGDHVLMTNT
       + ***   *    +* *  ******** **    *  *+      *+  ** ++
      -YSRRGNPTRDVVQRALAELEGGAGAVLTNTGMSAIHLVTTVFLKPGDLLVAPHD


      AYEPSQDFCSKILSKLGVTTSWF-DPLIGADIVKHLQPNTKIVFLESPGSITMEV
       *  *       * * *    * *        +   *    *+* +***      *
      CYGGSYRLFDS-LAKRGCYRVLFVDQGDEQALRAALAEKPKLVLVESPSNPLLRV


      HDVPAIVAAVRSVVPDAIIMIDNTWAAGVLFKALDFGIDVSIQAATKYLVGHSDA
       *+  *     * *   *+  +***      *    *   * *+ +   **** ****
      VDIAKICHLAREVG--AVSVVDNTFLSPALQNPLALGADLVLHSCTKYLNGHSDV


      MIGTAVCNARCWEQLRENAYL---MGQMVDADTAYITSRGLRTLGVRLRQHHESS
       *  *  *+  + + * *       *     *   *+  ****** *
      VAG--VVIAKDPDVVTELAWWANNIGVTGGAFDSYLLLRGLRTLVPRMELAQRNA


      LKVAEWLAEHPQVARVNHPALPGSKGHEFWKRDFTGSSGLFSFVLKKKLNNEELA
       +   *    * * ++ ** **    ***    *    *      ** *        *
      QAIVKYLQTQPLVKKLYHPSLPENQGHEIAARQQKGFGAMLSFELDGDEQT--LR


      NYLDNFSLFSMAYSWGGYESLILANQPEHIAAIRPQGEIDF--SGTLIRLHIGLE
       *    ***+ * * ** ****        *    *        * ****+  *+*
      RFLGGLSLFTLAESLGGVESLISHAATMTHAGMAPEARAAAGISETLLRISTGIE


      DDLIADLDAGFARIV
      +******+ **
      EDLIADLENGFRAANKG
```

Figure 1. Comparison of the *E. coli* β-cystathionase (C) and cystathionine γ-synthase (B). The sequences are presented in the one-letter code and have been aligned by introducing gaps (-) in order to maximize identities. Asterisks indicate identical residues and + signs indicate accepted replacements (I-L-V, D-E, R-K, T-S).

(From J. Belfaiza, C. Parsot, A. Martel, C. Bouthier de la Tour, D. Margarita, G. N. Cohen & I. Saint-Girons, *Proc. Natl. Acad. Sci. US* **83**, 867-871 (1986).)

Did the ancestral enzyme perform the two reactions or was it involved in the direct synthesis of homocysteine? The *in vitro* reaction of the present-day cystathionine synthase with O-succinylhomoserine and H_2S supports the second proposal. Moreover, indirect suppressors of

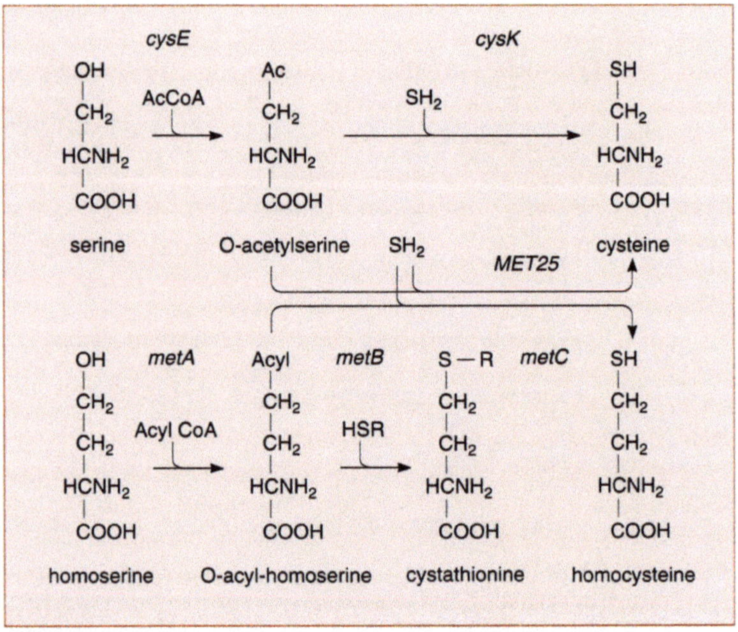

Figure 2. Some of the reactions involved in the cysteine and methionine biosynthetic pathways. The names of the *E. coli* genes encoding the relevant enzymes are indicated above the catalyzed reactions:
***cysE*, serine acetyltransferase; *cysK*, O-acetylserine sulfhydrylase; *metA*, homoserine transsuccinylase; *metB*, cystathionine γ-synthase; *metC*, cystathionine β-lyase (cystathionase). The reactions catalyzed by the *S. cerevisiae MET25* gene product (O-acetylhomoserine, O-acetylserine sulfhydrylase) are also shown. The radical R represents cysteine, and acyl can be either succinyl or acetyl.**
(From C. Parsot, I. Saint-Girons & G. N. Cohen, *Microbiological Sciences* **4**, 258-262 (1987). With permission from Blackwell Scientific Publications Ltd .)

E. coli metC mutants have been selected (*metQ*) which lead to the synthesis (or the overproduction) of an unidentified sulfur compound which is used by cystathionine synthase to synthesize homocysteine directly. This "new reaction" could represent a throwback to the ancestral pathway where only one enzyme was responsible for the synthesis of homocysteine from O-succinylhomoserine.

The contemporary cystathionine γ-synthase uses either O-succinylhomoserine or O-acetylhomoserine, depending on the microorganism from which it is extracted. The ancestral enzyme was probably able to utilize both compounds. Thus, there is a striking similarity between the biosynthetic pathways of homocysteine and cysteine, from homoserine and serine, respectively, as already noticed by Ycas (fig. 2).

Cysteine biosynthesis is catalyzed by O-acetylserine sulfhydrylase, the *E. coli cysK* gene product. This is also a pyridoxal phosphate-dependent enzyme, composed of two identical subunits. It was accordingly supposed that this enzyme would belong to the same family as cystathionine γ-synthase and cystathionine β-lyase; however, when the sequence of O-acetylserine sulfhydrylase was established, this appeared not to be the case (except in the vicinity of the pyridoxal phosphate binding site). Rather, the C-terminal moiety of O-acetylsulfhydrylase was found to be similar to the β-subunit of tryptophan synthase ! An evolutionary relationship between *E. coli* enzymes involved in the cysteine and methionine biosynthetic pathways is most likely if we consider these pathways in the yeast *Saccharomyces cerevisiae.* As indicated in figure 2, there is only one enzyme (encoded by the *MET25* gene) which is involved in the synthesis of homocysteine from O-acetylhomoserine in yeast. This pyridoxal-phosphate-dependent enzyme also catalyses the synthesis of cysteine from O-acetylserine, and is designated as O-acetylhomoserine, O-acetylserine sulfhydrylase. Its nucleotide sequence has been determined and its deduced amino acid sequence has been compared with those of the *E. coli metB* and *metC* gene products. Among about 100 identical residues detected by comparison of pairs of protein sequences, 56 residues are strictly conserved between the three polypeptides. The lysine residues that bind pyridoxal phosphate in cystathionine-γ-synthase and in cystathionine-β-lyase are aligned with a lysine residue in the sequence of O-acetylhomoserine, O-acetylserine sulfhydrylase. The common origin of these three enzymes, which are nowadays involved in the cysteine and/or the methionine biosynthetic pathways, appears to be well-documented. The existence of only one enzyme with broad specificity in *S. cerevisiae* most likely corresponds to a primitive state relative to *E. coli* for which there are at present three different enzymes: the homologous *metB* and *metC* gene products, and the *cysK* gene product, which derives from another ancestor. In *E. coli,* evolution has selected gene duplications and subsequent mutations leading to specialization in the substrate specificity of the encoded proteins and to different controls of their activity and expression.

In mammals, the transsulfuration pathway runs in the other direction from methionine to cysteine; in spite of this, the mammalian cystathionase (whose substrates are homocysteine and serine) has clearly evolved from the same ancestral gene: it shows 39% identity with the *metB* gene product. In contrast, the mammalian cystathionine synthase shows little similarity with the *metB* and *metC* gene products and appeared to have evolved from the cysteine biosynthetic enzymes.

Threonine, isoleucine, cysteine, and tryptophan biosynthetic pathways

Threonine is synthesized from homoserine by a two-step pathway which involves homoserine kinase and threonine synthase. These two enzymes are encoded by the *thrB* and *thrC* genes respectively. Threonine is also the precursor of isoleucine through five enzymatic reactions, of which only the first is specific for the isoleucine pathway, the four latter steps being achieved by enzymes that are also involved in valine biosynthesis (see p. 231-233). The reaction specific to the isoleucine pathway is catalyzed by threonine dehydratase, encoded by *ilvA*.

From a *B. subtilis ilvA* strain, auxotrophic for isoleucine, suppressor mutations allowing growth on homoserine or threonine have been isolated. These mutations affect the expression or the structure of threonine synthase, the *thrC* gene product. This led to the discovery that threonine synthase is endowed with a low-level threonine dehydratase activity. This activity is sufficient to compensate for the defect in the *ilvA*-encoded threonine dehydratase, once *thrC* expression is derepressed. The second type of mutation gives rise to an altered threonine synthase from which the reaction intermediate, α-aminocrotonate, is eliminated and then spontaneously hydrolyzes to α-ketobutyrate and ammonia. These two compounds actually correspond to the products of the reaction normally catalyzed by threonine dehydratase. As shown in figure 3, the reaction mechanisms of threonine synthase and threonine dehydratase share the same α-aminocrotonate intermediate.

These observations have led to the investigation of the sequence of the *B. subtilis thrC* gene. Comparison of *B. subtilis* threonine synthase and *S. cerevisiae* threonine dehydratase revealed 21% identity and 33% homology between the two polypeptide sequences. These data constitute strong evidence in favor of a common evolutionary origin for threonine synthase and threonine dehydratase. An important similarity was also detected between the sequences of these two enzymes, as well as that of *E. coli* threonine synthase and of *E. coli* serine dehydratase. All of these enzymes are pyridoxal phosphate-dependent. The lysine residue of D-serine dehydratase identified as covalently bound to pyridoxal phos-

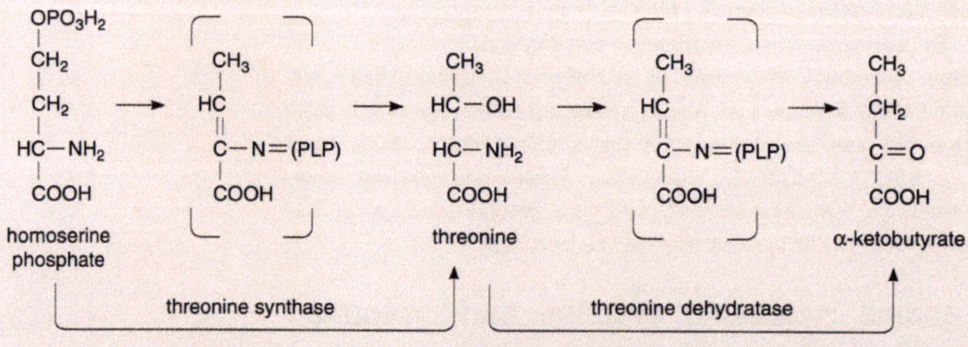

Figure 3. Structure of the substrates and products of the reactions catalyzed by threonine synthase and threonine dehydratase. The structure of the Schiff base between PLP and α-aminocrotonate, the common intermediate of the two reactions, is also shown. Inorganic phosphate, ammonia and water, which participate in the reactions, have not been indicated for the sake of clarity. PLP stands for pyridoxal phosphate.

(From C. Parsot, I. Saint-Girons & G. N. Cohen, *Microbiological Sciences* **4**, 258-262 (1987). With permission from Blackwell Scientific Publications Ltd.)

phate (after reduction of the Schiff base by sodium borohydride) is aligned in figure 4 with similarly positioned lysine residues lying in conserved regions of the threonine synthases, threonine dehydratase, and D-serine dehydratase. Similarity between threonine synthase and threonine dehydratase is further supported by comparison of the available sequences of the products of the *E. coli ilvA* and *tdc* genes which encode the biosynthetic and the catabolic threonine dehydratases.

Three points from these sequence comparisons must be emphasized. Firstly, the homology detected between threonine dehydratase and D-serine dehydratase reveals some interesting perspectives concerning the origin of the enzymes involved in the metabolism of D-amino acids. Secondly, the entire isoleucine biosynthetic pathway now appears to have arisen from the recruitment of the last enzyme of the threonine biosynthetic pathway and the exploitation of the substrate ambiguity of the enzymes acting in valine biosynthesis. Thirdly, the biosynthetic threonine dehydratases from *E. coli* and *S. cerevisiae* both have a C-terminal fragment of about 150 amino acid residues which is not present in the catabolic threonine dehydratase, in the D-serine dehydratase and in the threonine synthases. This extension could correspond to the acquisition

```
THREONINE SYNTHASE                   (thrC , E.coli) :          ..CLELFHGPTLAF K DFGGRFH..  (114)
                                                                         **  *     **  *
THREONINE SYNTHASE                   (thrC , B.subtilis) :      ..VKTEGVNPTGSF K DRGHVLA..  ( 66)
                                                                  ** * * *****    ** *

THREONINE DEHYDRATASE (Biosynthetic) (ilvA , E.coli) :          ..VKREDRQPVHSF K LRGAYAH..  ( 69)
                                                                  *****  *  **   **** **
THREONINE DEHYDRATASE (Biosynthetic) (ILV1 , S.cerevisiae) :    ..LKREDLLPVFSF K LRGAYHH..  (116)
                                                                  ********  **   ****  *
THREONINE DEHYDRATASE (Degradative)  (tdc  , E.coli) :          ..LKFENHQRTGSF(K)IRGAFNK..  ( 65)
                                                                  ** * *  ****   ****

TRYPTOPHAN SYNTHASE (Beta chain)     (trpB , E.coli) :          ..LKREDLLHGCAH(K)THQVLGQ..  ( 94)
                                                                  *******  *         **
TRYPTOPHAN SYNTHASE (Beta chain)     (trpB , B.subtilis) :      ..LKREDLNHTGSH K INNALGQ..  ( 97)
                                                                  ******   ***   *  ***
TRYPTOPHAN SYNTHASE (Beta chain)     (trpB , P.aeruginosa) :    ..LKREELNHTGAH K INNCIGQ..  ( 99)
                                                                  ******   **    *   **
TRYPTOPHAN SYNTHASE                  (TRP5 , S.cerevisiae) :    ..LKREDLNHTGSH K INNALAQ..  (391)
                                                                  ******   ***   *  ***

D-SERINE DEHYDRATASE                 (dsdA , E.coli) :          ..LKKDSHLPSGSI(K)ARGGIYE..  (125)
                                                                  ****  **|***    ****
                                                                          I
```

Figure 4. Comparison of the amino acid sequences around the proposed pyridoxal phosphate-binding site in threonine synthase, threonine dehydratase, tryptophan synthase and D-serine dehydratase. Only a portion of the sequences is shown. Numbers indicate the position of the last presented residue in the original sequences of threonine dehydratase from *E. coli* and *B. subtilis*, biosynthetic threonine dehydratase from *E. coli* and *S. cerevisiae*, degradative threonine synthatase from *E. coli*, β-chain of tryptophan syntetase from *E. coli*, *B. subtilis*, and *P. aeruginosa*, tryptophan synthetase from *S. cerevisiae* and D-serine dehydratase from *E. coli*. The lysine residues which have been experimentally demonstrated as bound to pyridoxal phosphate are enclosed. Asterisks indicate identical or similar residues present in at least two sets of sequences (accepted replacements are I-L-V-M, D-E, K-R, S-T, G-A). The arrow indicates the insertion of an isoleucine residue (I) in the sequence of D-serine dehydratase (between P and S) relative to the other sequences.
(From C. Parsot, I. Saint-Girons & G. N. Cohen, *Microbiological Sciences* **4**, 258-262 (1987). With permission from Blackwell Scientific Publications Ltd.)

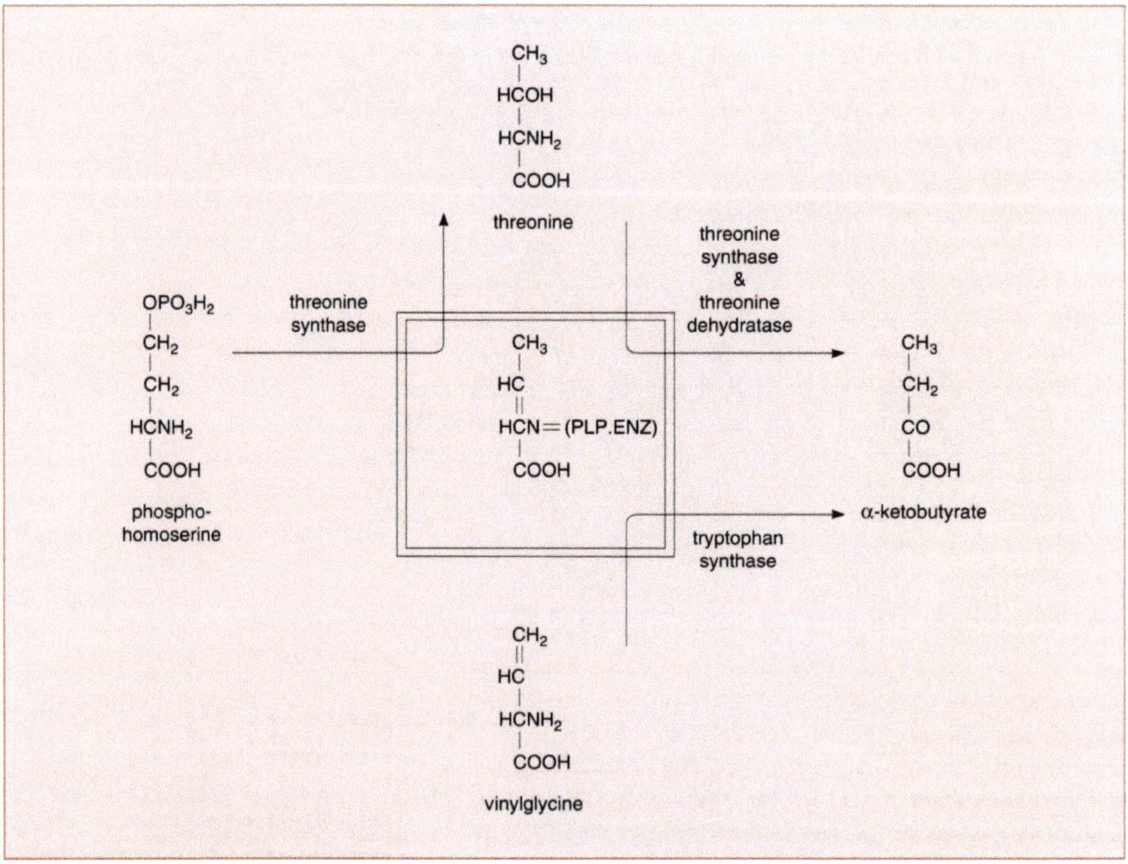

Figure 5. Schematic representation of the substrates and products, and of one of the reaction intermediates in some of the reactions catalyzed by threonine synthase and tryptophan synthase. The structure of the pyridoxal phosphate Schiff base of α-aminocrotonate, an intermediate in each of the reactions catalyzed by threonine synthase and dehydrase (see fig. 3) and tryptophan synthase β-chain is shown in the central box[46]. For the sake of clarity, the pyridoxal phosphate tautomerization of the double bond from β-γ to α-β carbon atoms involved in some of these reactions, inorganic phosphate, and water, are not presented.
(From C. Parsot, *Proc. Natl. Acad. Sci. US* 84, 5207-5210 (1987).)

of a regulatory domain by the biosynthetic threonine dehydratases, which are indeed inhibited by isoleucine.

Threonine dehydratase is also endowed with serine dehydratase activity. This is shared by another pyridoxal phosphate-dependent enzyme, namely the β_2 subunit of tryptophan synthase, which catalyzes the last step of the tryptophan biosynthetic pathway: the $\alpha_2\beta_2$ heterotetramer of tryptophan synthase catalyzes the hydrolysis of indole-3-glycerol phosphate to glyceraldehyde-3-phosphate and indole, as well as the sub-

46. Remember that the β_2 dimer of tryptophan synthase catalyzes the deamination of serine producing pyruvate and ammonia, from enzyme-bound PLP and L-serine, with the formation of an α-aminoacrylate Schiff base intermediate. It can also catalyze the conversion of 2-amino-3-butenoic acid (vinylglycine) to α-ketobutyrate and ammonia *via* the formation of the α-aminocrotonate Schiff base intermediate.

sequent condensation of indole onto L-serine to yield L-tryptophan. As we have seen (p. 249) in the absence of indole, the β_2 dimer deaminates L-serine, producing pyruvate and ammonia. Figure 5 illustrates the similar mechanism, and figure 6 the common evolutionary origin of threonine synthase (and therefore of threonine dehydratase) and of the β subunit of tryptophan synthase.

There are numerous identical residues scattered throughout the entire length of the sequences, with the exception of the C-terminal extension of biosynthetic threonine dehydratases discussed above, which is not present in tryptophan synthases from *B. subtilis* or from other organisms. In addition, the lysine residue of the tryptophan synthase β chain identified as bound to pyridoxal phosphate is aligned with the lysine residues of threonine synthases and threonine dehydratases which were proposed to bind pyridoxal phosphate. The divergence between threonine and tryptophan synthases is probably very ancient since there are no differences in the biosynthetic pathways of threonine or of tryptophan in eubacteria, lower eukaryotes, or plants.

The determination of the nucleotide sequence of most of the genes involved in the tryptophan, threonine, methionine, lysine, and isoleucine biosynthetic pathways, and the subsequent comparison of the encoded proteins, does not support Horowitz's hypothesis of retrograde evolution in biosynthetic pathways. The results presented above (and others) are in keeping with the proposal put forward by Ycas and then by Jensen concerning the substrate ambiguity of primitive enzymes. On the basis of the substrate ambiguity exhibited by some contemporary enzymes, these authors have suggested that ancestral enzymes possessed a broad specificity, allowing them to utilize a wide range of structurally related substrates and thereby yielding small amounts of related products. This process would have provided a biochemically diverse environment in which individual enzyme recruitment would have improved the function of slow – but already existing – ambiguous pathways. The specialization of the proteins encoded by *E. coli metB* and *metC* versus the substrate ambiguity exhibited by the *S. cerevisiae MET25* enzyme seems particularly illustrative.

A word of caution is necessary: a similar approach was attempted in the case of enzymes of the arginine pathway, namely N-acetylglutamate kinase and N-acetylglutamate-γ-semialdehyde dehydrogenase, functionally analogous to kinases and dehydrogenases of other pathways, which by their successive action also achieve the conversion of a carboxylate into an aldehyde function. To this effect, the corresponding genes of *E. coli*, *argB* and *argC* have been sequenced and the derived amino acid sequences have been compared with the known sequences of analogous enzymes active in the proline and homoserine biosynthetic pathways and in glycolysis. Figure 7 illustrates the series of reactions and the corresponding enzymes considered.

The sequences of the four kinases on the one hand, of the four

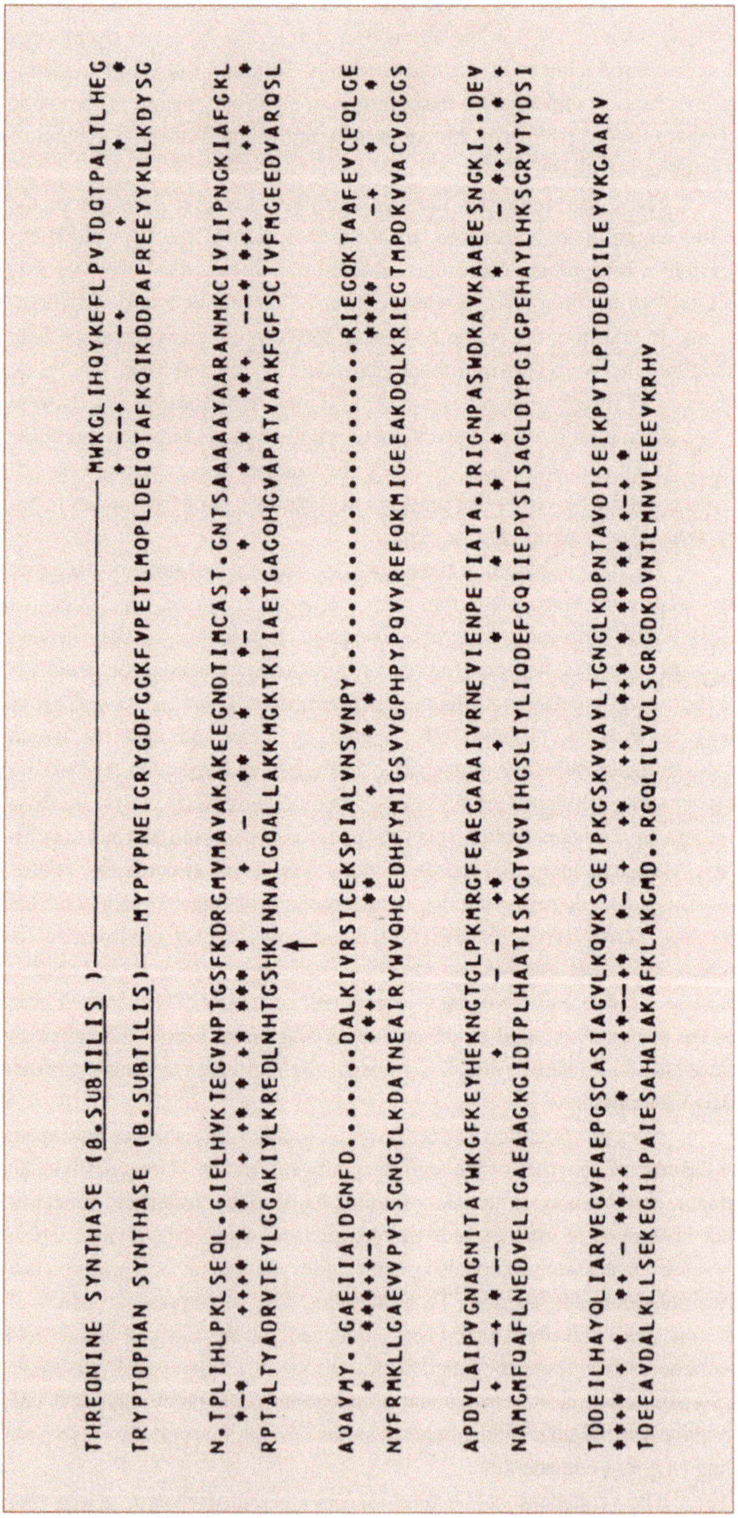

<

Figure 6. Comparison of *B. subtilis* tryptophan synthase and threonine synthase. The entire amino acid sequences are presented in the one letter code. They have been aligned by introducing gaps to maximize the similarity. Asterisks between the sequences indicate identical residues and + signs indicate conservative replacements: R-K, S-T, D-E, I-L-V-M-F. Hyphens indicate that the corresponding residue of tryptophan synthase from *E. coli*, *S. cerevisiae* and *Pseudomonas aeruginosa* is identical to the residue of *B. subtilis* tryptophan synthase. The ↑ sign indicates the lysine residue of tryptophan synthase which binds pyridoxal phosphate.

(From C. Parsot, *Proc. Natl. Acad. Sci. US* **84**, 5207-5210 (1987).)

dehydrogenases on the other, have been compared. In addition, each sequence has been compared to all the protein sequences present in the data banks. No convincing similarity was detected using any of these sequences as a probe (except for short identical stretches along the dehydrogenase polypeptides (obviously the 3-phosphoglyceraldehyde dehydrogenases from other organisms show the expected homology with the homologous *E. coli* enzyme). These negative results contrast with the strong similarities detected when comparing other analogous enzymes operating in different pathways (aspartate and ornithine transcarbamylases, cystathionine synthase and cystathionase, threonine synthase, threonine dehydratase, D-serine dehydratase and the β-chain of tryptophan synthase. Therefore, in the present case, the possibility (distasteful to the author of this book) of convergent evolution should be considered. It remains to be seen whether the tridimensional structure of the two groups of enzymes display similarities indicating homology. In any case, the concept of enzyme recruitment of Ycas and Jensen may be more difficult to prove than suggested by the few supporting examples available.

1) N-acetylglutamate →(*argB*) N-acetylglutamylphosphate →(*argC*) N-acetylglutamate semialdehyde

2) glutamate →(*proB*) glutamylphosphate →(*proA*) glutamate semialdehyde

3) aspartate →(*lysC*) aspartylphosphate →(*asd*) aspartate semialdehyde

4) 3-phosphoglycerate →(*pgk*) 3-phosphoglyceroylphosphate →(*gap*) 3-phosphoglyceraldehyde

Figure 7. Series of reactions converting a carboxyl group to an aldehyde group. Only the relevant substrate transformations are shown, with omission of ATP, ADP, NADPH, $NADP^+$, inorganic phosphate, and water. For each reaction, the name of the corresponding gene (according to the *E. coli* nomenclature) is indicated above the arrow.

Although similarities in the reactions catalyzed by enzymes involved in different metabolic pathways must focus attention on their possible evolutionary relationship, no conclusion can be drawn before sequence analyses are completed. Last but not least, it should be kept in mind that even if ancestral organisms possessed a smaller repertoire of enzymes, it does not mean that their metabolism was simpler and led to fewer end products.

Evolutionary pathway leading to the three isofunctional aspartokinases in *E. coli*

The biosynthesis of the amino acids deriving part or all of their carbon atoms from aspartic acid has been thoroughly studied in *E. coli.*

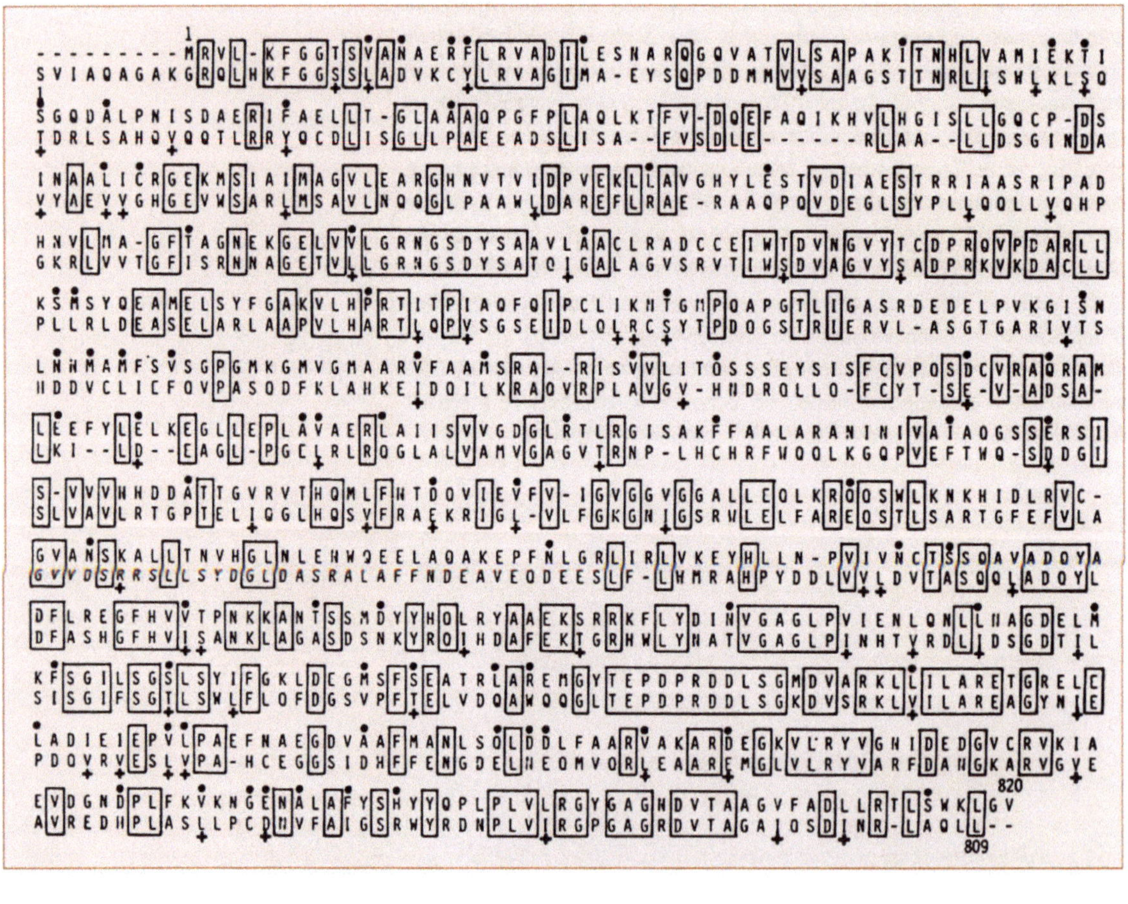

Figure 8. Comparison between the amino acid sequences of aspartokinase I-homoserine dehydrogenase I and aspartokinase II-homoserine dehydrogenase II of *E. coli*. Top line: AKI-HDHI; bottom line: AKII-HDHII. Identical residues have been boxed. Homologies resulting from a single base change in the DNA sequence are marked by a black dot; those corresponding to a conservative amino acid substitution are indicated by a cross.
(From M. M. Zakin, N. Duchange, P. Ferrara & G. N. Cohen (1983). With permission from the *Journal of Biological Chemistry.*)

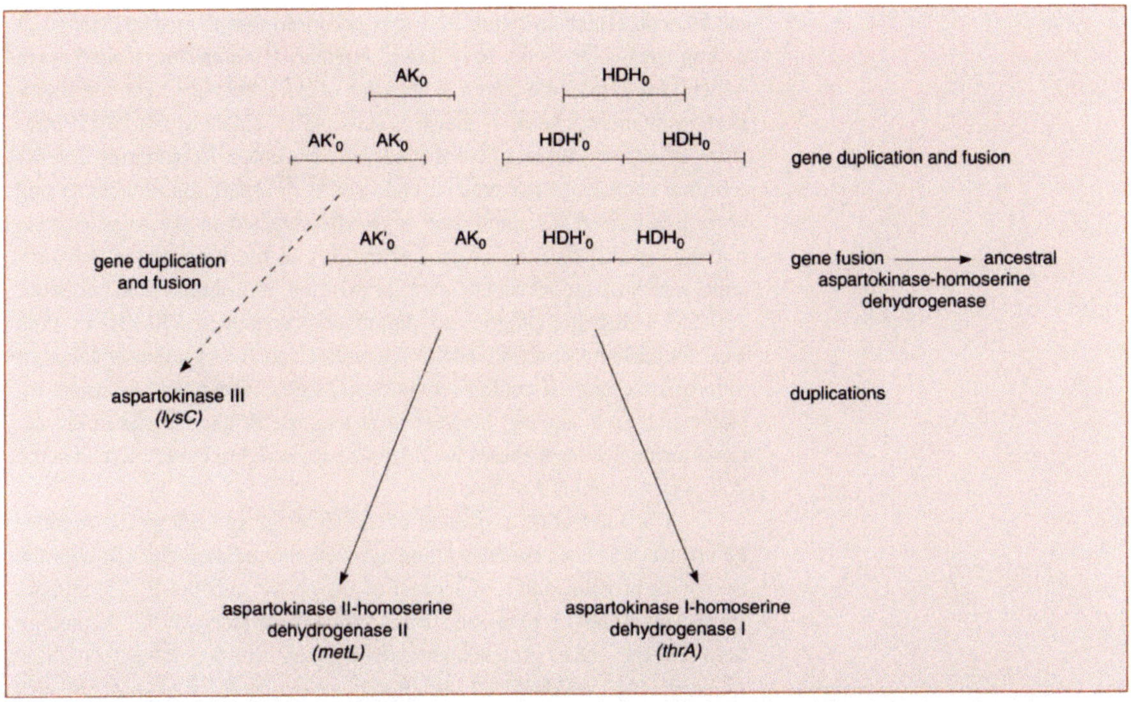

Figure 9. Hypothetical pathway of evolution of *E. coli* aspartokinase-homoserine dehydrogenases. This pathway, proposed in 1984, was proved to be wrong after the sequence of AKIII had been determined.
(From P. Ferrara, N. Duchange, M. M. Zakin & G. N. Cohen (1984).)

The first reaction of the common pathway, leading to diaminopimelate, lysine, methionine, threonine, and isoleucine, is catalyzed by three distinct isofunctional aspartokinases, differing in the way their synthesis and activities are regulated by the respective concentrations of the end metabolites of the aspartic acid family branched pathway. In addition, two of these aspartokinases are bifunctional and also carry homoserine dehydrogenase activities; these proteins are the tetrameric aspartokinase I-homoserine dehydrogenase I and the dimeric aspartokinase II-homoserine dehydrogenase II, respectively encoded by the *thrA* and *metL* genes (see p. 168 & 178). Comparison of their sequences reveal extensive homology which provides unequivocal evidence that the two present genes derive from a common ancestor (fig. 8).

A more detailed examination of each of the protein sequences revealed internal homologies both in the N- and C-terminal parts of the molecules, two segments of about 130 amino acids are repeated in the first one-third of the polypeptide chain, and two segments of about 250 amino acids and covering the COOH-terminal 500 amino acids also present a significant homology. These observations and data obtained from limited proteolysis experiments (see p. 169-170) suggested that the genes coding for the aspartokinase and homoserine dehydrogenase activities may have evolved independently by a process of gene duplication and fusion prior to the appearance of an ancestral fused gene, coding for a bifunctional aspartokinase-homoserine dehydrogenase enzyme, which was itself later duplicated to give rise to the original *thrA* and *metL* genes. As a part of this model represented in figure 9, it was proposed that the aspartokinase III coding sequence could have been duplicated before the fusion event which gave rise to the common ancestor of *thrA* and *metL*. The main supporting arguments were 1) that it is possible to isolate a proteolytic fragment of aspartokinase I-homoserine dehydrogenase I containing 250 N-terminal residues of the protein, endowed with aspartokinase activity and corresponding to the duplicated segments detected in the aspartokinase domain, and 2) that the internal homology in the homoserine dehydrogenase part of the molecule starts at position 330, amino acid sequence 330-565 being homologous to amino acid sequence 570-810 in both aspartokinase-homoserine dehydrogenases I and II. As a consequence, if the aspartokinase III coding sequence had been duplicated before the fusion event, the homology between aspartokinase III and the aspartokinase-homoserine dehydrogenases would not be expected to cover more than the 260-300 N-terminal residues.

The nucleotide sequence of the *lysC* gene, coding for aspartokinase III, was later elucidated and allowed comparison (fig. 10) with the two other proteins and a direct test of the above hypothesis. Three main points must be emphasized from the comparison: 1) the homology between the three proteins extends through the whole aspartokinase sequence (449 residues); 2) the identities calculated for pairs of proteins show that the aspartokinase III sequence is more similar to the aspartokinase I sequence than to the aspartokinase II; 3) as previously noticed in the comparison between the aspartokinase segments of the two bifunctional enzymes (fig. 8) and extended by the comparison of the three sequences, the homology is not evenly distributed along the sequences but clearly shows two regions of higher conservation (residues 16-63 and 209-315), the other regions being less homologous with most of the gaps.

The fact that the whole sequence of aspartokinase III is homologous to the N-terminal part of the aspartokinases-homoserine dehydrogenases seems well documented for the N-terminal and the central regions by the long stretches of homology between the three proteins, and for the C-terminal region by the presence of 12 identical residues between the last 36 positions of aspartokinase III and the corresponding part of AKI-HDHI. As this last segment is part of the duplicated segment of the de-

>

Figure 10. Comparison of the three aspartokinases of *E. coli*. The deduced amino acid sequence of AKIII and part of AKI-HDHI and AKII-HDHII are presented in the one-letter code. Sequences have been aligned in order to maximize homologies by introducing gaps indicated by a dash (–). Identical residues are indicated by asterisks under the sequences. Numbers refer to the amino acid positions in the alignment and numbers in parentheses refer to the position of the last presented amino acid in the original sequences. Brackets delimitate the most conserved sequences and are discussed in the text.

(From M. Cassan, C. Parsot, G. N. Cohen & J.-C. Patte (1986). With permission from the *Journal of Biological Chemistry*.)

```
                                                                                  70
AKIII       : MSEIVVS--------KFGGTSVADFDAMNRSADIVLSDANVRLV--VLSASAGITNLLVALAEGLEPCER
              ** *           *********     * ***  * *    *  **** * *** ***  *
AKI-HDHI    : MRVL-----------KFGGTSVANAERFLRVADILESNARQGQVATVLSAPAKITNHLVAMIEKTISGQ-
              * *            ********     ******  * * *  *  **** * *** ***  *
AKII-HDHII  : MSVIAQAGAKGRQLHKFGGSSLADVKCYLRVAGIMA-EYSQPDDHMVVSAAGSTTNRLISWLKLSQT---
              ****           **** * **    **** *      *     * **    ** *

                                                                                 140
AKIII       : FEKLDAIRNIQFA--ILERLRYPNVIREEIERLLENIT------------VLAEAAALATSP--ALTDEL
                 ***  *** *  *   *       **   *    *            **  ** *   *  *   **
AKI-HDHI    : ----DALPNISDAERIFAELLTGLAAAQPGFPLAQLKTFV-DQEFAQIKHVLHGISLLGQCP-DSINAAL
                  **  **  *  **   *  **  *     *    *** * *      **    **   * *   * *
AKII-HDHII  : -DRLSAHQVQQTLRRYQCDLISGLLPAEEADSLIS--AFVSDLE--------RLAALLDSGINDAVYAEV
                 * *    *   *    *  **  ***   *     ** * *          ****     **  **

                                                                                 210
AKIII       : VSHGELMSTLLFVEILRERDVQAQWFDVRKVMRTNDRFGRAEPDIAALAELAALQLLPRLNEGLVITQGF
              * *** **  *    *  *   * * * *   *   *      ***    * *  * *     ** * **
AKI-HDHI    : ICRGEKMSIAIMAGVLEARGHNVTVIDPVEKLLAVGHYLESTVDIAESTRRIAASRIPADHMVLMA--GF
                 ** **   *  **  **      * * * *         **** *    *    *     *    **
AKII-HDHII  : VGHGEVWSARLMSAVLNQQGLPAAWLDAREFLRA-ERAAQPQVDEGLSYPLLQQLLVQHPGKRLVVT-GF
              * ***  *  **  **   *  * * * ** ***  *     **   *  *    *       ** * **

                                                                                 280
AKIII       : IGSENKGRTTTLGRGGSDYTAALLAEALHASRVDIWTDVPGIYTTDPRVVSAAKRIDEIAFAEAAEMATF
              *   *** *  *** **** ** **  * **** ***** * ** *** *  *         ** * * *
AKI-HDHI    : TAGNEKGELVVLGRNGSDYSAAVLAACLRADCCEIWTDVNGVYTCDPRQVPDARLLKSMSYQEAMELSYF
                 * *** * *********** *** * *    ***** **** *** * ** **      ** **  *
AKII-HDHII  : ISRNNAGETVLLGRNGSDYSATQIGALAGVSRVTIWSDVAGVYSADPRKVKDACLLPLLRLDEASELARL
              *  ** **** **********    *    *** ** ** ***  *** * ** **      ** ***

                                                                                 350
AKIII       : GAKVLHPATLLPAVRSDIPVFVGSSKDPRAGGTLVCNKT-ENPPLFRALALRRNQTLLTLHSLNMLHSRG
              ******* ** *   * **     *  * *****      *     *      *  *       *    *
AKI-HDHI    : GAKVLHPRTITPIAQFQIPCLIKNTGNPQAPGTLIGASRDEDELPVKGISNLNNMAMFSVSGPCMKGMVG
              *********  *     **        * * ****     *            *        * *    *
AKII-HDHII  : AAPVLHARTLQPVSGSEIDLQLRCSYTPDQGSTRIERVL-ASGTGARIVTSHDDVCLIEFQVPASQDFKL
               * *** *** *   * *      *  *  * * *           *         *     *

                                                                                 420
AKIII       : FLAEVFGILARHNISVDLIT--TSEVSVALTLDTTGSTSTGD-TLLTQSLLMELSALCRVEVEEGLALVA
                **** ** *  *** ***   ** *      *          *    ** *   *    * *******
AKI-HDHI    : MAARVFAAMSRARISVVLITQSSSEYSISFCVPQSDCVRAERAMLEEFYLELKEGLLEPLAVAERLAIIS
                * **    ** *** ***   ** *  **   *         *    *    *** * * * * **
AKII-HDHII  : AHKEIDQILKRAQVRPLAVGVHNDRQLLQFC-YTSEVADSAL------KILDEAGLPGELRLRQGLALVA
                 *   ** **                 **  **               * * **   *    ******

AKIII       : LIGNDL--SKACAVGKEVFGVLEPFNIRMICYGASSHNLCFLVPGEDAEQVVQKLHSNLFE (449)
                *  *               *   **  *  * *   *   *   ***  *  **  **
AKI-HDHI    : VVGDGLRTLRGISAKFFAALARANINIVAIAQGSSERSISVVVNNDDATTGVRVTHQMLFN (461)
               ** **   *               **  * ** * * *  **   **   *   **  **
AKII-HDHII  : MVGACV--TRNPLHCHRFWQQLKGQPVEFTWQSDDGISLVAVLRTGPTESLIQGLHQSVFR (453)
               ** *    *           *         *     **  *      *     ***  *
```

hydrogenase moiety of AKI-HDHI, the homology between the C-terminal part of AKIII and the central region of AKI-HDHI seems to rule out the separation of the ancestor of *lysC* prior to the fusion event between the aspartokinase and dehydrogenase fusion. It is therefore proposed that the separation of the *lysC* ancestral gene arose *after* the fusion of the aspartokinase and homoserine dehydrogenase coding sequences and that *lysC* could derive from *thrA*. However, this evolutionary pathway supposes that the mutation rates of the three genes were not very different and that no genetic conversion occurred, these assumptions being beyond any experimental verification. Whether the duplication event giving rise to *lysC* encompassed part or all of the homoserine dehydrogenase coding portion of *thrA* is not determinable; there is no homology between the 3'-flanking region of *lysC* and the corresponding region of *thrA,* but even the nucleotide coding sequences show only about 40% identity, indicating that selective pressure has operated in order to conserve amino acid rather than nucleotide sequences.

The high degree of conservation of the amino acid sequences observed for segments 16-63 and 209-315 probably reflects structural and/or functional constraints. Attempts to utilize these sequences as probes to explore sequences of proteins endowed with kinase activity or concerned with aspartate metabolism failed to detect any significant region of homology that could lead to an educated guess for the localization of the aspartate or ATP-binding sites. On the other hand, data obtained from limited proteolysis on aspartokinase-homoserine dehydrogenases I and II have led to the conclusion that the native polypeptides have a triglobular structure: an N-terminal domain (corresponding to residues 1-245) carries the aspartokinase activity; a central domain (correponding to residues 250-500) is involved in protein association contacts; and a C-terminal domain (residues 500-820) is endowed with the homoserine dehydrogenase activity. Thus, part of the second very conserved region in the primary sequence (residues 209-315) spans the central domain observed by limited proteolysis (residues 250-500) and is likely to be involved in protein association contacts leading to the polymeric state, dimeric in the case of AKIII and AKII-HDHII and tetrameric in the case of AKI-HDHI.

>

Figure 11. Comparison of *E. coli* and *B. subtilis* homoserine dehydrogenase amino acid sequences. The entire sequence of *B. subtilis* homoserine dehydrogenase (HDH, *hom*) has been aligned with parts of the sequences of *E. coli* AKI-HDHI and AKII-HDHII (HDHI and HDHII). Symbols = and – indicate identical and similar (I-L-V-M, R-K, D-E, G-A, S-T, F-Y) residues, respectively. Asterisks above the alignment indicate positions strictly conserved in the three sequences and arrows refer to the glycine residues of a proposed nucleotide binding domain. The most conserved region is boxed. Numbers in parentheses refer to the position of the relevant residue in the original sequence. (From C. Parsot & G. N. Cohen (1988). With permission from the *Journal of Biological Chemistry.*)

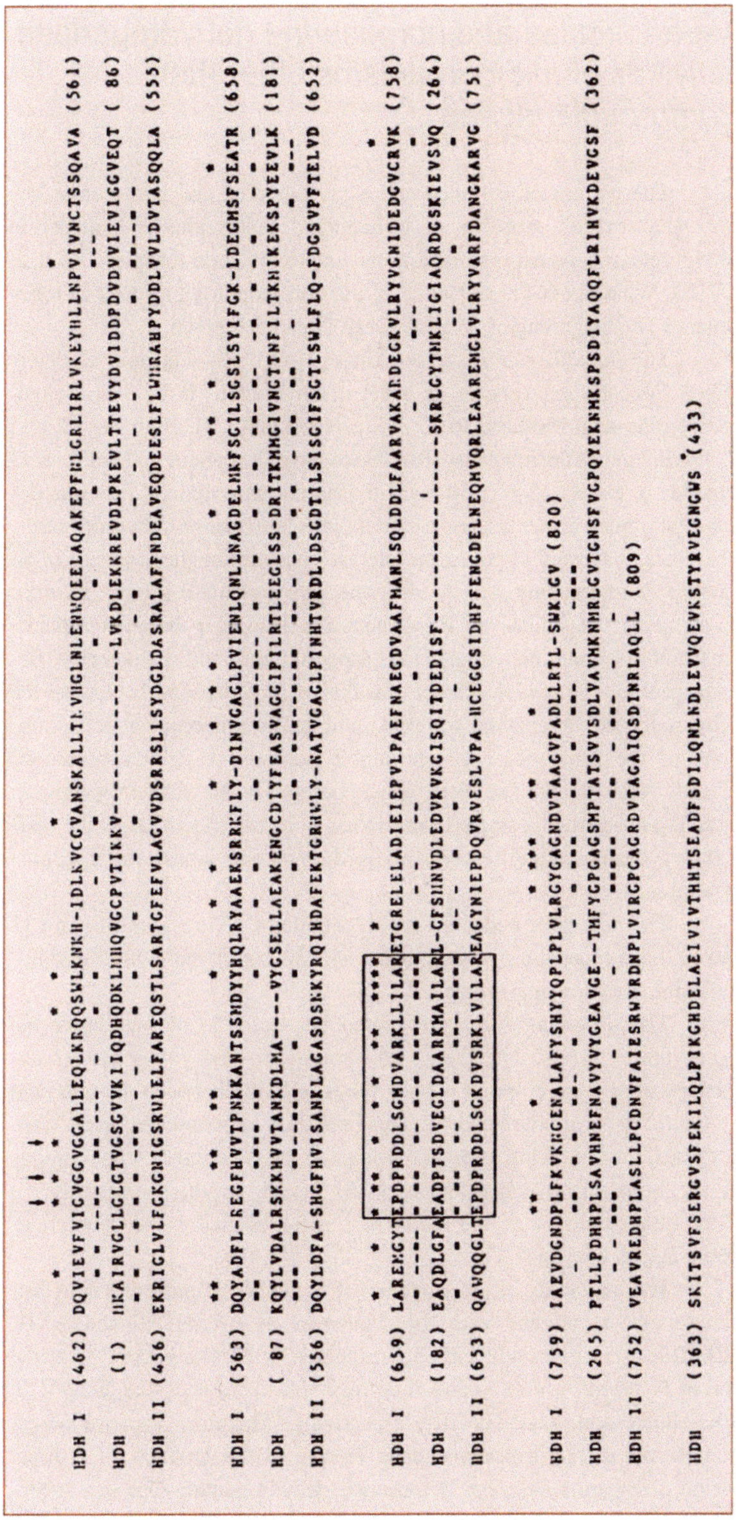

Aspartokinase and homoserine dehydrogenase activities of microorganisms other than *Enterobacteriaceae*

The pattern of isofunctional aspartokinases and homoserine dehydrogenases and probably of bifunctional aspartokinases-homoserine dehydrogenases seems restricted to the enteric bacteria (chapters XVII & XVIII). What about the sequence of the correponding enzymes in other genera? The beginning of an answer can already be given.

The *Bacillus subtilis* aspartokinase II (p. 198) sequence, as determined from the sequence of its gene, is homologous in its entire length (408 amino acid residues) to *E. coli* aspartokinase III. Both *E. coli* and *B. subtilis* monofunctional aspartokinases have thus about 300 amino acid residues in excess of the 250-residue aspartokinase domain that was defined by limited proteolysis on aspartokinase-homoserine dehydrogenases. Therefore, it seemed of particular interest to determine the sequence of an enzyme endowed only with homoserine dehydrogenase activity in order to complete the studies on the structural relationship between aspartokinase-homoserine dehydrogenase, aspartokinase and homoserine dehydrogenase. The obvious choice was that of *Bacillus subtilis* homoserine dehydrogenase, an enzyme which is inhibited by threonine at concentrations of the same order which inhibit homoserine dehydrogenase of *E. coli* AKI-HDHI. The *hom* gene coding for the dehydrogenase in *B. subtilis* is part of a single transcription unit that encompasses the *thrB* and *thrC* genes, coding respectively for homoserine kinase and threonine synthase.

The *hom* gene codes for a polypeptide of 433 residues. Figure 11 shows the comparison of its sequence with its counterparts in *E. coli* homoserine dehydrogenases I and II.

The following observations can be made: 1) the similarity between the *B. subtilis* and the *E. coli* homoserine dehydrogenases commences at the exact point where the similarity between *E. coli* and *B. subtilis* aspartokinase ends; 2) some residues are conserved in all three enzymes, thus allowing an hypothesis as to the function of some regions in the homoserine dehydrogenases; 3) the *B. subtilis* enzyme has a C-terminal extension of 108 amino acid residues that has no counterpart in the *E. coli* enzymes.

The beginning of the sequence of *B. subtilis* homoserine dehydrogenase can be aligned with the sequences of AKI-HDHI and AKII-HDHII from residues 462 and 456, respectively. Similarly, the C-terminal end of *B. subtilis* aspartokinase II corresponds to residues 468 and 461 in AKI-HDHI and AKII-HDHII, respectively. Therefore aspartokinases-homoserine dehydrogenases can be viewed as the products of a direct fusion of aspartokinase and homoserine dehydrogenase. One can speculate that, in an ancestral chromosome, genes coding for aspartokinase and homoserine dehydrogenase were in tandem. By the simple loss of a stop codon at the end of the aspartokinase gene, a fused gene coding for a bifunctional protein could have been created. This could have happened after the emergence of the *Enterobacteriaceae,* since bifunctional AK-HDHs have been detected only in these organisms. A comparable artificial fusion between the *S. typhimurium his D* and *hisC* genes has been obtained under appropriate selection conditions, showing that such a scheme is not unreasonable.

The comparison of homoserine dehydrogenases indicates that 52 amino acid residues are strictly conserved between the three sequences (indicated by asterisks in fig. 11). Among these highly-conserved residues, one finds the motif Gly-X-Gly-X-X-Gly, which is encountered in all known nucleotide binding domains. Around these glycine residues, the small, hydrophobic amino acid residues, predicted to be of structural importance for the β-α-β folding on this domain, are also present at the correct positions in the sequences of the three homoserine dehydrogenases. (This motif is present at similar positions in the homoserine dehydrogenases of *Corynebacterium glutamicum* and *Brevibacterium l actofermentum.)* It thus seems likely that this region constitutes the NADPH binding domain of homoserine dehydrogenase.

The most conserved region among the three homoserine dehydrogenases lies between positions 200 and 221 of *B. subtilis* homoserine dehydrogenase, where 12 out of 22 residues are strictly conserved (see the *boxed* region of fig. 11). Interestingly, there are seven charged residues among these 12 conserved positions, whereas there are only 12 charged residues that are conserved in the full length of the polypeptides. It seems reasonable to speculate that this 22-amino acid region could be involved in substrate binding or catalytic activity. Such a region is also detected by comparison with the homoserine dehydrogenase of *Corynebacterium glutamicum* and *Brevibacterium lactofermentum.*

The comparison between the three enzymes indicates that the *B. subtilis* homoserine dehydrogenase has a 108-amino acid residue C-terminal extension relative to the *E. coli* enzymes. Exploration of protein sequence libraries failed to detect any similarity between this extra segment and any other protein sequence. The following observations suggest that this part of the *B. subtilis* dehydrogenase could constitute a domain involved in the regulation of enzyme activity. First, since this region is not present in the *E. coli* enzymes, it cannot be directly involved in the catalytic activity (even though a truncated polypeptide, lacking the last 80 residues is not endowed with detectable activity (it may be that it is incorrectly folded). Second, numerous lines of evidence indicate that the threonine sensitivity of *E. coli* is intrinsically asociated with a threonine-sensitive kinase domain and thus appears to be a consequence of the fusion of homoserine dehydrogenase to aspartokinase. Therefore, the threonine sensitivity of the homoserine dehydrogenase from *B. subtilis* is likely to be achieved by a mechanism different from the mechanism used for regulation of the homoserine dehydrogenase in the *E. coli* bifunctional enzyme. This indirect evidence led the authors of this work to propose that the C-terminal sequence detected in *B. subtilis* homoserine dehydrogenase could constitute a regulatory domain involved in the sensitivity of this enzyme to threonine. This hypothesis provides an attractive scheme for the fusion of two enzymes that do not catalyze two consecutive reactions of a metabolic pathway, namely aspartokinase and homoserine dehydrogenase. One can speculate that the ancestral homoserine dehydrogenase was not regulated and that two different strategies have arisen to ensure regulation of this enzyme activity, *i.e.*, fusion to the C-terminal end of an already regulated aspartokinase or addition of a regulatory domain to the C-terminus of homoserine dehydrogenase. This hypothesis is conforted by the fact that the homoserine dehydrogenases of *Corynebacterium glutamicum* and of *Brevibacterium lactofermentum,* also species where separate proteins catalyze threonine-sensitive aspartokinase and homoserine dehydrogenase, possess also a C-terminal extension of about 100 residues at their COOH-terminal, for which there is no counterpart in either of the *E. coli* proteins.

Transmembrane facilitators

Many transport proteins of bacteria and eukaryotes are thought to possess a common structural motif of 12 transmembrane-spanning α-helical segments, similar to the lactose permease we have examined in some detail. Using statistical methods, Marger and Saier have found that five families of proteins which facilitate uniport, symport and antiport, comprise a single superfamily. These five families are drug-resistance proteins, prokaryotic and eukaryotic sugar facilitators (among them, glucose transporters, galactose, lactose, and maltose permeases from various yeasts), bacterial citrate transporters, bacterial sugar-phosphate transporters and, finally, some disaccharide symporters, including LacY, the product of enterobacterial Lac permease. A total of more than 50 sequences have been compared. The intrafamily and interfamily comparisons show a degree of homology sufficiently high to exclude chance occurrence. The authors propose that the superfamily arose before the appearance of eukaryotes on earth, the duplication events having occurred some 2-3 billion years ago.

DNA-binding regulator proteins

Great progress in computer-aided search of the data bases has made it possible to find homologies between bacterial regulators homologous to the Gal and Lac repressors. Each member of the GalR-LacI family has an amino-terminal DNA binding domain and several regions involved in effector recognition and oligomerization in the carboxyl terminal part of the protein. The DNA-binding domains, which can assume the form of a helix-turn-helix, are the most conserved portions of these proteins. This family includes, so far, some 20 members belonging to diverse Gram-positive and negative organisms, that regulate a variety of biosynthetic and transport functions. A very illuminating alignment was presented by Weickert and Adhya in 1992; it was shown that these proteins demonstrate a very high degree of similarity (60%) throughout the entire sequence. Since a portion of the operator sequences occupied by these proteins is also conserved, a similar DNA structure may be required for specific recognition of DNA by members of the family. The compilation and alignment presented should simplify the study by mutagenesis of current and new member proteins by identifying, in advance, residues which are more likely to be involved in the critical functions of DNA binding, inducer interactions and oligomerization. It may also give some hints for protein evolution.

SELECTED REFERENCES

The Evolution of Metabolic Function, R. P. Mortlock ed., 330 p. CRC Press (1992).

W. F. Loomis, *Four Billion Years, An Essay on the Evolution of Genes and Organisms*, 286 p. Sinauer Associates, Inc., Sunderland, Mass (1988).

Two different theories on the evolution of biosynthetic pathways

N. H. Horowitz, *Proc. Natl. Acad. Sci. US* **31**, 153-157 (1945).

N. H. Horowitz, in *Evolving Genes and Proteins*, V. Bryson & H. J. Vogel eds., pp. 16-23. Academic Press, New York (1965).

M. Ycas, *J. Theoret. Biol.* **44**, 145-160 (1974).

R. A. Jensen, *Ann. Rev. Microbiol.* **30**, 409-25 (1976).

C. Parsot, I. Saint-Girons & G. N. Cohen, *Microbiol. Sciences* **4**, 258-262 (1987).

Common origin of cystathionine-γ-ysnthase and cystathionase

J. Belfaiza, C. Parsot, A. Martel, C. Bouthier de la Tour, D. Margarita, G. N. Cohen & I. Saint Girons, *Proc. Natl. Acad. Sci. US* **83**, 867-871 (1986).

Common origin of threonine synthase, threonine dehydratase, D-serine dehydratase, and the B chain of tryptophan synthase

C. Parsot, *The EMBO Journal* **5**, 3013-3019 (1986).

C. Parsot, *Proc. Natl. Acad. Sci. US* **84**, 5207-5210 (1987).

Comparison of arg genes with homologous and analogous enzymes

C. Parsot, A. Boyen, G. N. Cohen & N. Glansdorff, *Gene* **68**, 275-283 (1988).

Evolution of the E. coli *aspartokinases and homoserine dehydrogenases*

M. M. Zakin, N. Duchange, P. Ferrara & G. N. Cohen, *J. Biol. Chem.* **258**, 3028-3031 (1983).

P. Ferrara, N. Duchange, M. M. Zakin & G. N. Cohen, *Proc. Natl. Acad. Sci. US* **81**, 3019-3023 (1984).

M. Cassan, C. Parsot, G. N. Cohen & J.-C. Patte, *J. Biol. Chem.* **261**, 1052-1057 (1986).

Structural and evolutionary relationships between E. coli *aspartokinase-homoserine dehydrogenases and monofunctional homoserine dehydrogenases*
C. Parsot & G. N. Cohen, *J. Biol. Chem.* **263**, 14654-14660 (1988).

Superfamily of transmembrane facilitators
M. D. Marger & M. H. Saier, *J. Trends Biochem. Sci.* **18**, 13-20 (1993).

DNA-binding regulator proteins
M. J. Weickert & S. Adhya, J. Biol. Chem. 267, 15869-15874 (1992).

Abbreviations

ACP	acyl carrier protein
ADP	adenosine diphosphate
ADPG	adenosinediphosphoglucose
AICAR	5-aminoimidazole-4-carboxamido-1-β-ribofuranosyl-5'-phosphate
AIR	aminoimidazole ribonucleotide
AK	aspartokinase
ALA	δ-aminolevulinic acid
AMP	adenosine monophosphate
APS	adenosine-5'-phosphosulfate
AS	anthranilate synthetase
ATCase	aspartate transcarbamylase
ATP	adenosine triphosphate
BCCP	biotin carboxyl carrier protein
CAIR	5-amino-4-imidazole carboxylate ribonucleotide
cAMP	cyclic AMP
cDNA	complementary DNA
CDRP	anthranilate deoxyribulotide
CAP	catabolite activator protein
CH_2THF	5-10 methylene tetrahydrofolate
CMP	cytidine monophosphate
CDP	cytidine diphosphate
CTP	cytidine triphosphate
CoA	coenzyme A
CRM	cross-reacting material
DAHP	3-deoxy-D-arabino-heptulosonate-7-phosphate
DAP	diaminopimelate
DAPA	7-8 diaminopelargonate
dADP	deoxyadenosine diphosphate
dATP	deoxyadenosine triphosphate
dCDP	deoxycytidine diphosphate
dCMP	deoxycytidine monophosphate
dCTP	deoxycytidine triphosphate
dGDP	deoxyguanosine diphosphate
dGTP	deoxyguanosine triphosphate
DHNA	1,4-dihydroxy-2-naphthoic acid
DMK	demethylmenaquinone
dNDP	deoxynucleoside diphosphate
dNTP	deoxynucleoside triphosphate
dTDP	deoxythymidine diphosphate
dTMP	deoxythymidine monophosphate
dTTP	deoxythymidine triphosphate
dUDP	deoxyuridine diphosphate

dUMP	deoxyuridine monophosphate
DEAE	dimethylaminoethyl-
DHQ	dehydroquinate
DNA	deoxyribonucleic acid
dTMP	deoxythymidine monophosphate
E4P	erythrose-4-phosphate
F^-	female
FAD	flavin adenine dinucleotide
FAICAR	5'-phosphoribosyl-4-carboxamide-5-formamidoimidazole
FeMoco	iron-molybdenum cofactor
FGAM	formylglycinamidine ribonucleotide
FGAR	formylglycinamide ribonucleotide
FMN	flavin mononucleotide
FPP	farnesylpyrophosphate
GAR	glycinamide ribonucleotide
GAT	glutamine amidotransferase
GMP	guanosine monophosphate
GDP	guanosine diphosphate
GS	glutamine synthetase
GSH	glutathione
GSSG	oxidized glutathione
GTP	guanosine triphosphate
HDH	homoserine dehydrogenase
Hfr	high frequency of recombination
HPr	heat-stable protein of the PTS system
IGP	indoleglycerol phosphate or imidazoleglycerol phosphate (cf. context)
IMP	inosinic acid
IPTG	isopropylthiogalactoside
7-KAP	7-keto-8-aminopelargonate
Lac	lactose
LAO	lysine-arginine-ornithine permeation system
Lrp	leucine-responsive regulatory protein
MBP	maltose-binding protein
^{5}N-methyl-H_4PtGlu	^{5}N-methyltetrahydropteroylglutamate
MK	menaquinone
mRNA	messenger RNA
NAD^+	nicotinamide adenine dinucleotide
NADH	reduced form of NAD^+
$NADP^+$	nicotinamide adenine dinucleotide phosphate
NADPH	reduced form of $NADP^+$
NDP	nucleoside diphosphate
NTP	nucleoside triphosphate
NEM	N-ethylmaleimide
NMR	nuclear magnetic resonance
OMHMB	2-octaprenyl-3-methyl-5-hydroxy-6-methoxy-1,4-benzoquinone
OMP	orotidine-5'-phosphate

ONPF	*o*-nitrophenylfucoside
ONPG	*o*-nitrophenylgalactoside
ORF	open reading frame
OSB	*o*-succinylbenzoate
OTCase	ornithine transcarbamylase
PII	protein interacting with glutamine synthetase adenylyltransferase
p-AB	*p* -aminobenzoic acid
p-ABG	*p* -aminobenzoylglutamic acid
PALA	N-phosphonacetyl-L-aspartate
PAPS	3'-phosphoadenosine-5'-phosphosulfate
PEP	phosphoenolpyruvate
Pi	inorganic phosphate
PLP	pyridoxal phosphate
PPi	pyrophosphate
PQQ	pyrroloquinoline quinone
PR-	phosphoribosyl-
PRFAR	PR-formimino-5-aminoimidazole-4-carboxamide ribonucleotide
PRPP	5-phosphoribosyl-1-pyrophosphate
PRT	anthranilate phosphoribosyltransferase
PTS	phosphotransferase system
RNA	ribonucleic acid
SAH	S-adenosylhomocysteine
SAICAR	5-amino-4-imidazole-N-succinylcarboxamide ribonucleotide
SAM	S-adenosylmethionine
TDG	thiodigalactoside
TMG	thiomethylgalactoside
TPEG	thiophenylethylgalactoside
TPG	thiophenylgalactoside
TPP	thiamine pyrophosphate
TQQ	tryptophyl tryptophanquinone
tRNA	transfer RNA
UDP	uridine diphosphate
UDPG	uridine diphosphoglucose
UMP	uridine monophosphate
UR	uridylylremoving enzyme
UTase	uridylyltransferase
UTP	uridine triphosphate
XMP	xanthylic acid

One-letter code abbreviations for amino acids

A	alanine
C	cysteine
D	aspartic acid
E	glutamic acid
F	phenylalanine
G	glycine
H	histidine
I	isoleucine
K	lysine
L	leucine
M	methionine
N	asparagine
P	proline
Q	glutamine
R	arginine
S	serine
T	threonine
V	valine
W	tryptophan
Y	tyrosine

Gene index

GENE	PRODUCT OR ACTIVITY
adk	adenylate kinase
argA	N-acetylglutamate synthase
argB	acetylglutamate kinase
argC	N-acetyl-γ-glutamylphosphate reductase
argD	acetylornithine aminotransferase
argE	acetylornithine deacetylase
argF	ornithine transcarbamylase
argG	argininosuccinate synthetase
argH	argininosuccinate lyase
argI	duplicate ornithine transcarbamylase
argR	arginine repressor
aroA	3-enolpyruvylshikimate-5-phosphate synthetase
aroB	dehydroquinate synthase
aroC	chorismate synthetase
aroD	5-dehydroquinate dehydratase
aroE	dehydroshikimate reductase
aroF	DAHP synthetase(tyr)
aroG	DAHP synthetase(phe)
aroH	DAHP synthetase(trp)
aroL	shikimate kinase
asd	aspartate semialdehyde dehydrogenase
asnA	asparagine synthetase (ammonia dependent)
asnB	asparagine synthetase(glutamine dependent)
asnC	asparagine synthesis modulator
aspC	aspartate aminotransferase
bioA	7,8-diaminopelargonic acid synthetase
bioB	biotin synthetase
bioC	unidentified step in pimeloyl CoA synthesis
bioD	dethiobiotin synthetase
bioF	7-keto-8-aminopelargonic acid synthetase
bioH	unidentified step in pimeloyl CoA synthesis
carA	carbamylphosphate synthetase, light subunit (glutamine)
carB	carbamylphosphate synthetase, heavy subunit (ammonia)
cfa	cyclopropane fatty acid synthase
coaA	pantothenate kinase
cobA	S-adenosylmethionine: uroporphyrinogenIII methyltransferase
cobB	cobyrinic acid *a, c*-diamide synthase
cobC	addition of 1-amino-2-propanol to cobyric acid
cobD	addition of 1-amino-2-propanol to cobyric acid
cobF	synthesis of intermediate between precorrin-3 and precorrin-6x
cobG	elimination of C20 (contraction of the tetrapyrrole macrocycle)

cobH	mutase catalyzing the methyl migration from C11 to C12
cobI	methylation at C20, precorrin-2 $\rightarrow$ precorrin-3
cobJ	synthesis of intermediate between precorrin-3 and precorrin-6x
cobK	reduction of precorrin-6x to precorrin-6y
cobL	precorrin 6y $\rightarrow$ precorrin-8x
cobM	synthesis of intermediate between precorrin-3 and precorrin-6x
cobN	cobaltochelatase, in conjunction with *cobS* and *cobT*
cobO	cob(I)alamine adenosyltransferase
cobP	cobinamide kinase-cobinamide phosphate guanylyltransferase
cobQ	further amidation of cobyrinic acid *a, c*-diamide
cobS	cobaltochelatase, in conjunction with *cobN* and *cobT*
cobT	cobaltochelatase, in conjunction with *cobN* and *cobS*
cobU	nicotinate mononucleotide: 5, 6-dimethylbenzimidazole phosphoribosyltransferase
cobV	incorporation of 5, 6-dimethylbenzimidazole-ribose-5'-phosphate
cobW	reduction of cobalt leading to Co(I) corrinoids
codA	cytosine deaminase
cpc1 operon	*Calothrix* genes encoding phycocyanin genes regardless of the light spectrum
cpc2 operon	*Calothrix* genes encoding phycocyanin genes expressed only in red light
cpc3 operon	silent *Calothrix* phycocyanin genes, expressed upon sulfur starvation
cpcE	subunit of phycocyanobilin lyase, *Synechococcus*
cpcF	subunit of phycocyanobilin lyase, *Synechococcus*
crp	cyclic AMP receptor protein
crr	glucose phosphocarrier protein IIIglc (PTS system)
cyaA	adenylate cyclase
cysA	sulfate permease
cysB	regulatory gene for cysteine biosynthesis
cysC	adenylylsulfate kinase
cysD	sulfate adenylyltransferase
cysE	serine transacetylase
cysG	pleiotropic mutation involved in sulfite reduction
cysH	adenylylsulfate reductase (PAPS sulfotransferase)
cysI	hemoprotein of sulfite reductase
cysJ	flavoprotein of sulfite reductase
cysK	O-acetylserine sulfhydrylase (cysteine synthase)
dapA	dihydrodipicolinate synthase
dapB	dihydrodipicolinate reductase
dapC	tetrahydrodipicolinate succinylase
dapD	succinyldiaminopimelate aminotransferase
dapE	N-succinyldiaminopimelate deacylase
dapF	diaminopimelate epimerase
dpp	dipeptide permease
dsdA	D-serine deaminase
envZ	gene regulating the production of porins
fabA	β-hydroxydecanoylthioester dehydrase
fadD	acyl CoA synthetase
FAS1	β-subunit of yeast fatty acid synthetase

FAS2	α-subunit of yeast fatty acid synthetase
FAS3	yeast acetyl CoA carboxylase
fdp	fructose diphosphatase
gal	genes involved in galactose degradation
gcv	glycine cleavage
gdh	glutamate dehydrogenase
glnA	glutamine synthetase
glnB	PII protein
glnD	uridylyltransferase-uridylyl removing enzyme
glnE	glutamine synthetase adenylyltransferase
glnG	NR_I protein
glnL	NR_{II} ; NR_I kinase
gltB	large subunit of glutamate synthase
gltD	small subunit of glutamate synthase
glyA	serine hydroxymethyltransferase
gpt	guanine phosphoribosyltransferase
guaA	GMP synthetase
guaB	IMP dehydrogenase
hemA	δ-aminolevulinate synthase
hemB	δ-aminolevulinate dehydratase
hemC	hydroxymethylbilane synthase (porphobilinogen deaminase)
hemD	uroporphyrinogen III synthase (cosynthase)
hemE	uroporphyrinogen III decarboxylase
hemF	coproporphyrinogen III oxidase
hemG	protoporphyrinogen oxidase
hemH	ferrochelatase
hisA	N-(5'-phospho-L-ribosylformimino)-5-amino-1-(5'-phosphoribosyl)-4-imidazolecarboxamide isomerase
hisB	imidazoleglycerol phosphate dehydratase and histidinol phosphatase
hisC	histidinol phosphate aminotransferase
hisD	L-histidinol: NAD^+ oxidoreductase
hisE	phosphoribosyl-ATP pyrophosphohydrolase
hisF	ammonia-dependent cyclase
hisG	phosphoribosylphosphate-ATP pyrophosphorylase
hisH	glutamine-dependent amidotransferase(forms imidazole glycerol phosphate in cooperation with the *hisF* product)
hisI	phosphoribosyl-AMP cyclohydrolase
hisJ	histidine periplasmic binding protein
hisR	histidinyl-tRNA
hisS	histidinyl-tRNA synthetase
hisT	pseudouridylate synthetase
hisU	maturation of His tRNA ? identical with a gyrase subunit?
hisW	maturation of His tRNA ? identical with a gyrase subunit?
hom	*B. subtilis* homoserine dehydrogenase
hpt	hypoxanthine phosphoribosyltransferase
ILV1	yeast threonine dehydratase

ilvA	threonine dehydratase(biosynthetic)
ilvB	α-aceto-α-hydroxyacid synthase
ilvC	α-aceto-α-hydroxyacid reductoisomerase
ilvD	dihydroxy acid dehydratase
ilvE	branched chain amino acid aminotransferase
lacA (or lac ac)	thiogalactoside acetyltransferase
lacI	Lac repressor
lacY	galactoside permease
lacZ	β-D-galactosidase
leuA	α-isopropylmalate synthase
leuB	β-isopropylmalate dehydrogenase
leuC	subunit of α-isopropylmalate isomerase
leuD	subunit of α-isopropylmalate isomerase
liv	high affinity branched-chain amino acid transport system
lrp	leucine-responsive regulatory protein
lysA	diaminopimelate decarboxylase
lysC	aspartokinase III
malE	periplasmic maltose-binding protein MBP
malF	transmembrane protein involved in maltose transport
malG	transmembrane protein involved in maltose transport
malK	membrane ATPase involved in maltose transport
menA	conversion of 1, 4-dihydroxynaphthoate to demethyl menaquinone and prenylation
menB	1, 4-dihydroxynaphthoate synthase
menC	conversion of chorismate to isochorismate
menD	*o*-succinylbenzoate synthase
menE	*o*-succinylbenzoylcoenzyme A synthetase
MET25	yeast O-acetylhomoserine, O-acetylserine sulfhydrylase
metA	homoserine transsuccinylase
metB	cystathionine-γ-synthase
metC	cystathionine-β-lyase (cystathionase)
metE	homocysteine: tetrahydropteroyltriglutamate methyltransferase
metF	5, 10-methylenetetrahydrofolate reductase
metH	B_{12}-dependent homocysteine transmethylase
metJ	methionine repressor
metK	S-adenosylmethionine synthetase
metL	aspartokinase II-homoserine dehydrogenase II
metQ	direct synthesis of homocysteine from cystathionine
metR	activation of *metE* and *metH*
metX	S-adenosylmethionine synthetase, expressed in rich media
nadA	quinolinate synthase
nadB	aspartate oxidase
nadC	quinolinate phosphoribosyltransferase
nadE	nicotinamide adenine dinucleotide synthetase
nadR	repressor of the synthesis of nicotinamide derivatives
ndk	nucleoside diphosphate kinase
nifHDK operon	operon involved in nitrogen fixation

nrdA	large subunit of ribonucleoside diphosphate reductase
nrdB	small subunit of ribonucleoside diphosphate reductase
ompC	outer membrane protein (porin)
ompF	outer membrane protein (porin)
ompR	positive regulation of *ompC* and *ompF*
opp	oligopeptide permease
panB	ketopantoate hydroxymethyltransferase
panC	pantothenate synthetase
panD	aspartate-1-decarboxylase
panE	ketopantoate reductase
pncB	nicotinate phosphoribosyltransferase
pdx	pyridoxine requirement
pheA	chorismate mutase-prephenate dehydratase
pqq	genes involved in pyrroloquinoline quinone synthesis
proA	γ-glutamylphosphate reductase
proB	γ-glutamate kinase
proC	pyrroline-5-carboxylate reductase
prs	phosphoribosylpyrophosphate synthetase
purA	adenylosuccinate synthetase
purB	adenylosuccinate lyase
purC	phosphoribosylaminoimidazole-succinocarboxamide synthetase
purD	phosphoribosylglycinamide synthetase
purE	phosphoribosylaminoimidazole carboxylase
purF	phosphoribosylpyrophosphate amidotransferase
purG	phosphoribosylformylglycinamidine synthetase
purH	phosphoribosylaminoimidazolecarboxamide formyltransferase
purI	phosphoribosylaminoimidazole synthetase
purJ	IMP cyclohydrolase
purL	phosphoribosylformylglycinamide synthetase
purR	repressor of the *pur* regulon genes
purT	same reaction as *purL*, but formate is the direct formyl donor
purU	source of formate for *purT*
putA	proline oxidase and repressor of proline utilization
putP	proline transporter
pyrB	catalytic subunit of aspartate transcarbamylase
pyrC	dihydroorotase
pyrD	dihydroorotate oxidase
pyrE	orotate phosphoribosyltransferase
pyrF	orotidine-5'-phosphate decarboxylase
pyrG	cytidine triphosphate synthetase
pyrH	uridine monophosphate kinase
pyrI	regulatory subunit of aspartate transcarbamylase
ribC	*B. subtilis* repressor of riboflavin synthesis
ribO	possible repressor binding site of the RibC repressor
serA	phosphoglycerate dehydrogenase
serB	phosphoserine phosphatase

serC	phosphoserine aminotransferase
Sms and Smr	alleles determining streptomycin sensitivity or resistance
speA	arginine decarboxylase
speB	agmatine amidinohydrolase
speC	ornithine decarboxylase
speD	S-adenosylmethionine decarboxylase
suc	succinyl CoA synthesis; mutations can be phenotypically repaired by providing lysine plus methionine
T6 (tsx)	gene determining the sensitivity or resistance to phage T_6
tdc	threonine dehydratase (degradative)
tdh	threonine dehydrogenase
thrA	aspartokinase I-homoserine dehydrogenase I
thrB	homoserine kinase
thrC	threonine synthase
thrL	threonine leader sequence
thrP	threonine promoter
TRP5	yeast tryptophan synthase
trpA	tryptophan synthase, component α
trpB	tryptophan synthase, component β
trpC	phosphoribosylanthranilate isomerase-indoleglycerol phosphate synthase
trpD	glutamine amidotransferase-anthranilate phosphoribosyltransferase
trpE	anthranilate synthetase(ammonia as the amino donor)
trpR	Trp repressor
tsx	see T_6
tyrA	chorismate mutase-prephenate dehydrogenase
tyrB	aromatic aminotransferase
tyrR	repressor of *aroF, tyrA* and *tyrB*
ubiA	3-octaprenyl-4-hydroxybenzoate synthesis from *p*-hydroxybenzoate
ubiC	chorismate(pyruvate) lyase
ubiD	3-octaprenyl-4-hydroxybenzoate decarboxylase
ubiG	2-octaprenyl-3-methyl-5-hydroxy-6-methoxy-1, 4-benzoquinone methyltransferase
upp	uracil phosphoribosyltransferase
vnfDGK operon	operon involved in nitrogen fixation through non-molybdenum vanadium-containing nitrogenase

Subject index

Direction éditoriale : Patricia Kerserho
Direction artistique : Sylvia Winter

Printed in France,
Imprimerie Nouvelle, Saint-Jean-de-Braye
Numéro d'édition : 6260
Dépôt légal : troisième trimestre 1994
HERMANN, PUBLISHERS IN ART AND SCIENCES

MIX
Papier aus verantwortungsvollen Quellen
Paper from responsible sources
FSC® C105338

If you have any concerns about our products,
you can contact us on
ProductSafety@springernature.com

In case Publisher is established outside the EU,
the EU authorized representative is:
Springer Nature Customer Service Center GmbH
Europaplatz 3, 69115 Heidelberg, Germany

Printed by Libri Plureos GmbH
in Hamburg, Germany